Ajith Abraham, Aboul-Ella Hassanien,
Patrick Siarry, and Andries Engelbrecht (Eds.)

Foundations of Computational Intelligence Volume 3

Studies in Computational Intelligence, Volume 203

Editor-in-Chief
Prof. Janusz Kacprzyk
Systems Research Institute
Polish Academy of Sciences
ul. Newelska 6
01-447 Warsaw
Poland
E-mail: kacprzyk@ibspan.waw.pl

Ajith Abraham, Aboul-Ella Hassanien,
Patrick Siarry, and Andries Engelbrecht (Eds.)

Foundations of Computational Intelligence Volume 3

Global Optimization

 Springer

Dr. Ajith Abraham
Machine Intelligence Research Labs
(MIR Labs)
Scientific Network for Innovation
and Research Excellence
P.O. Box 2259 Auburn,
Washington 98071-2259
USA
E-mail: ajith.abraham@ieee.org
http://www.mirlabs.org
http://www.softcomputing.net

Dr. Aboul-Ella Hassanien
College of Business Administration
Quantitative and Information System
Department
Kuwait University
P.O. Box 5486
Safat, 13055
Kuwait
E-mail: abo@cba.edu.kw

Dr. Patrick Siarry
Université Paris XII
Fac. Sciences, LERISS
61 avenue du Général de Gaulle
Building P2 - Room 350
94010 Créteil
France
E-mail: siarry@univ-paris12.fr

Dr. Andries Engelbrecht
University of Pretoria
Department of Computer Science
Pretoria 0002
South Africa
E-mail: engel@driesie.cs.up.ac.za

ISBN 978-3-642-10165-6   e-ISBN 978-3-642-01085-9

DOI 10.1007/978-3-642-01085-9

Studies in Computational Intelligence   ISSN 1860949X

Typeset & Cover Design: Scientific Publishing Services Pvt. Ltd., Chennai, India.

Printed in acid-free paper

9 8 7 6 5 4 3 2 1

springer.com

Preface

Foundations of Computational Intelligence

Volume 3: Global Optimization: Theoretical Foundations and Applications

Global optimization is a branch of applied mathematics and numerical analysis that deals with the task of finding the absolutely best set of admissible conditions to satisfy certain criteria / objective function(s), formulated in mathematical terms. Global optimization includes nonlinear, stochastic and combinatorial programming, multiobjective programming, control, games, geometry, approximation, algorithms for parallel architectures and so on. Due to its wide usage and applications, it has gained the attention of researchers and practitioners from a plethora of scientific domains. Typical practical examples of global optimization applications include: Traveling salesman problem and electrical circuit design (minimize the path length); safety engineering (building and mechanical structures); mathematical problems (Kepler conjecture); Protein structure prediction (minimize the energy function) etc.

Global Optimization algorithms may be categorized into several types: Deterministic (example: branch and bound methods), Stochastic optimization (example: simulated annealing). Heuristics and meta-heuristics (example: evolutionary algorithms) etc. Recently there has been a growing interest in combining global and local search strategies to solve more complicated optimization problems.

This edited volume comprises 17 chapters, including several overview Chapters, which provides an up-to-date and state-of-the art research covering the theory and algorithms of global optimization. Besides research articles and expository papers on theory and algorithms of global optimization, papers on numerical experiments and on real world applications were also encouraged. The book is divided into 2 main parts.

Part-I: Global Optimization Algorithms: Theoretical Foundations and Perspectives

In Chapter 1, Snasel et al. [1] introduce the fundamentals of genetic algorithm and illustrate a Higher Level Chromosome Genetic Algorithm (HLCGA) for solving combinatorial optimization problems. The developed HLCGA is applied for Turbo code interleaver optimization process aiming to leverage the efficiency of turbo code based digital communications.

Bacterial foraging optimization algorithm (BFOA) has been widely accepted as a global optimization algorithm for distributed optimization and control. Das et al. [2] in Chapter 2 provide all the related work on BFOA, which ranges from the foundational aspects, mathematical model, hybridization and adaptation to novel applications.

In the Third Chapter, Geem [3] presents the theoretical foundations of the Harmony Search (HS) algorithm, which mimics music improvisation where musicians try to find better harmonies based on randomness or their experiences, which can be expressed as a novel stochastic derivative rather than a calculus-based gradient derivative. The chapter also presents three applications that demonstrate the global optimization feature of the HS algorithm.

Festa and Resende [4] in the Fourth Chapter give an excellent overview of different ways to hybridize Greedy Randomized Adaptive Search Procedures (GRASP) to create new and more effective metaheuristics. Several types of hybridizations are considered, involving different constructive procedures, enhanced local search algorithms, and memory structures.

In the Fifth Chapter, Pant et al. [5] present the foundations of Particle Swarm Optimization (PSO) and some of the recent modified variants. The main focus is on the design and implementation of the modified PSO based on diversity, mutation, crossover and efficient initialization using different distributions and Low-discrepancy sequences.

Habet [6] in the Sixth Chapter presents a nice overview of Tabu Search (TS) metaheuristic algorithm to solve various combinatorial optimization problems. The TS algorithm is illustrated to solve a real-life optimization problem under constraints.

In the Seventh Chapter, Liberti et al. [7] introduce Mathematical Programming (MP) for describing optimization problems. MP is based on parameters, decision variables and objective function(s) subject to various types of constraints. A reformulation of a mathematical program P is a mathematical program Q obtained from P via symbolic transformations applied to the sets of variables, objectives and constraints. This chapter presents a survey of existing reformulations interpreted along these lines with some example applications.

Shcherbina [8] in the Eighth chapter provides a review of structural decomposition methods in discrete optimization and gives a unified framework in the form of Local Elimination Algorithms (LEA). Different local elimination schemes and related notions are considered. The connection of LEA schemes and a way of transforming the directed acyclic graph of computational LEA procedure to the tree decomposition are also presented.

In the Ninth Chapter, Avdagic et al. [9] present the general problem of decision making in unknown, complex or changing environment by an extension of static multiobjective optimization problem. Implementation of multiobjective genetic algorithm is used for solving such problems and the population of potential solutions to the problem for different test cases, such as homogeneous, - non-homogeneous, and the problem with changing number of objectives and decision making is also illustrated.

Abraham and Liu [10] in the Tenth Chapter illustrate the problem of premature convergence for the conventional PSO algorithm for multi-modal problems involving high dimensions. Analysis of the behavior of the PSO algorithm reveals that such premature convergence is mainly due to the decrease of velocity of particles in the search space that leads to a total implosion and ultimately fitness stagnation of the swarm. This paper introduces Turbulence in the Particle Swarm Optimization (TPSO) algorithm to overcome the problem of stagnation. The parameters of the TPSO are adapted by a fuzzy logic controller.

Part-II: Global Optimization Algorithms: Applications

In the Eleventh Chapter, Stoean et al. [11] propose an evolutionary algorithm approach for solving the central optimization problem of determining the equation of the hyper plane deriving from support vector learning. This approach helps to open the 'black-box' of support vector training and breaks the limits of the canonical solving component.

In the Twelfth Chapter, Baragona and Battaglia [12] Illustrate how evolutionary computation techniques have influenced the statistical theory and practice concerned with multivariate data analysis, time series model building and optimization. Chapter deals with variable selection in linear regression models, non linear regression, time series model identification and estimation, detection of outlying observations in time series with respect to location and type identification, cluster analysis and grouping problems, including clusters of directional data and clusters of time series.

Baron et al. [13] in the Thirteenth Chapter introduce a heuristic based on ant colony optimization and evolutionary algorithm and further hybridized with a Tabu search and a greedy algorithm to accelerate the convergence and to reduce the cost engendered by the evaluation process. Experimental results reveal that it is possible to offer the decision maker a reduced number of more accurate solutions in order to choose one according to technical, economic and financial criteria.

Elizabeth and Goldbarg [14] in the Fourteenth Chapter present the outlines for the development of Transgenetic Algorithms and reported the mplementation of these algorithms to a single and to a bi-objective combinatorial problem. The mono objective problem is the uncapacitated version of Traveling Purchaser Problem, where the proposed algorithm managed to find nine new best solutions for benchmark instances. The proposed approach is described and a didactic example with the well-known Traveling Salesman Problem illustrates its basic components. Applications of the proposed technique are reported for two NP-hard combinatorial problems: the Traveling Purchaser Problem and the Bi-objective Minimum Spanning Tree Problem.

Abdelsalam [15] in the Fifteenth Chapter presents a model that aims to support the optimal formulation and assignment of multi-functional teams in integrated product development (IPD) organizations - or any project-based organization. The model accounts for limited availability of personnel, required skills, team homogeneity, and, further, maximizes organization's payoff by formulating and assigning teams to projects with higher expected payoffs. A Pareto multi-objective particle swarm optimization approach was used to solve the model. The model was applied a hypothetical example that demonstrates the efficiency of the proposed solution algorithm and it allows personnel to work in several concurrent projects and considers both person-job and person-team fit.

In the Sixteenth Chapter, Omara and Arafa [16] illustrate two variants of genetic algorithms with some heuristic principles for task scheduling in distributed systems. In the first variant, two fitness functions have been applied one after another. The first fitness function is concerned with minimizing the total execution time (schedule length) and the second one is concerned with the load balance satisfaction. The second variant of genetic algorithm is based on task duplication technique.

Estimation of distribution algorithms (EDAs), are evolutionary algorithms that try to estimate the probability distribution of the good individuals in the population. Mohammed and Kamel [17] in the last Chapter present a new PSO algorithm that borrows ideas from EDAs. This algorithm is implemented and compared to previous PSO and EDAs hybridization approaches using a suite of well-known benchmark optimization functions.

We are very much grateful to the authors of this volume and to the reviewers for their great effort by reviewing and providing useful feedback to the authors. The editors would like to express thanks to Dr. Thomas Ditzinger (Springer Engineering Inhouse Editor, Studies in Computational Intelligence Series), Professor Janusz Kacprzyk (Editor-in-Chief, Springer Studies in Computational Intelligence Series) and Ms. Heather King (Editorial Assistant, Springer Verlag, Heidelberg) for the editorial assistance and excellent collaboration to produce this important scientific work. We hope that the reader will share our joy and will find the volume useful

December 2008   Ajith Abraham, Norway
                                                              Aboul Ella Hassanien, Egypt
                                                                    Patrick Siarry, France
                                                      Andries Engelbrecht, South Africa

References

[1] Snasel, V., Platos, J., Kromer, P., Ouddane, N.: Genetic Algorithms for the Use in Combinatorial Problems
[2] Das, S., Biswas, A., Dasgupta, S., Abraham, A.: Bacterial Foraging Optimization Algorithm: Theoretical Foundations, Analysis, and Applications
[3] Geem, Z.W.: Global Optimization Using Harmony Search: Theoretical Foundations and Applications
[4] Festa, P., Resende, M.G.C.: Hybrid GRASP heuristics
[5] Pant, M., Thangaraj, R., Abraham, A.: Particle Swarm Optimization: Performance Tuning and Empirical Analysis

Contents

Part II: Global Optimization Algorithms: Applications

Part I
Global Optimization Algorithms: Theoretical Foundations and Perspectives

Genetic Algorithms for the Use in Combinatorial Problems

Václav Snášel, Jan Platoš, Pavel Krömer, and Nabil Ouddane

Abstract. Turbo code interleaver optimization is a NP-hard combinatorial optimization problem attractive for its complexity and variety of real world applications. In this paper, we investigate the usage and performance of recent variant of genetic algorithms, higher level chromosome genetic algorithms, on the turbo code optimization task. The problem as well as higher level chromosome genetic algorithms, that can be use for combinatorial optimization problems in general, is introduced and experimentally evaluated.

1 Introduction

Evolutionary algorithms (EAs) are a family of iterative, stochastic search and soft optimization methods based on mimicking successful optimization strategies observed in nature [6, 10, 12, 5]. The essence of EAs lies in the emulation of Darwinian evolution utilizing the concepts of Mendelian inheritance for use in computer science and applications [5]. Along with fuzzy sets, neural networks and fractals, evolutionary algorithms are among the fundamental members of the class of soft computing methods.

Genetic algorithms (GA), introduced by John Holland and extended by David Goldberg, are a widely applied and highly successful EA variant based on computer emulation of genetic evolution. Many variants of standard generational GA have been proposed. The differences are mostly in particular selection, crossover, mutation and replacement strategy [10].

Václav Snášel, Jan Platoš, and Pavel Krömer
Department of Computer Science, Faculty of Electrical Engineering and
Computer Science, VŠB - Technical University of Ostrava, 17. listopadu 15, 708 33
Ostrava - Poruba, Czech Republic
e-mail: {vaclav.snasel,jan.platos,pavel.kromer.fei}@vsb.cz

Nabil Ouddane
Department of Telecommunications, Faculty of Electrical Engineering and
Computer Science, VŠB – Technical University of Ostrava, Czech Republic
e-mail: nabil.ouddane.st1@vsb.cz

A. Abraham et al. (Eds.): Foundations of Comput. Intel. Vol. 3, SCI 203, pp. 3–22.
springerlink.com   © Springer-Verlag Berlin Heidelberg 2009

Genetic algorithms have been successfully used to solve non-trivial multi-modal optimization problems. They inherit the robustness of emulated natural optimization processes and excel in browsing huge, potentially noisy problem domains. Their clear principles, ease of interpretation, intuitive and reusable practical use and significant results made genetic algorithms the method of choice for industrial applications while carefully elaborated theoretical foundations attracted the attention of the academy.

This paper presents an innovative variant of genetic algorithms designed and developed in order to solve combinatorial problems. The method, called Higher Level Chromosome Genetic Algorithms (HLCGA), will be thoroughly described and experimentally evaluated on an appealing combinatorial optimization problem. Turbo code interleaver optimization is an optimization process aiming to leverage the efficiency of turbo code based digital communications.

2 Evolutionary Optimization

Evolutionary algorithms are generic and reusable population-based meta-heuristic optimization methods [2, 10, 12]. EAs operate with a population (also known as pool) of artificial individuals (also referred to as items or chromosomes) encoding possible problem solutions. Encoded individuals are evaluated using a carefully selected domain specific objective function which assigns a fitness value to each individual. The fitness value represents the quality (ranking) of each individual as a solution to a given problem. Competing individuals explore the problem domain towards an optimal solution [10].

2.1 Evolutionary Search Process

For the purpose of EAs is necessary proper encoding, representing solutions of given problem as encoded chromosomes suitable for evolutionary search process. Finding proper encoding is a non-trivial, problem dependent task affecting the performance and results of evolutionary search in given problem domain. The solutions might be encoded into binary strings, real vectors or more complex, often tree-like, hierarchical structures (subject of genetic programming [11]). The encoding selection is based on the needs of particular application area.

The iterative phase of evolutionary search process starts with an initial population of individuals that can be generated randomly or seeded with potentially good solutions. Artificial evolution consists of iterative application of so called genetic operators, introducing to the algorithm evolutionary principles such as inheritance, survival of the fittest and random perturbations. Iteratively, the current population of problem solutions is modified with the aim to form new and hopefully better population to be used in then next generation. The evolution of problem solutions ends after satisfying specified termination criteria and especially the criterion of finding optimal or near-optimal solution. However, the decision whether a problem solution is best (i.e. global optimum was reached) is in many problem areas hard or impossible. After the termination of the search process, evolution winner is decoded and presented as the most optimal solution found.

2.2 Genetic Operators

Genetic operators and termination criteria are the most influential parameters of every evolutionary algorithm. All bellow presented operators have several implementations performing differently in various application areas.

- *Selection* operator is used for selecting chromosomes from population. Through this operator, selection pressure is applied on the population of solutions with the aim to pick promising solutions to form following generation. Selected chromosomes are usually called parents.
- *Crossover* operator modifies the selected chromosomes from one population to the next by exchanging one or more of their subparts. Crossover is used for emulating sexual reproduction of diploid organisms with the aim to inherit and increase the good properties of parents for offspring chromosomes. The crossover operator is applied with probability P_C called croosover probability or crossover value.
- *Mutation* operator introduces random perturbation in chromosome structure; it is used for changing chromosomes randomly and introducing new genetic material into the population. Mutation operator is applied with probability P_M (mutuation probability, mutation value).

There are several strategies and implementations of every introduced genetic operator. The *roulette wheel selection* is an example of fitness proportionate selection. The probabitily p_i of selecting i-th chromosome as parent in population of N chromosomes is defined by 1, where f_i stands for the fitness value of the i-th chromosome.

$$p_i = \frac{f_i}{\sum_{k=1}^{N} f_k} \tag{1}$$

In the *truncation selection*, the population is sorted according to the fitness value of the chromosomes. The N_T most fitted chromosomes are then taken as candidates for parenthood with the same probability. Some selection strategies, such as truncation selection, might introduce loss of variability, stagnation or premature convergence to the population. In order to increase the convergence speed of the algorithm, the concept of *elitism* is often introduced. In elitary genetic algorithms, the fittest chromosomes are allways considered for reproduction. Elitism can potentially lead to a loss of diversity in the population so its application must be carefully considered.

There are several widely used implementations of the crossover operator. In *one-point crossover*, a random position is selected within the parent chromosomes and their subparts are swapped. The *multi-point crossover*
operator divides the chromosomes in multiple parts and every second segment is exchanged between the two parents. The *segment crossover* is a
multi-point crossover variant in which the number of crossover points varies
among the generations. When applying the *uniform crossover*, every gene
(i.e. every value in the chromosome) is considered for exchange between the
parents.

Common mutation operators are i.e. *one-point mutation* in which one gene
in the chromosome is randomly changed. The *random selection* is a radical
version of the mutation operator in which is whole chromosome replaced by
a randomly generated alternative.

Naturally, the operators are subject to domain specific modifications
and tuning. Besides genetic operators, termination criteria are important
factor affecting the search process. Widely used termination criteria
are i.e.:

- Reaching optimal solution (which is often hard or impossible to recognize)
- Processing certain number of generations
- Processing certain number of generations without significant improvement
  in the population

EAs are successful general adaptable soft computing concept with good
results in many areas. The class of evolutionary techniques consists of more
particular algorithms having numerous variants, forged and tuned for specific
problem domains. The family of evolutionary algorithms consists of genetic
algorithms, genetic programming, evolutionary strategies and evolutionary
programming.

2.3   Genetic Algorithms

Genetic algorithms (GA) introduced by John Holland and extended by David
Goldberg are wide applied and highly successful EA variant. Basic workflow
of original (standard) generational GA (GGA) is:

1. Define objective function
2. Encode initial population of possible solutions as fixed
   length binary strings and evaluate chromosomes in initial
   population using objective function
3. Create new population (evolutionary search for better
   solutions)
   a. Select suitable chromosomes for reproduction (parents)
   b. Apply crossover operator on parents with respect to
      crossover probability to produce new chromosomes
      (known as offspring)

     c. Apply mutation operator on offspring chromosomes with respect to mutation probability. Add newly constituted chromosomes to new population

     d. Until the size of new population is smaller than size of current population go back to (a).

     e. Replace current population by new population

4. Evaluate current population using objective function

5. Check termination criteria; if not satisfied go back to (3).

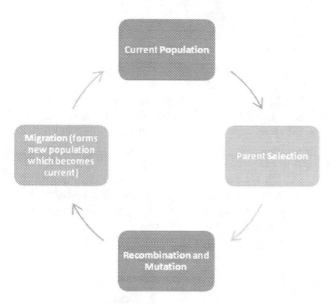

Fig. 1 Iterative phase of Genetic Algorithm

Many variants of standard generational GA have been proposed. The differences are mostly in particular selection, crossover, mutation and replacement strategy [10]. Different high-level approach is represented by steady-state genetic algorithms (SSGA). In GGA, in one iteration is replaced whole population [6] or fundamental part of population [16] while SSGA replace only few individuals at time and never whole population. This method is more accurate model of what happens in the nature and allows exploiting promising individuals as soon as they are created. However, no evidence that SSGA are fundamentally better than GGA was found [16].

3 Crossover Challenging Problems

There is an indispensable class of problem domains introducing various challenges to GA based solutions. Tasks such as solving combinatorial problems

harass specially the crossover operator. It is not easy to find suitable encoding
that will enable the use of fully featured (i.e. crossover enabled) genetic algorithms. The loss of crossover might be considered as significant weakening
of the algorithm.

3.1   The Role of Crossover in GA

Crossover operator is the main operator of genetic algorithms distinguishing
it from other stochastic search methods [12]. Its role in the GA process has
been intensively investigated and its omitting or traversing is expected to
affect the efficiency of GA solution significantly.

Crossover operator is primarily a creative force in the evolutionary search
process. It is supposed to propagate building blocks (low order, low defining-
length schemata with above average fitness) from one generation to another
and create new (higher order) building blocks by combining low order building blocks. It is intended to introduce to the population large changes with
small disruption of building blocks [17]. In contrast, mutation is expected to
insert new material to the population by random perturbation of chromosome structure. By this, however, can be new building blocks created or old
disrupted [17].

An experimental study on crossover and mutation in relation to frequency
and lifecycle of building blocks in chromosome population showed that the
ability of mutation and one point crossover to create new building blocks is
almost the same. However, crossover is irreplaceable to spread newly found
building blocks among the population (which can lead to loss of diversity in
the population) [17].

3.2   Traditional Approaches to Crossover Challenging Tasks

Crossover challenging problems, such as combinatorial problems, were genetically solved using several strategies. In general, they are:

- Averse scoring of invalid individuals
- Mutation only genetic algorithm
- Random key encoding
- Post-processing to fix corrupted chromosomes

Averse scoring of invalid individuals assigns to offspring chromosomes that
were corrupted by crossover extremely bad fitness value and continues following classic GA pattern. The issue of this approach is usually immense
increase of the solution space dimension (i.e. all invalid individuals are added
to the solution space). The Genetic Algorithm then usually deals most of the
time with irrelevant solutions instead of browsing the space of valid problem
solutions.

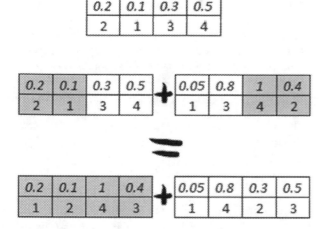

Fig. 2 Random key encoding examples

Mutation only genetic algorithm avoids the use of crossover operator totally. This, as already mentioned, can be seen as significant weakening of the algorithm.

Random key encoding is a strategy available for problems involving permutation evolution [15]. In random key encoding, the permutation is represented as a string of real numbers (random keys), whose position after ordering corresponds to the permutation index. Random key encoded chromosome and crossover of randomly encoded permutations is illustrated in Fig. 2.

Randomly encoded permutations can be rarely, when there are two identical random keys in the population, corrupted by crossover as well.

The post-processing of corrupted chromosomes hacks the evolutionary tendencies in chromosome population and eliminates the impact and creative power of crossover operator. The building blocks that crossover considers to be significant, though they make the chromosome invalid, are disrupted by the post-processing procedure.

In order to design more straightforward and less intrusive GA variant, the task of combinatorial optimization will be analyzed and novell GA version introduced in section 3. First, let us briefly introduce the real world optimization task that initiated the design and developement of higher level genetic algorithms.

4 Turbo Codes

The turbo codes are among the most promising innovative techniques in digital communications. As a very first channel encoding method, they allowed to approach the Shannon limit, a theoretical maximum information transfer rate of a channel for certain noise level [3].

The turbo codes were introduced by C. Berrou and A. Glavieux in 1993 [4] and they have become a hot topic soon after their introduction. Prior to the turbo codes, $3dB$ or more separated the spectral efficiency of real world channel encoding systems from the theoretical maximum described by Shannon theorem [3]. Turbo coding brought to the world of channel encoding one important principle: the feedback concept, exploited heavily in electronics, to be utilized in decoding of concatenated codes. And it was indeed the iterative decoding that helped the turbo codes to achieve its impressive near-optimum performance [3].

The turbo codes are an implementation of parallel concatenation of two circular recursive systematic convolutional (CRSC) codes based on a pseudo-random permutation (the interleaver). This is specially important since random codes were used by Shannon to calculate the theoretical potential of channel coding [3]. In general, more than two CRSC encoders can be paralelly concatenated but the quasi optimum performance can be achived with just two encoders as in the classical turbo code [3]. The general scheme of a classic turbo encoder is shown in 3.

The classic turbo decoding is based on the cooperation of two soft-in/soft-out (SISO) decoders, (also reffered to as probabilistic decoders). Each decoder processes its own data, and passes the extrinsic information to the other decoder. As the probabilistic decoders work together on the estimation of a common set of symbols, both machines have to give the same decision, with the same probability, about each symbol, as a single (global) decoder would [3, 4].

The improvements obtained by turbo coding are supported by the fact that each of the encoders typically produces a high-weight code word for most inputs, but it produces a low-weight (i.e. unsuitable) code word for only few inputs. The interleaver makes it more unlikely that both encoders will output a bad code word for the same input, increasing the chance that the decoder will be able to extract the correct information.

Turbo codes offer the best compromise between structure (concatenation) and randomness created by the interleaver. Its characteristic iterative decoding process is among the principal performance factors of the turbo codes. The significant characteristics of turbo codes are small bit error rate (BER) achieved even at low signal to noise ratio $\frac{E_b}{N_0}$ and the error floor at moderate and high values of $\frac{E_b}{N_0}$.

Previous studies proved that a random interleaver (random permutation of the input fame) can be in certain cases (i.e. for $BER > 10^{-5}$) more efficient than other channel encoding schemes [9]. In this paper, a genetically evolved turbo code interleaver will be compared to random interleaver by the means of BER to evaluate its efficiency. The increase of the interleaver size gives better performance and better interleaving gain while worsening latency. The relation 2 illustrates the influence on the latency:

$$t_d = \frac{K_f}{R_b} N_i \tag{2}$$

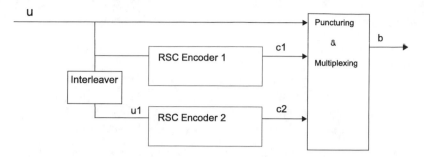

Fig. 3 A general scheme of turbo encoder

In 2, R_b is the code bit rate, K_f stands for the frame size and N_i is the number of the decoding stages. The performance of the turbo codes depends on two principal parameters, first is the code spectrum, and the second is decorrelation between the external information at the same number of iterations. The optimization process can be used for the amelioration of performance and the diminution of the matrix stature with safe performance. The latter is very interesting for multimedia real-time transmission systems over satellite because the interleaving matrix makes a considerable diminution of the codec complexity and delay. Interleaver matrix sizes vary from tens to ten-thousands of bits. When optimizing, it is highly inefficient, if not impossible, to test all the possible input vectors $(2N)$ with all the possible interleaver matrices $(N!)$, requiring $2N.N!$ tests in total. Therefore, advanced interleaver optimization methods are required.

The encoder processes a N-bit long information frame u. The input frame is interleaved by the N-bit interleaver to form permuted frame u_1. Original input frame is encoded by the encoder $RSC1$ and interleaved frame is encoded by $RSC2$. Hereafter, the two encoded frames c_1 and c_2 are merged together and with the original input sequence u following some puncturing and multiplexing scheme. The rate of the code is defined as $r = \frac{k}{n} = \frac{Input Symbols}{Output Symbols}$

4.1 Interleaver Evaluation

As noted earlier, the performance of a turbo code can be evaluated by the means of bit error rate, i. e. the ratio of the number of incorrectly decoded bits to the number of all bits transmitted information during some period. Unfortunatelly, it is rather hard to compute the BER for a turbo code and the simulations can be for large interleavers inaccurate.

The error floor of a $C(n,k)$ code can be analytically estimated:

$$BER \approx \frac{w_{free}}{2k} erfc(\sqrt{d_{free}\frac{k}{n}\frac{E_b}{N_0}}) \qquad (3)$$

To estimate BER, the following code properties must be known [8]:

- d_{free} - the free distance, i.e. the minimum number of different bits in any pair of codewords
- N_{free} - the free distance multiplicity, i.e. the number of input frames generating codewords with dfree
- w_{free} - the information bit multiplicity, i.e. the sum of the Hamming weights of the input frames generating the codewords with dfree

There are several algorithms for free distance evaluation. Garello et. al. [8] presented an algorithm designed to effectively compute free distances of large interleavers with unconstrained input weight based on constrained subcodes.

This work presents interleaver optimization and computational experiments driven by both, computer simulations used to estimate BER over an Additive White Gaussian Noise (AWGN) channel and algebraical estimation of maximum d_{free} evaluated using analytical approach by Garello et. al. [8].

5 Genetic Algorithms for Linear Ordering Problem

In this section, we introduce a general idea of genetic algorithms designed to solve linear ordering problem, as well as other combinatorial optimization problems.

Consider a permutation of N symbols $\Pi(N) = (i_1, i_2, \ldots, i_N)$, where $i_k \in [1, N]$ and $i_m \neq i_n$ for all $m \neq n \in [1, N]$. A sample permutation of 3 symbols ($N = 3$) is shown in (4).

$$\Pi_3 = (3, 1, 2) \tag{4}$$

The genetic encoding of a permutation might straightforwardly copy the notion of Π_N from 4 and the chromosome then consists of an integer vector $(i_1, i_2, \ldots, i_N)$ where $i_1 \neq i_2 \neq \cdots \neq i_N$ and $i_n \in [1, N]$. Imagine a binary data frame $I_5 = (0, 1, 0, 1, 1, 1)$ to be encoded by a turbo code system. The effect of a sample permutation $\Pi_5 = (5, 3, 4, 1, 2)$ is shown in 5:

$$O_5 = \Pi_5(I_5) = (1, 0, 1, 0, 1) \tag{5}$$

Unfortunatelly, the straightforwardly encoded permutations are unsuitable for crossover operator. The permutations encoded using integer vector $(i_1, i_2, \ldots, i_N)$ subject to $i_m \neq i_n$ for all $m \neq n \in [1, N]$ produce when applying common common crossover operators (such as one-point crossover or two-point crossover) invalid offspring.

Mutation operator is in genetic permutation evolution implemented as mutual replacement of two randomly chosen indices from permutation chromosome. The fitness function is problem specific. For the turbo code interleaver optimization task is the fitness function straightforward permutation of input bit word as illustrated in 5.

Since the permutation evolution is a fundamentally crossover challenging task, a novell variant of GA called higher level chomosome genetic algorithms was developed.

5.1 Higher Level Chromosome Genetic Algorithms

To enable the application of crossover for interleaver optimization, expecting performance increase, we have investigated the effect of uniform crossover on convergence ability of the classical interleaver optimizing GA (a mutation-only implementation). In the second phase, we have designed modified GA allowing the use of virtually any crossover operator for permutation evolution without breaking the chromosomes. New crossover friendly GA is based on separation of chromosomes into groups of the same size called higher level chromosomes (HLCs) that act on the GA level as regular (traditional) chromosomes and the contained primitive chromosomes act as genes (i.e. the crossover and mutation is applied on HLCs and some problem specific functions are used to transfer the effect of the operator on the primitive chromosomes assigned to the particular HLC). The idea of higher level chromosome genetic algorithms (HLCGA) is illustrated in Figure 4.

In order to speed up the algorithm convergence, a *semi-elitary selection* schema was used. In semi-elitary selection, one parrent is chosen by elitary manners (i.e. the currently fittest chromosome is allways taken as first parent and the second parent was selected using roulette wheel selection). We have tested all above introduced techniques on benchmarking problem consisting of search for an identity matrix. The results, summarized for 512 bit benchmark in Figure 5, have shown that GA with semi-elitary selection and HLCs performs best.

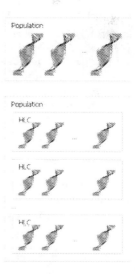

Fig. 4 Traditional population compared to population with HLCs

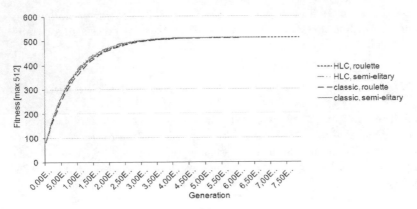

Fig. 5 Performance comparison of GA in benchmarking problem

HLCGA was implemneted for turbo code interleaver evolution. The following section describes in detail the experiments, particular algorithm setup and obtained results.

6 HLCGA Experiments

Genetic algorithms have been previously used for interleaver matrix optimization. Durand et al. [7] used customized GA to optimize the interleaver of the size 105, comparing their results to previous interleaver design techniques. Their genetic algorithm was fully based on mutation and the crossover operator was due to complications omitted.

Rekh et al. [14] presented another GA for the interleaver optimization, introducing two-point crossover to interleaver evolution process. Nevertheless, the crossover impact was limited by necessary correction of errors created during crossover application. The fitness criterion was BER and the size of optimized interleaver 50.

Two fitness functions were investigated for turbo code optimization task.

6.1 *Fitness Function Based on Average BER*

A simulation framework built upon the IT++ library[1] was used to experimentally evaluate proposed interleaver optimization method. IT++ is a robust and efficient C++ library of mathematic and coding algorithms and classes. It provides high performance of native code and excellent abstraction of well-defined object oriented framework. We have implemented an experimental framework allowing to simulate the transmission of binary data over an additive white Gaussian noise (AWGN) channel and Rayleigh fading channel.

---

[1] IT++ is available at http://itpp.sourceforge.net/

The AWGN channel is a good model for satellite and deep space communication links but not an appropriate model for terrestrial links. Rayleigh fading channel was used as a reasonable model for tropospheric and ionospheric signal propagation as well as the effect of heavily builtup urban environments on radio signals [13].

The evolved interleavers were evaluated by simulated transmission over AWGN channel for $\frac{E_b}{N_0} \in [0, 4]$ and flat Rayleigh fading channel for $\frac{E_b}{N_0} \in [0, 6]$
.

We have experimented with 64, 128, 512 and 1024 bit interleavers aiming to optimize in the future as large interleaver as possible.

The settings for all optimization experiments were:

- HLCGA with semi-elitary selection
- 1000 generations
- Probability of crossover 0.8
- Probability of mutation 0.2
- Population of 5 high level chromosomes per 6 genes
- Fitness criterion was minimal BER after simulated submission of 100 random frames of weight up to 6
- Simulations performed over additive white Gaussian noise (AWGN) channel and Rayleigh fading channel.

Experimental optimization results are summarized in Figures 6, 7, 8 and 9 respectively. In all of them is used the following notation: the curve denoted as $O1$ corresponds to the best interleaver found by GA with classic population, $O2$ describes performance of best interleaver found using HLCGA and *Rand* denotes a reference random block interleaver. *AWGN* marked curves illustrate experiments over additive white Gaussian noise channel and *Rayleigh* curves represent the experimental results measured over Rayleigh channel. In all figures can be seen that optimized interleavers perform better than reference random interleaver.

Figure 6 illustrates the binary error rate for an interleaver with the length of 64 bits; we can observe that an improvement for AWGN channel begin to appear from $\frac{E_b}{N_0} = 2dB$ and becomes more significant for larger $\frac{E_b}{N_0}$ values, especially for the interleaver obtained by second optimization method. Both optimized interleavers overperformed the random interleaver.

For $BER = 10^{-3}$ we have an $\frac{E_b}{N_0}$ of approximately $3.25dB$ for the random interelaver and $2.75dB$ for the second optimized interleaver achieving gain of $0.5dB$. The trend is valid for Rayleigh channel experiments as well and the supremacy of interleaver $O2$ is even more evident.

For 128 bits interleaver, shown in Figure 7, can be observed that for AWGN, the amelioration begins to be significant between the second optimization and the random interleaver from $\frac{E_b}{N_0} = 2.25dB$, it means for a larger signal noise rate values. For $BER = 10^{-3}$ we have $\frac{E_b}{N_0}$ equal $2.25dB$ for the second optimization and $2.5dB$ for the random interelaver having $0.25dB$ of gain. For the Rayleigh channel transmissions, the better performance of $O1$

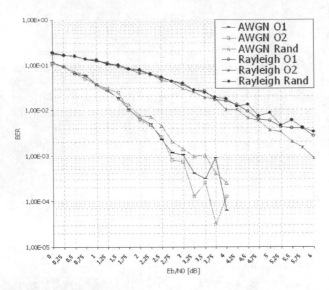

Fig. 6 64bit interleavers over AWGN channel and Rayleigh fading channel

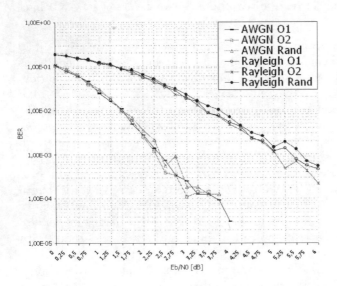

Fig. 7 128bit interleavers over AWGN channel and Rayleigh fading channel

and $O2$ comparing to random interleaver becomes to be clear for greater $\frac{E_b}{N_0}$ values ($> 5dB$) and $O2$ is again best performing among the three.

The gain becomes more considerable for interleaver length of 512bits as shown on Figure 8, for example in AWGN, we have for $BER = 10^{-4}$ the $\frac{E_b}{N_0} = 2.75dB$ for the second optimization method while having $3.5dB$ for the random interelaver. This indicates $0.75dB$ gain for 512bits length interleaver,

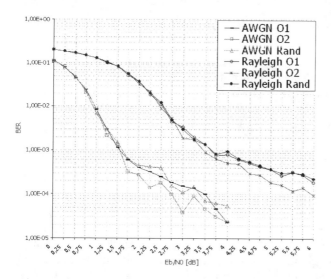

Fig. 8 512bit interleavers over AWGN channel and Rayleigh fading channel

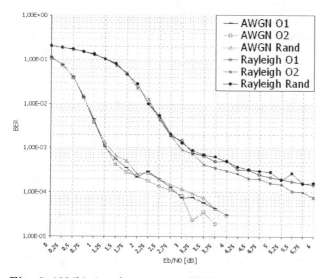

Fig. 9 1024bit interleavers over AWGN channel and Rayleigh fading channel

which is for this interleaver length a remarkable result. In Rayleigh fading channel, the initial BER values are for all three compared interleavers almost the same while for higher $\frac{E_b}{N_0}$, interleaver $O2$ achieves permanent gain over similarly performing $O1$ and random interleaver.

Similarly, optimized 1024 bit inteleavers, highlighting specially interleaver $O2$, shown at Figure 9 outperform reference random interleaver for both, AWGN channel and Rayleigh fading channel.

6.2 Fitness Function Based on Maximum Free Distance

The BER error floor specified in Equation 3 and the definitions of d_{free}, w_{free} and N_{free} suggest another way of designing the fitness function for TC interleaver optimization. It is desired to maximize d_{free} and minimize N_{free} and w_{free}. The fitness function f was defined as follows:

$$f = A \cdot d_{free} - B \cdot N_{free} - C \cdot w_{free} \tag{6}$$

where the coefficients A, B and C were fixed to 100, 10 and 1 respectively.

We have performed a series of experiments to investigate the ability of HLCGA with free distance as fitness function to resolve above average interleavers. Garello et. al. [8] provided an overview of the distribution of free distance for several turbo code configurations and frame lengths. We have confronted the results (i.e. maximum and average free distance) obtained by HLCGA with known maximum d_{free}, average d_{free} and d_{free} variance for rate $\frac{1}{3}$ turbo codes with 8 state constituent encoders. Table 1 shows these values and the results of performed experiments.

Values shown in Table 1 are illustrated in Figure 10 while Figure 11 provides zoomed view on the small interleaver ($N \in [2, 10]$) area from Figure 10. Both figures show best and average interlever d_{free} according to [8] and best and average interleaver d_{free} found by presented genetic algorithm.

When evolving small interleavers (i.e. crawling relatively small fitness landscapes), the evolved interleavers feature very good free distance. For interleaver lengths 2 to 6, the algorithm reached in all experimental runs the best known d_{free}. For interleaver lengths 8 to 10 found the algorithm best known dfree in some of experimental runs. Only for interleaver length 7 was not reached the best d_{free}. However, best and average free distances found by HLCGA (which was equal to 10) were superior to average d_{free} for interleaver length 7 (which was equal to 7).

Frankly, the interleaver lengths 2 to 10 are only of limited use. Therefore, the results of the HLCGA evolution of higher length interleavers are of much bigger interest.

HLCGA performed well when evolving the interleaver sizes 20, 40, 80, 160, 320, 640 and 1280. For frame size 20, the average dfree of an evolved interleaver was 12 while the average d_{free} was 10.2. For frame size 160 was the average evolved d_{free} 16.1 beating the average d_{free} 12.4 by 3.7.

The first drop in the value of best and average evolved free distance was observed for the last tested interleaver size 2560. The best evolved d_{free} was 19 while the best evolved d_{free} for the interleaver size 1280 was 21. Similarly, the average evolved d_{free} for $N = 2560$ was 18.1 while the average evolved d_{free} for $N = 1280$ was 18.6. This should not be seen as a disappointing result, since the setup of the algorithm (i.e. maximum number of generations,

Table 1 Comparing average error of random initial factors and CAS suggested initial factors

N	Avg. d_{free}	Max. d_{free}	d_{free} variance	Avg. evolved d_{free}	Max. evolved d_{free}
2	8	8	0	8	8
3	8	8	0	8	8
4	8.125	9	0.109	9	9
5	8.667	10	0.522	10	10
6	8.636	10	0.504	10	0
7	8.835	11	0.604	10	10
8	9.018	11	0.676	10.4	11
9	9.170	11	0.701	10.3	11
10	9.300	11	0.707	11	11
20	10.248	13	0.780	12	12
40	10.939	14	0.666	13.1	14
80	11.602	15	1.130	14.8	15
160	12.412	17	1.720	16.1	17
320	13.144	19	1.824	17.8	18
640	13.715	22	1.842	18.1	19
1280	14.306	22	1.889	18.6	21
2560	14.260	22	1.888	18.1	19

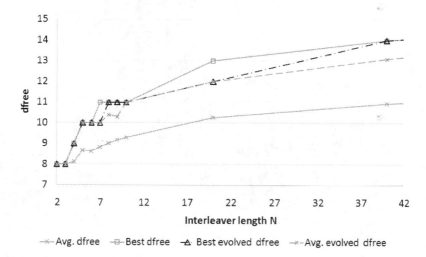

Fig. 10 Average and maximum d_{free} for interleaver sizes 2 to 40

population size) was fixed while the searched domain was growing rapidly. Anyway, the algorithm provided even for interleaver size 2560 average evolved d_{free} 18.1 clearly dominating the average d_{free} of 14.3.

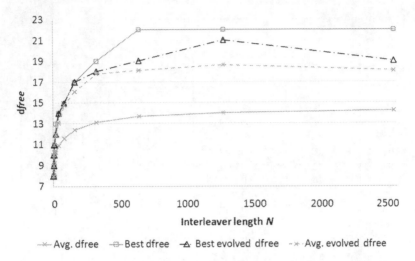

Fig. 11 Average and maximum d_{free} for interleaver sizes 40 to 2560

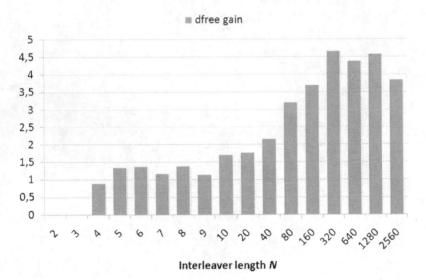

Fig. 12 The d_{free} gain observed when comparing average d_{free} with average d_{free} of an evolved interleaver

7   Conclusions

In this paper we discussed the problem of efficient genetic algorithms for the use in combinatorial problems. A novell genetic method called higher level genetic algorithm was designed to overcome some challenges (especially crossover obligations) raised by the nature of combinatorial problems.

HLCGA was implemented and used for the turbo code interleaver optimization task.

Two variants of HLCGA for turbo code interleaver optimization task were implemented. In the first case, a simulated data transmissions were used to estimate bit error rate that was used as fitness function. The interleavers evolved by HLCGA have shown better performance (i.e. lower BER) than interleavers evolved by mutation-only genetic algorithms and random interleavers. The drawback of this approach is the unstability of simulation as basis for BER evaluation. In any simulation, the BER for a turbo code system could be wrongly evaluated just because 'bad luck'.

The second set of experiments featured an analytical estimate of turbo code free distance as the basis of fitness function. The free distance is known as a very reliable and accurate measure estimating the error floor of a turbo code system. The algorithm was tested in a computer framework and the outputs (i.e. the average evolved d_{free} and the maximum evolved d_{free}) for some interleaver sizes were compared to known properties of the free distance distribution for compatible turbo code system.

Experimental results have shown that the average free distance of an evolved interleaver is higher than the average free distance for interleavers of the same size. The free distance gain, summarized in Figure 12, is especially for larger interleaver sizes significant. Optimized interleavers are, according to computer experiments, better than random interleavers or interleavers deleloped by previous methods. A hardware implementtion of the optimized turbo code systems is currently in progress to provide a final verification of the success of HLCGA in the field of digital communications.

The HLCGA is a general high level variant of genetic algorithms. Its succesfull deployment in the area of turbo code optimization suggests that it might be considered as a general metaheuristic solver also for other combinatorial optimization problems such as the linear ordering problem [1] or the travelling salesman problem. The problems, infamous for its NP-hardness, have a number of real world applications and any contriution in this field is acknowledged.

References

1. Schiavinotto, T., Stützle, T.: The Linear Ordering Problem: Instances, Search Space Analysis and Algorithms. Journal of Mathematical Modelling and Algorithms 4(3), 367–402 (2004)
2. Bäck, T., Hammel, U., Schwefel, H.-P.: Evolutionary computation: comments on the history and current state. IEEE Transactions on Evolutionary Computation 1(1), 3–17 (1997)
3. Berrou, C.: The ten-year-old turbo codes are entering into service. IEEE Communications Magazine 41(8), 110–116 (2003)

4. Berrou, C., Glavieux, A., Thitimajshima, P.: Near Shannon limit error-correcting coding and decoding: turbo codes. In: Proc. Int. Conf. on Commun., pp. 1064–1070 (1993)
5. Bodenhofer, U.: Genetic Algorithms: Theory and Applications. Lecture notes. Fuzzy Logic Laboratorium Linz-Hagenberg (Winter, 2003/2004)
6. Dianati, M., Song, I., Treiber, M.: An introduction to genetic algorithms and evolution strategies. Technical report, University of Waterloo, Ontario, N2L 3G1, Canada (July 2002)
7. Durand, N., Alliot, J., Bartolom, B.: Turbo codes optimization using genetic algorithms. In: Angeline, P.J., Michalewicz, Z., Schoenauer, M., Yao, X., Zalzala, A. (eds.) Proceedings of the Congress on Evolutionary Computation, Mayflower Hotel, Washington D.C., USA, vol. 2, pp. 816–822. IEEE Press, Los Alamitos (1999)
8. Garello, R., Chiaraluce, F., Pierleoni, P., Scaloni, M., Benedetto, S.: On error floor and free distance of turbo codes. In: IEEE International Conference on Communications (ICC 2001), vol. 1, pp. 45–49 (2001)
9. Hokfelt, J., Maseng, T.: Methodical interleaver design for turbo codes. In: International Symposium on Turbo Codes
10. Jones, G.: Genetic and evolutionary algorithms. In: von Rague, P. (ed.) Encyclopedia of Computational Chemistry. John Wiley and Sons, Chichester (1998)
11. Koza, J.: Genetic programming: A paradigm for genetically breeding populations of computer programs to solve problems. Technical Report STAN-CS-90-1314, Dept. of Computer Science, Stanford University (1990)
12. Mitchell, M.: An Introduction to Genetic Algorithms. MIT Press, Cambridge (1996)
13. Proakis, J.G.: Digital Communications, 4th edn. McGraw-Hill, New York (2001)
14. Rekh, S., Rani, S., Hordijk, W., Gift, P., Shanmugam: Design of an interleaver for turbo codes using genetic algorithms. The International Journal of Artificial Intelligence and Machine Learning 6, 1–5 (2006)
15. Snyder, L.V., Daskin, M.S.: A random-key genetic algorithm for the generalized traveling salesman problem. European Journal of Operational Research 174(1), 38–53 (2006)
16. Townsend, H.A.R.: Genetic Algorithms - A Tutorial (2003)
17. Wu, A.S., Lindsay, R.K., Riolo, R.: Empirical observations on the roles of crossover and mutation. In: Bäck, T. (ed.) Proc. of the Seventh Int. Conf. on Genetic Algorithms, pp. 362–369. Morgan Kaufmann, San Francisco (1997)

Bacterial Foraging Optimization Algorithm: Theoretical Foundations, Analysis, and Applications

Swagatam Das, Arijit Biswas, Sambarta Dasgupta, and Ajith Abraham

Abstract. Bacterial foraging optimization algorithm (BFOA) has been widely accepted as a global optimization algorithm of current interest for distributed optimization and control. BFOA is inspired by the social foraging behavior of *Escherichia coli*. BFOA has already drawn the attention of researchers because of its efficiency in solving real-world optimization problems arising in several application domains. The underlying biology behind the foraging strategy of *E.coli* is emulated in an extraordinary manner and used as a simple optimization algorithm. This chapter starts with a lucid outline of the classical BFOA. It then analyses the dynamics of the simulated chemotaxis step in BFOA with the help of a simple mathematical model. Taking a cue from the analysis, it presents a new adaptive variant of BFOA, where the chemotactic step size is adjusted on the run according to the current fitness of a virtual bacterium. Nest, an analysis of the dynamics of reproduction operator in BFOA is also discussed. The chapter discusses the hybridization of BFOA with other optimization techniques and also provides an account of most of the significant applications of BFOA until date.

1 Introduction

Bacteria Foraging Optimization Algorithm (BFOA), proposed by Passino [1], is a new comer to the family of nature-inspired optimization algorithms. For over the last five decades, optimization algorithms like Genetic Algorithms (GAs) [2], Evolutionary Programming (EP) [3], Evolutionary Strategies (ES) [4], which draw their inspiration from evolution and natural genetics, have been dominating the realm of optimization algorithms. Recently natural swarm inspired algorithms like Particle Swarm Optimization (PSO) [5], Ant Colony Optimization (ACO) [6] have

Swagatam Das, Arijit Biswas, and Sambarta Dasgupta
Department of Electronics and Telecommunication Engineering,
Jadavpur University, Kolkata, India

Ajith Abraham
Norwegian University of Science and Technology, Norway
e-mail: ajith.abraham@ieee.org

A. Abraham et al. (Eds.): Foundations of Comput. Intel. Vol. 3, SCI 203, pp. 23–55.
springerlink.com © Springer-Verlag Berlin Heidelberg 2009

found their way into this domain and proved their effectiveness. Following the same trend of swarm-based algorithms, Passino proposed the BFOA in [1]. Application of group foraging strategy of a swarm of *E.coli* bacteria in multi-optimal function optimization is the key idea of the new algorithm. Bacteria search for nutrients in a manner to maximize energy obtained per unit time. Individual bacterium also communicates with others by sending signals. A bacterium takes foraging decisions after considering two previous factors. The process, in which a bacterium moves by taking small steps while searching for nutrients, is called chemotaxis and key idea of BFOA is mimicking chemotactic movement of virtual bacteria in the problem search space.

Since its inception, BFOA has drawn the attention of researchers from diverse fields of knowledge especially due to its biological motivation and graceful structure. Researchers are trying to hybridize BFOA with different other algorithms in order to explore its local and global search properties separately. It has already been applied to many real world problems and proved its effectiveness over many variants of GA and PSO. Mathematical modeling, adaptation, and modification of the algorithm might be a major part of the research on BFOA in future.

This chapter is organized as follows: Section 2 provides the biological motivation behind the BFOA algorithm and outlines the algorithm itself in a comprehensive manner. Section 3 provides a simple mathematical analysis of the computational chemotaxis of BFOA in the framework of the classical gradient descent search algorithm. A mathematical model of reproduction operator is furnished in section 4. Section 5 discusses the hybridization of BFOA with other soft computing algorithms. Section 6 provides an overview of the applications of BFOA in different fields of science and engineering. The chapter is finally summarized in Section 7.

2 The Bacteria Foraging Optimization Algorithm

During foraging of the real bacteria, locomotion is achieved by a set of tensile flagella. Flagella help an *E.coli* bacterium to tumble or swim, which are two basic operations performed by a bacterium at the time of foraging [7, 8]. When they rotate the flagella in the clockwise direction, each flagellum pulls on the cell. That results in the moving of flagella independently and finally the bacterium tumbles with lesser number of tumbling whereas in a harmful place it tumbles frequently to find a nutrient gradient. Moving the flagella in the counterclockwise direction helps the bacterium to swim at a very fast rate. In the above-mentioned algorithm the bacteria undergoes chemotaxis, where they like to move towards a nutrient gradient and avoid noxious environment. Generally the bacteria move for a longer distance in a friendly environment. Figure 1 depicts how clockwise and counter clockwise movement of a bacterium take place in a nutrient solution.

When they get food in sufficient, they are increased in length and in presence of suitable temperature they break in the middle to from an exact replica of itself. This phenomenon inspired Passino to introduce an event of reproduction

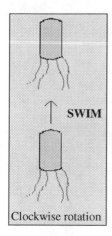 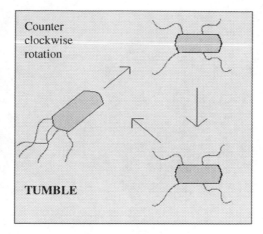

Fig. 1 Swim and tumble of a bacterium

in BFOA. Due to the occurrence of sudden environmental changes or attack, the chemotactic progress may be destroyed and a group of bacteria may move to some other places or some other may be introduced in the swarm of concern. This constitutes the event of elimination-dispersal in the real bacterial population, where all the bacteria in a region are killed or a group is dispersed into a new part of the environment.

Now suppose that we want to find the minimum of $J(\theta)$ where $\theta \in \mathfrak{R}^p$ (i.e. θ is a p-dimensional vector of real numbers), and we do not have measurements or an analytical description of the gradient $\nabla J(\theta)$. BFOA mimics the four principal mechanisms observed in a real bacterial system: chemotaxis, swarming, reproduction, and elimination-dispersal to solve this non-gradient optimization problem. A virtual bacterium is actually one trial solution (may be called a search-agent) that moves on the functional surface (see Figure 2) to locate the global optimum.

Let us define a chemotactic step to be a tumble followed by a tumble or a tumble followed by a run. Let j be the index for the chemotactic step. Let k be the index for the reproduction step. Let l be the index of the elimination-dispersal event. Also let

> p: Dimension of the search space,
> S: Total number of bacteria in the population,
> Nc: The number of chemotactic steps,
> N_s: The swimming length.
> N_{re}: The number of reproduction steps,
> N_{ed}: The number of elimination-dispersal events,
> P_{ed}: Elimination-dispersal probability,
> $C(i)$: The size of the step taken in the random direction specified by the tumble.

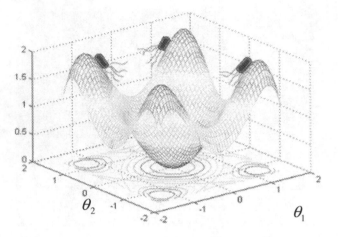

Fig. 2 A bacterial swarm on a multi-modal objective function surface

Let $P(j,k,l) = \{\theta^i(j,k,l) \mid i = 1,2,\dots,S\}$ represent the position of each member in the population of the S bacteria at the j-th chemotactic step, k-th reproduction step, and l-th elimination-dispersal event. Here, let $J(i,j,k,l)$ denote the cost at the location of the i-th bacterium $\theta^i(j,k,l) \in \Re^p$ (sometimes we drop the indices and refer to the i-th bacterium position as θ^i). Note that we will interchangeably refer to J as being a "cost" (using terminology from optimization theory) and as being a nutrient surface (in reference to the biological connections). For actual bacterial populations, S can be very large (e.g., S =109), but p = 3. In our computer simulations, we will use much smaller population sizes and will keep the population size fixed. BFOA, however, allows $p > 3$ so that we can apply the method to higher dimensional optimization problems. Below we briefly describe the four prime steps in BFOA.

i) **Chemotaxis:** This process simulates the movement of an *E.coli* cell through swimming and tumbling via flagella. Biologically an *E.coli* bacterium can move in two different ways. It can swim for a period of time in the same direction or it may tumble, and alternate between these two modes of operation for the entire lifetime. Suppose $\theta^i(j,k,l)$ represents i-th bacterium at j-th chemotactic, k-th reproductive and l-th elimination-dispersal step. $C(i)$ is the size of the step taken in the random direction specified by the tumble (run length unit). Then in computational chemotaxis the movement of the bacterium may be represented by

$$\theta^i(j+1,k,l) = \theta^i(j,k,l) + C(i)\frac{\Delta(i)}{\sqrt{\Delta^T(i)\Delta(i)}}, \tag{1}$$

where Δ indicates a vector in the random direction whose elements lie in [-1, 1].

ii) **Swarming:** An interesting group behavior has been observed for several motile species of bacteria including *E.coli* and *S. typhimurium*, where intricate and stable spatio-temporal patterns (swarms) are formed in semisolid nutrient medium. A group of *E.coli* cells arrange themselves in a traveling ring by moving up the nutrient gradient when placed amidst a semisolid matrix with a single nutrient chemo-effecter. The cells when stimulated by a high level of *succinate*, release an attractant *aspertate*, which helps them to aggregate into groups and thus move as concentric patterns of swarms with high bacterial density. The cell-to-cell signaling in *E. coli* swarm may be represented by the following function.

$$J_{cc}(\theta, P(j,k,l)) = \sum_{i=1}^{S} J_{cc}(\theta, \theta^i(j,k,l))$$

$$= \sum_{i=1}^{S} [-d_{attractant} \exp(-w_{attractant} \sum_{m=1}^{p} (\theta_m - \theta_m^i)^2)] + \sum_{i=1}^{S} [h_{repellant} \exp(-w_{repellant} \sum_{m=1}^{p} (\theta_m - \theta_m^i)^2)]$$

$$(2)$$

where $J_{cc}(\theta, P(j,k,l))$ is the objective function value to be added to the actual objective function (to be minimized) to present a time varying objective function, S is the total number of bacteria, p is the number of variables to be optimized, which are present in each bacterium and $\theta = [\theta_1, \theta_2, ..., \theta_p]^T$ is a point in the p-dimensional search domain. $d_{aatractant}, w_{attractant}, h_{repellant}, w_{repellant}$ are different coefficients that should be chosen properly [1, 9].

iii) **Reproduction:** The least healthy bacteria eventually die while each of the healthier bacteria (those yielding lower value of the objective function) asexually split into two bacteria, which are then placed in the same location. This keeps the swarm size constant.

iv) **Elimination and Dispersal:** Gradual or sudden changes in the local environment where a bacterium population lives may occur due to various reasons e.g. a significant local rise of temperature may kill a group of bacteria that are currently in a region with a high concentration of nutrient gradients. Events can take place in such a fashion that all the bacteria in a region are killed or a group is dispersed into a new location. To simulate this phenomenon in BFOA some bacteria are liquidated at random with a very small probability while the new replacements are randomly initialized over the search space.

The pseudo-code as well as the flow-chart (Figure 3) of the complete algorithm is presented below:

The BFOA Algorithm

Parameters:
[Step 1] Initialize parameters p, S, N_c, N_s, N_{re}, N_{ed}, P_{ed}, $C(i)(i=1,2...S)$, θ^i.

Algorithm:

[Step 2] Elimination-dispersal loop: $l=l+1$

[Step 3] Reproduction loop: $k=k+1$

[Step 4] Chemotaxis loop: $j=j+1$
    [a] For $i =1,2...S$ take a chemotactic step for bacterium i as follows.
    [b] Compute fitness function, $J(i, j, k, l)$.

$$J(i,j,k,l) = J(i,j,k,l) + J_{cc}(\theta^i(j,k,l),P(j,k,l))$$ (i.e. add

on the cell-to cell attractant–repellant profile to simulate the swarming behavior)

where, J_{cc} is defined in (2).

    [c] Let $J_{last}=J(i, j, k, l)$ to save this value since we may find a better cost via a run.
    [d] Tumble: generate a random vector $\Delta(i) \in \Re^p$ with each element

$$\Delta_m(i), m = 1,2,..., p,$$ a random number on [-1, 1].

    [e] Move: Let

$$\theta^i(j+1,k,l) = \theta^i(j,k,l) + C(i)\frac{\Delta(i)}{\sqrt{\Delta^T(i)\Delta(i)}}$$

This results in a step of size $C(i)$ in the direction of the tumble for bacterium i.
    [f] Compute $J(i, j+1,k,l)$ and let

$$J(i,j+1,k,l) = J(i,j,k,l) + J_{cc}(\theta^i(j+1,k,l),P(j+1,k,l)).$$

    [g] Swim
        i) Let $m=0$ (counter for swim length).
        ii) While $m< N_s$ (if have not climbed down too long).
        • Let $m=m+1$.
        • If $J(i, j+1,k,l) < J_{last}$ (if doing better), let $J_{last} = J(i, j+1,k,l)$ and let

$$\theta^i(j+1,k,l) = \theta^i(j,k,l) + C(i)\frac{\Delta(i)}{\sqrt{\Delta^T(i)\Delta(i)}}$$

And use this $\theta^i(j+1, j, k)$ to compute the new J $(i, j+1, k, l)$ as we did in [f]

• Else, let $m = N_s$. This is the end of the while statement.

[h] Go to next bacterium $(i+1)$ if $i \neq S$ (i.e., go to [b] to process the next bacterium).

[**Step 5**] If $j < N_c$, go to step 4. In this case continue chemotaxis since the life of the bacteria is not over.

[**Step 6**] Reproduction:

[a] For the given k and l, and for each $i = 1,2,..., S$, let

$$J^i_{health} = \sum_{j=1}^{N_c+1} J(i, j, k, l) \tag{3}$$

be the health of the bacterium i (a measure of how many nutrients it got over its lifetime and how successful it was at avoiding noxious substances). Sort bacteria and chemotactic parameters $C(i)$ in order of ascending cost J_{health} (higher cost means lower health).

[b] The S_r bacteria with the highest J_{health} values die and the remaining S_r bacteria with the best values split (this process is performed by the copies that are made are placed at the same location as their parent).

[**Step 7**] If $k < N_{re}$, go to step 3. In this case, we have not reached the number of specified reproduction steps, so we start the next generation of the chemotactic loop.

[**Step 8**] Elimination-dispersal: For $i = 1,2..., S$ with probability P_{ed}, eliminate and disperse each bacterium (this keeps the number of bacteria in the population constant). To do this, if a bacterium is eliminated, simply disperse another one to a random location on the optimization domain. If $l < N_{ed}$, then go to step 2; otherwise end.

In Figure 4 we illustrate the behavior of a bacterial swarm on the constant cost contours of the two dimensional sphere model: $f(x_1, x_2) = x_1^2 + x_2^2$. Constant cost contours are curves in $x_1 - x_2$ plane along which $f(x_1, x_2) = x_1^2 + x_2^2 = $ constant.

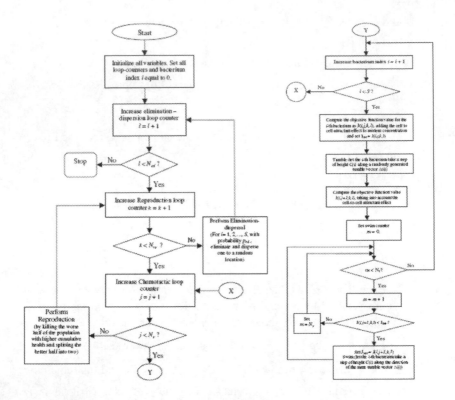

Fig. 3 Flowchart of the Bacterial Foraging Algorithm

3 Analysis of the Chemotactic Dynamics in BFOA

Let us consider a single bacterium cell that undergoes chemotactic steps according to (1) over a single-dimensional objective function space. Since each dimension in simulated chemotaxis is updated independently of others and the only link between the dimensions of the problem space are introduced via the objective functions, an analysis can be carried out on the single dimensional case, without loss of generality. The bacterium lives in continuous time and at the t-th instant its position is given by $\theta(t)$. Next we list a few simplifying assumptions that have been considered for the sake of gaining mathematical insight.

i) The objective function $J(\theta)$ is continuous and differentiable at all points in the search space.

The function is uni-modal in the region of interest and its one and only optimum (minimum) is located at $\theta = \theta_0$. Also $J(\theta) \neq 0$ for $\theta \neq \theta_0$.

ii) The chemotactic step size C is smaller than 1 (Passino himself took $C = 0.1$ in [8]).

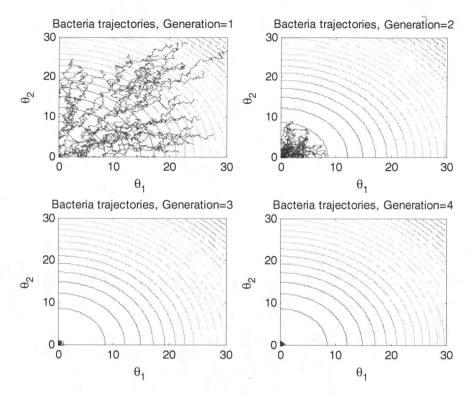

Fig. 4 Convergence behavior of virtual bacteria on the two-dimensional constant cost contours of the sphere model

iii) The analysis applies to the regions of the fitness landscape where gradients of the function are small i.e. near to the optima.

3.1 Derivation of Expression for Velocity

Now, according to BFOA, the bacterium changes its position only if the modified objective function value is less than the previous one i.e. $J(\theta) > J(\theta + \Delta\theta)$ i.e. $J(\theta) - J(\theta + \Delta\theta)$ is positive. This ensures that bacterium always moves in the direction of decreasing objective function value. A particular iteration starts by generating a random number, which assumes only two values with equal probabilities. It is termed as the *direction of tumble* and is denoted by Δ. It can assume only two values 1 or −1 with equal probabilities. For one-dimensional optimization problem Δ is of unit magnitude. The bacterium moves by an amount of $C\Delta$ if objective function value is reduced for new location. Otherwise, its position will not change at all. Assuming uniform rate of position change, if the bacterium moves $C\Delta$ in unit time, its position is changed by $(C\Delta)(\Delta t)$ in Δt sec. It decides to move in the direction in which concentration of nutrient increases or in other words objective function decreases i.e. $J(\theta) - J(\theta + \Delta\theta) > 0$. Otherwise it remains immobile. We have assumed that Δt is an infinitesimally small positive quantity, thus sign of the quantity $J(\theta) - J(\theta + \Delta\theta)$ remains unchanged if Δt divides it. So, bacterium will change its position if and only if $\dfrac{J(\theta) - J(\theta + \Delta\theta)}{\Delta t}$ is positive. This crucial decision making (i.e. whether to take a step or not) activity of the bacterium can be modeled by a unit step function (also known as Heaviside step function [10, 11]) defined as,

$$u(x) = 1, \text{ if } x > 0;$$
$$= 0, \text{ otherwise.} \tag{3}$$

Thus, $\Delta\theta = u(\dfrac{J(\theta) - J(\theta + \Delta\theta)}{\Delta t}).(C.\Delta)(\Delta t)$, where value of $\Delta\theta$ is 0 or $(C\Delta)(\Delta t)$ according to value of the unit step function. Dividing both sides of above relation by Δt we get,

$$\Rightarrow \frac{\Delta\theta}{\Delta t} = u[-\frac{\{J(\theta + \Delta\theta) - J(\theta)\}}{\Delta t}]C.\Delta \tag{4}$$

Velocity is given by, $V_b = \underset{\Delta t \to 0}{Lim}\dfrac{\Delta\theta}{\Delta t} = \underset{\Delta t \to 0}{Lim}[u\{-\dfrac{J(\theta + \Delta\theta) - J(\theta)}{\Delta t}\}.C.\Delta]$

$$\Rightarrow V_b = \underset{\Delta t \to 0}{Lim}[u\{-\frac{J(\theta + \Delta\theta) - J(\theta)}{\Delta\theta}\frac{\Delta\theta}{\Delta t}\}.C.\Delta]$$

as $\Delta t \to 0$ makes $\Delta\theta \to 0$, we may write,

$$V_b = [u\{-\left(\underset{\Delta\theta \to 0}{Lim}\frac{J(\theta + \Delta\theta) - J(\theta)}{\Delta\theta}\right)\left(\underset{\Delta t \to 0}{Lim}\frac{\Delta\theta}{\Delta t}\right)\}.C.\Delta]$$

Again, $J(x)$ is assumed to be continuous and differentiable. $\underset{\Delta\theta \to 0}{Lim}\dfrac{J(\theta + \Delta\theta) - J(\theta)}{\Delta\theta}$ is the value of the gradient at that point and may be denoted by $\dfrac{dJ(\theta)}{d\theta}$ or G. Therefore we have:

$$V_b = u(-GV_b)C\Delta \tag{5}$$

where, $G = \dfrac{dJ(\theta)}{d\theta} = $ gradient of the objective function at θ.

In (5) argument of the unit step function is $-GV_b$. Value of the unit step function is 1 if G and V_b are of different sign and in this case the velocity is $C\Delta$. Otherwise, it is 0 making bacterium motionless. So (5) suggests that bacterium will move the direction of negative gradient. Since the unit step function $u(x)$ has a jump discontinuity at $x = 0$, to simplify the analysis further, we replace $u(x)$ with the continuous logistic function $\phi(x)$, where

$$\phi(x) = \frac{1}{1 + e^{-kx}}$$

We note that,
$$u(x) = \underset{k \to \infty}{Lt}\ \phi(x) = \underset{k \to \infty}{Lt}\ \frac{1}{1 + e^{-kx}} \tag{6}$$

Figure 5 illustrates how the logistic function may be used to approximate the unit step function used for decision-making in chemotaxis. For analysis purpose k cannot be infinity. We restrict ourselves to moderately large values of k (say $k = 10$) for which $\phi(x)$ fairly approximates $u(x)$. Thus, for moderately high values of k $\phi(x)$ fairly approximates $u(x)$. Hence from (5),

$$V_b = \frac{C\Delta}{1 + e^{kGV_b}} \tag{7}$$

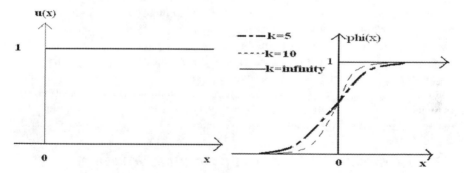

Fig. 5 The unit step and the logistic functions

According to assumptions (ii) and (iii), if C and G are very small and $k \sim 10$, then also we may have $|kGV_b| \ll 1$. In that case we neglect higher order terms in the expansion of e^{kgv_b} and have $e^{kgv_b} \approx 1 + kGV_b$. Substituting it in (7) we obtain,

$$\Rightarrow V_b = \frac{C.\Delta}{2} \frac{1}{1 + \dfrac{kGV_b}{2}}$$

$$\Rightarrow V_b = \frac{C.\Delta}{2}(1 - \frac{kGV_b}{2}) \quad [\because | \frac{kGV_b}{2} | \ll 1, \text{ neglecting higher}$$

$$\text{terms, } (1 + \frac{kGV_b}{2})^{-1} \approx (1 - \frac{kGV_b}{2})]$$

After some manipulation we have,

$$\Rightarrow V_b = \frac{C\Delta}{2} \cdot \frac{1}{1 + \dfrac{kCG\,\Delta}{4}} \tag{8}$$

$$\Rightarrow V_b = \frac{C\Delta}{2}(1 - \frac{kGC\,\Delta}{4}) \quad [\because | \frac{kGC\Delta}{4} | = | \frac{kGC}{4} | \ll 1, \text{ as } |\Delta| = 1 \text{ and}$$

$$\text{neglecting the higher order terms.}]$$

$$\Rightarrow V_b = \frac{C\Delta}{2} - \frac{kGC^2\Delta^2}{8}$$

$$\Rightarrow V_b = -\frac{kC^2}{8}G + \frac{C\Delta}{2} \quad [\because \Delta^2 = 1] \tag{9}$$

Equation (9) is applicable to a single bacterium system and it does not take into account the cell-to-cell signaling effect. A more complex analysis for the two-bacterium system involving the swarming effect has been included at the appendix. It indicates that, a complex perturbation term is added to the dynamics of each bacterium due to the effect of the neighboring bacteria cells. However, the term becomes negligibly small for small enough values of C (~0.1) and the dynamics under these circumstances get practically reduced to that described in equation (9). In what follows, we shall continue the analysis for single bacterium system for better understanding of the chemotactic dynamics.

3.2 Experimental Verification of Expression for Velocity

Characteristic equation of chemotaxis (9) represents the dynamics of bacterium taking chemotactic steps. In order to verify how reliably the equation represents the motion of the virtual bacterium compare results obtained from (10) with that of according to BFOA. First the equation is expressed in iterative form, which is,

$$V_b(n) = \theta(n) - \theta(n-1) = -\frac{kC^2}{8}G(n-1) + \frac{C\Delta(n)}{2}$$

$$\Rightarrow \theta(n) = \theta(n-1) - \frac{kC^2}{8}G(n-1) + \frac{C\Delta(n)}{2} \tag{10}$$

where n is the iteration index. The tumble vector is also a function of iteration count (i.e. chemotactic step number) i.e. it is generated repeatedly for successive iterations. We have taken $J(\theta) = \theta^2$ as objective function for this experimentation. Bacterium was initialized at –2 i.e. $\theta(0) = -2$ and C is taken as 0.2. Gradient of $f(x)$ is $2x$. Therefore $G(n-1)$ may be replaced by $2\theta(n-1)$. Finally for this specific case we get,

$$\theta(n) = (1 - \frac{kC^2}{4})\theta(n-1) + \frac{C\Delta(n)}{2} \tag{11}$$

We compute values of $\theta(n)$ for successive iterations according to above iterative relation. Also values of positions are noted following guidelines of BFOA. With current position is changed by $C\Delta$ if objective function value decreases for new position. Results have been presented in Figure 6. Figure 6 (a) shows position in successive iteration according to BFOA and as obtained from (11). Here also we have assumed position of bacterium changes linearly between two consecutive iterations. Mismatch between actual and predicted values is also shown. In Figure 6 (b) actual and predicted values of velocity is shown. Velocity is assumed to be constant between two successive iterations. According to BFOA magnitude of velocity is either C (0.2 in this case) or 0. Difference between actual and predicted velocity is shown as error. Time lapsed between two consequent iterations is spent for computation and is termed as unit time. This may be perceived as the time required by a bacterium to measure nutrient content of a new point on fitness landscape. Actually it is the time taken by the processor to perform numerical computations.

3.3   Chemotaxis and the Classical Gradient Decent Search

From expression (9) of Section 3.1, we get

$$V_b = -\frac{kC^2}{8}G + \frac{C\Delta}{2} \Rightarrow \frac{d\theta}{dt} = -\alpha'G + \beta' \tag{12}$$

where α' is $\frac{-kC^2}{8}$ and β' is $\frac{C\Delta}{2}$. The classical gradient descent search algorithm is given by the following dynamics in single dimension [12]:

$$\frac{d\theta}{dt} = -\alpha.G + \beta \tag{13}$$

where, α is the learning rate and β is the momentum. Similarity between equations (12) and (13) suggests that chemotaxis may be considered a modified gradient descent search, where α', a function of chemotactic step-size can be identified as the learning rate parameter.

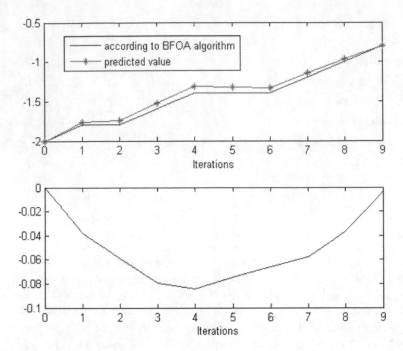

(a) Graphs showing actual, predicted positions of bacterium and error in estimation over successive iterations.

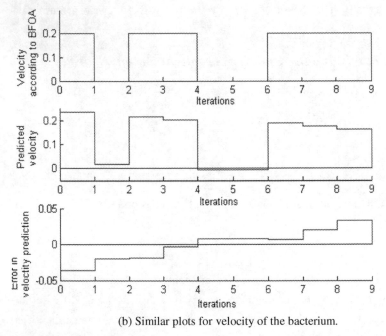

(b) Similar plots for velocity of the bacterium.

Fig. 6 Comparison between actual and predicted motional state of the bacterium

Already we have discussed that magnitude of gradient should be small within the region of our analysis. For chemotaxis of BFOA, when G becomes very mall, the gradient descent term $\alpha' G$ of equation (12) becomes ineffective. But the random search term $\dfrac{C\Delta}{2}$ plays an important role in this context. From equation (12), considering $G \to 0$, we have

$$\frac{d\theta}{dt} = \frac{C.\Delta}{2} \neq 0 \qquad (14)$$

So there is a convergence towards actual minima. The random search or momentum term $C\Delta \big/ 2$ in the RHS of equation (13) provides an additional feature to the classical gradient descent search. When gradient becomes very small, the random term dominates over   gradient decent term and the bacterium changes its position. But random search term may lead to change in position in the direction of increasing objective function value. If it happens then again magnitude of gradient increases and dominates the random search term.

3.4 Oscillation Problem: Need for Adaptive Chemotaxis

If magnitude of the gradient decreases consistently, near the optima or very close to the optima $\alpha' G$ of expression (12) becomes comparable to β. Then gradually β becomes dominant. When $|G| \mapsto 0, |\dfrac{d\theta}{dt}| \approx |\beta| = \dfrac{C\Delta}{2} = \dfrac{C}{2}$ $\because |\Delta| = 1$. Let us assume the bacterium has reached close to the optimum. But since we obtain $|\dfrac{d\theta}{dt}| = \dfrac{C}{2}$, the bacterium does not stop taking chemotactic steps and oscillates about the optima. This crisis can be remedied if step size C is made adaptive according to the following relation,

$$C = \frac{|J(\theta)|}{|J(\theta)|+\lambda} = \frac{1}{1+ \lambda \big/ |J(\theta)|}, \qquad (15)$$

where λ is a positive constant. Choice of a suitable value for λ has been discussed in the next subsection. Here we have assumed that the global optimum of the cost function is 0. Thus from (25) we see, if $J(\theta) \to 0$, then $C \to 0$. So there would be no oscillation if the bacterium reaches optima because random search term vanishes as $C \to 0$. The functional form given in equation (15) causes C to vanish nears the optima. Besides, it plays another important role described below. From (15), we have, when $J(\theta)$ is large $\dfrac{\lambda}{|J(\theta)|} \to 0$ and consequently $C \to 1$.

The adaptation scheme presented in equation (15) has an important physical significance. If magnitude of cost function is large for an individual bacterium, it is in the vicinity of noxious substance. It will then try to move to a place with better nutrient concentration by taking large steps. On the other hand the bacterium, when in nutrient rich zone i.e. with small magnitude of the objective function value, tries to retain its position. Naturally, its step size becomes small.

The BFOA is made adaptive according to the above rule and its performance improved with respect to speed of convergence, quality of solution and rate of success rate.

3.5 A Special Case

If the optimum value of the objective function is not exactly zero, step-size adapted according to (15) may not vanish near optima. Step-size would shrink if the bacterium comes closer to the optima, but it may not approach zero always. To get faster convergence for such functions it becomes necessary to modify the adaptation scheme. Use of gradient information in the adaptation scheme i.e. making step-size a function of the function-gradient (say $C = C(J(\theta), G)$) may not be practical enough, because in real-life optimization problems, we often deal with discontinuous and non-differentiable functions. In order to make BFOA a general black-box optimizer, our adaptive scheme should be a generalized one performing satisfactorily in these situations too.   Therefore to accelerate the convergence under these circumstances, we propose an alternative adaptation strategy in the following way:

$$C = \frac{\left|J(\theta) - J_{best}\right|}{\left|J(\theta) - J_{best}\right| + \lambda} \tag{16}$$

J_{best} is the objective function value for the globally best bacterium (one with lowest value of objective function). $\left|J(\theta) - J_{best}\right|$ is the deviation in fitness value of an individual bacterium from global best. Expression (16) can be rearranged to give,

$$C = \frac{1}{1 + \dfrac{\lambda}{\left|J(\theta) - J_{best}\right|}}. \tag{17}$$

If a bacterium is far apart from the global best, $\left|J(\theta) - J_{best}\right|$ would be large making $C \approx 1 \because \dfrac{\lambda}{\left|J(\theta) - J_{best}\right|} \rightarrow 0$. On the other hand if another bacterium is very close to it, step size of that bacterium will almost vanish, because $\left|J(\theta) - J_{best}\right|$ becomes small and denominator of (17) grows very large.   The scenario is

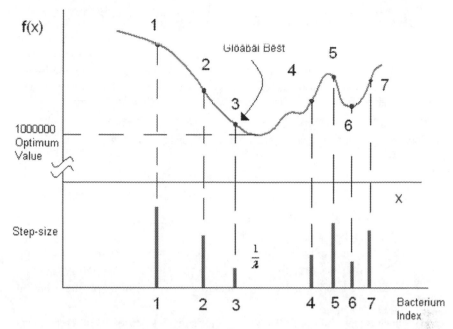

Fig. 7 An objective function with optimum value much greater than zero and a group of seven bacteria are scattered over the fitness landscape. Their step height is also shown

depicted in Figure 7. BFOA with adaptive scheme of equation (15) is referred as ABFOA1 and the BFOA with adaptation scheme described in (17) is referred as ABFOA2.

Figure 7 shows how the step-size becomes large as objective function value becomes large for an individual bacterium. The bacterium with better function value tries to take smaller step and to retain its present position. For best bacterium of the swarm $|J(\theta) - J_{best}|$ is 0. Thus, from (17) its step-size is $\frac{1}{\lambda}$, which is quite small. The adaptation scheme bears a physical significance too. A bacterium located at relatively less nutrient region of fitness landscape will take large step sizes to attain better fitness. Whereas, another bacterium located at a location, best in regard to nutrient content, is unlikely to move much.

In real world optimization problems optimum value of objective function is very often found to be zero. In those cases adaptation scheme of (15) works satisfactorily. But for functions, which do not have a moderate optimum value, (16) should be used for better convergence. Note that neither of two proposed schemes contains derivative of objective function so they can be used for discontinuous and non-differentiable functions as well. In [13], Dasgupta *et al.* have established the efficacy of the adaptive BFO variants by comparing their performances with classical BFOA, its other state-of-the-art variants, a recently proposed variant of PSO and a standard real-coded GA on numerical benchmarks as well as one engineering optimization problem.

4   Analysis of the Reproduction Step in BFOA

This section presents a simple mathematical analysis of the reproduction operator
of BFOA for a two-bacterium system [14]. Let us consider a small population of
two bacteria that sequentially undergoes the four basic steps of BFOA over a one-
dimensional objective function. The bacteria live in continuous time and at the t-th
instant its position is given by $\theta(t)$. Below we list a few assumptions that were
considered for the sake of gaining mathematical insight into the dynamics of reproduction.

Assumptions:

    i)   The objective function $J(\theta)$ is continuous and differentiable at all points
        in the search space.

    ii)   The analysis applies to the regions of the fitness landscape where gradi-
        ents of the function are small i.e., near to the optima. The region of
        fitness landscapes between θ_1 and θ_2 is monotonous at the time of
        reproduction.

    iii) During reproduction, two bacteria remain close to each other and one
        of them must not superpose on another (i.e. $|\theta_2 - \theta_1| \mapsto 0$ may happen
        due to reproduction but $\theta_2 \neq \theta_1$. Suppose P and Q represent the respec-
        tive positions of the two bacteria as shown in fig.6). At the start of
        reproduction θ_1 and θ_2 remain apart from each other but as the process
        progresses they come close to each other gradually.

    iv) The bacterial system lives in continuous time.

4.1   Analytical Treatment

In our two bacterial system, $\theta_1(t)$ and $\theta_2(t)$ represent the position of the two
bacteria at time t and $J(\theta_1), J(\theta_2)$ denote the cost function values at those positions respectively. During reproduction, the virtual bacterium with a relatively
larger value of the cost function (for a minimization problem) is liquidated while
the other is split into two. These two offspring bacteria start moving from the
same location. Hence in effect, through reproduction the least healthy bacteria
shift towards the healthier bacteria. Health of a bacterium is measured in terms of
the accumulated cost function value, possessed by the bacterium until that time
instant. The accumulated cost may be mathematically modeled as $\int_0^t J(\theta_1(t))dt$.

For a minimization problem, higher accumulated cost represents that a bacterium

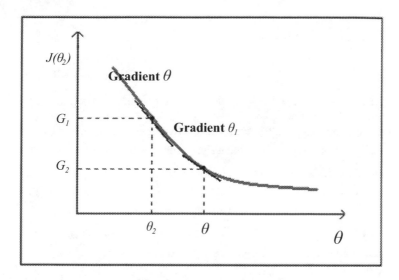

Fig. 8 A two-bacterium system on arbitrary fitness landscape

did not get as many nutrients during its lifetime of foraging and hence is not as "healthy" and thus unlikely to reproduce .The two-bacterial system working on a single-dimensional fitness landscape has been depicted in Figure 8.

To simulate the bacterial reproduction we have to take a decision on which bacterium will split in next generation and which one will die. This decision may be modeled with the help of the well-known unit step function $u(x)$ defined in equation (3). In what follows, we shall denote $\theta_1(t)$ and $\theta_2(t)$ as θ_1 and θ_2 respectively. Now if we consider that $\Delta\theta_1$ is the infinitesimal displacement $(\Delta\theta_1 \to 0)$ of the first bacterium in infinitesimal time Δt $(\Delta t \to 0)$ towards the second bacterium in favorable condition i.e. when the second is healthier than the first one, then the instantaneous velocity of the first one is given by, $\dfrac{\Delta\theta_1}{\Delta t}$.

Now when we are trying to model reproduction we assume the instantaneous velocity of the worse bacterium to be proportional with the distance between the two bacteria, i.e. as they come closer their velocity decreases but this occurs unless we incorporate the decision making part. So, if the first bacterium is the worse one then,

$$\frac{\Delta\theta_1}{\Delta t} \infty (\theta_2 - \theta_1)$$

$$\Rightarrow \frac{\Delta\theta_1}{\Delta t} = \bar{k}(\theta_2 - \theta_1) \qquad \text{[Where, } \bar{k} \text{ is the proportionality constant]}$$

$$\Rightarrow \frac{\Delta\theta_1}{\Delta t} = 1.(\theta_2 - \theta_1) = (\theta_2 - \theta_1) \tag{18}$$

[If we assume that $\overline{k} = 1 \ \sec^{-1}$]

Since we are interested in modeling a dynamics of the reproduction operation, the decision making i.e. whether one of the bacteria will move towards the other, can not be discrete i.e. it is not possible to check straightaway whether the other bacterium is at a better position or not. So a bacterium (suppose θ_1) will be checking whether a position situated at an infinitesimal distance from θ_1 is healthier or not and then it will move (see Figure 9). The health of first bacterium is given by the integral of $J(\theta_1)$ from zero to time t and the same for the differentially placed position is given by the integral of $J(\theta_1 + \Delta\theta_1)$ from zero to time t. Then we may model the decision making part with the unit step function in the following way:

$$\frac{\Delta\theta_1}{\Delta t} = u[\int_0^t J(\theta_1)dt - \int_0^t J(\theta_1 + \Delta\theta_1)dt].(\theta_2 - \theta_1) \tag{19}$$

Similarly, when we consider the second bacterium, we get,

$$\frac{\Delta\theta_2}{\Delta t} = u[\int_0^t J(\theta_2)dt - \int_0^t J(\theta_2 + \Delta\theta_2)dt].(\theta_1 - \theta_2) \tag{20}$$

In equation (19), $\int_0^t J(\theta_1)dt$ represents the health of the first bacterium at the time

instant t and $\int_0^t J(\theta_1 + \Delta\theta_1)dt$ represents the health corresponding to $(\theta_1 + \Delta\theta_1)$

at the time instant t. We are going to carry out calculations with the equation for bacterium 1 only, as the results for other bacterium can be obtained in a similar fashion.

Fig. 9 Change of position of the bacteria during reproduction

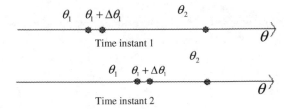

Since we are considering only the monotonous part of any function, so if θ_2 is at a better position, then any position, in-between θ_1 and θ_2, has a lesser objective function value compared to θ_1. So we may conclude $J(\theta_1 + \Delta\theta_1)$ is less

than $J(\theta_1)$. In that case we can imagine that $\int_0^t J(\theta_1 + \Delta\theta_1).dt$ is less than

$\int_0^t J(\theta_1).dt$ as t is not too high, the functional part under consideration is mo-

notonous and change of $\theta_1 + d\theta_1$ with respect to t is same as that of θ_1. We re-
write the equation (19) corresponding to bacterium 1 as,

$$\Rightarrow \frac{\Delta\theta_1}{\Delta t} = u[-\int_0^t \frac{J(\theta_1 + \Delta\theta_1) - J(\theta_1)}{\Delta t} dt](\theta_2 - \theta_1)$$

[∵ $\Delta t > 0$. We know for a positive constant Δt, $u(\frac{x}{\Delta t}) = u(x)$ as x and

$\frac{x}{\Delta t}$ are of same sign and unit step function depends only upon sign of the
argument.]

$$\Rightarrow \underset{\substack{\Delta t \to 0 \\ \Delta\theta_1 \to 0}}{Lt} \frac{\Delta\theta_1}{\Delta t} = \underset{\substack{\Delta t \to 0 \\ \Delta\theta_1 \to 0}}{Lt} u[-\int_0^t \frac{J(\theta_1 + \Delta\theta_1) - J(\theta_1)}{\Delta t} dt].(\theta_2 - \theta_1)$$

$$\Rightarrow \underset{\substack{\Delta t \to 0 \\ \Delta\theta_1 \to 0}}{Lt} \frac{\Delta\theta_1}{\Delta t} = \underset{\substack{\Delta t \to 0 \\ \Delta\theta_1 \to 0}}{Lt} u[-\int_0^t \frac{J(\theta_1 + \Delta\theta_1) - J(\theta_1)}{\Delta\theta_1} \frac{\Delta\theta_1}{\Delta t} dt].(\theta_2 - \theta_1) \tag{21}$$

Again, $J(x)$ is assumed to be continuous and differentiable.
$\underset{\Delta\theta \to 0}{Lim} \frac{J(\theta_1 + \Delta\theta_1) - J(\theta_1)}{\Delta\theta_1}$ is the value of the gradient at that point and may be

denoted by $\frac{dJ(\theta_1)}{d\theta_1}$ or G_1. So we write,

$$\Rightarrow \frac{d\theta_1}{dt} = u[-\int_0^t (\frac{dJ}{d\theta_1} \frac{d\theta_1}{dt}) dt].(\theta_2 - \theta_1) \text{ [Where } \frac{d\theta_1}{dt} \text{ is the instantaneous}$$

velocity of the first bacterium]

$$\Rightarrow v_1 = u[-\int_0^t G_1 v_1 dt].(\theta_2 - \theta_1) \tag{22}$$

[Where $v_1 = \frac{d\theta_1}{dt}$ and G_1 is the gradient of J at $\theta = \theta_1$.]

Now in equation (19) we have not yet considered the fact that the event of reproduction is taking place at t=1 only. So we must introduce a function of

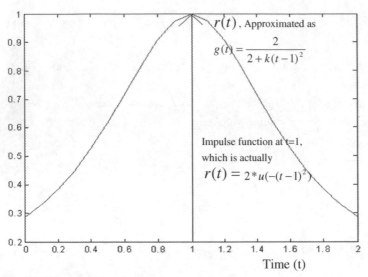

Fig. 10 Function r (t) and g(t)

time $r(t) = 2*u(-(t-1)^2)$ (unit step) ($u(-(t-1)^2)$ is multiplied with 2 for getting $r(t) = 1$, not 0.5, when t=1) in product with the right hand side of equation (19). This provides a sharp impulse of strength 1 unit at time t = 1. Now it is well known that $u(x)$ may be approximated with the continuous logistic function $\phi(x)$, where $\phi(x) = \dfrac{1}{1+e^{-kx}}$.

We note that,

$$u(x) = \underset{k\to\infty}{Lt}\ \phi(x) = \underset{k\to\infty}{Lt}\ \frac{1}{1+e^{-kx}} \tag{23}$$

Following this we may write:

$$r(t) = 2*u(-(t-1)^2) \approx \frac{2}{1+e^{k(t-1)^2}}$$

For moderately large value of k, since $t \to 1$, we can have $\left|k(t-1)^2\right| \ll 1$ and thus $e^{k(t-1)^2} \approx 1+k(t-1)^2$. Using this approximation of the exponential term we may replace the unit step function $r(t)$ with another continuous function g(t) where

$$g(t) = \frac{2}{2+k(t-1)^2}, \qquad \text{(we can take } k = 10)$$

which is not an impulsive function just at $t = 1$ rather a continuous function as shown in Figure 10. Higher value of k will produce more effective result. Due to the presence of this function we see that $v_1 (i.e., \dfrac{d\theta_1}{dt})$ will be maximum at $t = 1$ and decreases drastically when we move away from $t = 1$ in both sides.

So equation (22) is modified and becomes,

$$v_1 = u[-\int_0^t G_1 v_1 dt](\theta_2 - \theta_1).\frac{2}{2 + k(t-1)^2} \qquad (24)$$

For ease of calculation we denote the term within the unit step function as $M = -\int_0^t G_1 v_1 dt$ to obtain,

$$v_1 = u(M)(\theta_2 - \theta_1).\frac{2}{2 + k(t-1)^2} \qquad (25)$$

Since $u(M) = \underset{\alpha \to \infty}{\text{Lt}} \dfrac{1}{1 + e^{-\alpha M}}$

We take a smaller value of α for getting into the mathematical analysis (say $\alpha = 10$). Since, we have the region, under consideration with very low gradient and the velocity of the particle is low, (so product $G_1 v_1$ is also small enough), and the time interval of the integration is not too large (as the time domain under consideration is not large), so we can write, by expanding the exponential part and neglecting the higher order terms:

$$u(M) = \frac{1}{1 + (1 - \alpha M)}$$

$$= \frac{1}{2(1 - \alpha M / 2)}$$

Putting this expression in equation (25) we get,

$$v_1 = \frac{1}{2(1 - \alpha M/2)}(\theta_2 - \theta_1)\frac{2}{2(1 + (k/2)(t-1)^2)}$$

$$\Rightarrow \frac{v_1}{\theta_2 - \theta_1}((1 + (k/2)(t-1)^2) = \frac{1}{2}(1 + \frac{\alpha M}{2}) \qquad (26)$$

$$[\because |\theta_2 - \theta_1| \mapsto 0 \text{ but } |\theta_2 - \theta_1| \neq 0$$

$$\text{also} \because \left|\frac{\alpha M}{2}\right| \ll 1, \text{ neglecting higher order terms, } (1 - \frac{\alpha M}{2})^{-1} \approx (1 + \frac{\alpha M}{2})]$$

Now the equation given by (26) is true for all values possible values of t, so we can differentiate both sides of it with respect to t and get,

$$\Rightarrow \frac{(\theta_2 - \theta_1)\frac{dv_1}{dt} - v_1(\frac{d\theta_2}{dt} - \frac{d\theta_1}{dt})}{(\theta_2 - \theta_1)^2}((1 + (k/2)(t-1)^2) + \frac{v_1}{\theta_2 - \theta_1}k(t-1) = \frac{1}{4}\frac{d(\alpha M)}{dt}$$

(27)

Now, $\dfrac{d(CM)}{dt} = \dfrac{d(-\alpha \int_0^t v_1 G_1 dt)}{dt} = -\alpha v_1 G_1$ [By putting the expression for M

and applying the Leibniz theorem for differentiating integrals]

So from (27), we get,

$$\frac{(\theta_2 - \theta_1)\frac{dv_1}{dt} - v_1(\frac{d\theta_2}{dt} - \frac{d\theta_1}{dt})}{(\theta_2 - \theta_1)^2}((1 + (k/2)(t-1)^2) + \frac{v_1}{\theta_2 - \theta_1}k(t-1) = -\frac{1}{4}\alpha v_1 G_1$$

Putting $\dfrac{d\theta_1}{dt} = v_1$ and $\dfrac{d\theta_2}{dt} = v_2$ after some further manipulations (where we need to cancel out $(\theta_2 - \theta_1)$, which we can do as $|\theta_2 - \theta_1| \mapsto 0$ towards the end of reproduction but never $|\theta_2 - \theta_1| \neq 0$ according to assumption (iii)), we get,

$$\frac{dv_1}{dt} = -\frac{v_1^2}{\theta_2 - \theta_1} - v_1[\frac{k(t-1)}{1 + (k/2)(t-1)^2} + \frac{\alpha G_1(\theta_2 - \theta_1)}{4(1 + (k/2)(t-1)^2)} - \frac{v_2}{\theta_2 - \theta_1}]$$

$$\Rightarrow \frac{dv_1}{dt} = -Pv_1^2 - Qv_1$$

(28)

Where, $P = \dfrac{1}{\theta_2 - \theta_1}$ and $Q = (\dfrac{k(t-1)}{1 + (k/2)(t-1)^2} + \dfrac{\alpha G_1(\theta_2 - \theta_1)}{4(1 + (k/2)(t-1)^2)} - \dfrac{v_2}{\theta_2 - \theta_1})$

The above equation is for the first bacterium and similarly we can derive the equation for the second bacterium, which looks like,

$$\frac{dv_2}{dt} = -P'v_2^2 - Q'v_2 ,$$

(29)

where, $P' = \dfrac{1}{\theta_1 - \theta_2}$ and $Q' = (\dfrac{k(t-1)}{1 + (k/2)(t-1)^2} + \dfrac{\alpha G_2(\theta_1 - \theta_2)}{4(1 + (k/2)(t-1)^2)} - \dfrac{v_1}{\theta_1 - \theta_2})$

4.2 Physical Significance

A possible way to visualize the effect of the dynamics presented in equations (28) and (29) is to see how the velocities of the bacteria vary over short time intervals over which the coefficients P and Q can be assumed to remain fairly constant. The velocity of a bacterium (which is at the better place) has been plotted over five short time intervals in Figure 11 (P and Q are chosen arbitrarily in those intervals). Note that at the time of reproduction ($t = 1$) the graph is highly steep indicating sharp decrease in velocity.

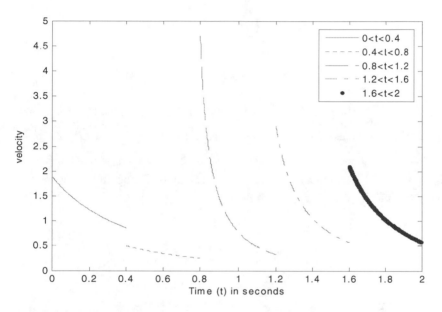

Fig. 11 Piece-wise change in velocity over small time intervals

Now if we study the second term in the expression of Q from equation (28) i.e. the term $\dfrac{\alpha G_1(\theta_2 - \theta_1)}{4(1 + (k/2)(t-1)^2)}$, as $G_1 \to 0$, $(\theta_2 - \theta_1)$ is also small and α is not taken to be very large. At the denominator also we have got some divisors greater than 1. So the term becomes insignificantly small and all we can neglect it from Q. In equation (29) also we can similarly neglect the term $\dfrac{\alpha G_2(\theta_1 - \theta_2)}{4(1 + (k/2)(t-1)^2)}$ from Q'. Again we assume, the velocity of both the particles to be negative for the time being. So we can replace, $v_1 = -|v_1|$ and

$v_2 = -|v_2|$ in Q and Q' in equations (28) and (29). After doing all this simplifications for getting a better mathematical insight, equations (28) and (29) become,

$$\frac{dv_1}{dt} = -Pv_1^2 - Qv_1 \,,$$ (30)

where, $P = \dfrac{1}{\theta_2 - \theta_1}$ and $Q = (\dfrac{k(t-1)}{1 + (k/2)(t-1)^2} + \dfrac{|v_2|}{\theta_2 - \theta_1})$

$$\frac{dv_2}{dt} = -P'v_2^2 - Q'v_2 \,,$$ (31)

where, $P' = \dfrac{1}{\theta_1 - \theta_2}$ and $Q' = (\dfrac{k(t-1)}{1 + (k/2)(t-1)^2} + \dfrac{|v_1|}{\theta_1 - \theta_2})$.

Now, for $\theta_2 > \theta_1$ P and Q are both positive. That means the first bacterium
slows down very quickly. Whereas the second particle has P' and Q' (assuming
the other term independent of $(\theta_1 - \theta_2)$ in Q' is lesser than this) both negative.
That means this bacterium accelerates. This acceleration is hopefully towards the
first bacterium.

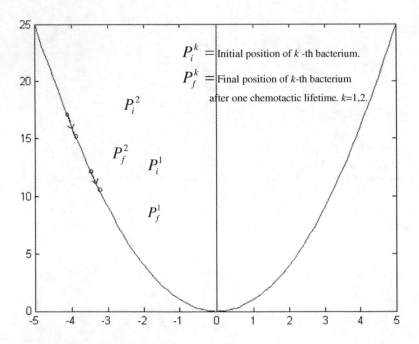

Fig. 12 Initial and final positions of the two bacteria (after one chemotactic lifetime)

Since the rate of change of velocity of bacterium 1 and 2 are dependent on $(\theta_2 - \theta_1)$ and $(\theta_1 - \theta_2)$ respectively, it is evident that the distance between the two bacteria guides their dynamics. If we assume, $\theta_2 > \theta_1$ and they don't traverse too long, the first bacterium is healthier (less accumulated cost) than the second one, when the function is decreasing monotonically in a minimization problem and also the time rate change of first bacterium is less than that of the second (as depicted in Figure 12 clearly, where we take $J(\theta) = \theta^2$).

So at the time of reproduction, in a two bacteria system, the healthier bacterium when senses that it is in a better position compared to its fellow bacterium, it hopes that the optima might be very near so it slows down and its search becomes more fine-tuned. This can be compared with the real bacterium involved in foraging. Whenever it senses that food might be nearby then it obviously slows down and searches that place thoroughly at cost of some time [15 - 17].

The second bacterium moves away from that place with a high acceleration quite naturally getting the information from the first bacterium that the fitter place is away from its present position. In biological system for grouped foraging when one member of the group share information from its neighbors it tries to move towards the best position found out by the neighboring members [15].

Thus we see that reproduction was actually included in BFOA in order to facilitate grouped global search, which is explained from our small analysis.

4.3 Avoiding Premature Convergence

Again if we observe the bacterium at the better position more carefully we will be seeing, that this has a tendency to decelerate at a very high rate and it becomes at rest very quickly. Now when it is near the optima, we can conclude that as $t \to \infty$, $v_{better} \to 0$ (velocity of the better one). Thus as it reaches the optima it stabilize without any further oscillation. Thus reproduction helps the better bacterium to stabilize at the optima.

But the darker side of this fact lies in premature convergence i.e. the best bacterium can converge towards a local optima and the search process gets disturbed. So we understand that at the start of search process reproduction can cause premature convergence but the same can lead to a stable system if applied near the global optima. So we suggest an adaptive scheme related to reproduction operator. The reproduction rate should be made adaptive and it should be increased gradually towards the end of this search process. This has been proved experimentally.

5 Hybridization of BFOA with Other Approaches

We have a handful of optimization algorithms for applying in practical problems but as we know from NFL (No Free Lunch theorem) [18] that no algorithm can perform satisfactorily well over every possible optimization problems. Some algorithms are inspired by natural evolution whereas some are by natural flocking of birds or swarming of bees. Some algorithm can have an extremely good local search behavior while some other can have a good global search property. This may be the reason why hybridization of different algorithms can give better performance as compared to the parent algorithms.

In 2007, Kim *et al.* proposed a hybrid approach involving genetic algorithm (GA) and BFOA for function optimization [19]. The proposed algorithm outperformed both GA and BFOA over a few numerical benchmarks and a practical PID tuner design problem.

Biswas *et al.* coupled BFOA and PSO to develop a new algorithm called BSO (Bacterial Swarm Optimization) [20]. This algorithm provided some very good results when tested over a set of benchmark problems and a difficult engineering problem of spread spectrum radar poly-phase code design. BSO performs local search through the chemotactic movement operation of BFOA whereas a PSO operator accomplishes the global search over the entire search space. In this way it balances between exploration and exploitation, enjoying best of both the worlds. In BSO, after undergoing a chemo-tactic step, a PSO operator also mutates each bacterium. In this phase, the bacterium is stochastically attracted towards the globally best position found so far in the entire population at current time and also towards its previous heading direction. The PSO operator uses only the globally best position found by the entire population to update the velocities of the bacteria and eliminates term involving the personal best position as the local search in different regions of the search space is already taken care of by the chemo-tactic operator of BFOA.

The chemotaxis step of BFOA have been hybridized with another powerful optimization algorithm of current interest called the Differential Evolution (DE) [21] and gave rise to an algorithm known as CDE (Chemotactic Differential Evolution) [22]. Biswas *et al.* proved efficiency of this algorithm too on a set of optimization problems, both numerical benchmark and practical. In this algorithm a bacterium undergoes a differential mutation step just after one chemotaxis step and the rest is kept similar to that of the original BFOA algorithm. Thus each of the bacteria explores the fitness landscape more carefully.

6   Applications of BFOA

Ulagammai *et al.* applied BFOA to train a Wavelet-based Neural Network (WNN) and used the same for identifying the inherent non-linear characteristics of power system loads [23]. In [24], BFOA was used for the dynamical resource allocation in a multiple input/output experimentation platform, which mimics a temperature grid plant and is composed of multiple sensors and actuators organized in zones. Acharya *et al.* proposed a BFOA based Independent Component Analysis (ICA) [25] that aims at finding a linear representation of non-gaussian data so that the components are statistically independent or as independent as possible. The proposed scheme yielded better mean square error performance as compared to a CGAICA (Constrained Genetic Algorithm based ICA). Chatterjee *et al.* reported

Table 1 A Summary of State-of-the-art research works on BFOA

Area of research	Sub-topic	Researchers	References
Hybridization	BFOA-GA, BFOA-PSO, BFOA-DE Hybridization	Dong Hwa Kim, Jae Hoon Cho, Ajith Abraham, Swagatam Das, Arijit Biswas, Sambarta Dasgupta,	[19], [20], [22]
Mathematical Analysis	Chemotaxis, Reproduction, modeling in varying and dynamics environment	Swagatam Das, Sambarta Dasgupta, Arijit Biswas, Ajith Abraham, W. J. Tang, Q. H. Wu, J. R. Saunders	[13], [14], [30], [31]
Modification of BFOA	Adaptive chemotactic step size, modified step size using Hybrid least square-Fuzzy Logic, advanced BFOA using fuzzy logic and clonal selection, BFOA in dynamic environments, BFOA with varying population, cooperative approach to BFOA	Kevin M. Passino, Sambarta Dasgupta, Arijit Biswas, Swagatam Das, Ajith Abraham, Dong Hwa Kim, Jae Hoon Cho, S. Mishra, W. J Tang, Q H Wu, J R Saunders, Carlos Fernandes, Vitorino Ramos, Agostinho C. Rosa, Hanning Chen, Yunlong Zhu, Kunyuan Hu.	[1], [9], [19], [29], [27], [28], [32], [37]
Application in the field of electrical engineering and Control	Optimization of real power loss and voltage stability and distribution static compensator, Harmonic estimation, Active power filter for load compensation, dynamic resource allocation in multi-zone temperature experimentation, PID controller design,	S. Mishra, M. Tripathi, C.N. Bhende, L.L Lai, Mario A. Munoz, Jesus A. Lopez, Eduardo Caicedo, Dong Hwa Kim	[27], [28], [32], [33], [34]
Filtering Problem	Application of BFOA to extended Kalman filter based simultaneous localization and mapping problems	Amitava Chatterjee, Fumitoshi Matsuno	[26]
Learning and Neural network problems	Wavelet neural network training, Optimal learning of Neuro fuzzy structure, Parameter optimization of extreme learning machine	M. Ulagammai, P. Venkatesh, P.S. Kannan, Narayan Prasad Padhy, D.H Kim, Jae-Hoon Cho, Dae-Jong Lee	[23], [35],
Pattern Recognition	Circle detection with Adaptive BFOA, Independent component analysis	Sambarta Dasgupta, Arijit Biswas, Swagatam Das, Ajith Abraham, D P Acharya, G Panda, S Mishra, Y V S Laxmi	[36], [25]
Scheduling Problem	BFOA for job shop scheduling	Chunguo Wu, Na Zhang, Jingqing Jiang, Jinhui Yang and Yanchun Liang	[38]

an interesting application of BFOA in [26] to improve the quality of solutions for the extended Kalman Filters (EKFs), such that the EKFs can offer to solve simultaneous localization and mapping (SLAM) problems for mobile robots and autonomous vehicles.

Tripathy and Mishra proposed an improved BFO algorithm for simultaneous optimization of the real power losses and Voltage Stability Limit (VSL) of a mesh power network [27]. In their modified algorithm, firstly, instead of the average value, the minimum value of all the chemotactic cost functions is retained for deciding the bacterium's health. This speeds up the convergence, because in the average scheme described by Passino [1], it may not retain the fittest bacterium for the subsequent generation. Secondly for swarming, the distances of all the bacteria in a new chemotactic stage are evaluated from the globally optimal bacterium to these points and not the distances of each bacterium from the rest of the others, as suggested by Passino [1]. Simulation results indicated the superiority of the proposed approach over classical BFOA for the multi-objective optimization problem involving the UPFC (Unified Power Flow Controller) location, its series injected voltage, and the transformer tap positions as the variables. Mishra and Bhende used the modified BFOA to optimize the coefficients of Proportional plus Integral (PI) controllers for active power filters [28]. The proposed algorithm was found to outperform a conventional GA with respect to the convergence speed.

Mishra, in [29], proposed a Takagi-Sugeno type fuzzy inference scheme for selecting the optimal chemotactic step-size in BFOA. The resulting algorithm, referred to as Fuzzy Bacterial Foraging (FBF), was shown to outperform both classical BFOA and a Genetic Algorithm (GA) when applied to the harmonic estimation problem. However, the performance of the FBF crucially depends on the choice of the membership function and the fuzzy rule parameters [29] and there is no systematic method (other than trial and error) to determine these parameters for a given problem. Hence FBF, as presented in [29], may not be suitable for optimizing any benchmark function in general. In Table 1 we summarize the current researches on different aspects and applications of BFOA.

7 Conclusions

Search and optimization problems are ubiquitous through the various realms of science and engineering. This chapter has provided a comprehensive overview of one promising real-parameter optimization algorithm called the Bacterial Foraging Optimization Algorithm (BFOA). BFOA is currently gaining popularity due to its efficacy over other swarm and evolutionary computing algorithms in solving engineering optimization problems. It mimics the individual as well as grouped foraging behavior of *E.coli* bacteria that live in our intestine.

The chapter first outlines the classical BFOA in sufficient details. It then develops a simple mathematical model of the simulated chemotaxis operation of BFOA. With the help of this model it analyses the chemotactic dynamics of a single bacterium moving over a one-dimensional fitness landscape. The analysis indicates that the chemotactic dynamics has some striking similarity with the classical gradient descent search although the former never uses an analytic expression of the derivative of the objective function. A problem of oscillations near the optimum is identified from the presented analysis and two adaptation rules for the chemotactic step-height have been proposed to promote the quick convergence of the algorithm near the global optimum of the search space. The chapter also provides an analysis of the reproduction step of BFOA for a two-bacterium system. The analysis reveals how the dynamics of reproduction helps in avoiding premature convergence.

In recent times, a symbiosis of swarm intelligence with other computational intelligence algorithms has opened up new avenues for the next generation computing systems. The chapter presents an account of the research efforts aiming at hybridizing BFOA with other popular optimization techniques like PSO, DE, and GA for improved global search and optimization. It also discusses the significant applications of BFOA in diverse domains of science and engineering. The content of the chapter reveals that engineering search and optimization problems including those from the fields of pattern recognition, bioinformatics, and machine intelligence will find new dimensions in the light of swarm intelligence techniques like BFOA.

References

[1] Passino, K.M.: Biomimicry of Bacterial Foraging for Distributed Optimization and Control. IEEE Control Systems Magazine, 52–67 (2002)
[2] Holland, J.H.: Adaptation in Natural and Artificial Systems. University of Michigan Press, Ann Harbor (1975)
[3] Fogel, L.J., Owens, A.J., Walsh, M.J.: Artificial Intelligence through Simulated Evolution. John Wiley, Chichester (1966)
[4] Rechenberg, I.: Evolutionsstrategie 1994. Frommann-Holzboog, Stuttgart (1994)
[5] Kennedy, J., Eberhart, R.: Particle swarm optimization. In: Proceedings of IEEE International Conference on Neural Networks, pp. 1942–1948 (1995)
[6] Dorigo, M., Gambardella, L.M.: Ant Colony System: A Cooperative Learning Approach to the Traveling Salesman Problem. IEEE Transactions on Evolutionary Computation 1(1), 53–66 (1997)
[7] Berg, H., Brown, D.: Chemotaxis in escherichia coli analysed by three-dimensional tracking. Nature 239, 500–504 (1972)
[8] Berg, H.: Random Walks in Biology. Princeton Univ. Press, Princeton (1993)
[9] Liu, Y., Passino, K.M.: Biomimicry of Social Foraging Bacteria for Distributed Optimization: Models, Principles, and Emergent Behaviors. Journal of Optimization Theory And Applications 115(3), 603–628 (2002)
[10] Abramowitz, M., Stegun, I.A. (eds.): Handbook of Mathematical Functions with Formulas, Graphs, and Mathematical Tables. Dover, New York (1972)
[11] Bracewell, R.: Heaviside's Unit Step Function, H(x), The Fourier Transform and Its Applications, 3rd edn., pp. 57–61. McGraw-Hill, New York (1999)
[12] Snyman, J.A.: Practical Mathematical Optimization: An Introduction to Basic Optimization Theory and Classical and New Gradient-Based Algorithms. Springer Publishing, Heidelberg (2005)
[13] Dasgupta, S., Das, S., Abraham, A., Biswas, A.: Adaptive Computational Chemotaxis in Bacterial Foraging Optimization: An Analysis. IEEE Transactions on Evolutionary Computation (in press, 2009)
[14] Abraham, A., Biswas, A., Dasgupta, S., Das, S.: Anaysis of Reproduction Operator in Bacterial Foraging Optimization. In: IEEE Congress on Evolutionary Computation CEC 2008, IEEE World Congress on Computational Intelligence, WCCI 2008, pp. 1476–1483. IEEE Press, USA (2008)
[15] Murray, J.D.: Mathematical Biology. Springer, New York (1989)
[16] Bonabeau, E., Dorigo, M., Theraulaz, G.: Swarm Intelligence: From Natural to Artificial Systems. Oxford Univ. Press, New York (1999)

[17] Okubo, A.: Dynamical aspects of animal grouping: swarms, schools, flocks, and herds. Advanced Biophysics 22, 1–94 (1986)

[18] Wolpert, D.H., Macready, W.G.: No Free Lunch Theorems for Optimization. IEEE Transactions on Evolutionary Computation 1(1), 67–82 (1997)

[19] Kim, D.H., Abraham, A., Cho, J.H.: A hybrid genetic algorithm and bacterial foraging approach for global optimization. Information Sciences 177(18), 3918–3937 (2007)

[20] Biswas, A., Dasgupta, S., Das, S., Abraham, A.: Synergy of PSO and Bacterial Foraging Optimization: A Comparative Study on Numerical Benchmarks. In: Corchado, E., et al. (eds.) Second International Symposium on Hybrid Artificial Intelligent Systems (HAIS 2007), Innovations in Hybrid Intelligent Systems, ASC. Advances in Soft computing Series, vol. 44, pp. 255–263. Springer, Germany (2007)

[21] Storn, R., Price, K.: Differential evolution – A Simple and Efficient Heuristic for Global Optimization over Continuous Spaces. Journal of Global Optimization 11(4), 341–359 (1997)

[22] Biswas, A., Dasgupta, S., Das, S., Abraham, A.: A Synergy of Differential Evolution and Bacterial Foraging Algorithm for Global Optimization. Neural Network World 17(6), 607–626 (2007)

[23] Ulagammai, L., Vankatesh, P., Kannan, P.S., Padhy, N.P.: Application of Bacteria Foraging Technique Trained and Artificial and Wavelet Neural Networks in Load Forecasting. Neurocomputing, 2659–2667 (2007)

[24] Munoz, M.A., Lopez, J.A., Caicedo, E.: Bacteria Foraging Optimization for Dynamical resource Allocation in a Multizone temperature Experimentation Platform. In: Anal. and Des. of Intel. Sys. using SC Tech., ASC, vol. 41, pp. 427–435 (2007)

[25] Acharya, D.P., Panda, G., Mishra, S., Lakhshmi, Y.V.S.: Bacteria Foaging Based Independent Component Analysis. In: International Conference on Computational Intelligence and Multimedia Applications. IEEE Press, Los Alamitos (2007)

[26] Chatterjee, A., Matsuno, F.: Bacteria Foraging Techniques for Solving EKF-Based SLAM Problems. In: Proc. International Control Conference (Control 2006), Glasgow, UK, August 30- September 01 (2006)

[27] Tripathy, M., Mishra, S.: Bacteria Foraging-Based to Optimize Both Real Power Loss and Voltage Stability Limit. IEEE Transactions on Power Systems 22(1), 240–248 (2007)

[28] Mishra, S., Bhende, C.N.: Bacterial Foraging Technique-Based Optimized Active Power Filter for Load Compensation. IEEE Transactions on Power Delivery 22(1), 457–465 (2007)

[29] Mishra, S.: A hybrid least square-fuzzy bacterial foraging strategy for harmonic estimation. IEEE Trans. on Evolutionary Computation 9(1), 61–73 (2005)

[30] Tang, W.J., Wu, Q.H., Saunders, J.R.: A Novel Model for Bacteria Foraging in Varying Environments. In: Gavrilova, M.L., Gervasi, O., Kumar, V., Tan, C.J.K., Taniar, D., Laganá, A., Mun, Y., Choo, H. (eds.) ICCSA 2006. LNCS, vol. 3980, pp. 556–565. Springer, Heidelberg (2006)

[31] Biswas, A., Das, S., Dasgupta, S., Abraham, A.: Stability Analysis of the Reproduction Operator in Bacterial foraging Optimization. In: IEEE/ACM International Conference on Soft Computing as Transdisciplinary Science and Technology (CSTST 2008), Paris, France, pp. 568–575. ACM Press, New York (2008)

[32] Fernandes, C., Ramos, V., Agostinho, C.: Varying the Population Size of Artificial Foraging Swarms on Time Varying Landscapes. In: Duch, W., Kacprzyk, J., Oja, E., Zadrożny, S. (eds.) ICANN 2005. LNCS, vol. 3696, pp. 311–316. Springer, Heidelberg (2005)

[33] Tripathy, M., Mishra, S., Lai, L.L., Zhang, Q.P.: Transmission Loss Reduction Based on FACTS and Bacteria Foraging Algorithm. In: PPSN, pp. 222–231 (2006)

[34] Mishra, S., Bhende, C.N.: Bacterial Foraging Technique-Based Optimized Active Power Filter for Load Compensation. IEEE Transactions on Power Delivery 22(1), 457–465 (2007)

[35] Kim, D.H., Cho, C.H.: Bacterial Foraging Based Neural Network Fuzzy Learning. In: IICAI 2005, pp. 2030–2036 (2005)

[36] Dasgupta, S., Biswas, A., Das, S., Abraham, A.: Automatic Circle Detection on Images with an Adaptive Bacterial Foraging Algorithm. In: 2008 Genetic and Evolutionary Computation Conference, GECCO 2008, pp. 1695–1696. ACM Press, New York (2008)

[37] Chen, H., Zhu, Y., Hu, K., He, X., Niu, B.: Cooperative Approaches to Bacterial Foraging Optimization. In: ICIC (2), pp. 541–548 (2008)

[38] Wu, C., Zhang, N., Jiang, J., Yang, J., Liang, Y.: Improved Bacterial Foraging Algorithms and Their Applications to Job Shop Scheduling Problems. In: Beliczynski, B., Dzielinski, A., Iwanowski, M., Ribeiro, B. (eds.) ICANNGA 2007. LNCS, vol. 4431, pp. 562–569. Springer, Heidelberg (2007)

Global Optimization Using Harmony Search: Theoretical Foundations and Applications

Zong Woo Geem

Abstract. This chapter presents the theoretical foundations of the music-phenomenon-mimicking optimization technique, harmony search (HS) algorithm. The HS algorithm mimics music improvisation where musicians try to find better harmonies based on randomness or their experiences, which can be expressed as a novel stochastic derivative rather than a calculus-based gradient derivative. The chapter also presents three applications that demonstrate the global optimization feature of the HS algorithm. For the water network design, HS reached global optimum faster than other algorithms such as genetic algorithm, simulated annealing, shuffled frog leaping algorithm, cross entropy algorithm, and scatter search; for the multiple dam scheduling, HS reached five different global optima with identical benefit while the genetic algorithm found a near-optimum under the same number of function evaluations; and HS reached global optimum for the tree-like network layout while genetic algorithm and evolutionary algorithm found near-optimum.

1 Introduction

The harmony search (HS) algorithm [1] mimics how musicians in the improvisation process enrich their experiences by practice in searching for better harmonies. This kind of computational intelligence process can be adopted in the optimization process. A musical instrument in improvisation corresponds to a decision variable in optimization; its pitch range corresponds to a value range; and a harmony corresponds to a solution vector. In the HS algorithm, a musician plays a pitch basically based upon one of three factors: randomness, experience, and variation of experience. This mechanism can be expressed as a novel stochastic derivative, instead of a calculus-based gradient derivative [2] used in mathematical optimization techniques.

The HS algorithm has so far been applied to various real-life, scientific, and engineering optimization problems. Examples include Sudoku puzzle solving [3], tour routing [4], music composition [5], multicast routing [6], web documents clustering [7], structural design [8-11], ground water modeling [12], soil stability analysis [13], vehicle routing [14], fluid network design [15], branched network layout [16], multiple dam operation [17], offshore structure mooring [18], satellite

Zong Woo Geem
Environmental Planning and Management, Johns Hopkins University

A. Abraham et al. (Eds.): Foundations of Comput. Intel. Vol. 3, SCI 203, pp. 57–73.
springerlink.com   © Springer-Verlag Berlin Heidelberg 2009

heat pipe design [19], leakage detection [20], transport energy modeling [21], energy system dispatch [22], heat exchanger design [23], photo-electronic detection [24], ecological conservation [25], model parameter calibration [26-27].

After many successes applying the HS algorithm in various problems, this chapter especially focuses on the global optimization feature of the HS algorithm by reviewing several combinatorial optimization problems. Literature shows that the HS algorithm reached global optimum after searching very few combinations ($8.36 \times 10^{-5}\%$ for water network design; $5.09 \times 10^{-29}\%$ for multiple dam scheduling; and $1.19 \times 10^{-21}\%$ for branched network layout) and performed better than other phenomenon-mimicking algorithms. Thus, this research will show how the HS algorithm works for the global optimization problems in a detailed manner, hoping other researchers will also use this algorithm in their own global optimization problems.

2 Harmony Search Algorithm

The music-inspired HS algorithm evolves a group of harmonies (= solution vectors) using the following steps:

2.1 Problem Formulation

The problem to be solved by HS can be formulated as an optimization problem as follows [28]:

$$\text{Optimize } z(\pmb{x}) \tag{1}$$

$$\text{Subject to } \mathbf{g}(\pmb{x}) \geq 0 \tag{2}$$

$$\mathbf{h}(\pmb{x}) = 0 \tag{3}$$

$$\pmb{x} \in S \tag{4}$$

where $z(\pmb{x})$ = an objective function; $\pmb{x}$ = a solution vector with n decision variables ($x_1, x_2, \ldots, x_n$); $\mathbf{g}(\pmb{x})$ = a vector of inequality constraints; $\mathbf{h}(\pmb{x})$ = a vector of equality constraints; and S = the solution space. Because the decision variable x_i is discrete in combinatorial optimization problems, $S = \{x_i(1), x_i(2), \ldots, x_i(k), \ldots, x_i(K_i)\}$.

2.2 Initialization of Harmony Memory

The HS algorithm starts with a group of harmonies. The value of the decision variable in each harmony can be randomly selected as follows:

$$x_i \leftarrow x \in S_i \tag{5}$$

Then, the random harmonies and corresponding objective function values are stored in harmony memory (HM) expressed as a matrix:

$$\mathbf{HM} = \begin{bmatrix} x_1^1 & x_2^1 & \cdots & x_n^1 & z(\mathbf{x}^1) \\ x_1^2 & x_2^2 & \cdots & x_n^2 & z(\mathbf{x}^2) \\ \vdots & \cdots & \cdots & \cdots & \vdots \\ x_1^{HMS} & x_2^{HMS} & \cdots & x_n^{HMS} & z(\mathbf{x}^{HMS}) \end{bmatrix} \tag{6}$$

In Eq. 6, each row represents each harmony, and the number of total harmonies is HMS (harmony memory size).

In order to start with more converged HM, random harmonies may be generated more than HMS. Then only better harmonies, in terms of the objective function as in Eq. 1, are included in the HM.

2.3 Improvisation of New Harmony

After the HM is prepared, a new harmony, $\mathbf{x}^{New} = (x_1^{New}, x_2^{New}, ..., x_n^{New})$, is improvised based on the HM or random basis.

2.3.1 Experience Consideration

A value of the decision variable x_i^{New} can be chosen from any candidate values stored in the HM with a probability of HMCR (harmony memory considering rate).

$$x_i^{New} \leftarrow x \in \{x_i^1, x_i^2, ..., x_i^{HMS}\} \quad \text{w.p.} \quad HMCR \tag{7}$$

2.3.2 Random Selection

Instead of the above-mentioned experience consideration, a value of the decision variable x_i^{New} can be chosen from any candidate values in total solution space S_i rather than from the HM.

$$x_i^{New} \leftarrow x \in S_i \quad \text{w.p.} \quad (1 - HMCR) \tag{8}$$

2.3.3 Pitch Adjustment

Only if a value of the decision variable x_i^{New} is obtained from the experience consideration operation (not from the random selection operation) in Eq. 7, the value can be further varied into its neighboring values with a probability of PAR (pitch adjusting rate).

$$x_i^{New} \leftarrow \begin{cases} x_i(k+m) & \text{w.p.} \quad PAR \\ x_i(k) & \text{w.p.} \quad (1-PAR) \end{cases} \tag{9}$$

where $x_i(k)$ ($x_i(k) \in S_i$) is identical to the value of the decision variable x_i^{New} obtained in the experience consideration operation; and neighboring index m is an element of the candidate set $\{\ldots, -2, -1, 1, 2, \ldots\}$. Normally, $m \in \{-1, 1\}$.

2.4 Stochastic Derivative

The above three operations (experience consideration, random selection, and pitch adjustment) can be re-written as a stochastic derivative [2] as follows:

$$
\begin{aligned}
\left. \frac{\partial f}{\partial x_i} \right|_{x_i = x_i(k)} &= \frac{1}{K_i} \cdot (1 - HMCR) \\
&+ \frac{n(x_i(k))}{HMS} \cdot HMCR \cdot (1 - PAR) \\
&+ \frac{n(x_i(k \mp m))}{HMS} \cdot HMCR \cdot PAR
\end{aligned}
\tag{10}
$$

The first term in the right hand side of Eq. 10 stands for the probability that $x_i(k)$ ($x_i(k) \in S_i$) is selected for the value of x_i^{New} in the random selection operation; the second term stands for the probability that $x_i(k)$ is selected in the experience consideration operation; and the third term stands for the probability that $x_i(k)$ is selected in the pitch adjustment operation.

Thus, the stochastic derivative for discrete variables in HS denotes the summation of the probabilities that a certain candidate value is selected for the value of x_i^{New} using one of three operations (experience consideration, random selection, and pitch adjustment). As iteration continues, the probability to choose the value of global optimum vector increases [2]. Here, the cumulative density function of total candidate values for each decision variable should be equal to one:

$$\sum_{k=1}^{K_i} \left. \frac{\partial f}{\partial x_i} \right|_{x_i = x_i(k)} = 1 \tag{11}$$

2.5 Optional Operations

Once a new harmony vector x^{New} is improvised based upon the above three basic operations, it can be further processed with other operations such as ensemble consideration and penalty consideration.

2.5.1 Ensemble Consideration

A value of the decision variable x_i^{New} can be chosen from the correlation among decision variables [29]. The value of x_i^{New} is determined based on the value of x_j^{New} where the pairs (x_i^l, x_j^l) in the HM have the highest correlation in terms of statistical determination coefficient.

$$x_i^{New} \leftarrow f(x_j^{New}) \quad \text{where} \quad \max_{i \neq j}\left\{\left[Corr(\boldsymbol{x}_i, \boldsymbol{x}_j)\right]^2\right\} \tag{12}$$

2.5.2 Penalty Consideration

This operation checks if the new harmony $\boldsymbol{x}^{New}$ violates any constraint in the optimization formulation. If $\boldsymbol{x}^{New}$ violates any constraint, a certain amount of penalty is added to the objective function value.

$$z(\boldsymbol{x}^{New}) \leftarrow z(\boldsymbol{x}^{New}) + \left\{\left(p_1 \cdot |\Delta|\right)^{p_2} + p_3\right\} \tag{13}$$

where Δ = quantitative amount of constraint violation; p^1, p^2, and p^3 = penalty coefficients. p^1 makes the penalty amount proportional to the violated amount; p^2 makes the penalty amount exponential to the violated amount; and p^3 contributes constant amount to the penalty amount in order to prevent any slight violation.

2.6 Update of Harmony Memory

If the newly generated harmony vector $\boldsymbol{x}^{New}$ is better than the worst harmony vector $\boldsymbol{x}^{Worst}$ in the HM in terms of the objective function value (including the penalty), the new harmony is included in the HM and the worst vector is excluded from the HM.

$$\left(\boldsymbol{x}^{New} \in \mathbf{HM}\right) \wedge \left(\boldsymbol{x}^{Worst} \notin \mathbf{HM}\right) \tag{14}$$

There may exist a maximum allowed number of identical vectors in order to prevent premature HM.

$$n\left(\boldsymbol{x}^{New} = \boldsymbol{x}^j, \; j \in \{1, 2, ..., HMS\}\right) \leq MaxVec - 1 \tag{15}$$

where $n(\cdot)$ = function which counts identical vectors in HM; and $MaxVec$ = maximum allowed number of identical vectors stored in HM.

2.7 Termination of Computation

If the number of harmony improvisations reaches MaxImp (maximum improvisations) or other criteria, the computation stops. Otherwise, the process described in sections 2.3 - 2.7 is repeated.

2.8 Pseudo Code of the Algorithm

The above-mentioned procedure of the HS algorithm can be expressed as a pseudo code as follows:

```
Procedure HS

  Initialize Harmony Memory (HM)

  While (Not Termination Condition)

    For I = 1 to N Do (N = # of Decision Variables)
      If (UN(1) < HMCR) (UN = Uniform Random Number)
        X[I] = x, x ∈ HM
        If (UN(2) < PAR)
          X[I] = X[I] ± Δ (Δ = Discrete Increment)
        End If
      Else
        X[I] = x, x ∈ Ω (Ω = Value Set)
      End If

      X[I] ∈ X_New

    End Do

    Consider Ensemble of X_New

    Add Penalty to z(X_New)

    If X_New is better than X_Worst
      (X_New ∈ HM) ∧ (X_Worst ∉ HM)
    End If

  End While

End Procedure
```

3   Examples of Global Optimization

In order to show the HS ability to find global optimum in real-world combinatorial optimization problems, three examples are demonstrated: water distribution network design [15], multiple dam scheduling [17], and fluid-transport pipeline layout [16].

3.1   Design of Water Distribution Networks

The HS algorithm was applied to the design of water distribution networks [15]. The water network design means finding minimal cost diameters for all pipes in the system while satisfying water quantity and pressure requirements. Alperovits and Shamir [30] proposed a mathematical technique for this optimization problem. However, because this approach resulted in unrealistic solutions (for example, continuous diameters), researchers have introduced various phenomenon-mimicking algorithms [32-36].

Figure 1 shows an example network which consists of one tank (fixed head = 210 m), six demand nodes, and eight distribution pipes with two loops [30].

The objective function of the problem is a design cost function including penalty function as follows:

$$\text{Minimize } z = \sum_{i=1}^{n} f(D_i) + \sum_{j=1}^{J} \Phi(H_j(\mathbf{D})) \tag{16}$$

where $f(\cdot)$ = cost function of diameter D_i for pipe i; and $\Phi(\cdot)$ = penalty function for deficit nodal pressure head H_j at demand node j. The nodal pressure head at each node is calculated based on pipe diameters.

Constraints for this problem include mass conservation, energy conservation, and minimum head requirement. The constraint of mass conservation at each node can be expressed as follows:

$$\sum Q_j^{in} - \sum Q_j^{out} = d_j \tag{17}$$

where Q_j^{in} = pipe flow into node j; Q_j^{out} = pipe flow out of node j; and d_j = nodal demand at node j.

The constraint of energy conservation at each loop can be expressed as follows:

$$\sum_{i} h_f = 0, \quad i \in k \tag{18}$$

where h_f is pressure head loss due to flow friction along pipe i, and pipe i is an element of loop set k.

Fig. 1 Schematic of Two-Loop Network

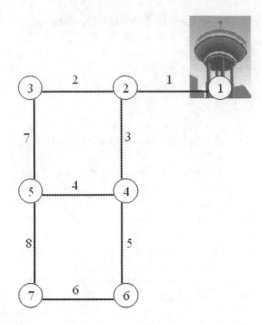

The constraint of minimum head requirement at each node can be expressed as follows:

$$H_j \geq H_j^{\min} \tag{19}$$

where $H_j^{\min}$ = minimum required head for node j.

While mass conservation and energy conservation constraints can be satisfied by using hydraulic software [31], the minimum head requirement constraint can be satisfied by introducing a penalty function added to the objective function as in Eq. 16.

Details of the penalty function are as follows:

$$\Phi(H_j) = p_1\{\max(0, \Delta)\}^{p_2} + p_3 \, \mathrm{sgn}\{\max(0, \Delta)\}, \Delta = H_j^{\min} - H_j \tag{20}$$

The decision variable for this problem is a diameter D_i (unit = inch) for each pipe, which has 14 commercial types as follows:

$$D_i \in S_i = \{1, 2, 3, 4, 6, 8, 10, 12, 14, 16, 18, 20, 22, 24\} \tag{21}$$

Thus, the number of total enumerations of candidate network designs becomes $14^8 = 1.48 \times 10^9$. However, the HS algorithm reached the global optimum [32], z(18, 10, 16, 4, 16, 10, 10, 1) = \$ 419,000, after testing only 1,234 designs, taking 0.5 second on Intel Celeron 1.8GHz CPU. When compared with other phenomenon-mimicking algorithms in terms of the number of iterations, HS (1,234 iterations) required less iterations than genetic algorithm (7,467 iterations) [33], simulated annealing (more than 25,000 iterations) [34], shuffled frog leaping algorithm (11,155 iterations) [35], cross entropy algorithm (35,000 iterations) [36], and scatter search (3,215 iterations) [32].

Figure 2 shows the changes of stochastic derivative values for pipe 1 [2]. Initially, the diameter of 3 inches had the highest chance (0.1601) to be selected. However, the highest chance occurred with 10-inch diameter at 26 iterations (0.2501), 16-inch diameter at 47 and 141 iterations (0.2096 and 0.5651), and finally 18-inch diameter, which is the element of global solution vector, at 585 and 1,234 iterations (0.6326 and 0.6641).

$$\left.\frac{\partial f}{\partial D_1}\right|_{D_1=18} = 0.6641 \tag{22}$$

Here, the stochastic derivative set for each candidate diameter is {1/0.0071, 2/0.0071, 3/0.0071, 4/0.0071, 6/0.0071, 8/0.0071, 10/0.0071, 12/0.0071, 14/0.0071, 16/0.0881, 18/0.6641, 20/0.1601, 22/0.0161, 24/0.0071}. The cumulative density function of total candidate diameters for pipe 1 becomes one:

$$\sum_{k=1}^{14} \left.\frac{\partial f}{\partial D_1}\right|_{D_1=D(k)} = 1 \tag{23}$$

The above computation was performed with HMS = 20, HMCR = 0.9, and PAR = 0.2. However, in order to obtain better algorithm parameters, further sensitivity analysis was performed with HMS = {10, 30, 50}, HMCR = {0.9, 0.95, 0.97}, and PAR = {0.05, 0.1, 0.3}. Out of 270 runs (10 random runs for each parameter combination), minimum and median numbers of function evaluations reaching global optimum were 374 and 1,958, respectively. Also, the analysis found one typical combination (HMS = 50, HMCR = 0.9, PAR = 0.3). With 100 random runs of the combination, global optimum was reached 13 times.

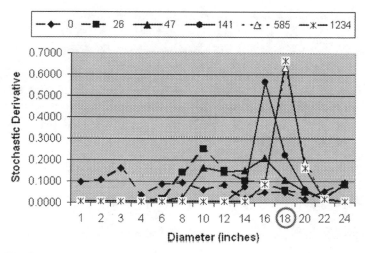

Fig. 2 Stochastic Derivative with Different Iterations

3.2 Scheduling of Multiple Dams

The operation of multiple dams, as shown in Figure 3, is a combinatorial optimization problem of complex decision making process. Traditionally researchers have used mathematical techniques for finding multiple dam schedules. However, because most of the mathematical models are suitable for simplified dam systems, genetic algorithm has gathered attention among researchers. Wardlaw and Sharif [37] proposed a multiple dam scheduling model with various schemes (real-value coding, tournament selection, uniform crossover, and modified uniform mutation).

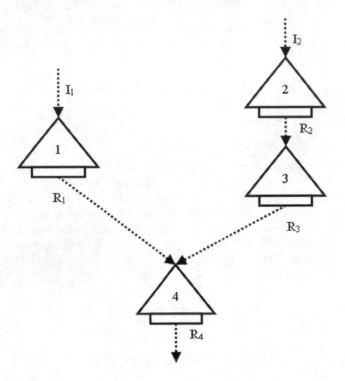

Fig. 3 Four-dam system

The objective function of the problem is both hydropower generation and irrigation benefits as follows:

$$\text{Maximize } z = \sum_i \sum_t p_i(t) \cdot R_i(t) + \sum_i \sum_t b_i(t) \cdot R_i(t) \tag{24}$$

where $R_i(t)$ = discrete water release at time t from dam i; $p_i(t)$ = unit benefit from hydropower generation; $b_i(t)$ = unit benefit from irrigation.

The continuity constraint is as follows:

$$\mathbf{S}_i(t \mid 1) = \mathbf{S}_i(t) + \mathbf{I}_i(t) + \mathbf{M} \cdot \mathbf{R}_t(t) \tag{25}$$

where $\mathbf{S}_i(t)$ = vector of dam storages at time t; $\mathbf{I}_i(t)$ = vector of inflows to dam i; $\mathbf{M}$ = dam connection matrix.

Here, the dam storage $S_i(t)$ should be placed between lower and upper limits as follows:

$$S_i^{MIN}(t) \leq S_i(t) \leq S_i^{MAX}(t) \tag{26}$$

Also, the decision variable $R_i(t)$ should have a discrete value as follows:

$$R_i(t) \in \left\{ R_i^1, R_i^2, \ldots, R_i^{K_i} \right\} \tag{27}$$

The HS algorithm tackled the multiple-dam operation and obtained five different global optima that have an identical objective function value (401.3) [17]. Figure 4 shows one of those global optima.

The HS algorithm tackled the multiple-dam operation and obtained five different global optima that have an identical objective function value (401.3) [17]. Figure 4 shows one of those global optima.

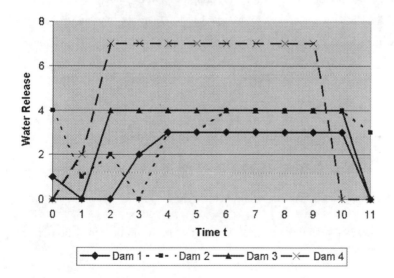

Fig. 4 Optimal Water Releases at Each Dam

While the HS algorithm (HMS = 30, HMCR = 0.95, PAR = 0.05, and MaxImp = 35,000) reached global optima five times, an improved genetic algorithm (binary, gray & real-value representations, tournament selection, three crossover strategies, and uniform & modified uniform mutation) in literature [37] obtained just near-optimum (400.5).

For this combinatorial optimization problem which has 48 decision variables, the number of total enumerations is 6.87×10^{34}. However, the HS algorithm reached five different global optima after testing only 35,000 operations. When a linear programming solver tackled this problem which does not have non-linear functions, it could reach one of global optima. However, it could not provide four alternative global optima that HS found.

3.3   Layout of Fluid-Transport Branched Pipelines

Networks deliver fluid, electricity, or vehicles. While some networks require redundancy, others do not. For example, while urban water supply systems require redundancy (loops) for reliability-purpose, irrigation and sewer systems whose geometric layouts are branched rather than looped rarely require the redundancy [38].

For the problem of a tree-like branched network layout, the minimal spanning tree (MST) algorithm is a popular approach. The MST algorithm finds a branched network which contains all nodes in the network and has no loop with minimum total length.

Figure 5 shows an example of MST. The first network from the left is a base graph which has four nodes and five arcs. The values assigned to the arcs are the lengths between two nodes. For example, the length between node 1 and node 2 is $\sqrt{2}$, and the length between node 1 and node 3 is 1. The other three networks are candidate tree-like layouts. The second one has a total length of $\sqrt{2}$ +2; the third one 3; and the fourth one $\sqrt{2}$ +2. Thus, the third one becomes MST because it has minimal total length.

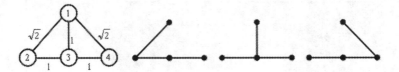

Fig. 5 Base Graph and Three Candidate Branched Layouts

However, this MST technique cannot be applicable to the branched layout for fluid networks because each arc has not only length but also capacity which should satisfy volumetric requirement at each node.

Figure 6 shows an example. The first network from the left is a base graph which has one supply node (node 1) and three demand nodes (node 2 ~ 4). The values assigned to the nodes are fluid demands: the demand for node 2 is 10; the demand for node 3 is 10; and the demand for node 4 is 100. The other three networks are candidate branched layouts.

For the second network, the arc capacity between node 1 and node 2 is 120 because the supply node should provide the total amount of fluid demands. After node 2 consumes 10 units of fluid, the arc capacity between node 2 and node 3

Fig. 6 Base Graph and Three Fluid-Transport Branched Layouts

becomes 110. Likewise, after the node 3 consumes 10 units, the arc capacity between node 2 and node 3 becomes 100. If the cost of each arc can be calculated by arc length × arc capacity, total cost for the second layout is approximately 380 (= $\sqrt{2}$ ×120 + 1 ×110 + 1 ×100), and those of the other two layouts become 230 and 200, respectively. Thus, the fourth layout appears to be the best one, rather than the third one in the MST algorithm.

When the HS algorithm improvises a candidate layout by growing from the source node to all the other nodes, the HS selects the next arc based on good memory rather than the shortest arc in MST algorithm. If a candidate arc is already stored in HM, it has a much higher chance to be selected as the next arc. If node 1 in Figure 7 has three candidate arcs, but only the arc between node 1 and node 4 was already stored in HM, that arc has a much higher chance of HMCR. The other two candidate arcs have chance of 1-HMCR.

Fig. 7 Tree Growing Concept in Harmony Search

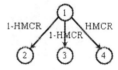

The HS model for finding the best branched layout starts with problem and algorithm initializations. First, the connectivity between node i and node j (node i ≠ node j) is established using the following function.

$$Conn(i, j) = \begin{cases} 0 & \text{if } i \text{ is not connected to } j \\ 1 & \text{if } i \text{ is connected to } j \end{cases} \qquad (28)$$

Once multiple random layouts are generated and stored in HM, one candidate tree layout is generated based on the HM. The candidate tree layout starts growing from the source node. First, the algorithm identifies how many demand nodes are connected to the source node, then it gives certain probability to each arc. If a certain arc is already stored in HM, the arc has a higher chance to be selected for the new layout, as shown in Figure 7. After determining the probabilities of all arcs connected to the source node, one downstream node can be chosen based upon its probability. Then, both the source node and the chosen node become root nodes, and a next downstream node, which does not form a loop, is chosen based on its probability. This procedure continues until all nodes are connected in tree-like manner.

The above-mentioned tree growing technique, rather than random generating technique [39], is a very efficient searching scheme because it is guaranteed to produce one of the complete branched candidate networks that occupy only a small amount of the total of all candidate networks.

Once a candidate layout is generated, the arc capacity between upstream node i and sequel downstream node j is calculated based on the following equation.

$$Capa(i, j) = \sum_{j \in DN} Q_j, \quad i \in UN \tag{29}$$

where DN is the set of downstream nodes; UN is the set of upstream (= root) nodes; and Q_j is the fluid demand at node j.

After each arc capacity is calculated, each arc cost is then calculated by multiplying both arc capacity and arc length as follows:

$$Cost(i, j) = Capa(i, j)^\alpha \times Length(i, j) \tag{30}$$

where α is a coefficient; and $Length(\cdot)$ is a distance function between nodes i and j.

The HS model was applied to the layout design of a branched fluid-transport network which has 64 nodes, as shown in Figure 8 [16].

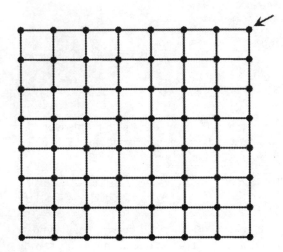

Fig. 8 Base Graph of 64-Node Network

The HS algorithm found the global optimum (layout cost = 5062.8) shown in Figure 9 while the evolutionary algorithm found 5095.2 (0.64% difference from the global optimum) [40] and genetic algorithm found 5218.0 (3.07% difference from the global optimum) [41]. HS reached the global optimum after testing only 1,500 layouts, while the number of total enumerations is 1.26×10^{26}.

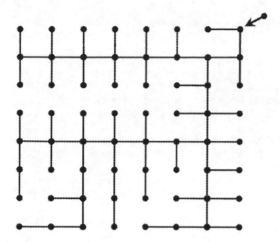

Fig. 9 Optimal Branched Layout for 64-Node Network

4 Conclusions

This chapter presented theoretical foundations of the harmony search algorithm for combinatorial optimization. It also reviewed three applications (water distribution network design, multiple dam scheduling, and fluid-transport branched network layout) where the HS algorithm has successfully found a global optimum solution.

HS, which was inspired from music improvisation, searches better solution vectors based upon factors of randomness, experience, and variation of experience. This stochastic process is represented using a novel stochastic derivative.

In the application of water network design, HS reached global optimum after testing 1,234 designs out of total 1.48×10^9 designs. Also, the stochastic derivative was visualized to show how the chance to select the value of global optimum increases with each iteration.

In the application of multiple dam operation, HS reached global optima after testing 35,000 operations out of total 6.87×10^{34} operations. While GA found a near-optimal solution, HS found five different global optima.

In the application of fluid network layout, HS reached global optimum after testing 1,500 layouts out of total 1.26×10^{26} layouts while evolutionary algorithm and genetic algorithm reached near-optima.

From these successful applications, it is expected that more researchers will use the HS algorithm for global optimization problems in their fields.

References

1. Geem, Z.W., Kim, J.H., Loganathan, G.V.: A new heuristic optimization algorithm: Harmony search. Simulation 76, 60–68 (2001)
2. Geem, Z.W.: Novel Derivative of Harmony Search Algorithm for Discrete Design Variables. Applied Mathematics and Computation 199, 223–230 (2008)

3. Geem, Z.W.: Harmony search algorithm for solving Sudoku. In: Apolloni, B., Howlett, R.J., Jain, L. (eds.) KES 2007, Part I. LNCS (LNAI), vol. 4692, pp. 371–378. Springer, Heidelberg (2007)
4. Geem, Z.W., Tseng, C.L., Park, Y.: Harmony search for generalized orienteering problem: Best touring in China. In: Wang, L., Chen, K., S. Ong, Y. (eds.) ICNC 2005. LNCS, vol. 3612, pp. 741–750. Springer, Heidelberg (2005)
5. Geem, Z.W., Choi, J.Y.: Music composition using harmony search algorithm. In: Giacobini, M. (ed.) EvoWorkshops 2007. LNCS, vol. 4448, pp. 593–600. Springer, Heidelberg (2007)
6. Forsati, R., Haghighat, A.T., Mahdavi, M.: Harmony Search based Algorithms for Bandwidth-Delay-Constrained Least-Cost Multicast Routing. Computer Communications 31, 2505–2519 (2008)
7. Mahdavi, M., Chehreghani, M.H., Abolhassani, H., Forsatia, R.: Novel Meta-Heuristic Algorithms for Clustering Web Documents. Applied Mathematics and Computation 201, 441–451 (2008)
8. Lee, K.S., Geem, Z.W.: A new structural optimization method based on the harmony search algorithm. Computers & Structures 82, 781–798 (2004)
9. Lee, K.S., Geem, Z.W., Lee, S.-H., Bae, K.-W.: The Harmony Search Heuristic Algorithm for Discrete Structural Optimization. Engineering Optimization 37, 663–684 (2005)
10. Saka, M.P.: Optimum Geometry Design of Geodesic Domes Using Harmony Search Algorithm. Advances in Structural Engineering 10, 595–606 (2007)
11. Erdal, F., Saka, M.P.: Effect of Beam Spacing in the Harmony Search Based Optimum Design of Grillages. Asian Journal of Civil Engineering (Building and Housing) 9, 215–228 (2008)
12. Ayvaz, M.T.: Simultaneous Determination of Aquifer Parameters and Zone Structures with Fuzzy C-Means Clustering and Meta-Heuristic Harmony Search Algorithm. Advances in Water Resources 30, 2326–2338 (2007)
13. Cheng, Y.M., Li, L., Lansivaara, T., Chi, S.C., Sun, Y.J.: An Improved Harmony Search Minimization Algorithm Using Different Slip Surface Generation Methods for Slope Stability Analysis. Engineering Optimization 40, 95–115 (2008)
14. Geem, Z.W., Lee, K.S., Park, Y.: Application of Harmony Search to Vehicle Routing. American Journal of Applied Sciences 2, 1552–1557 (2005)
15. Geem, Z.W.: Optimal cost design of water distribution networks using harmony search. Engineering Optimization 38, 259–280 (2006)
16. Geem, Z.W., Park, Y.: Harmony search for Layout of Rectilinear Branched Networks. WSEAS Transactions on Systems 6, 1349–1354 (2006)
17. Geem, Z.W.: Optimal scheduling of multiple dam system using harmony search algorithm. In: Sandoval, F., Prieto, A.G., Cabestany, J., Graña, M. (eds.) IWANN 2007. LNCS, vol. 4507, pp. 316–323. Springer, Heidelberg (2007)
18. Ryu, S., Duggal, A.S., Heyl, C.N., Geem, Z.W.: Offshore mooring cost optimization via harmony search. In: Proceedings of 26th International Conference on Offshore Mechanics and Arctic Engineering, ASME, San Diego, CA, USA (2007) CD-ROM
19. Geem, Z.W., Hwangbo, H.: Application of harmony search to multi-objective optimization for satellite heat pipe design. In: Proceedings of US-Korea Conference on Science, Technology, & Entrepreneurship (UKC 2006), Teaneck, NJ, USA (2006) CD-ROM
20. Kim, S.H., Yoo, W.S., Oh, K.J., Hwang, I.S., Oh, J.E.: Transient analysis and leakage detection algorithm using GA and HS algorithm for a pipeline system. Journal of Mechanical Science and Technology 20, 426–434 (2006)
21. Ceylan, H., Ceylana, H., Haldenbilena, S., Baskan, O.: Transport Energy Modeling with Meta-Heuristic Harmony Search Algorithm, an Application to Turkey. Energy Policy 36, 2527–2535 (2008)

22. Vasebi, A., Fesanghary, M., Bathaeea, S.M.T.: Combined Heat and Power Economic Dispatch by Harmony Search Algorithm. International Journal of Electrical Power & Energy Systems 29, 713–719 (2007)
23. Fesanghary, M., Damangir, E., Soleimani, I.: Design optimization of shell and tube heat exchangers using global sensitivity analysis and harmony search algorithm. Applied Thermal Engineering (2008), doi:10.1016/j.applthermaleng.2008.05.018
24. Dong, H., Bo, Y., Gao, M., Zhu, T.: Improved harmony search for detection with photon density wave. In: Proceedings of International Symposium on Photo-Electronic Detection and Imaging (ISPDI 2007), Beijing, China (2007) CD-ROM
25. Geem, Z.W., Williams, J.C.: Harmony Search and Ecological Optimization. International Journal of Energy and Environment 1, 150–154 (2007)
26. Kim, J.H., Geem, Z.W., Kim, E.S.: Parameter estimation of the nonlinear Muskingum model using harmony search. Journal of the American Water Resources Association 37, 1131–1138 (2001)
27. Zarei, O., Fesanghary, M., Farshi, B., Saffar, R.J., Razfar, M.R.: Optimization of Multi-Pass Face-Milling via Harmony Search Algorithm. Journal of Materials Processing Technology (2008), doi:10.1016/j.jmatprotec.2008.05.029
28. Mays, L.W., Tung, Y.K.: Hydrosystems engineering and management. McGraw-Hill, New York (1992)
29. Geem, Z.W.: Improved harmony search from ensemble of music players. In: Gabrys, B., Howlett, R.J., Jain, L.C. (eds.) KES 2006. LNCS (LNAI), vol. 4251, pp. 86–93. Springer, Heidelberg (2006)
30. Alperovits, E., Shamir, U.: Design of optimal water distribution systems. Water Resources Research 13, 885–900 (1977)
31. Rossman, L.A.: EPANET2 Users Manual. US Environmental Protection Agency. Cincinnati, OH, USA (2000)
32. Lin, M.D., Liu, Y.H., Liu, G.F., Chu, C.W.: Scatter Search Heuristic for Least-Cost Design of Water Distribution Networks. Engineering Optimization 39, 857–876 (2007)
33. Wu, Z.Y., Boulos, P.F., Orr, C.H., Ro, J.J.: Using genetic algorithms to rehabilitate distribution systems. Journal of the American Water Works Association 93, 74–85 (2001)
34. Cunha, M.C., Sousa, J.: Water distribution network design optimization: simulated annealing approach. ASCE Journal of Water Resources Planning and Management 125, 215–221 (1999)
35. Eusuff, M., Lansey, K.E.: Optimization of water distribution network design using the shuffled frog leaping algorithm. Journal of Water Resources Planning and Management, ASCE 129, 210–225 (2003)
36. Perelman, L., Ostfeld, A.: An adaptive heuristic cross-entropy algorithm for optimal design of water distribution systems. Engineering Optimization 39, 413–428 (2007)
37. Wardlaw, R., Sharif, M.: Evaluation of genetic algorithms for optimal reservoir system operation. Journal of Water Resources Planning and Management, ASCE 125, 25–33 (1999)
38. Geem, Z.W.: Geometry Layout for Real-World Tree Networks Using Harmony Search. In: Proceedings of the 3rd Indian International Conference on Artificial Intelligence (IICAI 2007), Pune, India, pp. 268–277 (2007)
39. Hassanli, A.M., Dandy, G.C.: Optimal Layout and Hydraulic Design of Branched Networks Using Genetic Algorithms. Applied Engineering in Agriculture, ASAE 21, 55–62 (2005)
40. Walters, G., Smith, D.: Evolutionary design algorithm for optimal layout of tree networks. Engineering Optimization 24, 261–281 (1995)
41. Walters, G., Lohbeck, T.: Optimal layout of tree networks using genetic algorithms. Engineering Optimization 22, 27–48 (1993)

Hybrid GRASP Heuristics

Paola Festa and Mauricio G.C. Resende

Abstract. Experience has shown that a crafted combination of concepts of different metaheuristics can result in robust combinatorial optimization schemes and produce higher solution quality than the individual metaheuristics themselves, especially when solving difficult real-world combinatorial optimization problems. This chapter gives an overview of different ways to hybridize GRASP (Greedy Randomized Adaptive Search Procedures) to create new and more effective metaheuristics. Several types of hybridizations are considered, involving different constructive procedures, enhanced local search algorithms, and memory structures.

1 Introduction

Combinatorial optimization problems involve a finite number of alternatives: given a finite solution set X and a real-valued objective function $f : X \rightarrow R$, one seeks a solution $x^* \in X$ with $f(x^*) \leq f(x)$, $\forall \ x \in X$. Several combinatorial optimization problems can be solved in polynomial time, but many of them are computationally intractable in the sense that no polynomial time algorithm exists for solving it unless P = NP [27]. Due to the computational complexity of hard combinatorial problems, there has been an extensive research effort devoted to the development of approximation and heuristic algorithms, especially because many combinatorial optimization problems, including routing, scheduling, inventory and production planning, and

Paola Festa
Dept. of Mathematics and Applications, University of Napoli FEDERICO II,
Naples, Italy
e-mail: paola.festa@unina.it

Mauricio G.C. Resende
Algorithms and Optimization Research Dept., AT&T Labs Research, Florham Parl,
NJ USA
e-mail: mgcr@research.att.com

A. Abraham et al. (Eds.): Foundations of Comput. Intel. Vol. 3, SCI 203, pp. 75–100.
springerlink.com

facility location, arise in real-world situations such as in transportation (air, rail, trucking, shipping), energy (electrical power, petroleum, natural gas), and telecommunications (design, location).

To deal with hard combinatorial problems, heuristic methods are usually employed to find good, but not necessarily guaranteed optimal solutions. The effectiveness of these methods depends upon their ability to adapt to a particular realization, avoid entrapment at local optima, and exploit the basic structure of the problem. Building on these notions, various heuristic search techniques have been developed that have demonstrably improved our ability to obtain good solutions to difficult combinatorial optimization problems. One of the most promising of such techniques are usually called *metaheuristics* and include, but are not restricted to, simulated annealing [43], tabu search [28, 29, 32], evolutionary algorithms like genetic algorithms [36], ant colony optimization [19], scatter search [35, 45, 47], path-relinking [30, 31, 33, 34], iterated local search [8, 49], variable neighborhood search [37], and GRASP (Greedy Randomized Adaptive Search Procedures) [21, 22].

Metaheuristics are a class of methods commonly applied to suboptimally solve computationally intractable combinatorial optimization problems. The term metaheuristic derives from the composition of two Greek words: *meta* and *heuriskein*. The suffix 'meta' means 'beyond', 'in an upper level', while 'heuriskein' means 'to find'. In fact, metaheuristics are a family of algorithms that try to combine basic heuristic methods in higher level frameworks aimed at efficiently exploring the set of feasible solution of a given combinatorial problem. In [72] the following definition has been given:

"A metaheuristic is an *iterative master process* that guides and modifies the operations of *subordinate heuristics* to efficiently produce high-quality solutions. It may manipulate a complete (or incomplete) single solution or a collection of solutions at each iteration. The subordinate heuristics may be high (or low) level procedures, or a simple local search, or just a construction method."

Osman and Laporte [52] in their metaheuristics bibliography define a metaheuristics as follows:

"A metaheuristic is formally defined as an *iterative generation process* which guides a *subordinate heuristic* by combining intelligently different concepts for exploring and exploiting the search space, learning strategies are used to structure information in order to find efficiently near-optimal solutions. "

In the last few years, many heuristics that do not follow the concepts of a single metaheuristic have been proposed. These heuristics combine one or more algorithmic ideas from different metaheuristics and sometimes even from outside the traditional field of metaheuristics. Experience has shown that a crafted combination of concepts of different metaheuristics can result in robust combinatorial optimization schemes and produce higher solution quality than the individual metaheuristics themselves. These approaches combining different metaheuristics are commonly referred to as *hybrid metaheuristics*.

This chapter gives an overview of different ways to hybridize GRASP to create new and more effective metaheuristics. Several types of hybridizations are considered, involving different constructive procedures, enhanced local search algorithms, and memory structures.

In Section 2 the basic GRASP components are briefly reviewed. Hybrid construction schemes and hybridization with path-relinking are considered in Sections 3 and 4, respectively.

Hybridization schemes of GRASP with other metaheuristics are explained in Section 5. Concluding remarks are given in the last section.

2   A Basic GRASP

A basic GRASP metaheuristic [21, 22] is a multi-start or iterative method. Given a finite solution set X and a real-valued objective function $f : X \to R$ to be minimized, each GRASP iteration is usually made up of a construction phase, where a feasible solution is constructed, and a local search phase which starts at the constructed solution and applies iterative improvement until a locally optimal solution is found. Repeated applications of the construction procedure yields diverse starting solutions for the local search and the best overall solution is kept as the result.

The construction phase builds a solution x. If x is not feasible, a repair procedure is invoked to obtain feasibility. Once a feasible solution x is obtained, its neighborhood is investigated by the local search until a local minimum is found. The best overall solution is kept as the result. An extensive

```
procedure GRASP(f(·), g(·), MaxIterations, Seed)
1     x_best:=∅; f(x_best):=+∞;
2     for k = 1, 2, . . . ,MaxIterations→
3         x:=ConstructGreedyRandomizedSolution(Seed, g(·));
4         if (x not feasible) then
5             x:=repair(x);
6         endif
7         x:=LocalSearch(x, f(·));
8         if (f(x) < f(x_best)) then
9             x_best:=x;
10        endif
11    endfor;
12    return(x_best);
end GRASP
```

Fig. 1 Pseudo-code of a basic GRASP for a minimization problem

```
procedure ConstructGreedyRandomizedSolution(Seed, g(·))
1    x:=∅;
2    Sort the candidate elements i according to their incremental
     costs g(i);
3    while (x is not a complete solution)→
4        RCL:=MakeRCL();
5        v:=SelectIndex(RCL, Seed);
6        x := x ∪ {v};
7        Resort remaining candidate elements j according to their
         incremental costs g(j);
8    endwhile;
9    return(x);
end ConstructGreedyRandomizedSolution;
```

Fig. 2 Basic GRASP construction phase pseudo-code

survey of the literature is given in [26]. The pseudo-code in Figure 1 illustrates the main blocks of a GRASP procedure for minimization, in which MaxIterations iterations are performed and Seed is used as the initial seed for the pseudorandom number generator.

Starting from an empty solution, in the construction phase, a complete solution is iteratively constructed, one element at a time (see Figure 2). The basic GRASP construction phase is similar to the semi-greedy heuristic proposed independently by [39]. At each construction iteration, the choice of the next element to be added is determined by ordering all candidate elements (i.e. those that can be added to the solution) in a candidate list C with respect to a greedy function $g : C \rightarrow R$. This function measures the (myopic) benefit of selecting each element. The heuristic is adaptive because the benefits associated with every element are updated at each iteration of the construction phase to reflect the changes brought on by the selection of the previous element. The probabilistic component of a GRASP is characterized by randomly choosing one of the best candidates in the list, but not necessarily the top candidate. The list of best candidates is called the *restricted candidate list* (RCL). In other words, the RCL is made up of elements $i \in C$ with the best (i.e., the smallest) incremental costs $g(i)$. There are two main mechanisms to build this list: a *cardinality-based* (CB) and a *value-based* (VB) mechanism. In the CB case, the RCL is made up of the k elements with the best incremental costs, where k is a parameter. In the VB case, the RCL is associated with a parameter $\alpha \in [0, 1]$ and a threshold value $\mu = g_{min} + \alpha(g_{max} - g_{min})$, where g_{min} and g_{max} are the smallest and the largest incremental costs, respectively, i.e.

$$g_{min} = \min_{i \in C} g(i), \qquad g_{max} = \max_{i \in C} g(i). \tag{1}$$

```
procedure LocalSearch(x, f(·))
1      Let N(x) be the neighborhood of x;
2      H:={y ∈ N(x) | f(y) < f(x)};
3      while (|H| > 0)→
4          x:=Select(H);
5          H:={y ∈ N(x) | f(y) < f(x)};
6      endwhile
7      return(x);
end LocalSearch
```

Fig. 3 Pseudo-code of a generic local search procedure

Then, all candidate elements i whose incremental cost $g(i)$ is no greater than the threshold value are inserted into the RCL, i.e. $g(i) \in [g_{min}, \mu]$. Note that, the case $\alpha = 0$ corresponds to a pure greedy algorithm, while $\alpha = 1$ is equivalent to a random construction.

Solutions generated by a GRASP construction are not guaranteed to be locally optimal with respect to simple neighborhood definitions. Hence, it is almost always beneficial to apply a local search to attempt to improve each constructed solution. A local search algorithm iteratively replaces the current solution by a better solution in the neighborhood of the current solution. It terminates when no better solution is found in the neighborhood. The *neighborhood structure* N for a problem relates a solution s of the problem to a subset of solutions $N(s)$. A solution s is said to be *locally optimal* if in $N(s)$ there is no better solution in terms of objective function value. The key to success for a local search algorithm consists of the suitable choice of a neighborhood structure, efficient neighborhood search techniques, and the starting solution. Figure 3 illustrates the pseudo-code of a generic local search procedure for a minimization problem.

It is difficult to formally analyze the quality of solution values found by using the GRASP methodology. However, there is an intuitive justification that views GRASP as a repetitive sampling technique. Each GRASP iteration produces a sample solution from an unknown distribution of all obtainable results. The mean and variance of the distribution are functions of the restrictive nature of the candidate list, as experimentally shown by Resende and Ribeiro in [56].

An especially appealing characteristic of GRASP is the ease with which it can be implemented either sequentially or in parallel, where only a single global variable is required to store the best solution found over all processors. Moreover, few parameters need to be set and tuned, and therefore development can focus on implementing efficient data structures to assure quick GRASP iterations.

3   Hybrid Construction Mechanisms

In this Section, we briefly describe enhancements and alternative techniques for the construction phase of GRASP.

Reactive GRASP

Reactive GRASP is the first enhancement that incorporate a learning mechanism in the memoryless construction phase of the basic GRASP.

The value of the RCL parameter α is selected at each iteration from a discrete set of possible values with a probability that depends on the solution values found along the previous iterations. One way to accomplish this is to use the rule proposed in [53]. Let $\mathcal{A} - \{\alpha_1, \alpha_2, \ldots, \alpha_m\}$ be the set of possible values for α. At the first GRASP iteration, all m values have the same probability to be selected, i.e.

$$p_i = \frac{1}{m}, \qquad i = 1, 2, \ldots, m. \tag{2}$$

At any subsequent iteration, let $\hat{z}$ be the incumbent solution and let A_i be the average value of all solutions found using $\alpha = \alpha_i$, $i = 1, \ldots, m$. The selection probabilities are periodically reevaluated as follows:

$$p_i = \frac{q_i}{\sum_{j=1}^{m} q_j}, \tag{3}$$

where $q_i = \frac{\hat{z}}{A_i}$, $i = 1, \ldots, m$.

Reactive GRASP has been successfully applied in solving several combinatorial optimization problems arising in real-world applications [11, 18].

Cost perturbations

Another step toward an improved and alternative solution construction mechanism is to allow *cost perturbations*. The idea to introduce some "noise" in the original costs in a fashion resembles the noising method of Charon and Hudry [15, 16] and can be usefully applied in all cases when the construction algorithm is not very sensitive to randomization or for the problem to be solved there is available no greedy algorithm for randomization.

Experimental results in the literature have shown that embedding a strategy of costs perturbation into a GRASP framework improves the best overall results. The hybrid GRASP with path-relinking proposed for the Steiner problem in graphs by Ribeiro et al. in [62] uses this cost perturbation strategy and is among the most effective heuristics currently available. Path-relinking will be in detail described in Section 4.

Bias functions

Another construction mechanism has been proposed by Bresina [12]. Once the RCL is built, instead of choosing with equal probability one candidate among the RCL elements, Bresina introduced a family of probability distributions to bias the selection toward some particular candidates. A bias function is based on a rank $r(x)$ assigned to each candidate x according to its greedy function value and is evaluated only for the elements in RCL. Several different bias functions have been introduced:

i.   random bias: bias$(r(x)) = 1$;
ii. linear bias: bias$(r(x)) = \frac{1}{r(x)}$;
iii. log bias: bias$(r(x)) = \log^{-1}[r(x) + 1]$;
iv. exponential bias: bias$(r(x)) = e^{-r}$;
v.   polynomial bias of order n: bias$(r(x)) = r^{-n}$.

Let bias$(r(x))$ be one of the bias function defined above. Once these values have been evaluated for all elements of the RCL, the probability p_x of selecting element x is

$$p_x = \frac{\text{bias}(r(x))}{\sum_{y \in RCL} \text{bias}(r(y))}. \tag{4}$$

A successful application of Bresina's bias function can be found in [10], where experimental results show that the evaluation of bias functions may be restricted only to the elements of the RCL.

Other hybrid construction proposals

Resende and Werneck [57] proposed the following further construction methods:

i.   *Sample greedy construction.*
     Instead of randomizing the greedy algorithm, a greedy algorithm is applied to each solution in a random sample of candidates. At each step, a fixed-size subset of the candidates is sampled and the incremental contribution to the cost of the partial solution is computed for each sampled element. An element with the best incremental contribution is selected and added to the partial solution. This process is repeated until, as before, the construction terminates when no further candidate exists. Resende and Werneck in [57] proposed for the p-median problem a sample greedy construction scheme, whose general framework for a minimization problem is shown in Figure 4.
ii. *Random plus greedy construction.* A partial random solution is built and a greedy algorithm is then applied to complete the construction. The size k of the randomly built portion determines how greedy or random the construction will be. The pseudo-code is reported in Figure 5.
iii. *Proportional greedy construction.*
     In each iteration of proportional greedy, the incremental cost $g(c)$ for every candidate element $c \subset C$ is computed and then a candidate is picked at random, but in a biased way. In fact, the probability of a given candidate $v \in C$ being selected is inversely proportional to $g(v) - \min\{g(c) \mid c \in C\}$.

```
procedure ConstructSampleGreedySolution(Seed, g(·), k)
Let C be the set of candidate elements.
1    x:=∅;
2    while (x is not a complete solution)→
3       RCL:=select-randomly(Seed,k,C);   /*k candidates at random*/
4       Evaluate incremental costs of candidates in RCL;
5       v :=argmin{g(i) | i ∈ RCL};
6       x := x ∪ {v};
7       C := C \ {v};
8    endwhile;
9    return(x);
end ConstructSampleGreedySolution;
```

Fig. 4 Sample greedy GRASP construction phase pseudo-code

```
procedure ConstructRand+GreedySolution(Seed, g(·), k)
Let C be the set of candidate elements.
1    x:=select-randomly(Seed,k,C);   /*k candidates at random*/
2    C := C \ x;
3    while (x is not a complete solution)→
4       Evaluate incremental costs of candidates in C;
5       v :=argmin{g(i) | i ∈ C};
6       x := x ∪ {v};
7       C := C \ {v};
8    endwhile;
9    return(x);
end ConstructRand+GreedySolution;
```

Fig. 5 Random plus greedy GRASP construction phase pseudo-code

4 GRASP and Path-Relinking

Path-relinking is a heuristic proposed in 1996 by Glover [30] as an intensification strategy exploring trajectories connecting elite solutions obtained
by tabu search or scatter search [31, 33, 34]. It can be traced back to the
pioneering work of Kernighan and Lin [42].

The result of the combination of the basic GRASP with path-relinking is
a hybrid technique, leading to significant improvements in solution quality.
The first proposal of a hybrid GRASP with path-relinking is in 1999 due to
Laguna and Martí [46]. It was followed by several extensions, improvements,
and successful applications [5, 13, 24, 25].

Starting from one or more elite solutions, paths in the solution space leading towards other guiding elite solutions are generated and explored in the
search for better solutions. This is accomplished by selecting moves that

introduce attributes contained in the guiding solutions. At each iteration, all moves that incorporate attributes of the guiding solution are analyzed and the move that best improves (or least deteriorates) the initial solution is chosen.

Path-relinking is applied to a pair of solutions $\mathbf{x}, \mathbf{y}$, where one can be the solution obtained from the current GRASP iteration, and the other is a solution from an elite set of solutions. $\mathbf{x}$ is called the *initial solution* and $\mathbf{y}$ the *guiding solution*. The set $\mathcal{E}$ of elite solutions has usually a fixed size that does not exceed MaxElite. Given the pair $\mathbf{x}, \mathbf{y}$, their common elements are kept constant, and the space of solutions spanned by these elements is searched with the objective of finding a better solution. The size of the solution space grows exponentially with the the distance between the *initial* and *guiding* solutions and therefore only a small part of the space is explored by path-relinking. The procedure starts by computing the symmetric difference $\Delta(\mathbf{x}, \mathbf{y})$ between the two solutions, i.e. the set of moves needed to reach $\mathbf{y}$ (target solution) from $\mathbf{x}$ (initial solution). A path of solutions is generated linking $\mathbf{x}$ and $\mathbf{y}$. The best solution x^* in this path is returned by the algorithm.

Let us denote the set of solutions spanned by the common elements of the n-vectors $\mathbf{x}$ and $\mathbf{y}$ as

$$S(\mathbf{x}, \mathbf{y}) := \{w \text{ feasible} \mid w_i = x_i = y_i, i \notin \Delta(\mathbf{x}, \mathbf{y})\} \setminus \{\mathbf{x}, \mathbf{y}\}. \tag{5}$$

Clearly, $|S(\mathbf{x}, \mathbf{y})| = 2^{n - d(\mathbf{x}, \mathbf{y})} - 2$, where $d(\mathbf{x}, \mathbf{y}) = |\Delta(\mathbf{x}, \mathbf{y})|$. The underlying assumption of path-relinking is that there exist good-quality solutions in $S(\mathbf{x}, \mathbf{y})$, since this space consists of all solutions which contain the common elements of two good solutions $\mathbf{x}$ and $\mathbf{y}$. Since the size of this space is exponentially large, a greedy search is usually performed where a path of solutions

$$\mathbf{x} = \mathbf{x}_0, \mathbf{x}_1, \dots, \mathbf{x}_{d(\mathbf{x}, \mathbf{y})}, \mathbf{x}_{d(\mathbf{x}, \mathbf{y})+1} = \mathbf{y}, \tag{6}$$

is built, such that $d(\mathbf{x}_i, \mathbf{x}_{i+1}) = 1$, $i = 0, \dots, d(\mathbf{x}, \mathbf{y})$, and the best solution from this path is chosen. Note that, since both $\mathbf{x}$ and $\mathbf{y}$ are, by construction, local optima in some neighborhood $N(\cdot)^1$, in order for $S(\mathbf{x}, \mathbf{y})$ to contain solutions which are not contained in the neighborhoods of $\mathbf{x}$ or $\mathbf{y}$, $\mathbf{x}$ and $\mathbf{y}$ must be sufficiently distant.

Figure 6 illustrates the pseudo-code of the path-relinking procedure applied to the pair of solutions $\mathbf{x}$ (starting solution) and $\mathbf{y}$ (target solution). In line 1, an initial solution $\mathbf{x}$ is selected at random among the elite set elements and it usually differs sufficiently from the guiding solution $\mathbf{y}$. The loop in lines 6 through 14 computes a path of solutions $\mathbf{x}_1, \mathbf{x}_2, \dots, \mathbf{x}_{d(\mathbf{x}, \mathbf{y})-2}$, and the solution x^* with the best objective function value is returned in line 15. This is achieved by advancing one solution at a time in a greedy manner. At each iteration, the procedure examines all moves $m \in \Delta(x, y)$ from the current

---

1 The same metric $d(\mathbf{x}, \mathbf{y})$ is usually used.

```
procedure Path-relinking(f(·), x, E)
1       Choose, at random, a pool solution y ∈ E to relink with x;
2       Compute symmetric difference Δ(x, y);
3       f* := min{f(x), f(y)};
4       x* := arg min{f(x), f(y)};
5       x := x;
6       while (Δ(x, y) ≠ ∅) →
7              m* := arg min{f(x ⊕ m) | m ∈ Δ(x, y)};
8              Δ(x ⊕ m*, y) := Δ(x, y) \ {m*};
9              x := x ⊕ m*;
10             if (f(x) < f*) then
11                    f* := f(x);
12                    x* := x;
13             endif;
14      endwhile;
15      x* := LocalSearch(x*, f(·));
16      return (x*);
end Path-relinking;
```

Fig. 6 Pseudo-code of a generic path-relinking for a minimization problem

solution x and selects the one which results in the least cost solution (line 7), i.e. the one which minimizes $f(x \oplus m)$, where $x \oplus m$ is the solution resulting from applying move m to solution x. The best move m^* is made, producing solution $x \oplus m^*$ (line 9). The set of available moves is updated (line 8). If necessary, the best solution x^* is updated (lines 10–13). $\Delta(x, \mathbf{y}) = \emptyset$. Since x^* is not guaranteed to be locally optimal, a local search is usually applied and the locally optimal solution is returned.

We now describe a possible way to hybridize the basic GRASP described in Section 2 with path-relinking. The integration of the path-relinking procedure with the basic GRASP is shown in Figure 7. The pool $\mathcal{E}$ of elite solutions is initially empty, and until it reaches its maximum size no path relinking takes place. After a solution $\mathbf{x}$ is found by GRASP, it is passed to the path-relinking procedure to generate another solution. The procedure $\texttt{AddToElite}(\mathcal{E}, x_p)$ attempts to add to the elite set of solutions the currently found solution. Since we wish to maintain a pool of good but diverse solutions, each solution obtained by path-relinking is considered as a candidate to be inserted into the pool if it is sufficiently different from every other solution currently in the pool. If the pool already has $\texttt{MaxElite}$ solutions and the candidate is better than the worst of them, then a simple strategy is to have the former replace the latter. Another strategy, which tends to increase the diversity of the pool, is to replace the pool element most similar to the candidate among all pool elements with cost worse than the candidate's.

More formally, in several papers, a solution x_p is added to the elite set $\mathcal{E}$ if either one of the following conditions holds:

```
procedure GRASP+PR(f(·), g(·), MaxIterations, Seed, MaxElite)
1     x_best:=∅; f(x_best):=+∞; E := ∅
2     for k = 1, 2, ..., MaxIterations→
3         x:=ConstructGreedyRandomizedSolution(Seed, g(·));
4         if (x not feasible) then
5             x:=repair(x);
6         endif
7         x:=LocalSearch(x, f(·));
8         if (k ≤MaxElite) then
9             E := E ∪ {x};
10            if (f(x) < f(x_best)) then
11                x_best:=x;
12            endif
13        else
14            x_p:=Path-relinking(f(·), x, E);
15            AddToElite(E, x_p);
16            if (f(x_p) < f(x_best)) then
17                x_best:=x_p;
18            endif
19        endif
20    endfor;
21    return(x_best);
end GRASP+PR
```

Fig. 7 Pseudo-code of a basic GRASP with path-relinking heuristic for a minimization problem

1. $f(x_p) < \min\{f(\mathbf{w}) : \mathbf{w} \in \mathcal{E}\}$,
2. $\min\{f(\mathbf{w}) : \mathbf{w} \in \mathcal{E}\} \leq f(x_p) < \max\{f(\mathbf{w}) : \mathbf{w} \in \mathcal{E}\}$ and $d(x_p, \mathbf{w}) > \beta n$, $\forall \mathbf{w} \in \mathcal{E}$, where β is a parameter between 0 and 1 and n is the number of decision variables.

If x_p satisfies either of the above, it then replaces an elite solution $\mathbf{z}$ no better than x_p and most similar to x_p, i.e. $\mathbf{z} = \text{argmin}\{d(x_p, \mathbf{w}) : \mathbf{w} \in \mathcal{E}$ such that $f(\mathbf{w}) \geq f(x_p)\}$.

Figure 7 shows the simplest way to combine GRASP with path-relinking, which is applied as an intensification strategy to each local optimum obtained after the GRASP local search phase.

In general, two basic strategies can be used:

i. path-relinking is applied as a post-optimization step to all pairs of elite solutions;
ii. path-relinking is applied as an intensification strategy to each local optimum obtained after the local search phase.

Applying path-relinking as an intensification strategy to each local optimum (strategy ii.) seems to be more effective than simply using it as a post-optimization step [58].

Several further alternatives have been recently considered and combined, all involving the trade-offs between computation time and solution quality. They include:

a. do not apply path-relinking at every GRASP iteration, but only periodically;
b. explore only one path, starting from either **x** (*forward path-relinking*) or **y** (*backward path-relinking*);
c. explore two different paths, using first **x**, then **y** as the initial solution (*forward and backward path-relinking*);
d. do not follow the full path, but instead only part of it (*truncated path-relinking*).

Ribeiro et al. [61] observed that exploring two different paths for each pair (**x**, **y**) takes approximately twice the time needed to explore only one of them, with very marginal improvements in solution quality. They have also observed that if only one path is to be investigated, better solutions are found when path-relinking starts from the best among **x** and **y**. Since the neighborhood of the initial solution is much more carefully explored than that of the guiding one, starting from the best of them gives the algorithm a better chance to investigate in more detail the neighborhood of the most promising solution. For the same reason, the best solutions are usually found closer to the initial solution than to the guiding solution, allowing pruning the relinking path before the latter is reached.

Resende and Ribeiro [55] performed extensive computational experiments, running implementations of GRASP with several different variants of path-relinking. They analyzed the results and illustrated the trade-offs between the different strategies.

5   GRASP and Other Metaheuristics

In this section, we describe and comment on some enhancements of the basic GRASP obtained by hybridization with other approaches and optimization strategies. We also report on experience showing that a crafted combination of concepts of different metaheuristics/techniques can result in robust combinatorial optimization schemes and produce higher solution quality than the individual metaheuristics themselves, especially when solving difficult real-world combinatorial optimization problems.

Most of the GRASP hybrid approaches involve other metaheuristics in the basic local search scheme described in Section 2. They include methods that explore beyond the current solution's neighborhood by allowing cost-increasing moves, by exploring multiple neighborhoods, and by exploring very large neighborhoods.

5.1 GRASP and Tabu Search

Tabu search (TS) is a metaheuristic strategy introduced by Glover [28, 29, 30, 32, 33] that makes use of memory structures to enable escape from local minima by allowing cost-increasing moves. During the search, short-term memory TS uses a special data structure called *tabu list* to store information about solutions generated in the last iterations[2]. The process starts from a given solution and, as any local search heuristic, it moves in iterations from the current solution s to some solution $t \in N(s)$. To avoid returning to a just-visited local minimum, reverse moves mov_t that lead back to that local minimum are forbidden, or made *tabu*, for a number of iterations that can be a priori fixed (fixed sized tabu list) or adaptively varying (variable sized tabu list).

---

procedure TS(x, $f(\cdot)$, k)
1   Let $N(x)$ be the neighborhood of x;
2    $s := x$; $T := \emptyset$; $x_b := x$;
3   **while** (stopping criterion not satisfied)$\rightarrow$
4        $\hat{N}(s) := N(s) \setminus T$;
5        $t :=$argmin$\{f(w) \mid w \in \hat{N}(s)\}$;
6   **if** ($|T| \geq k$) **then**
7   Remove from T the oldest entry;
8   **endif**
9        $T := T \cup \{t\}$;
10   **if** ($f(t) < f(x_b)$) **then**
11          $x_b := t$;
12   **endif**
13      $s := t$;
14   **endwhile**
15   **return**(x_b);
end TS

Fig. 8 Short memory TS pseudo-code for a minimization problem

Figure 8 shows pseudo-code for a short-term TS using a fixed k sized tabu list T, that, for ease of handling, stores the complete solutions t instead of the corresponding moves mov_t.

It is clear that TS can be used as a substitute for the standard local search in a GRASP. This type of search allows the exploration beyond the neighborhood of the greedy randomized solution. By using the number of cost-increasing moves as a stopping criterion one can balance the amount

---

[2] Usually, the tabu list stores all moves that reverse the effect of recent local search steps.

```
procedure simulated-annealing (x, f(·), T, Seed)
1      s := x; x_b := x;
2      while (T > 0 and stopping criterion not satisfied)→
3          t :=select-randomly(Seed, N(s));
4          if (f(t) − f(s) < 0) then
5              s := t;
6              if (f(t) < f(x_b)) then x_b := t;
7              endif
8          else s := t with probability e^{−(f(t)−f(s))/(K·T))};
9          endif
10         Decrement T according to a defined criterion;
11     endwhile
12     return (x_b);
end simulated-annealing
```

Fig. 9 SA pseudo-code for a minimization problem

of time that GRASP allocates to constructing a greedy randomized solution and exploring around that solution with tabu search.

Examples of GRASP with tabu search include [18] for the single source capacitated plant location problem, [1] for multi-floor facility layout, [71] for the capacitated minimum spanning tree problem, [48] for the m-VRP with time windows, and [20] for the maximum diversity problem.

5.2 GRASP and Simulated Annealing

Simulated annealing (SA) [43] is based on principles of mechanical statistics and on the idea of simulating the annealing process of a mechanical system.

It offers a further possibility to enhance the basic GRASP local search phase and pseudo-code in Figure 9 shows how SA can be used as a substitute for the standard local search in a GRASP.

As any stochastic local search procedure, SA is also given a starting solution x which is used to initialize the current solution s. At each iteration, it randomly selects a trial solution $t \in N(s)$. In perfect correspondence of mechanical systems state change rules, if t is an improving solution, then t is made the current solution. Otherwise, t is made the current solution with probability given by

$$e^{-\frac{f(t)-f(s)}{K \cdot T}}, \tag{7}$$

where $f(x)$ is interpreted as the energy of the system in state x, K is the Boltzmann constant, and T a control parameter called the *temperature*.

There are many ways to implement SA, depending on the adopted stopping criterion and on the rule (*cooling schedule*) applied to decrement the temperature parameter T (line 10). Note that, the higher is the temperature T the higher is the probability of moving on a not improving solution t.

Usually, starting from a high initial temperature T_0, at iteration k the cooling schedule changes the temperature by setting $T_{k+1} := T_k \cdot \gamma$, where $0 < \gamma < 1$.

Therefore, initial iterations can be thought of as a diversification phase, where a large part of the solution space is explored. As the temperature cools, fewer non-improving solutions are accepted and those cycles can be thought of as intensification cycles.

To make use of SA as a substitute for the standard local search in GRASP, one should limit the search to the intensification part, since the diversification is already guaranteed by the randomness of the GRASP construction phase. Limitation to only intensification part can be done by starting already with a cool temperature T_0.

Examples of hybrid GRASP with SA include [70] for a simplified fleet assignment problem and [17] for the rural postman problem.

5.3 GRASP, Genetic Algorithms, and Population-Based Heuristics

Evolutionary metaheuristics such as genetic algorithms (GA) [36], ant colony optimization [19], scatter search [35, 45, 47], and evolutionary path-relinking [57] require the generation of an initial population of solutions.

Rooted in the mechanisms of evolution and natural genetics and therefore derived from the principles of natural selection and Darwin's evolutionary theory, the study of heuristic search algorithms with underpinnings in natural and physical processes began as early as the 1970s, when Holland [40] first proposed genetic algorithms. This type of evolutionary technique has been theoretically and empirically proven to be a robust search method [36] having a high probability of locating the global solution optimally in a multimodal search landscape.

In nature, competition among individuals results in the fittest individuals surviving and reproducing. This is a natural phenomenon called *the survival of the fittest*: the genes of the fittest survive, while the genes of weaker individuals die out. The reproduction process generates diversity in the gene pool. Evolution is initiated when the genetic material (chromosomes) from two parents recombines during reproduction. The exchange of genetic material among chromosomes is called *crossover* and can generate good combination of genes for better individuals. Another natural phenomenon called *mutation* causes regenerating lost genetic material. Repeated selection, mutation, and crossover cause the continuous evolution of the gene pool and the generation of individuals that survive better in a competitive environment.

In complete analogy with nature, once encoded each possible point in the search space of the problem into a suitable representation, a GA transforms a population of individual solutions, each with an associated *fitness* (or objective function value), into a new generation of the population. By applying genetic operators, such as crossover and mutation [44], a GA

```
procedure GA(f(·))
1      Let N(x) be the neighborhood of a solution x;
2      k := 0;
3      Initialize population P(0); x_b :=argmin{f(x) | x ∈ P(0)};
4      while (stopping criterion not satisfied)→
5          k := k + 1;
6          Select P(k) from P(k − 1);
7          t :=argmin{f(x) | x ∈ P(k)};
8          if (f(t) < f(x_b)) then
9              x_b := t;
10         endif
11         Alter P(k);
12     endwhile
13     return(x_b);
end GA
```

Fig. 10 Pseudo-code of a generic GA for a minimization problem

successively produces better approximations to the solution. At each iteration, a new generation of approximations is created by the process of selection
and reproduction. In Figure 10 a simple genetic algorithm is described by the
pseudo-code, where $P(k)$ is the population at iteration k.

In solving a given optimization problem $\mathcal{P}$, a GA consists of the following
basic steps.

1. Randomly create an initial population $P(0)$ of individuals, i.e. solutions
   for $\mathcal{P}$.
2. Iteratively perform the following substeps on the current generation of the
   population until the termination criterion has been satisfied.

   a. Assign fitness value to each individual using the fitness function.
   b. Select parents to mate.
   c. Create children from selected parents by crossover and mutation.
   d. Identify the best-so-far individual for this iteration of the GA.

Scatter Search (SS) operates on a *reference set* of solutions, that are combined to create new ones. One way to obtain a new solution is to linearly
combine two reference set solutions. Unlike a GA, the reference set of solutions is relatively small, usually consisting of less than 20 solutions. At the
beginning, a starting set of solutions is generated to guarantee a critical level
of diversity and some local search procedure is applied to attempt to improve
them. Then, a subset of the best solutions is selected as reference set, where
the quality of a solution is evaluated both in terms of objective function and
diversity with other reference set candidates. At each iteration, new solutions
are generated by combining reference set solutions. One criterion used to

select reference solutions for combination takes into account the convex regions spanned by the reference solutions.

Evolutionary path-relinking (EvPR) has been introduced by Resende and Werneck [57] and applied as a post-processing phase for GRASP with PR. In EvPR, the solutions in the pool are evolved as a series of populations $P(1), P(2), \ldots$ of equal size. The initial population $P(0)$ is the pool of elite solutions produced by GRASP with PR. In iteration k, PR is applied between a set of pairs of solutions in population $P(k)$ and, with the same rules used to test for membership in the pool of elite solutions, each resulting solution is tested for membership in population $P(k+1)$. This evolutionary process is repeated until no improvement is seen from one population to the next.

As just described, all above techniques are evolutionary metaheuristics requiring the generation of an initial population of solutions. Usually, these initial solutions are randomly generated, but another way to generate them is to use a GRASP.

Ahuja et al. [4] used a GRASP to generate the initial population of a GA for the quadratic assignment problem. Alvarez et al. [6] proposed a GRASP embedded scatter search for the multicommodity capacitated network design problem. Very recently, Contreras and Díaz used GRASP to initialize the reference set of scatter search for the single source capacitated facility location problem. GRASP with EvPR has been recently used in [59] for the uncapacitated facility location problem and in [54] for the max-min diversity problem.

5.4 GRASP and Variable Neighborhood Search

Almost all randomization effort in implementations of the basic GRASP involves the construction phase. On the other hand, strategies such as Variable Neighborhood Search (VNS) and Variable Neighborhood Descent (VND) [38, 51] rely almost entirely on the randomization of the local search to escape from local optima. With respect to this issue, probabilistic strategies such as GRASP and VNS may be considered as complementary and potentially capable of leading to effective hybrid methods.

The variable neighborhood search (VNS) metaheuristic, proposed by Hansen and Mladenović [38], is based on the exploration of a dynamic neighborhood model. Contrary to other metaheuristics based on local search methods, VNS allows changes of the neighborhood structure along the search.

VNS explores increasingly distant neighborhoods of the current best found solution. Each step has three major phases: neighbor generation, local search, and jump. Let N_k, $k = 1, \ldots, k_{max}$ be a set of pre-defined neighborhood structures and let $N_k(x)$ be the set of solutions in the kth-order neighborhood of a solution x. In the first phase, a neighbor $x' \in N_k(x)$ of the current solution is applied. Next, a solution x'' is obtained by applying local search to x'. Finally, the current solution jumps from x to x'' in case the latter

```
procedure VNS(x, f(·), k_max, Seed)
1      x_b := x; k := 1;
2      while (k ≤ k_max)→
3          x' :=select-randomly(Seed, N_k(x));
4          x'' :=LocalSearch(x', f(·));
5          if (f(x'') < f(x')) then
6              x := x''; k := 1;
7              if (f(x'') < f(x_b)) then x_b := x'';
8              endif
9          else k := k + 1;
10         endif
11     endwhile
12     return(x_b);
end VNS
```

Fig. 11 Pseudo-code of a generic VNS for a minimization problem

improved the former. Otherwise, the order of the neighborhood is increased by one and the above steps are repeated until some stopping condition is satisfied.

Usually, until a stopping criterion is met, VNS generates at each iteration a solution x at random. In hybrid GRASP with VNS, where VNS is applied as local search, the starting solution is the output x of the GRASP construction procedure and the pseudo-code of a generic VNS local search is illustrated in Figure 11.

Examples of GRASP with VNS include [14] for the prize-collecting Steiner tree problem in graphs, [25] for the MAX-CUT problem, and [9] for the strip packing problem.

VND allows the systematic exploration of multiple neighborhoods and is based on the facts that a local minimum with respect to one neighborhood is not necessarily a local minimum with respect to another and that a global minimum is a local minimum with respect to all neighborhoods. VND also is based on the empirical observation that, for many problems, local minima with respect to one or more neighborhoods are relatively close to each other. Since a global minimum is a local minimum with respect to all neighborhoods, it should be easier to find a global minimum if more neighborhoods are explored.

Let $N_k(x)$, for $k = 1, \ldots, k_{max}$, be k_{max} neighborhood structures of solution x. The search begins with a given starting solution x which is made the current solution s. Each major iteration (lines 2–11) searches for an improving solution t in up to k_{max} neighborhoods of s. If no improving solution is found in any of the neighborhoods, the search ends. Otherwise, t is made the current solution s and the search is applied starting from s.

```
procedure VND(x, f(·), k_max)
1      x_b := x; s := x; flag:=true;
2      while (flag)→
3          flag:=false;
4          for k = 1, ..., k_max →
5              if (∃t ∈ N_k(s) | f(t) < f(s)) then
6                  if (f(t) < f(x_b)) then x_b := t;
7                  endif
8                  s := t; flag:=true; break;
9              endif
10         endfor
11     endwhile
12     return(x_b);
end VND
```

Fig. 12 Pseudo-code of a generic VND for a minimization problem

In hybrid GRASP with VND, where VND is applied as local search, the starting solution is the output x of the GRASP construction procedure and the pseudo-code of a generic VND local search is illustrated in Figure 12. A first attempt in the direction of hybridizing GRASP with VNS has been done by Martins et al. [50]. The construction phase of their hybrid heuristic for the Steiner problem in graphs follows the greedy randomized strategy of GRASP, while the local search phase makes use of two different neighborhood structures as a VND strategy. Their heuristic was later improved by Ribeiro, Uchoa, and Werneck [61], one of the key components of the new algorithm being another strategy for the exploration of different neighborhoods. Ribeiro and Souza [60] also combined GRASP with VND in a hybrid heuristic for the degree-constrained minimum spanning tree problem. In the more recent literature, Ribeiro and Vianna [64] and Andrade and Resende [7] proposed a hybrid GRASP with VND for the phylogeny problem and for PBX telephone migration scheduling problem, respectively.

5.5 GRASP and Iterated Local Search

Iterated Local Search (ILS) [49] is a multistart heuristic that at each iteration k finds a locally optimal solution searched in the neighborhood of an initial solution obtained by perturbation of the local optimum found by local search at previous iteration $k - 1$.

The efficiency of ILS strongly depends on the perturbation (line 3) and acceptance criterion (line 5) rules. A "light" perturbation may cause local search to lead back to the starting solution l, while a "strong" perturbation may cause the search to resemble random multi-start. Usually, the acceptance criterion

```
procedure ils (x, f(·), history)
1     t :=LocalSearch(x, f(·)); x_b := t;
2     while (stopping criterion not satisfied)→
3         s :=perturbation(t, history);
4         ŝ :=LocalSearch(s, f(·));
5         t :=AcceptanceCriterion(t, ŝ, history);
6         if (f(t) < f(x_b)) then x_b := t;
7         endif
8     endwhile
9     return (x_b);
end ils
```

Fig. 13 ILS pseudo-code for a minimization problem

resembles SA, i.e. $\hat{s}$ is accepted if it is an improving solution; otherwise, it is accepted with some positive probability.

ILS can be applied to enhance the basic GRASP local search phase and pseudo-code in Figure 13 shows how it can be used as a substitute for the standard local search in a GRASP. The procedure LocalSearch can also be the basic GRASP local search as defined in Figure 3.

Ribeiro and Urrutia [63] designed a hybrid GRASP with ILS for the mirrored traveling tournament problem, where the acceptance rule makes use of a threshold parameter β, initialized to 0.001. Then, each time the best solution changes (line 6), it is reinitialized to the same value, while it is doubled if the current solution does not chance after a fixed number of iterations. Finally, a solution $\hat{s}$ is accepted if $f(\hat{s}) < (1 + \beta) \cdot f(t)$ and the adopted stopping criterion has been to allow at most 50 cost-deteriorating moves without improvement in the current best solution.

5.6   GRASP and Very-Large Scale Neighborhood Search

As for any local search procedure, to efficiently search in the neighborhood of a solution, it is required that the neighborhood have a small size. Nevertheless, the larger the neighborhood, the better the quality of the locally optimal solution. Neighborhoods whose sizes grow exponentially as a function of problem dimension are called *very large scale neighborhoods* and they necessarily require efficient search techniques to be explored.

Ahuja et al. [2] presented a survey of methods called very-large scale neighborhood (VLSN) search. The following three classes of VLSN methods are described:

1. variable-depth methods where exponentially large neighborhoods are searched with heuristics;

2. a VLSN method that uses network flow techniques to identify improving neighborhood solutions;
3. a VLSN method that explores neighborhoods for NP-hard problems induced by restrictions of the problems that are solved in polynomial time.

In particular, with respect to class 2, they define special neighborhood structures called multi-exchange neighborhoods. The search is based on the cyclic transfer neighborhood structure that transforms a cost-reducing exchange into a negative cost subset-disjoint cycle in an *improving graph* and then a modified shortest path label-correcting algorithm is used to identify these negative cycles.

Ahuja et al. in [3] present two generalizations of the best known neighborhood structures for the capacitated minimum spanning tree problem. The new neighborhood structures defined allow cyclic exchanges of nodes among multiple subtrees simultaneously. To judge the efficacy of the neighborhoods, local improvement and tabu search algorithms have been developed. Local improvement uses a GRASP construction mechanism to generate repeated starting solutions for local improvement.

5.7 Other Hybridizations

In the previous sections of this chapter, we have reviewed some important hybridizations of GRASP, mostly involving the GRASP local search phase. More recently, several further hybridizations have been proposed. They include the use of GRASP in Branch & Bound framework and the combination of GRASP with data mining techniques.

GRASP and branch & bound

In 2004, Rocha et al. [66] proposed a hybridization of GRASP as an upper bound for a branch and bound (B&B) procedure to solve a scheduling problem with non-related parallel machines and sequence-dependent setup times. In 2007, Fernandes and Lourenço [23] proposed a hybrid GRASP and B&B for the job-shop scheduling problem. The B&B method is used within GRASP to solve subproblems of one machine scheduling subproblem obtained from the incumbent solution.

GRASP and data mining

In 2006, Jourdan et al. [41] presented a short survey enumerating opportunities to combine metaheuristics and data mining (DM) techniques. By using methods and theoretical results from statistics, machine learning, and pattern recognition, DM automatically explores large volumes of data (instances described according to several attributes) with the objective of discovering patterns. In fact, DM is also known as Knowledge-Discovery in Databases.

In GRASP with data mining (DM-GRASP), after executing a significant number of GRASP iterations, the data mining process extracts patterns from an elite set of solutions which will guide the GRASP construction procedure in the subsequent iterations. In fact, instead of building the randomized greedy solution from scratch, the construction procedure starts from a solution pattern (a partial solution) that was previously mined. Computational experiments have shown that the hybridization has benefited in both running time and quality of the solutions found.

DM-GRASP has been introduced in 2005 by Santos et al [69] for the maximum diversity problem. In 2006, Ribeiro et al. [65] also proposed a hybrid GRASP with DM and tested it the set packing problem as a case study and Santos et al. [68] solved a real world problem, called server replication for reliable multicast.

Very recently, s survey of applications of DM-GRASP has been published by Santos et al. [67].

6   Concluding Remarks

Simulated annealing, tabu search, ant colony, genetic algorithms, scatter search, path-relinking, GRASP, iterated local search, and variable neighborhood search are often listed as examples of "classical" metaheuristics. In the last few years, several different algorithms have been designed and proposed in the literature that do not purely apply the basic ideas of one single "classical" metaheuristic, but they combine various algorithmic ideas of different metaheuristic frameworks. The design and implementation of hybrid metaheuristics are emerging as one of the most exciting field.

In this chapter, we have surveyed hybridizations of GRASP and other metaheuristics. Among these, we highlight: path-relinking, tabu search, simulated annealing, genetic algorithms and population-based heuristics, variable neighborhood search and variable neighborhood descent, iterated local search, very large scale neighborhood local search, and very recent hybrids, such as GRASP with data mining and GRASP with branch and bound.

References

1. Abdinnour-Helm, S., Hadley, S.W.: Tabu search based heuristics for multi-floor facility layout. International J. of Production Research 38, 365–383 (2000)
2. Ahuja, R.K., Ergun, O., Orlin, J.B., Punnen, A.P.: A survey of very large-scale neighborhood search techniques. Discrete Appl. Math. 123, 75–102 (2002)
3. Ahuja, R.K., Orlin, J.B., Sharma, D.: Multi-exchange neighborhood structures for the capacitated minimum spanning tree problem. Mathematical Programming 91, 71–97 (2001)
4. Ahuja, R.K., Orlin, J.B., Tiwari, A.: A greedy genetic algorithm for the quadratic assignment problem. Computers and Operations Research 27, 917–934 (2000)

5. Aiex, R., Resende, M.G.C., Pardalos, P.M., Toraldo, G.: GRASP with path relinking for three-index assignment. INFORMS J. on Computing 17(2), 224–247 (2005)
6. Alvarez, A.M., Gonzalez-Velarde, J.L., De Alba, K.: GRASP embedded scatter search for the multicommodity capacitated network design problem. J. of Heuristics 11, 233–257 (2005)
7. Andrade, D.V., Resende, M.G.C.: A GRASP for PBX telephone migration scheduling. In: Eighth INFORMS Telecommunication Conference (April 2006)
8. Baum, E.B.: Iterated descent: A better algorithm for local search in combinatorial optimization problems. Technical report, California Institute of Technology (1986)
9. Beltrán, J.D., Calderón, J.E., Cabrera, R.J., Pérez, J.A.M., Moreno-Vega, J.M.: GRASP/VNS hybrid for the strip packing problem. In: Proceedings of Hybrid Metaheuristics (HM 2004), pp. 79–90 (2004)
10. Binato, S., Hery, W.J., Loewenstern, D., Resende, M.G.C.: A greedy randomized adaptive search procedure for job shop scheduling. In: Ribeiro, C.C., Hansen, P. (eds.) Essays and surveys on metaheuristics, pp. 58–79. Kluwer Academic Publishers, Dordrecht (2002)
11. Binato, S., Oliveira, G.C.: A Reactive GRASP for transmission network expansion planning. In: Ribeiro, C.C., Hansen, P. (eds.) Essays and surveys on metaheuristics, pp. 81–100. Kluwer Academic Publishers, Dordrecht (2002)
12. Bresina, J.L.: Heuristic-biased stochastic sampling. In: Proceedings of the Thirteenth National Conference on Artificial Intelligence (AAAI 1996), pp. 271–278 (1996)
13. Canuto, S.A., Resende, M.G.C., Ribeiro, C.C.: Local search with perturbations for the prize-collecting Steiner tree problem in graphs. Networks 38, 50–58 (2001)
14. Canuto, S.A., Resende, M.G.C., Ribeiro, C.C.: Local search with perturbations for the prize-collecting Steiner tree problem in graphs. Networks 38, 50–58 (2001)
15. Charon, I., Hudry, O.: The noising method: A new method for combinatorial optimization. Operations Research Letters 14, 133–137 (1993)
16. Charon, I., Hudry, O.: The noising methods: A survey. In: Ribeiro, C.C., Hansen, P. (eds.) Essays and surveys on metaheuristics, pp. 245–261. Kluwer Academic Publishers, Dordrecht (2002)
17. de la Peña, M.G.B.: Heuristics and metaheuristics approaches used to solve the rural postman problem: A comparative case study. In: Proceedings of the Fourth International ICSC Symposium on Engineering of Intelligent Systems (EIS 2004) (2004)
18. Delmaire, H., Díaz, J.A., Fernández, E., Ortega, M.: Reactive GRASP and tabu search based heuristics for the single source capacitated plant location problem. INFOR 37, 194–225 (1999)
19. Dorigo, M., Stützle, T.: Ant Colony Optimization. MIT Press, Cambridge (2004)
20. Duarte, A., Martí, R.: Tabu search and GRASP for the maximum diversity problem. European J. of Operational Research 178(1), 71–84 (2007)
21. Feo, T.A., Resende, M.G.C.: A probabilistic heuristic for a computationally difficult set covering problem. Operations Research Letters 8, 67–71 (1989)
22. Feo, T.A., Resende, M.G.C.: Greedy randomized adaptive search procedures. J. of Global Optimization 6, 109–133 (1995)

23. Fernandes, S., Lourenço, H.R.: A GRASP and Branch-and-Bound Metaheuristic for the Job-Shop Scheduling. In: Cotta, C., van Hemert, J. (eds.) EvoCOP 2007. LNCS, vol. 4446, pp. 60–71. Springer, Heidelberg (2007)
24. Festa, P., Pardalos, P.M., Pitsoulis, L.S., Resende, M.G.C.: GRASP with path-relinking for the weighted MAXSAT problem. ACM J. of Experimental Algorithmics 11, 1–16 (2006)
25. Festa, P., Pardalos, P.M., Resende, M.G.C., Ribeiro, C.C.: Randomized heuristics for the MAX-CUT problem. Optimization Methods and Software 7, 1033–1058 (2002)
26. Festa, P., Resende, M.G.C.: GRASP: An annotated bibliography. In: Ribeiro, C.C., Hansen, P. (eds.) Essays and Surveys on Metaheuristics, pp. 325–367. Kluwer Academic Publishers, Dordrecht (2002)
27. Garey, M.R., Johnson, D.S.: Computers and intractability: A guide to the theory of NP-completeness. W.H. Freeman and Company, New York (1979)
28. Glover, F.: Tabu search – Part I. ORSA J. on Computing 1, 190–206 (1989)
29. Glover, F.: Tabu search – Part II. ORSA J. on Computing 2, 4–32 (1990)
30. Glover, F.: Tabu search and adaptive memory programing – Advances, applications and challenges. In: Barr, R.S., Helgason, R.V., Kennington, J.L. (eds.) Interfaces in Computer Science and Operations Research, pp. 1–75. Kluwer, Dordrecht (1996)
31. Glover, F.: Multi-start and strategic oscillation methods – Principles to exploit adaptive memory. In: Laguna, M., Gonzáles-Velarde, J.L. (eds.) Computing Tools for Modeling, Optimization and Simulation: Interfaces in Computer Science and Operations Research, pp. 1–24. Kluwer, Dordrecht (2000)
32. Glover, F., Laguna, M.: Tabu Search. Kluwer Academic Publishers, Dordrecht (1997)
33. Glover, F., Laguna, M.: Tabu search. Kluwer Academic Publishers, Dordrecht (1997)
34. Glover, F., Laguna, M., Martí, R.: Fundamentals of scatter search and path relinking. Control and Cybernetics 39, 653–684 (2000)
35. Glover, F., Laguna, M., Martí, R.: Scatter Search. In: Ghosh, A., Tsutsui, S. (eds.) Advances in Evolutionary Computation: Theory and Applications, pp. 519–537. Kluwer Academic Publishers, Dordrecht (2003)
36. Goldberg, D.E.: Genetic algorithms in search, optimization and machine learning. Addison-Wesley, Reading (1989)
37. Hansen, P., Mladenović, N.: An introduction to variable neighborhood search. In: Voss, S., Martello, S., Osman, I.H., Roucairol, C. (eds.) Meta-heuristics, Advances and trends in local search paradigms for optimization, pp. 433–458. Kluwer Academic Publishers, Dordrecht (1998)
38. Hansen, P., Mladenović, N.: Developments of variable neighborhood search. In: Ribeiro, C.C., Hansen, P. (eds.) Essays and Surveys in Metaheuristics, pp. 415–439. Kluwer Academic Publishers, Dordrecht (2002)
39. Hart, J.P., Shogan, A.W.: Semi-greedy heuristics: An empirical study. Operations Research Letters 6, 107–114 (1987)
40. Holland, J.H.: Adaptation in Natural and Artificial Systems. Univ. of Michigan Press, Ann Arbor (1975)
41. Jourdan, L., Dhaenens, C., Talbi, E.-G.: Using datamining techniques to help metaheuristics: A short survey. In: Almeida, F., Blesa Aguilera, M.J., Blum, C., Moreno Vega, J.M., Pérez Pérez, M., Roli, A., Sampels, M. (eds.) HM 2006. LNCS, vol. 4030, pp. 57–69. Springer, Heidelberg (2006)

42. Kernighan, B.W., Lin, S.: An efficient heuristic procedure for partitioning problems. Bell System Technical Journal 49(2), 291–307 (1970)
43. Kirkpatrick, S.: Optimization by simulated annealing: Quantitative studies. J. of Statistical Physics 34, 975–986 (1984)
44. Koza, J.R., Bennett III, F.H., Andre, D., Keane, M.A.: Genetic Programming III, Darwinian Invention and Problem Solving. Morgan Kaufmann Publishers, San Francisco (1999)
45. Laguna, M.: Scatter Search. In: Pardalos, P.M., Resende, M.G.C. (eds.) Handbook of Applied Optimization, pp. 183–193. Oxford University Press, Oxford (2002)
46. Laguna, M., Martí, R.: GRASP and path relinking for 2-layer straight line crossing minimization. INFORMS J. on Computing 11, 44–52 (1999)
47. Laguna, M., Martí, R.: Scatter Search: Methodology and Implementations in C. Kluwer, Dordrecht (2003)
48. Lim, A., Wang, F.: A smoothed dynamic tabu search embedded GRASP for m-VRPTW. In: Proceedings of ICTAI 2004, pp. 704–708 (2004)
49. Lourenço, H.R., Martin, O.C., Stützle, T.: Iterated local search. In: Glover, F., Kochenberger, G. (eds.) Handbook of Metaheuristics, pp. 321–353. Kluwer Academic Publishers, Dordrecht (2003)
50. Martins, S.L., Resende, M.G.C., Ribeiro, C.C., Pardalos, P.: A parallel GRASP for the Steiner tree problem in graphs using a hybrid local search strategy. Journal of Global Optimization 17, 267–283 (2000)
51. Mladenović, N., Hansen, P.: Variable neighborhood search. Computers and Operations Research 24, 1097–1100 (1997)
52. Osman, I.H., Laporte, G.: Metaheuristics: A bibliography. Annals of Operations Research 63, 513 (1996)
53. Prais, M., Ribeiro, C.C.: Reactive GRASP: An application to a matrix decomposition problem in TDMA traffic assignment. INFORMS J. on Computing 12, 164–176 (2000)
54. Resende, M.G.C., Martí, R., Gallego, M., Duarte, A.: GRASP and path relinking for the max-min diversity problem. Technical report, AT&T Labs Research, Florham Park, NJ–USA (2008)
55. Resende, M.G.C., Ribeiro, C.C.: A GRASP with path-relinking for private virtual circuit routing. Networks 41, 104–114 (2003)
56. Resende, M.G.C., Ribeiro, C.C.: Greedy randomized adaptive search procedures. In: Glover, F., Kochenberger, G. (eds.) Handbook of Metaheuristics, pp. 219–249. Kluwer Academic Publishers, Dordrecht (2003)
57. Resende, M.G.C., Ribeiro, C.C.: A hybrid heuristic for the p-median problem. J. of Heuristics 10, 59–88 (2004)
58. Resende, M.G.C., Ribeiro, C.C.: GRASP with path-relinking: Recent advances and applications. In: Ibaraki, T., Nonobe, K., Yagiura, M. (eds.) Metaheuristics: Progress as Real Problem Solvers, pp. 29–63. Springer, Heidelberg (2005)
59. Resende, M.G.C., Werneck, R.F.: A hybrid multistart heuristic for the uncapacitated facility location problem. European J. of Operational Research 174, 54–68 (2006)
60. Ribeiro, C.C., Souza, M.C.: Variable neighborhood search for the degree constrained minimum spanning tree problem. Discrete Applied Mathematics 118, 43–54 (2002)
61. Ribeiro, C.C., Uchoa, E., Werneck, R.F.: A hybrid GRASP with perturbations for the Steiner problem in graphs. INFORMS Journal on Computing 14, 228–246 (2002)

62. Ribeiro, C.C., Uchoa, E., Werneck, R.F.: A hybrid GRASP with perturbations for the Steiner problem in graphs. INFORMS J. on Computing 14, 228–246 (2002)
63. Ribeiro, C.C., Urrutia, S.: Heuristics for the mirrored traveling tournament problem. European J. of Operational Research 179, 775–787 (2007)
64. Ribeiro, C.C., Vianna, D.S.: A GRASP/VND heuristic for the phylogeny problem using a new neighborhood structure. Intl. Trans. in Op. Res. 12, 325–338 (2005)
65. Ribeiro, M.H., Plastino, A., Martins, S.L.: Hybridization of GRASP metaheuristic with data mining techniques. J. of Mathematical Modelling and Algorithms 5, 23–41 (2006)
66. Rocha, P.L., Ravetti, M.G., Mateus, G.R.: The metaheuristic GRASP as an upper bound for a branch and bound algorithm in a scheduling problem with non-related parallel machines and sequence-dependent setup times. In: Proceedings of the 4th EU/ME Workshop: Design and Evaluation of Advanced Hybrid Meta-Heuristics, vol. 1, pp. 62–67 (2004)
67. Santos, L.F., Martins, S.L., Plastino, A.: Applications of the DM-GRASP heuristic: A survey. In: International Transactions on Operational Research (2008)
68. Santos, L.F., Milagres, R., Albuquerque, C.V., Martins, S., Plastino, A.: A hybrid GRASP with data mining for efficient server replication for reliable multicast. In: 49th Annual IEEE GLOBECOM Technical Conference (2006)
69. Santos, L.F., Ribeiro, M.H., Plastino, A., Martins, S.L.: A hybrid GRASP with data mining for the maximum diversity problem. In: Blesa, M.J., Blum, C., Roli, A., Sampels, M. (eds.) HM 2005. LNCS, vol. 3636, pp. 116–127. Springer, Heidelberg (2005)
70. Sosnowska, D.: Optimization of a simplified fleet assignment problem with metaheuristics: Simulated annealing and GRASP. In: Pardalos, P.M. (ed.) Approximation and complexity in numerical optimization. Kluwer Academic Publishers, Dordrecht (2000)
71. Souza, M.C., Duhamel, C., Ribeiro, C.C.: A GRASP heuristic for the capacitated minimum spanning tree problem using a memory-based local search strategy. In: Resende, M.G.C., Sousa, J. (eds.) Metaheuristics: Computer Decision-Making, pp. 627–658. Kluwer Academic Publishers, Dordrecht (2004)
72. Voss, S., Martello, S., Osman, I.H., Roucairo, C. (eds.): Meta-heuristics: Andvances and trends in local search paradigms for optimization. Kluwer Academic Publishers, Dordrecht (1999)

Particle Swarm Optimization: Performance Tuning and Empirical Analysis

Millie Pant, Radha Thangaraj, and Ajith Abraham

Abstract. This chapter presents some of the recent modified variants of Particle Swarm Optimization (PSO). The main focus is on the design and implementation of the modified PSO based on diversity, Mutation, Crossover and efficient Initialization using different distributions and Low-discrepancy sequences. These algorithms are applied to various benchmark problems including unimodal, multimodal, noisy functions and real life applications in engineering fields. The effectiveness of the algorithms is discussed.

1 Introduction

The concept of PSO was first suggested by Kennedy and Eberhart [1]. Since its development is 1995, PSO has emerged as one of the most promising optimizing technique for solving global optimization problems. Its mechanism is inspired by the social and cooperative behavior displayed by various species like birds, fish etc including human beings. The PSO system consists of a population (swarm) of potential solutions called particles. These particles move through the search domain with a specified velocity in search of optimal solution. Each particle maintains a memory which helps it in keeping the track of its previous best position. The positions of the particles are distinguished as personal best and global best. PSO has been applied to solve a variety of optimization problems and its performance is compared with other popular stochastic search techniques like Genetic algorithms, Differential Evolution, Simulated Annealing etc. [2], [3], [4]. Although PSO has shown a very good performance in solving many test as well as real life optimization problems, it suffers from the problem of premature convergence like most of the stochastic search techniques, particularly in case of multimodal optimization problems. The *curse* of premature convergence greatly affects the performance of algorithm and many times lead to a sub optimal solution [5]. Aiming at this shortcoming of PSO algorithms, many variations have been

Millie Pant and Radha Thangaraj
Department of Paper Technology, IIT Roorkee, India
email: millifpt@iitr.ernet.in, t.radha@ieee.org

Ajith Abraham
Q2S, Norwegian University of Science and Technology, Norway
email: ajith.abraham@ieee.org

A. Abraham et al. (Eds.): Foundations of Comput. Intel. Vol. 3, SCI 203, pp. 101–128.
springerlink.com   © Springer-Verlag Berlin Heidelberg 2009

developed to improve its performance. Some of the interesting modifications that helped in improving the performance of PSO include introduction of inertia weight and its adjustment for better control of exploration and exploitation capacities of the swarm [6] [7], introduction of constriction factor to control the magnitudes of velocities [8], impacts of various neighborhood topologies on the swarm [9], extension of PSO via genetic programming [10], use of various mutation operators into PSO [11] – [13]. In the present study ten recent versions of PSO are considered. Out of the ten chosen versions, five versions are based on the efficient initialization of swam, three versions are diversity guided and the remaining versions makes use of cross-over operator to improve the performance of PSO.

The present article has seven sections including the introduction. In the next section, a brief description of the basic PSO is given. Section 3 is divided into three subsections; in 3.1, PSO versions with different initialization schemes are described; in section 3.2 three diversity guided PSO are given and in Section 3.3 PSO with crossover operator is described. Section 4 is devoted to numerical problems consisting of ten popular bench mark problems and two real life problems. In Section 5 and Section 6, describe the experimental settings and numerical results respectively. The chapter finally concludes with Section 7.

2 Particle Swarm Optimization

The working of the Basic Particle Swarm Optimization (BPSO) may be described as: For a D-dimensional search space the position of the ith particle is represented as $X_i = (x_{i1}, x_{i2}, \ldots x_{iD})$. Each particle maintains a memory of its previous best position $P_{besti} = (p_{i1}, p_{i2} \ldots p_{iD})$. The best one among all the particles in the population is represented as $P_{gbest} = (p_{g1}, p_{g2} \ldots p_{gD})$. The velocity of each particle is represented as $V_i = (v_{i1}, v_{i2}, \ldots v_{iD})$. In each iteration, the P vector of the particle with best fitness in the local neighborhood, designated g, and the P vector of the current particle are combined to adjust the velocity along each dimension and a new position of the particle is determined using that velocity. The two basic equations which govern the working of PSO are that of velocity vector and position vector given by:

$$v_{id} = wv_{id} + c_1 r_1 (p_{id} - x_{id}) + c_2 r_2 (p_{gd} - x_{id}) \tag{1}$$

$$x_{id} = x_{id} + v_{id} \tag{2}$$

The first part of equation (1) represents the inertia of the previous velocity, the second part is the cognition part and it tells us about the personal experience of the particle, the third part represents the cooperation among particles and is therefore named as the social component. Acceleration constants c_1, c_2 and inertia weight w are the predefined by the user and r_1, r_2 are the uniformly generated random numbers in the range of [0, 1].

3 Modified Version of Particle Swarm Optimization

Empirical studies have shown that the basic PSO has a tendency of premature convergence [518], [559], [602], [606], [649] and the main reason   for this behavior is due to the loss of diversity in successive iterations. It has been observed that the presence of a suitable operator may help in improving the performance of PSO quite significantly. This chapter concentrates on two things, first is on the efficient generation of population using different initialization schemes and second is the use of diversity to guide the swarm using different operations like repulsion, mutation and crossover.

3.1 Efficient Initialization Particle Swarm Optimization

PSO (and other search techniques, which depend on the generation of random numbers) works very well for problems having a small search area (i.e. a search area having low dimension), but as the dimension of search space is increased, the performance deteriorates and many times converge prematurely giving a suboptimal result [5]. This problem becomes more persistent in case of multimodal functions having several local and global optima. One of the reasons for the poor performance of a PSO may be attributed to the dispersion of initial population points in the search space i.e. to say, if the swarm population does not cover the search area efficiently, it may not be able to locate the potent solution points, thereby missing the global optimum [14]. This difficulty may be minimized to a great extent by selecting a well-organized distribution of random numbers.

This section analyzes the behavior of some simple variations of PSO where only the initial distribution of random numbers is changed. Initially in the algorithms the initial uniform distribution is replaced by other probability distributions like exponential, lognormal and Gaussian distributions. It is interesting to see that even a small change in the initial distribution produces a visible change in the numerical results. After that more specialized algorithms are designed which use low discrepancy sequences for the generation of random numbers. A brief description of the algorithms is given in the subsequent sections.

The most common practice of generating random numbers is the one using an inbuilt subroutine (available in most of the programming languages), which uses a uniform probability distribution to generate random numbers. It has been shown that uniformly distributed particles may not always be good for empirical studies of different algorithms. The uniform distribution sometimes gives a wrong impression of the relative performance of algorithms as shown by Gehlhaar and Fogel [15].

3.1.1 Initializing the Swarm Using Different Probability Distributions [16]

Different Probability Distributions like Exponential and Gaussian have already been used for the fine tuning of PSO parameters [17], [18]. But for initializing the swarm most of the approaches use uniformly distributed random numbers. Pant et al. [16] investigated the possibility of having a different probability distribution (Gaussian, Exponential, Lognormal) for the generation of random number other than the uniform distribution. Empirical results showed that distributions other than uniform distribution are equally competent and in most of the cases are better than uniform distribution. The algorithms GPSO, EPSO and LNPSO use

Gaussian, exponential and lognormal distributions respectively. The algorithms follow the steps of BPSO given in Section 2 except for the fact that they use mentioned distributions in place of uniform distributions.

3.1.2 Initializing the Swarm Using Low-Discrepancy Sequences [19]

Theoretically, it has been proved that low discrepancy sequences are much better than the pseudo random sequences because they are able to cover the search space more evenly in comparison to pseudo random sequences (please see Figures 1(a) and 1(b)). Some previous instances where low discrepancy sequences have been used to improve the performance of optimization algorithms include [20] – [24]. In [22] – [24] authors have made use of Sobol and Faure sequences. Similarly, Nguyen et al. [21] have shown a detailed comparison of Halton Faure and Sobol sequences for initializing the swarm. In the previous studies, it has already been shown that the performance of Sobol sequence dominates the performance of Halton and Faure sequences. The performance of PSO using Van der Corput sequence called VCPSO along with PSO with Sobol sequence called SOPSO (which is said be superior than other low discrepancy sequences according to the previous studies) for swarm initialization is scrutinized and tested them for solving global optimization problems in large dimension search spaces by Pant et al. [19].

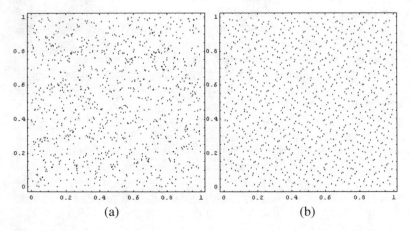

(a)   (b)

Fig. 1(a) Sample points generated using a pseudo random sequence. 1(b) Sample points generated using a quasi random sequence

Brief description of the sequences used in VCPSO and SOPSO:

Van der Corput Sequence
A Van der Corput sequence is a low-discrepancy sequence over the unit interval first published in 1935 by the Dutch mathematician J. G. Van der Corput. It is a digital $(0, 1)$-sequence, which exists for all bases $b \geq 2$. It is defined by the *radical inverse function* $\varphi_b : N_0 \rightarrow [0, 1)$. If $n \in N_0$ has the b-adic expansion

$$n = \sum_{j=0}^{T} a_j b^{j-1} \tag{3}$$

with $a_j \in \{0,...,b-1\}$, and $T = \lfloor \log_b n \rfloor$ then φ_b is defined as

$$\varphi_b(n) = \sum_{j=0}^{T} \frac{a_j}{b^j} \tag{4}$$

In other words, the jth b-adic digit of n becomes the jth b-adic digit of $\varphi_b(n)$ behind the decimal point. The Van der Corput sequence in base b is then defined as $(\varphi_b(n))_{n \geq 0}$.

The elements of the Van der Corput sequence (in any base) form a dense set in the unit interval: for any real number in [0, 1] there exists a sub sequence of the Van der Corput sequence that converges towards that number. They are also uniformly distributed over the unit interval.

Sobol Sequence
The construction of the Sobol sequence [25] uses linear recurrence relations over the finite field, F2, where F2 = {0, 1}. Let the binary expansion of the non-negative integer n be given by $n = n_1 2^0 + n_2 2^1 + ... + n_w 2^{w-1}$. Then the nth element of the jth dimension of the Sobol Sequence, $X_n^{(j)}$, can be generated by:

$$X_n^{(j)} = n_1 v_1^{(j)} \oplus n_2 v_2^{(j)} \oplus ...... \oplus n_w v_w^{(j)}$$

where $v_i^{(j)}$ is a binary fraction called the ith direction number in the jth dimension. These direction numbers are generated by the following q-term recurrence relation:

$$v_i^{(j)} = a_1 v_{i-1}^{(j)} \oplus a_2 v_{i-2}^{(j)} \oplus ... \oplus a_q v_{i-q+1}^{(j)} \oplus v_{i-q}^{(j)} \oplus (v_{i-q}^{(j)}/2^q) \quad i > q,$$ and the bit, a_i,

comes from the coefficients of a degree-q primitive polynomial over F2.

VC-PSO and SO-PSO Algorithm
It has been shown that uniformly distributed particles may not always be good for empirical studies of different algorithms. The uniform distribution sometimes gives a wrong impression of the relative performance of algorithms as shown by Gehlhaar and Fogel [15].

The quasi random sequences on the other hand generates a different set of random numbers in each iteration, thus providing a better diversified population of solutions and thereby increasing the probability of getting a better solution.

Keeping this fact in mind we decided to use the Vander Corput sequence and Sobol sequence for generating the swarm. The swarm population follows equation (1) and (2) for updating the velocity and position of the swarm. However for the generation of the initial swarm Van der Corput Sequence and Sobol Sequences have been used for VC-PSO and SO-PSO respectively.

3.2 Diversity Guided Particle Swarm Optimization

Diversity may be defined as the dispersion of potential candidate solutions in the search space. Interested readers may please refer to [26] for different formulae used for calculating diversity. One of the drawbacks of most of the population based search techniques is that they work on the principle of contracting the search domain towards the global optima. Due to this reason after a certain number of iterations all the points get accumulated to a region which may not even be a region of local optima, thereby giving suboptimal solutions [5]. Thus without a suitable diversity enhancing mechanism it is very difficult for an optimization algorithm to reach towards the true solution. The problem of premature convergence becomes more persistent in case of highly multimodal functions like Rastringin and Griewank having several local minima. This section presents three algorithms Attraction Repulsion PSO (ATREPSO), Gaussian Mutation PSO (GMPSO) and Quadratic Interpolation PSO (QIPSO) which use different diversity enhancing mechanisms to improve the performance of the swarm. All the algorithms described in the given sub sections use diversity threshold values d_{low} and d_{high} to guide the movement of the swarm. The threshold values are predefined by the user. In ATREPSO, the swarm particles follow the mechanism of repulsion so that instead of converging towards a particular location the particles are diverged from that location. In case of GMPSO and QIPSO evolutionary operators like mutation and crossover are induced in the swarm to perturb the population. These algorithms are described in the following subsections.

3.2.1 Attraction Repulsion Particle Swarm Optimization Algorithm [27]

The Attraction Repulsion Particle Swarm Optimization Algorithm (ATREPSO) of Pant et al. [27] is a simple extension of the Attractive and Repulsive PSO (ARPSO) proposed by Vesterstorm [28], where a third phase called *in between* phase or the phase of *positive conflict* is added In ATREPSO, the swarm particles switches alternately between the three phases of attraction, repulsion and an 'in between' phase which consists of a combination of attraction and repulsion. The three phases are defined as:

Attraction phase (when the particles are attracted towards the global optimal)

$$v_{id} = wv_{id} + c_1 r_1 (p_{id} - x_{id}) + c_2 r_2 (p_{gd} - x_{id}) \tag{5}$$

Repulsion phase (particles are repelled from the optimal position)

$$v_{id} = wv_{id} - c_1 r_1 (p_{id} - x_{id}) - c_2 r_2 (p_{gd} - x_{id}) \tag{6}$$

In-between phase (neither total attraction nor total repulsion)

$$v_{id} = wv_{id} + c_1 r_1 (p_{id} - x_{id}) - c_2 r_2 (p_{gd} - x_{id}) \tag{7}$$

In the in-between phase, the individual particle is attracted by its own previous best position p_{id} and is repelled by the best known particle position p_{gd}. In this way there is neither total attraction nor total repulsion but a balance between the two.

The swarm particles are guided by the following rule

$$v_{id} = \begin{cases} wv_{id} + c_1 r_1 (p_{id} - x_{id}) + c_2 r_2 (p_{gd} - x_{id}), div > d_{high} \\ wv_{id} + c_1 r_1 (p_{id} - x_{id}) - c_2 r_2 (p_{gd} - x_{id}), d_{low} < div < d_{high} \\ wv_{id} - c_1 r_1 (p_{id} - x_{id}) - c_2 r_2 (p_{gd} - x_{id}), div < d_{low} \end{cases} \qquad (8)$$

3.2.2 Gaussian Mutation Particle Swarm Optimization Algorithm [29]

The concept of mutation is quite common to Evolutionary Programming and Genetic Algorithms. The idea behind mutation is to increase of diversity of the population. There are several instances in PSO also where mutation is introduced in the swarm. Some mutation operators that have been applied to PSO include Gaussian [244], Cauchy [655], [656], Nonlinear [589], Linear [589] etc. The Gaussian Mutation Particle Swarm Optimization (GMPSO) algorithm given in this section is different from the above mentioned algorithms as it uses the threshold values to decide the activation of mutation operator. The concept is similar to that of ATREPSO i.e. to use diversity to decide the movement of the swarm. The algorithm uses the general equations (1) and (2) for updating the velocity and position vectors. At the same time a track of diversity is also kept which starts decreasing slowly and gradually after a few iterations because of the fast information flow between the swam particles leading to clustering of particles. It is at this stage that the Gaussian mutation operator given as $X_{t+1}[i] = X_t[i] + \eta *$ *Rand()*, where *Rand* is a random number generated by Gaussian distribution, is activated with the hope to increase the diversity of the swarm population. Here η is a scaling parameter.

3.2.3 Quadratic Interpolation Particle Swarm Optimization Algorithm [30]

As mentioned in the previous section, there are several instances available in literature on the use of mutation operator however there are not much references on the use of reproduction operator. One of the earlier references on the use of reproduction operator can be found in Clerc [101]. The Quadratic Interpolation Particle Swarm Optimization (QIPSO) algorithm described in this chapter uses concept of reproduction. It uses diversity as a measure to guide the swarm. When the diversity becomes less than d_{low}, then the quadratic crossover operator is activated to generate a new potential candidate solution. The process is repeated iteratively till the diversity reaches the specified threshold d_{high}. The quadratic crossover operator used in this paper is a nonlinear crossover operator which makes use of three particles of the swarm to produce a particle which lies at the point of minima of the quadratic curve passing through the three selected particles.

It uses a = X_{min}, (best particle with minimum function value) and two other randomly selected particles {b, c} (a, b and c are different particles) from the swarm to determine the coordinates of the new particle $\tilde{x}^i = (\tilde{x}^1, \tilde{x}^2, \ldots \ldots \tilde{x}^n)$, where

$$\tilde{x}^i = \frac{1}{2} \frac{(b^{i^2} - c^{i^2}) * f(a) + (c^{i^2} - a^{i^2}) * f(b) + (a^{i^2} - b^{i^2}) * f(c)}{(b^i - c^i) * f(a) + (c^i - a^i) * f(b) + (a^i - b^i) * f(c)} \qquad (9)$$

The nonlinear nature of the quadratic crossover operator used in this work helps in finding a better solution in the search space.

3.3   Crossover Based Particle Swarm Optimization

In this section two more modifications applied to the QIPSO given in Section 3.2 are described.

3.3.1   QIPSO-1 [31] and QIPSO-2 [32] Algorithms

The basic idea of QIPSO-1 and QIPSO-2 are modified versions of QIPSO algorithm given in section 3.2, which differ from each other in selection criterion of the individual. In QIPSO-1, the new point generated by the quadratic interpolation given by equation (9) is accepted in the swarm only if it is better than the worst particle of the swarm, where as in QIPSO-2, the particle is accepted if it is better than the global best particle.

4   Numerical Problems

One of the shortcomings of population based search techniques is that there are not many concrete proofs available to establish their authority for solving a wide range of problems. Therefore the researchers often depend on empirical studies to scrutinize the behavior of an algorithm. The numerical problems may be divided into two classes; benchmark problems and real life problems. For the present article ten standard benchmark functions and two real life problems described in the following subsections are taken.

4.1   Benchmark Problems

A collection of ten benchmark problems given in Table 1 is taken for the present study to analyze the behavior of algorithms taken in this study. These problems may not be called exhaustive but they provide a good launch pad for testing the credibility of an optimization algorithm. The first eight problems are scalable i.e. the problems can be tested for any number of variables. However for the present study medium sized problems of dimension 20 are taken. The three dimensional graphs of the test functions are depicted in Figures 2(a) to (i).

The special properties of the benchmark functions taken in this study may be described as:

- The first function f_1, commonly known as Rastringin function, is a highly multimodal function where the degree of multimodality increases with the increase in the dimension of the problem.
- The second function f_2, also known as spherical function is a continuous, strictly convex and unimodal function and usually do not pose much difficulty for an optimization algorithm.

- Griewank function is the third function. It is highly multimodal function having several local minima.
- The search space of the fourth function is dominated by a large gradual slope. Despite the apparent simplicity of the problem it is considered difficult for search algorithms because of its extremely large search space combined with relatively small global optima.
- f_5 is a noisy function where a uniformly distributed random noise is added to the objective function. Due to the presence of noise the objective function keeps changing from time to time and it becomes a challenge for an optimization algorithm to locate the optimum.
- Functions f_6 to f_8 are again multimodal functions having several optima. Such functions provide a suitable platform for testing the credibility of an optimization algorithm.
- Function f_9 and f_{10} are two dimensional functions. Function f_{10} is although simple in appearance but it an interesting and challenging function having 786 local minima and 18 global minima.

Table 1 Numerical Benchmark Problems [3]

Name of function	Function Definition	Range	Minimum Value
Rastrigin Function	$f_1(x) = \sum_{i=1}^{n}(x_i^2 - 10\cos(2\pi x_i) + 10)$	[-5.12,5.12]	0
Spherical Function	$f_2(x) = \sum_{i=1}^{n} x_i^2$	[-5.12,5.12]	0
Griewank Function	$f_3(x) = \frac{1}{4000}\sum_{i=0}^{n-1} x_i^2 + \sum_{i=0}^{n-1}\cos(\frac{x_i}{\sqrt{i+1}}) + 1$	[-600,600]	0
Rosenbrock Function	$f_4(x) = \sum_{i=0}^{n-1} 100(x_{i+1} - x_i^2)^2 + (x_i - 1)^2$	[-30,30]	0
Noisy Function	$f_5(x) = (\sum_{i=0}^{n-1}(i+1)x_i^4) + rand[0,1]$	[-1.28,1.28]	0
Schewefel Function	$f_6(x) = -\sum_{i=1}^{n} x_i \sin(\sqrt{\lvert x_i \rvert})$	[-500,500]	-8329.658
Ackley Function	$f_7(x) = 20 + e - 20\exp(-0.2\sqrt{\frac{1}{n}\sum_{i=1}^{n} x_i^2})$ $-\exp(\frac{1}{n}\sum_{i=1}^{n}\cos(2\pi x_i))$	[-32,32]	0
Sine Function	$f_8(x) = -\sum_{i=1}^{n}\sin(x_i)(\sin(i\frac{x_i^2}{\pi}))^{2m}$, $m = 10$	[-π,π]	---
Himmelblau Function	$f_9(x) = (x_2 + x_1^2 - 11)^2 + (x_1 + x_2^2 - 7)^2 + x_1$	[-5,5]	-3.78396
Shubert Function	$f_{10}(x) = \sum_{j=1}^{5} j\cos((j+1)x_1 + j)\sum_{j=1}^{5} j\cos((j+1)x_2 + j)$	[-10,10]	-186.7309

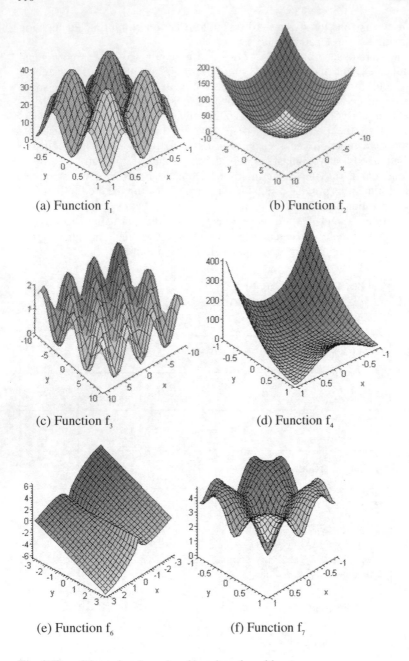

(a) Function f_1 (b) Function f_2

(c) Function f_3 (d) Function f_4

(e) Function f_6 (f) Function f_7

Fig. 2 Three Dimensional graphs of benchmark problems

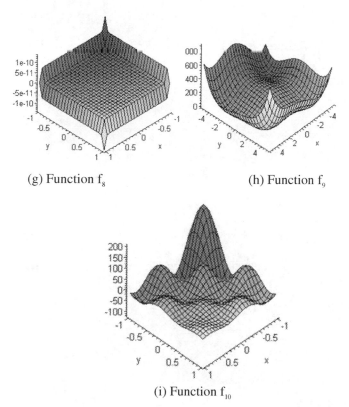

(g) Function f_8 (h) Function f_9

(i) Function f_{10}

Fig. 2 (*continued*)

4.2 Real Life Problems

Two real life engineering design problems are considered to depict the effectiveness of the algorithms discussed in the present article. These are nonlinear problems and both the problems are common in the field of electrical engineering. Mathematical model of the problems are given as:

4.2.1 Design of a Gear Train [33]

The first problem is to optimize the gear ratio for the compound gear train. This problem shown in Figure 3 was introduced by Sandgren [34]. It is to be designed such that the gear ratio is as close as possible to 1/6.931. For each gear the number of teeth must be between 12 and 60. Since the number of teeth is to be an integer, the variables must be integers. The mathematical model of gear train design is given by,

$$\text{Min} \quad f = \left\{ \frac{1}{6.931} - \frac{T_d T_h}{T_a T_f} \right\}^2 = \left\{ \frac{1}{6.931} - \frac{x_1 x_2}{x_3 x_4} \right\}^2$$

Subject to: $12 \leq x_i \leq 60 \quad i = 1,2,3,4$

$[x_1, x_2, x_3, x_4] = [T_d, T_b, T_a, T_f]$, x_i's should be integers. T_a, T_b, T_d, and T_f are the number of teeth on gears A, B, D and F respectively.

Fig. 3 Compound Gear Train

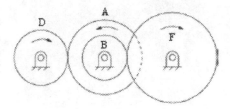

4.2.2 Transistor Modeling [35]

The second problem is a transistor modeling problem. The mathematical model of the transistor design is given by,

Minimize $f(x) = \gamma^2 + \sum_{k=1}^{4}(\alpha_k^2 + \beta_k^2)$

Where

$\alpha_k = (1 - x_1 x_2) x_3 \{\exp[x_5(g_{1k} - g_{3k}x_7 \times 10^{-3} - g_{5k}x_8 \times 10^{-3})] - 1\} g_{5k} + g_{4k}x_2$

$\beta_k = (1 - x_1 x_2) x_4 \{\exp[x_6(g_{1k} - g_{2k} - g_{3k}x_7 \times 10^{-3}$

$+ g_{4k}x_9 \times 10^{-3})] - 1\} g_{5k}x_1 + g_{4k}$.

$\gamma = x_1 x_3 - x_2 x_4$

Subject to: $x_i \geq 0$

The numerical constants g_{ik} are given by the matrix

$$\begin{bmatrix} 0.485 & 0.752 & 0.869 & 0.982 \\ 0.369 & 1.254 & 0.703 & 1.455 \\ 5.2095 & 10.0677 & 22.9274 & 20.2153 \\ 23.3037 & 101.779 & 111.461 & 191.267 \\ 28.5132 & 111.8467 & 134.3884 & 211.4823 \end{bmatrix}$$

This objective function provides a least-sum-of-squares approach to the solution of a set of nine simultaneous nonlinear equations, which arise in the context of transistor modeling.

5 Experimental Settings

Like all Evolutionary Algorithms, PSO has a set of parameters which is to be defined by the user. These parameters are population size, inertia weight, acceleration constants etc. these parameters may be varied as per the complexity of the problem. For example the population size in PSO related literature has been suggested as 2n to 5n, where n is the number of decision variables or a fixed population size. In the present study a fixed population size of thirty is taken for all the problems, which gave reasonably good results. Similarly various examples are available on the variations done in inertia weight and acceleration constants. For the present study, which consist of small moderate size problems of dimension 2 and 20, the list parameters which gave sufficiently good results is summarized below.

Common Parameters:
Population Size (NP):   Number of variables 30
Inertia weight: Linearly decreasing (0.9 – 0.4)
Acceleration Constants: c1 = c2 = 2.0
Stopping Criterion: Maximum number of generations = 10000

Probability Distributions for initializing the swarm:
Gaussian distribution:

$$f(x) = \frac{1}{\sqrt{2\pi}} e^{-\frac{x^2}{2}} \tag{10}$$

with mean 0 and standard deviation 1, i.e. N (0,1).
Exponential distribution:

$$f(x) = \frac{1}{2b} \exp(-|x - a|/b), \quad -\infty \le x < \infty, \tag{11}$$

with a, b > 0.It is evident that one can control the variance by changing the parameters a and b.
Log-normal distribution:

$$f(x) = \frac{e^{-(\ln x)^2 / 2}}{x\sqrt{2\pi}} \tag{12}$$

with mean 0 and standard deviation 1.

Diversity Measurement

$$Diversity(S(t)) = \frac{1}{n_s} \sum_{i=1}^{n_s} \sqrt{\sum_{j=1}^{n_x} (x_{ij}(t) - \overline{x_j(t)})^2}$$

Threshold values: $d_{high} = 0.25$, $d_{low} = 5.0*10^{-6}$

Repair method for points violating the boundary conditions

Hardware Settings
All algorithms are executed on a P-IV PC. Programming language used is DEV C++

Table 2 Comparison results of PSO, GPSO, EPSO, and LNPSO (Mean/diversity/standard deviation)

Function	PSO	GPSO	EPSO	LNPSO
f_1	22.339158	**9.750543**	12.173973	23.507127
	0.000115	0.364310	5.380822e-05	0.264117
	15.932042	5.433786	9.274301	15.304573
f_2	1.167749e-45	**1.114459e-45**	1.167749e-45	**1.114459e-45**
	2.426825e-23	3.168112e-23	3.909771e-23	2.778935e-23
	5.222331e-46	4.763499e-46	5.222331e-46	4.763499e-46
f_3	0.031646	**0.004748**	0.011611	0.011009
	0.000710	1.631303e-08	0.001509	0.000877
	0.025322	0.012666	0.019728	0.019186
f_4	22.191725	9.992837	8.995165	**4.405738**
	2.551408	2.527997	1.8737	2.904427
	1.615544e+04	3.168912	3.959364	4.121244
f_5	8.681602	0.636016	**0.380297**	0.537461
	0.340871	0.210458	0.237913	0.254176
	9.001534	0.296579	0.281234	0.285361
f_6	-6178.559896	**-6354.119792**	-6306.353646	-6341.4000
	0.072325	0.059937	0.026106	0.034486
	489.3329	483.654032	575.876696	568.655436
f_7	3.483903e-18	**3.136958e-18**	3.368255e-18	3.368255e-18
	3.651635e-18	5.736429e-13	3.903653e-18	3.846865e-18
	8.359535e-19	8.596173e-19	8.596173e-19	8.596173e-19
f_8	-18.1594	-18.5162	**-18.675**	-18.3944
	1.17699	0.603652	2.63785	0.221685
	1.05105	0.907089	1.06468	1.02706
f_9	-3.331488	**-3.63828**	**-3.63828**	-3.49261
	2.747822e-05	1.71916e-009	1.65462e-009	1.4056e-009
	1.24329	0.346782	0.346782	0.445052
f_{10}	-186.730941	**-186.731**	**-186.731**	**-186.731**
	0.362224	1.0056	0.19356	0.82143
	1.424154e-05	3.3629e-014	3.11344e-014	2.00972e-014

6 Numerical Results

A comparative analysis of the algorithms described is given Tables 2 to 9. Each algorithm was executed 100 times and the average fitness function value, diversity and standard deviation are reported. Tables 2 to 5 give the numerical results for benchmark problems whereas, the numerical results of real life problems are given in Tables 6 to 9.

Table 2, gives the numerical results of PSO versions initialized with Gaussian, exponential and lognormal probability distributions. From the numerical results it can be seen that the PSO using Gaussian mutation, GPSO, gave the best

Table 3 Comparison results of PSO, VC-PSO and SO-PSO (Mean/diversity/standard deviation)

Function	PSO	VC-PSO	SO-PSO	Function	PSO	VC-PSO	SO-PSO
f_1	22.3391 0.00011 15.9320	9.99929 1.00441 4.08386	**8.95459** 0.319194 2.65114	f_6	-6178.559 0.072325 4.893e+02	**-6503.05** 3.469e-06 477.252	-6252.51 2.478e-06 472.683
f_2	1.16e-45 2.42e-23 5.22e-46	**1.17e-108** 7.15e-054 4.36e-108	1.51e-108 6.36e-055 4.46e-108	f_7	3.483e-18 3.651e-18 8.359e-19	5.473e-19 5.039e-19 1.776e-18	**4.585e-19** 6.506e-17 1.538e-18
f_3	0.03164 0.00071 0.02532	**0.00147** 1.233e-08 0.00469	0.001847 9.940e-09 0.004855	f_8	-18.1594 1.17699 1.05105	-18.2979 0.0306 0.8902	**-18.70665** 0.0316574 1.028749
f_4	22.1917 2.55140 1.61e+04	**6.30326** 2.01591 3.99428	6.81079 2.61624 3.76973	f_9	-3.331488 2.747e-05 1.24329	-3.58972 1.439e-09 0.388473	**-3.78396** 3.946e-09 1.47699
f_5	8.68160 0.34087 9.00153	**0.410042** 0.230096 0.294763	0.806175 0.191133 0.868211	f_{10}	-186.730 0.36222 1.424e-05	**-186.731** 1.10502 2.770e-14	**-186.731** 0.32435 3.595e-14

performance in comparison to other versions, followed by EPSO and LNPSO. For the first function, f_1, GPSO gave the best function value of approximately 10.00 which is much better than the values obtained by the other algorithms. For f_2, which is a simple spherical function all the algorithms gave more or less similar results. However GPSO and LNPSO gave a slightly better performance. For f_3, once again the average fitness function value obtained by GMPSO is much better than the average fitness function value obtained by EPSO and LNPSO. For f_6 and f_7 once again GMPSO outperformed the other algorithms given in Table 2. For f9, both GMPSO and EPSO gave same result, which is better than the other two algorithms. Whereas for f_{10}, GMPSO, EPSO and LNPSO gave same result which is marginally better than the result obtained by basic PSO. In all, out of the 10 test functions GPSO outperformed others in 7 test cases. EPSO gave better results in 4 cases and LNPSO performed better in 3 cases. In all the cases the results were better than Basic PSO using uniform distribution.

Table 3, gives the comparison of PSO versions initialized with low discrepancy sequences with the basic PSO. It can be observed that PSO initialized with Sobol sequence (SOPSO) gave slightly better results than PSO initialized with Van der corput sequence. But a notable thing is that although SOPSO outperformed VCPSO in most of the test cases, the percentage of improvement is only marginal. Whereas, if we compare these results with basic PSO then the quality of solutions obtained by SOPSO and VCPSO is significantly better than the solutions obtained by basic PSO. For example, in f_1, which is a highly multimodal function the optimum function value obtained by VCPSO and SOPSO is approximately 10.00 and 9.00 respectively where as the optimum function value obtained by basic PSO is approximately 22.00. Likewise, there is a significant improvement in the

Table 4 Comparison results of PSO, QIPSO, ATREPSO and GMPSO (Mean/diversity/ standard deviation)

Function	PSO	QIPSO	ATREPSO	GMPSO
f_1	22.339158	**11.946888**	19.425979	20.079185
	0.000115	0.015744	7.353246	7.143211e-05
	15.932042	9.161526	14.349046	13.700202
f_2	1.167749e-45	**0.000000**	4.000289 e-17	7.263579e-17
	2.426825e-23	0.000000	8.51205	0.00026
	5.222331e-46	0.000000	0.000246	6.188854e-17
f_3	0.031646	**0.01158**	0.025158	0.024462
	0.000710	3.391647e-05	0.000563	0.000843
	0.025322	0.01285	0.02814	0.039304
f_4	22.191725	**8.939011**	19.49082	14.159547
	2.551408	1.983866	1.586547	6.099418e-05
	1.615544e+04	3.106359	3.964335e+04	4.335439e+04
f_5	8.681602	**0.451109**	8.046617	7.160675
	0.340871	0.0509	2.809409	0.29157
	9.001534	0.328623	8.862385	7.665802
f_6	-6178.559896	**-6355.58664**	-6183.6776	-6047.670898
	0.072325	0.00881	199.95052	0.062176
	489.3329	477.532584	469.611104	482.926738
f_7	3.483903e-18	**2.461811e-24**	0.018493	1.474933e-18
	3.651635e-18	0.000127	42.596802	0.061308
	8.359535e-19	0.014425	0.014747	1.153709e-08
f_8	-18.1594	-18.4696	**-18.9829**	-18.3998
	1.17699	1.2345	0.39057	1.63242
	1.05105	0.092966	0.272579	0.403722
f_9	-3.331488	**-3.783961**	-3.751458	-3.460233
	2.747822e-05	0.637823	3.214462	9.066805e-06
	1.24329	0.190394	0.174460	0.45782
f_{10}	-186.730941	**-186.730942**	-186.730941	**-186.730942**
	0.362224	2.169003	5.410105	0.239789
	1.424154e-05	3.480934e-14	1.424154e-05	1.525879e-05

function value for functions f_4 and f_5. For f_4, VCPSO and SOPSO gave function values as 6.00 and 7.00 approximately and basic PSO gave an average fitness function value of 22.00. In f_5, VCPSO and SOPSO converged to 0.4 and 0.8 respectively while basic PSO converged to an optimum function value of 8.00.

In Table 4, the results of diversity guided PSO algorithms are given. From the numerical results it is evident that the PSO assisted with quadratic crossover operator, QIPSO is a clear winner. QIPSO gave significantly better performance than ATREPSO, GMPSO and basic PSO in 9 out of 10 test cases taken for the present study. Second place goes to ATREPSO and third to GMPSO.

Table 5, gives the comparison of modified versions of QIPSO with basic PSO. If a comparison is done between QIPSO1 and QIPSO2, than from the numerical

Table 5 Comparison results of PSO, QIPSO-1 and QIPSO-2 (Mean best fitness)

Fun ction	BPSO Mean Best Fitness	QIPSO-1 Mean Best Fitness	QIPSO-2 Mean Best Fitness
$f1$	22.339158	0.994954	**5.97167e-01**
$f2$	1.167749e-45	**2.523604e-45**	8.517991e-43
$f3$	0.031646	0.015979	**2.940000e-02**
$f4$	22.191725	77.916591	**51.0779**
$f5$	8.681602	0.454374	**4.540630e-01**
$f6$	-6178.559896	**-9185.074692**	-9.185054e+03
f7	3.483903e-18	**5.89622e-10**	6.300262e-09
$f8$	-18.1594	**-27.546**	**-27.546**
$f9$	-3.331488	-3.58972	**-3.78396**
$f10$	-186.730941	**-186.731**	**-186.731**

results it can be seen that QIPSO-2, in which the new particle is accepted in the swarm only if it is better than the global best particle is better than QIPSO-1 in terms of average fitness function value. However once again it can be observed that the improvement of QISPO-2 over QIPSO-1 is only marginal whereas both the algorithms performed much better than the basic PSO. Empirical results are graphically illustrated in Figures 4-7.

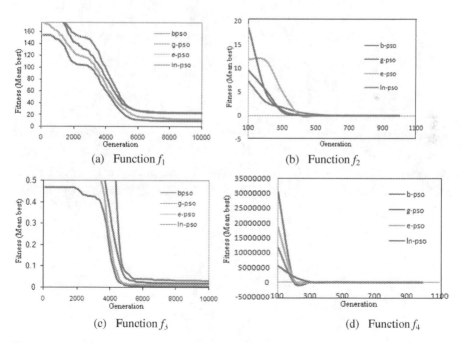

(a) Function f_1

(b) Function f_2

(c) Function f_3

(d) Function f_4

Fig. 4 Performance for BPSO, GPSO, EPSO and LNPSO

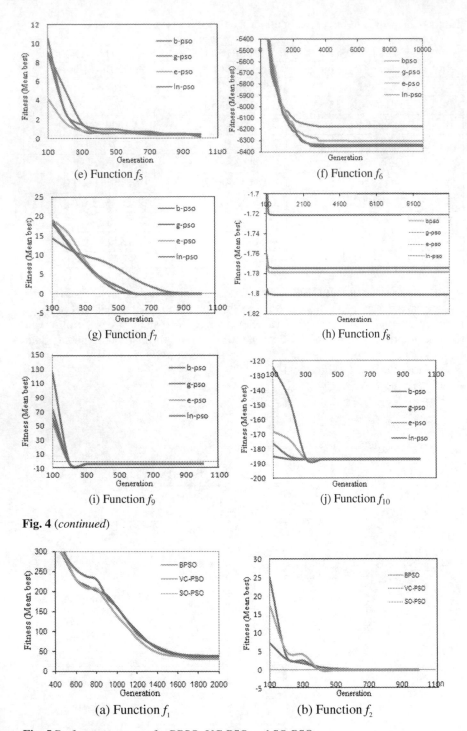

Fig. 4 (*continued*)

Fig. 5 Performance curves for BPSO, VC-PSO and SO-PSO

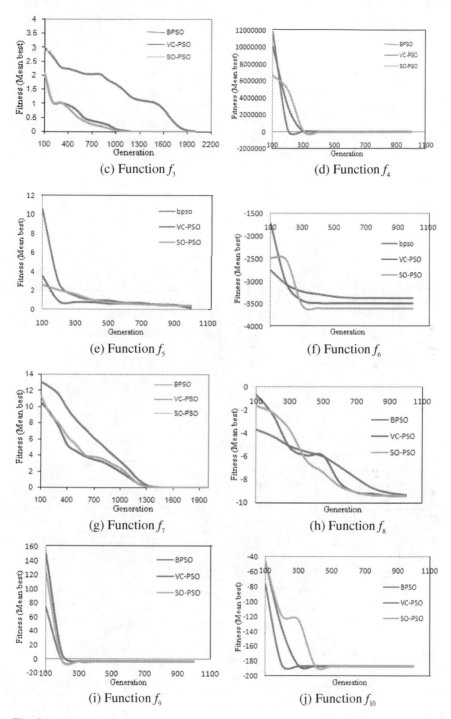

(c) Function f_3

(d) Function f_4

(e) Function f_5

(f) Function f_6

(g) Function f_7

(h) Function f_8

(i) Function f_9

(j) Function f_{10}

Fig. 5 (*continued*)

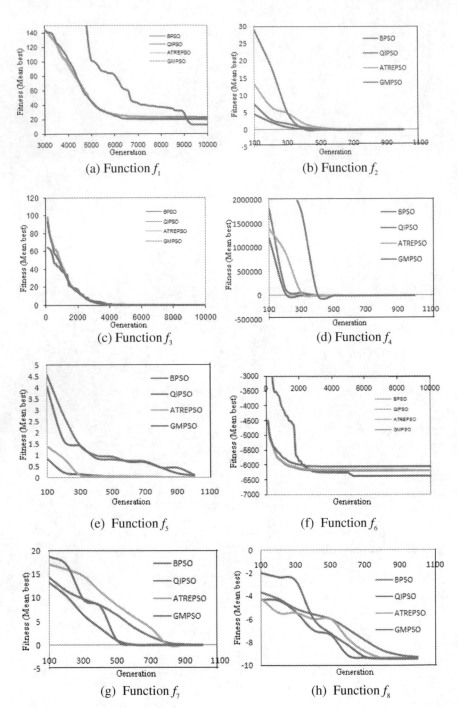

(a) Function f_1

(b) Function f_2

(c) Function f_3

(d) Function f_4

(e) Function f_5

(f) Function f_6

(g) Function f_7

(h) Function f_8

Fig. 6 Performance curves for BPSO, QIPSO, ATREPSO and GMPSO

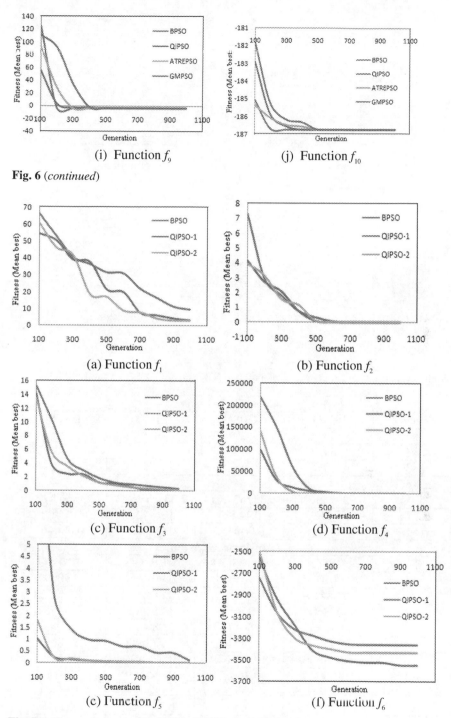

(i) Function f_9 (j) Function f_{10}

Fig. 6 (*continued*)

(a) Function f_1 (b) Function f_2

(c) Function f_3 (d) Function f_4

(e) Function f_5 (f) Function f_6

Fig. 7 Performance curves for BPSO, QIPSO-1and QIPSO-2

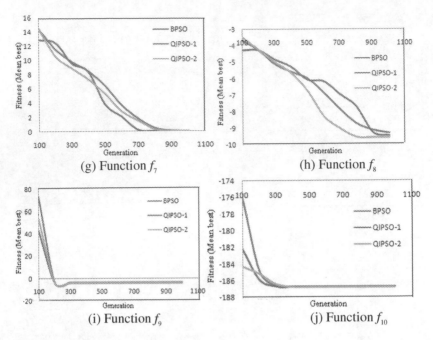

Fig. 7 (*continued*)

Table 6 Comparison results of real life problems (BPSO, GPSO, EPSO and LNPSO)

Gear Train Design				
Item	BPSO	GPSO	EPSO	LNPSO
x1	13	20	20	20
x2	31	13	13	13
x3	57	53	53	53
x4	49	34	34	34
f (x)	9.989333e-11	**2.331679e-11**	**2.331679e-11**	2.33168e-011
Gear Ratio	0.14429	0.14428	0.14428	0.14428
Error (%)	0.007398	0.000467	0.000467	0.000467
Transistor Modeling				
Item	BPSO	GPSO	EPSO	LNPSO
x1	0.901019	0.901241	0.901279	0.90097
x2	0.88419	0.883919	0.888237	0.880522
x3	4.038604	3.756517	3.854668	3.94582
x4	4.148831	3.861717	3.986954	4.081
x5	5.243638	5.387461	5.338548	5.28292
x6	9.932639	10.551659	10.410919	9.95503
x7	0.100944	0.26037	0.091619	0.221577
x8	1.05991	1.077294	1.083181	1.05418
x9	0.80668	0.764622	0.752615	0.825799
f(x)	0.069569	**0.058406**	0.05974	0.06292

Table 7 Comparison results of real life problems (BPSO, VC-PSO and SO-PSO)

Gear Train Design			
Item	BPSO	VC-PSO	SO-PSO
x1	13	16	16
x2	31	19	19
x3	57	49	49
x4	49	43	43
f (x)	9.989333e-11	**2.782895e-12**	2.7829e-012
Gear Ratio	0.14429	0.14428	0.14428
Error (%)	0.007398	0.000467	0.000467
Transistor Modeling			
Item	BPSO	VC-PSO	SO-PSO
x1	0.901019	0.900433	0.901031
x2	0.88419	0.52244	0.885679
x3	4.038604	1.07644	4.05936
x4	4.148831	1.949464	4.17284
x5	5.243638	7.853698	5.23002
x6	9.932639	8.836444	9.88428
x7	0.100944	4.771224	0.025906
x8	1.05991	1.007446	1.06251
x9	0.80668	1.854541	0.802467
f(x)	0.069569	**0.011314**	0.067349

Table 8 Comparison results of real life problems (BPSO, QIPSO, ATREPSO and GMPSO)

Gear Train Design				
Item	BPSO	QIPSO	ATREPSO	GMPSO
x1	13	15	19	19
x2	31	26	16	16
x3	57	51	43	43
x4	49	53	49	49
f (x)	9.98e-11	**2.33e-11**	2.78e-12	2.78e-12
Gear Ratio	0.14429	0.14428	0.14428	0.14428
Error (%)	0.007398	0.000467	0.000467	0.000467
Transistor Modeling				
Item	BPSO	QIPSO	ATREPSO	GMPSO
x1	0.901019	0.90104	0.900984	0.90167
x2	0.88419	0.884447	0.886509	0.877089
x3	4.038604	4.004119	4.09284	3.532352
x4	4.148831	4.123703	4.201832	3.672409
x5	5.243638	5.257661	5.214615	5.512315
x6	9.932639	9.997876	9.981726	10.80285
x7	0.100944	0.096078	5.69e-06	0.56264
x8	1.05991	1.062317	1.061709	1.074696
x9	0.80668	0.796956	0.772014	0.796591
f(x)	0.069569	0.066326	0.066282	**0.065762**

Table 9 Comparison results of real life problems (BPSO, QIPSO-1 and QIPSO-2)

Gear Train Design			
Item	BPSO	QIPSO-1	QIPSO-2
x1	13	19	13
x2	31	16	20
x3	57	43	34
x4	49	49	53
f (x)	9.989333e-11	**2.7829e-012**	2.33168e-011
Gear Ratio	0.14429	0.14428	0.14428
Error (%)	0.007398	0.000467	0.000467
Transistor Modeling			
Item	BPSO	QIPSO-1	QIPSO-2
x1	0.901019	0.901952	0.90107
x2	0.88419	0.895188	0.653572
x3	4.038604	3.66753	1.42074
x4	4.148831	3.67355	2.0913
x5	5.243638	5.44219	7.29961
x6	9.932639	11.2697	10.00
x7	0.100944	0.097903	4.09852
x8	1.05991	1.10537	1.00974
x9	0.80668	0.679967	1.59885
f(x)	0.069569	0.061881	**0.0514062**

Numerical results of real life problems are given in Tables 6 – 9. From these Tables, it is very difficult to claim the superiority of a particular algorithm over the others because the optimum function value obtained by all the algorithms is more or less similar. Although in some cases modified algorithms gave slightly better results than the basic PSO. This is probably due to the fact that both the real life problems, though nonlinear in nature, are small in size and do not pose any severe challenge for an optimization algorithm.

7 Conclusions

This article presents some recent simple and modified versions PSO. The algorithms considered may be divided into two classes; (1) algorithms without having any special operator but simply changing the initial configuration of the swarm and (2) algorithms having some special operator .

In all nine modified versions of PSO are presented in this chapter. These are:

- Gaussian Particle Swarm Optimization (GPSO)
- Exponential Particle Swarm Optimization (EPSO)

- Lognormal Particle Swarm Optimization (LNPSO)
- Sobol Particle Swarm Optimization (SOPSO)
- Van der Corput Particle Swarm Optimization (VCPSO)
- Attraction and Repulsion Particle Swarm Optimization (ATREPSO)
- Gaussian Mutation Particle Swarm Optimization (GMPSO)
- Quadratic Interpolation Particle Swarm Optimization (QIPSO)

The first five algorithms namely GPSO, EPSO, LNPSO, SOPSO and VCPSO described in the chapter use different initialization schemes for generating the swarm population. These schemes include Gaussian, exponential and lognormal probability distributions and quasi random sequences Sobol and Vander Corput to initialize the swarm. As expected, PSO algorithms initialized with quasi random sequences performed much better than the PSO initiated with the usual computer generated random numbers having uniform distribution (Please also see Table 3). However the interesting part of the study is that PSO initiated with Gaussian, exponential and lognormal distribution improved its performance quite significantly (Please also see Table 2).

The second part of the research consisted of modified PSO versions assisted with special operators like repulsion, mutation and crossover. In this part three algorithms called ATREPSO, GMPSO and QIPSO are given. The QIPSO is further modified into two versions QIPSO1 and QIPSO2. The common feature of these operator assisted PSO algorithms is that they all use diversity as guide to implement the operators. All the nine algorithms are applied on ten standard benchmark problems and two real life problems. The results obtained by these algorithms on the ten benchmark problems and two real life problems were either superior or at par with the basic PSO having uniform probability distribution to initialize the swarm. We did not compare the algorithms without any special operator with the ones having special operator because it will not be a fair comparison. However, among other algorithms PSO assisted with crossover operator QIPSO and its versions gave the best results. The present study may further be extended to solve the constrained optimization problems. Another interesting thing will be to combine PSO algorithms having different initialization scheme with PSO assisted with some special operator.

References

1. Kennedy, J., Eberhart, R.C.: Particle Swarm Optimization. In: IEEE International Conference on Neural Networks (Perth, Australia), IEEE Service Center, Piscataway, NJ, pg. IV, pp. 1942–1948 (1995)
2. Angeline, P.J.: Evolutionary Optimization versus Particle Swarm Optimization: Philosophy and Performance Difference. In: The 7th Annual Conference on Evolutionary Programming, San Diego, USA (1998)
3. Vesterstrom, J., Thomsen, R.: A Comparative study of Differential Evolution, Particle Swarm optimization, and Evolutionary Algorithms on Numerical Benchmark Problems. In: Proc. IEEE Congr. Evolutionary Computation, Portland, OR, June 20-23, pp. 1980–1987 (2004)

4. Vesterstrøm, J.S., Riget, J., Krink, T.: Division of Labor in Particle Swarm Optimisation. In: Proceedings of the Fourth Congress on Evolutionary Computation (CEC 2002), vol. 2, pp. 1570–1575 (2002)
5. Liu, H., Abraham, A., Zhang, W.: A Fuzzy Adaptive Turbulent Particle Swarm Optimization. International Journal of Innovative Computing and Applications 1(1), 39–47 (2007)
6. Shi, Y., Eberhart, R.C.: A Modified Particle Swarm Optimizer. In: Proc. IEEE Congr. Evolutionary Computation, pp. 69–73 (1998)
7. Eberhart, R.C., Shi, Y.: Particle Swarm Optimization: Developments, Applications and Resources. In: Proc. IEEE Congr. Evolutionary Computation, vol. 1, pp. 27–30 (2001)
8. Clerc, M.: The Swarm and the Queen: Towards a Deterministic and adaptive Particle Swarm Optimization. In: Proc. of the IEEE Congress on Evolutionary Computation, vol. 3, pp. 1951–1957 (1999)
9. Kennedy, J.: Small Worlds and Mega-Minds: Effects of Neighborhood Topology on Particle Swarm Performance. In: Proc. of the IEEE Congress on Evolutionary Computation, vol. 3, pp. 1931–1938 (1999)
10. Poli, R., Langdon, W.B., Holland, O.: Extending Particle Swarm Optimization via Genetic Programming. In: Keijzer, M., Tettamanzi, A.G.B., Collet, P., van Hemert, J., Tomassini, M. (eds.) EuroGP 2005. LNCS, vol. 3447, pp. 291–300. Springer, Heidelberg (2005)
11. Ting, T.-O., Rao, M.V.C., Loo, C.K., Ngu, S.-S.: A New Class of Operators to Accelerate Particle Swarm Optimization. In: Proceedings of the IEEE Congress on Evolutionary Computation, vol. (4), pp. 2406–2410 (2003)
12. Paquet, U., Engelbrecht, A.P.: A New Particle Swarm Optimizer for Linearly Constrained Optimization. In: Proceedings of the IEEE Congress on Evolutionary Computation, vol. (1), pp. 227–233 (2003)
13. Parsopoulos, K.E., Plagianakos, V.P., Magoulus, G.D., Vrahatis, M.N.: Objective Function "Strectching" to Alleviate Convergence to Local Minima. Nonlinear Analysis, Theory, Methods and Applications 47(5), 3419–3424 (2001)
14. Grosan, C., Abraham, A., Nicoara, M.: Search Optimization Using Hybrid Particle Sub-Swarms and Evolutionary Algorithms. International Journal of Simulation Systems, Science & Technology, UK 6(10&11), 60–79 (2005)
15. Gehlhaar, Fogel: Tuning Evolutionary programming for conformationally flexible molecular docking. In: Proceedings of the fifth Annual Conference on Evolutionary Programming, pp. 419–429 (1996)
16. Pant, M., Radha, T., Singh, V.P.: Particle Swarm Optimization: Experimenting the Distributions of Random Numbers. In: 3rd Indian Int. Conf. on Artificial Intelligence (IICAI 2007), India, pp. 412–420 (2007)
17. Krohling, R.A., Coelho, L.S.: PSO-E: Particle Swarm with Exponential Distribution. In: IEEE Congress on Evolutionary Computation, Canada, pp. 1428–1433 (2006)
18. Krohling, R.A., Swarm, G.: A Novel Particle Swarm Optimization Algorithm. In: Proc. of the 2004 IEEE Conference on Cybernetics and Intelligent Systems, Singapore, pp. 372–376 (2004)
19. Pant, M., Thangaraj, R., Abraham, A.: Improved Particle Swarm Optimization with Low-discrepancy Sequences. In: IEEE Cong. on Evolutionary Computation (CEC 2008), Hong Kong (accepted, 2008)
20. Kimura, S., Matsumura, K.: Genetic Algorithms using low discrepancy sequences. In: Proc of GEECO 2005, pp. 1341–1346 (2005)

21. Nguyen, X.H., Nguyen, Q.U., Mckay, R.I., Tuan, P.M.: Initializing PSO with Randomized Low-Discrepancy Sequences: The Comparative Results. In: Proc. of IEEE Congress on Evolutionary Algorithms, pp. 1985–1992 (2007)
22. Parsopoulos, K.E., Vrahatis, M.N.: Particle Swarm Optimization in noisy and continuously changing environments. In: Proceedings of International Conference on Artificial Intelligence and soft computing, pp. 289–294 (2002)
23. Brits, R., Engelbrecht, A.P., van den Bergh, F.: A niching Particle Swarm Optimizater. In: Proceedings of the fourth Asia Pacific Conference on Simulated Evolution and learning, pp. 692–696 (2002)
24. Brits, R., Engelbrecht, A.P., van den Bergh, F.: Solving systems of unconstrained Equations using Particle Swarm Optimization. In: Proceedings of the IEEE Conference on Systems, Man and Cybernetics, vol. 3, pp. 102–107 (2002)
25. Chi, H.M., Beerli, P., Evans, D.W., Mascagni, M.: On the Scrambled Sobol Sequence. In: Sunderam, V.S., van Albada, G.D., Sloot, P.M.A., Dongarra, J. (eds.) ICCS 2005. LNCS, vol. 3516, pp. 775–782. Springer, Heidelberg (2005)
26. Engelbrecht, A.P.: Fundamentals of Computational Swarm Intelligence. John Wiley & Sons Ltd., Chichester (2005)
27. Pant, M., Radha, T., Singh, V.P.: A Simple Diversity Guided Particle Swarm Optimization. In: IEEE Cong. on Evolutionary Computation (CEC 2007), Singapore, pp. 3294–3299 (2007)
28. Riget, J., Vesterstrom, J.S.: A diversity-guided particle swarm optimizer – the arPSO. Technical report, EVAlife, Dept. of Computer Science, University of Aarhus, Denmark (2002)
29. Pant, M., Radha, T., Singh, V.P.: A New Diversity Based Particle Swarm Optimization using Gaussian Mutation. Int. J. of Mathematical Modeling, Simulation and Applications (accepted)
30. Pant, M., Thangaraj, R.: A New Particle Swarm Optimization with Quadratic Crossover. In: Int. Conf. on Advanced Computing and Communications (ADCOM 2007), India, pp. 81–86. IEEE Computer Society Press, Los Alamitos (2007)
31. Pant, M., Thangaraj, R., Abraham, A.: A New Particle Swarm Optimization Algorithm Incorporating Reproduction Operator for Solving Global Optimization Problems. In: 7th International Conference on Hybrid Intelligent Systems, Kaiserslautern, Germany, pp. 144–149. IEEE Computer Society press, USA (2007)
32. Millie Pant, T., Pant, M., Radha, T., Singh, V.P.: A New Particle Swarm Optimization with Quadratic Interpolation. In: Int. Conf. on Computational Intelligence and Multimedia Applications (ICCIMA 2007), India, vol. 1, pp. 55–60. IEEE Computer Society Press, Los Alamitos (2007)
33. Kannan, B.K., Kramer, S.N.: An Augmented Lagrange Multiplier Based Method for Mixed Integer Discrete Continuous Optimization and its Applications to Mechanical Design. J. of Mechanical Design, 116/405 (1994)
34. Sandgren, E.: Nonlinear Integer and Discrete Programming in Mechanical Design. In: Proc. of the ASME Design Technology Conference, Kissimme, Fl, pp. 95–105 (1988)
35. Price, W.L.: A Controlled Random Search Procedure for Global Optimization. In: Dixon, L.C.W., Szego, G.P. (eds.) Towards Global Optimization 2, vol. X, pp. 71–84. North Holland Publishing Company, Amsterdam (1978)
36. Secrest, B.R., lamont, G.B.: Visualizing Particle Swarm Optimization – Gaussian Particle Swarm Optimization. In: Proc. of IEEE Swarm Intelligence Symposium, pp. 198–204 (2003)

37. Stacey, A., Jancic, M., Grundy, I.: Particle Swarm Optimization with Mutation. In: Proc. of the IEEE Congress on Evolutionary Computation, vol. 2, pp. 1425–1430 (2003)
38. van der Bergh, F.: An Analysis of Particle Swarm Optimizers. PhD thesis, Department of Computer Science, University of Pretoria, Pretoria, South Africa (2002)
39. van der Bergh, F., Engelbrecht, A.P.: A New Locally Convergent Particle Swarm Optimizer. In: Proc. of the IEEE Int. Conf. on Systems, Man, and Cybernetics, pp. 96–101 (2002)
40. Xie, X., Zhang, W., Yang, Z.: A Dissipative Particle Swarm Optimization. In: Proc. of the IEEE Congress on Evolutionary Computation, vol. 2, pp. 1456–1461 (2002)
41. Higashi, H., Iba, H.: Particle Swarm Optimization with Gaussian Mutation. In: Proc. of IEEE Swarm Intelligence Symposium, pp. 72–79 (2003)
42. Yao, X., Liu, Y.: Fast Evolutionary Programming. In: Fogel, L.J., Angeline, P.J., Back, T.B. (eds.) Proc. of the 5th Annual Conf. Evolutionary Programming, pp. 451–460 (1996)
43. Yao, X., Liu, Y., Lin, G.: Evolutionary Programming made faster. IEEE Trans. On Evolutionary Computation 3(2), 82–102 (1999)
44. Ting, T.-O., Rao, M.V.C., Loo, C.K., Ngu, S.-S.: A new Class of Operators to accelerate Particle Swarm optimization. In: Proceedings of IEEE Congress on Evolutionary Computation, vol. 4(656), pp. 2406–2410 (2003)
45. Clerc, M.: Think Locally, Act Locally: The way of Life of Cheap-PSO, an Adaptive PSO. Technical report (2001), http://clerc.maurice.free.fr/PSO/
46. Rigit, J., Vesterstorm, J.S.: Controlling Diversity in Particle Swarm Optimization. Master's thesis, University of Aahrus, Denmark (487) (2002)
47. Rigit, J., Vesterstorm, J.S.: Particle Swarms: Extensions for improved local, multi modal, and dynamic search in Numerical optimization. Masters thesis, department of Computer Science, University of Aahrus (620) (2002)
48. Brits, R.: Niching Strategies for Particle swarm optimization. Masters thesis, Department of Computer Science, university of Pretoria (67) (2002)
49. Brits, R.E., Van den Bergh, F.: Solving unconstrained equations using Particle Swarm Optimization. In: Proceedings of the IEEE congress on systems, man and cybernetics, vol. 3(70), pp. 102–107 (2002)

Tabu Search to Solve Real-Life Combinatorial Optimization Problems: A Case of Study

Djamal Habet

Summay. Tabu Search (TS) is a very powerful metaheuristic that has shown its efficiency when solving various combinatorial optimization problems, ranging from academic to real-life ones. The idea behind TS is to include during the search a guidance based on adaptive memory and learning. We present in this paper a short overview on Tabu Search and detail its application to solve successfully a real-life optimization problem under constraints.

1 Introduction

The Tabu Search (TS) Algorithm is a metaheuristic approach designed for solving combinatorial optimization problems. TS is a conceptually simple and elegant method and has become an established optimization methodology that is rapidly spreading in various fields. Planning and scheduling, transportation, routing and network design, continuous and stochastic optimization, manufacturing and financial analysis are some examples of these applications [22, 25, 20, 23, 10, 11, 12, 15, 18].

TS is originally introduced by F. Glover (see for instance [16]). It is an intelligent algorithm based on adaptive memory and learning. Like other local search methods, Tabu visits the search space by jumping from one configuration to another one within its neighborhood. The search method is partially based on the hill-climbing method that explores the neighborhood, that it defines, by trying to reach at each iteration a better configuration (in respects to some quality criteria). TS succeed in escaping local minima by using a tabu list that records the forbidden moves. Moves are classified as

Djamal Habet
LSIS – UMR CNRS 6168
Avenue Escadrille Normandie-Niemen, 13397 Marseille Cedex 20
e-mail: Djamal.Habet@lsis.org

A. Abraham et al. (Eds.): Foundations of Comput. Intel. Vol. 3, SCI 203, pp. 129–151.
springerlink.com   © Springer-Verlag Berlin Heidelberg 2009

forbidden if certain conditions imposed on the moves are satisfied. The aim
of maintaining a tabu list is to force the search process to avoid cycling and
thus impose some diversification.

The application of the Tabu metaheuristic to solve a combinatorial optimization problem can be difficult. In fact, such task needs the definition
of many components of a Tabu-like algorithm such as the search space, the
neighborhood, the move heuristic, the tabu list management, ... For instance,
defining the tabu tenure can require several (and expensive) empirical experiments in order the be fixed in accordance with the problem specifications.
Such a drawback is generally shared by the other local search methods. This
remark is strengthed by the rich literature around Tabu Search. This literature consists of several cases of study showing that every problem requires
special treatment and fine tuning so that the resolution by the tabu method
(sharing the same principles) is effective.

The first purpose of this paper is to give a general and a short description
of the Tabu Search (any interested reader by more details can easily find
more information in several tutorials about TS [5, 21]). The second and the
major purpose of this work is to explain, meticulously and step by step, the
application of Tabu Search to treat a real-world problem named the problem of managing an Agile Earth Observing Satellite (AEOS). This problem
was the subject of an international competition (ROADEF'2003) and concerns the maximization of a gain function under some hard constraints. The
gain has the particularity to be described by a non-linear function which
increases the hardness of the problem. All the components of our tabu resolution will be described, including the required algorithms to construct the
neighborhood and evaluate it, the aspiration criteria, the diversification, the
intensification, ...

The paper is structured as follows. It starts with a quick overview of the
tabu search in Section 2. We describe the AEOS problem in Section 3. The
components of the proposed tabu search algorithm to solve this problem are
described in Section 4. The obtained results are discussed in Section 5 and
Section 6 concludes the paper.

2 Tabu Search Principles

Over the last years, incomplete methods have received a high intention and
have proved their ability to solve successfully NP-hard combinatorial optimization problems (and often issued from real-world). One of the most popular incomplete methods is the local search (LS). LS is an iterative search
procedure that starts from an initial solution and progressively improves it by
applying a series of local modifications (called *moves*). At each iteration, the
search moves to an improving solution (often according to the optimization
criterion) that differs slightly from the current one. The search terminates
when it encounters a local optimum with respect to the transformations that

it considers. One of the major drawbacks of the local search is that this local optimum can be a solution with a bad quality and the search process is trapped in the currently explored zone (as in a hill-climbing algorithm). To overcome this weakness, several techniques (commonly named by Metaheuristics) are designed including simulated annealing, ant systems, genetic algorithms, tabu search and random walk. Among this none-exhaustive list, Tabu Search (TS) is one of the methods which is popular and efficient to overcome local optima.

In particular, Tabu Search (TS) is a metaheuristic designed to tackle hard combinatorial optimization problems. By contrast with random approaches, TS is based on the belief that an intelligent search should include more systematic forms of guidance based on adaptive memory and learning. TS can be described as a form of neighborhood search with a set of critical and complementary components.

For a given optimization instance (S, f) characterized by a search space S and an objective function f, a neighborhood $\mathcal{N}$ is introduced. It associates to each configuration s in S (a configuration is an affectation of the problem variables by some values), a non-empty subset $\mathcal{N}(s)$ of S. A typical TS algorithm begins with an initial configuration $s_0 \in S$, then repeatedly visits a series of the best local configurations following the neighborhood function. At each iteration, one of the best neighbors $s' \in S$ is selected to become the current configuration, even if s' does not improve the current configuration in terms of the cost function.

To avoid the problem of cycles occurring and to allow the search to go beyond local optima, a tabu list is introduced. This adds a short term memory component to the method. A tabu list maintains a selective history H (short term memory), composed of previously encountered configurations or, more generally, pertinent attributes of such configurations. A simple TS strategy consists in preventing configurations of H from being considered on the next k iterations, called the tabu tenure. This can vary according to different attributes, and it is generally problem dependent. At each iteration, TS looks for the best neighbor from this dynamically modified neighborhood $\mathcal{N}(H, s)$, instead of $\mathcal{N}(s)$ itself. Such a strategy prevents the search from being trapped in short term cycling and makes the process more rigorous.

Standard tabu lists are usually implemented as circular lists of fixed length. It has been shown, however, that fixed-length tabus cannot always prevent cycling, and some authors have proposed varying the tabu list length during the search (see for instance [13, 14]). Another solution is to randomly generate the tabu tenure of each move within some specified interval: using this approach requires a somewhat different scheme for recording tabu move which are then usually stored by using tags in an array [9].

When attributes of configurations, instead of configurations themselves, are recorded in a tabu list, some unvisited, but nonetheless interesting configurations may be prevented from being considered.

It is thus necessary to use algorithmic devices that will allow one to revoke (cancel) the tabu status of such intersting configurations. These are called aspiration criteria. The simplest and most commonly used aspiration criterion (found in almost all TS implementations) consists in allowing a move, even if it is tabu, if it results in a configuration with an objective value better than that of the current best-known configuration (since the new configuration has obviously not been previously visited) [9]. Much more complicated aspiration criteria have been proposed and successfully implemented but they are rarely used [5, 21].

Two other important ingredients of TS are intensification and diversification. On the one hand, the intensification consists in focusing the search to exploit regions of the search space, or characteristics of solutions, that the search history suggests that they are promising. On the other hand, the diversification undertakes to explore regions that differ in significant respects from regions previously visited.

On the one hand, the idea behind the concept of search intensification is that, as an intelligent human being would probably do, one should explore more thoroughly the portions of the search space that seem promising in order to make sure that the best solutions in these areas are indeed found. From time to time, one would thus stop the normal searching process to perform an intensification phase. In general, intensification is based on some intermediate-term memory, such as a recency memory, in which one records the number of consecutive iterations that various solution componentshave been present in the current solution without interruption. A typical approach to intensification is to restart the search from the best currently known solution and to fix in it the components that seem more attractive. Another technique that is often used consists in changing the neighborhood structure to one allowing more powerful or more diverse moves. Intensification is used in many TS implementations, but it is not always necessary. This is because there are many situations where the search performed by the normal searching process is thorough enough. There is thus no need to spend time exploring more carefully the portions of the search space that have already been visited [9].

On the other hand, one of the main problems of all methods based on Local Search approaches, and this includes TS in spite of the beneficial impact of tabus, is that they tend to be too local: they tend to spend most, if not all, of their time in a restricted portion of the search space. The negative consequence of this fact is that, although good solutions may be obtained, one may fail to explore the most interesting parts of the search space and thus end up with solutions that are still pretty far from the optimal ones. Diversification is an algorithmic mechanism that tries to alleviate this problem by forcing the search into previously unexplored areas of the search space. It is usually based on some form of long-term memory of the search, such as a frequency memory, in which one records the total number of iterations (since the beginning of the search) that various solution components have been present in the current solution or have been involved in the selected moves. In cases where it is possible to identify useful regions of the search space, the frequency memory can be refined to track the number of iterations spent in these different regions.

There are two major diversification techniques. The first, called restart diversification, involves forcing a few rarely used components in the current solution (or the best known solution) and restarting the search from this point. The second diversification method, continuous one, integrates diversification considerations directly into the regular searching process [9].

Therefore, a TS is described by specifying its main elements: the space search and the neighborhood definition, the cost function to evaluate the configurations and the neighborhood, the tabu list management, the aspiration criteria and finally the intensification and diversification phases. In the next sections, we describe and precise the definition and the application of these elements to solve a real-life optimization problem which is the problem of selecting and scheduling photographs of an agile earth observing satellite. We starts bu describing this problem then the dedicated tabu algorithm to solve it.

3   AEOS Problem

Earth Observing Satellites (EOS) are platforms that orbit the planet and are equipped with optical instruments. A new generation of EOS are *Agile* Earth Observing Satellites (AEOS). This means that, while the single onboard camera remains fixed on the satellite, the whole satellite is mobile along three axes (roll, pitch and yaw). This mobility potentially increases the efficiency of the whole system. In the course of their mission, satellites take photographs of specific areas of the Earth in function of requests from a number of users including governments, research institutes, and companies. Each request generates a revenue or a gain, and typically the number of requests exceeds what can be feasibly accomplished during a mission.

An observation request concerns some specified areas modeled by polygons. Each one is cut into a set of contiguous strips (rectangular shape) covering the request. A strip can be acquired using two opposite azimuths according to the satellite rotation sense (direct and indirect). Moreover, some requests are mono while others are stereo. A mono request consists of a single shot of each strip in the polygon. A stereo request consists of two shots of each strip at different angles but in the same direction. Finally, for each pair of shots, the satellite requires a minimum transition time to maneuver the camera from the end of the first strip to the start of the second one [24].

The input of an AEOS management problem is a set of candidate strips, from the current set of requests, that could be acquired. The problem to be dealt with is twofold: to select a set of strip acquisitions that maximize the total gain, and to order them in time. Formally, the problem can be described as follows [1]:

Formal Description

A problem with n strips involves $2n$ possible acquisitions since for each strip i two shooting directions are possibles. These shots are numbered:

- $2i - 1$: an odd number for a shot acquired in the direct azimuth, and
- $2i$: an even number for a shot acquired in the indirect azimuth.

For each strip $i \in [1, n]$ let:

- $tw(i)$ be the index of its stereo twin strip, 0 if i is mono.
- $d(i)$ be its shooting duration.
- $su(i)$ be its corresponding surface.

For each shot (image) $j \in [1, 2n]$ let:

- $es(j)$ and $ls(j)$ be its earliest and latest start dates respectively.
- $ee(j)$ and $le(j)$ be its earliest and latest end dates respectively.

For each shot pair (i, j), $i \neq j \in [1, 2n]$, $t(i \to j)$ denotes the minimum transition time between the end of i to the beginning of j. We assume that $t(i \to j) \geq es(j) - le(i)$ (otherwise it is obviously underestimated). Four variables are associated to each shot $i \in [1, 2n]$:

- $x_i \in \{0, 1\}$ equals 1 if and only if shot i is selected.
- $y_{0 \to i} \in \{0, 1\}$ equals 1 if and only if shot i is the first of the selection.
- $y_{i \to 2n+1} \in \{0, 1\}$ equals 1 if and only if shot i is the last of the selection.
- t_i is the shooting start date of i. The value of this variable is irrelevant when $x_i = 0$. The t_i values allow to order and to schedule the shots in time

Let m be the number of polygons. The kth polygon is characterized by:

- $s(k)$ its total area surface.
- g_k its gain when fully acquired.
- $p(k) \subset [1, 2n]$ the set of shots that it contains.We recall that each polygon is divided into a set of strips. Moreover, two shots are possibles per strip according to the sense of its acquisition.

Two continuous variables are associated to each polygon $k \in [1, m]$:

- $S_k \in [0, 1]$ is the percentage of the surface covered by the selected strips.
- $G_k \in [0, 1]$ is the corresponding percentage of the polygon gain.

Finally, one binary variable is defined for each pair of acquisitions $i \neq j \in [1, 2n]$:

- $y_{i \to j} \in \{0, 1\}$ equals 1 if and only if the shot j is immediately acquired after the shot i.

Constraints

The equations constraining these variables are listed below. The acquisition of the same strip in both directions is forbidden by (1). Equation (2) states the stereo constraints: simultaneous selection with identical direction. The time window for the starting time of a shot is imposed by (3). Shooting dates of consecutive images must respect minimum transition times (4).

(1) $\forall j \in [1, n], x_{2j-1} + x_{2j} \leq 1$.

(2) $\forall j \in [1, n]$, if $tw(j) \neq 0$ then $(x_{2j-1} = x_{2tw(j)-1}$ and $x_{2j} = x_{2tw(j)})$.

(3) $\forall i \in [1, 2n]$, if $x_i = 1$ then $t_i \in [es(i), ls(i)]$.

(4) $\forall i \neq j \in [1, 2n], t_j - t_i \geq (d(i) + t(i \rightarrow j)).y_{i \rightarrow j} + (es(j) - ls(j)).(1 - y_{i \rightarrow j})$.
    We denote by $C4$ the set of shot pairs (i, j) that satisfy the condition (4).

Furthermore, there is at most one first shot and one last shot and each selected acquisition has exactly one predecessor and one successor. In the rest of this paper, we assume that these two constraints are always satisfied.

Objective Function

The criterion to maximize is a global gain G defined by the sum of the gains associated to the complete or partial acquisition of each polygon k and formulated by:

$$G = \sum_{k=1}^{m} g_k \times G_k \text{ such as:}$$

- $\forall k \in [1, m], S_k = \dfrac{1}{s(k)} \sum_{i \in p(k)} su(i).x_i$.
- $G_k = f(S_k)$.
- $f : [0, 1] \rightarrow [0, 1]$ is a non-linear function, piecewise linear and defined by the points $\{(0, 0), (0.4, 0.1), (0.7, 0.4), (1, 1)\}$.

Some Related Works

Compared to other optimization problems, the Earth Observing Satellite management problem has received a limited attention. For the AEOS problem and in the context of the ROADEF'2003 Challenge, various methods were proposed. In [26], J.F. Cordeau and G. Laporte have proposed a tabu search algorithm which borrows from the Unified Tabu Search Algorithm [4] developed for the Vehicle Routing Problem with Time Windows (VRPTW). An important feature of their algorithm is the possibility of exploring infeasible solutions during the search by allowing the violation of the time window constraints. Thus, the value of a solution is defined by $f(s) = G(s) - \alpha w(s)$, where $w(s)$ is the total time window constraints. The parameter α is initially

set at 1 and self-adjusts during the course of the search to allow a mixture
of feasible and infeasible solutions. The moves switch between the insertion
and the removal of mono and stereo shots with respect to all the constraints
except the time window one which is relaxed. Moreover, the algorithm uses
two diversification mechanisms. The first one is a continuous diversification
scheme which introduces penalties on poor solutions. These penalties drive
the search process toward the less explored regions of the search space whenever a local optimum is reached. The second diversification mechanism per-
turbs the solution under certain circumstances. If the best known solution
has not improved after a certain number of iterations then the search stops
and restarts from the best reached solution. However, this best solution is
perturbed by removing a portion of randomly selected shots.

E.J. Kuipers [26] proposed a local search algorithm based on two stages. In
the first one, the most promising solutions are constructed then further optimized in the second stage. These two parts use Simulated Annealing (SA)
algorithms. The neighborhood is constructed in two steps. In the first one, 1,
2, 3 or 4 requests (or parts of requests) are removed from the current solution, and in the second step 1, 2, 3 or 4 requests (or parts of requests) are
added to the solution resulting from the first one. The number of the added
or removed requests (1, 2, 3 or 4) in both steps is randomly chosen. Moreover,
a library is constructed to file the best ways of ordering a set of 5 strips in
terms of transition durations. This library is updated at each new insertion
and used by the two SA algorithms. Other additional schemes are added to
attempt the speeding up of the optimization process of the first stage (for instance, a limitation on the number of the strips treated according to their gain).
However, the second stage considers all the strips during the neighborhood
construction.

In addition, a number of variants of the AEOS problem have been studied
in the literature. The paper [17] presents the space mission problem which
consists of selecting and scheduling a set of jobs on a single machine among
a set of candidates jobs. Each candidate job has a fixed duration, a given
time window and a weight. The aim is to select a feasible sequence of jobs
that maximizes the sum of weights. The space mission problem is NP-hard
and it is very close to the AEOS management problem with the following
simplifications: null transition times, no unique strip acquisition and stereoscopic constraints, and a linear objective function. Several interesting greedy
algorithms and an optimal algorithm for solving the space mission problem
are proposed. The optimal algorithm is based on a Dynamic Programming
scheme. Upper bound formulations are presented, based on a preemptive
relaxation (a job can be fragmented) and a lagrangian relaxation. The proposed algorithms are tested on 30 randomly generated instances with up to
200 candidate jobs.

The selection and scheduling problem for the SPOT5 satellite concerns a
non-agile satellite with only one moving axis (rolling). A consequence of this
lack of manoeuvrability is that the starting time of each candidate image is

fixed. This feature would result in a very simple (polynomial) problem if there were only one imaging instrument on the board of the satellite, but there are 3 instruments. Nevertheless, it is possible to pre-compute binary and ternary constraints which model the compatibilities between candidate images. Each candidate image is weighted and the problem is to find a feasible subset of the candidate images maximizing the sum of weights. This optimization problem is NP-hard and can be formulated in the general Valued Constraint Satisfaction Problem model [27]. The paper [3] fully describes the problem and proposes some instances benchmark. Some results with dedicated exact and approximate methods are given in [2, 3] and the column generation technique has been used to compute upper bounds on the benchmark instances [7]. Very good results on these instances were obtained by M. Vasquez and J.K. Hao using a dedicated Tabu Search algorithm [28]. These authors have also obtained very good upper bounds by means of an original partition method assessing the quality of their previous results [29]. Note that for some instances the optima values are still unknown.

The work presented in [8] concerns a selection and scheduling problem of a semi-agile satellite. The differences with respect to the AEOS problem are as follows: (a) the criterion to be maximized is the number of selected images (images are not weighted), (b) the satellite is slightly mobile in two axes (pitch and roll), but remains fixed during an image acquisition; so, there is only one possible azimuth for an acquisition, (c) the satellite's kinematics do not allow for a given strip to be acquired twice during the same track, and (d) there are no stereoscopic requests, hence no corresponding stereo constraints.

We now describe some related problems. In the first one which is named the Maximum Shot Sequencing Problem (MSP), the aim is to select and to schedule a set of images over several consecutive tracks (a given image can be acquired from two or more tracks), so several possible disjoint time windows are given for each image. In the second problem, named the Maximum Shot Orbit Sequencing Problem (MSOP), only one track is processed, so a single time window is associated to each candidate image. MSOP and MSP are also NP-hards and several algorithms are proposed and based on graph theoretic concepts. In a first approach, the time is discretized. The constraints (given by the satellite kinematics) are such that MSOP amounts to a longest path problem. With time discretization, MSP amounts to find a maximal independent set in an incompatibility graph, and can be solved by an approximate algorithm based on a near-optimal partition into cliques. Due to the nature of the objective, an interesting upper bound is available. Exact and approximate algorithms are also presented for the continuous time model to solve both MSP and MSOP. The exact algorithm is a branch-and-bound algorithm with graph-based heuristics and bounds. The approximate algorithm is a kind of greedy algorithm also based upon graph properties. Experiments are conducted on a set of randomly generated instances, allowing the proposed algorithms to be assessed against upper bounds and the impact of the discretization to be measured.

The paper [19] describes a scheduling problem concerning a satellite equipped with a radar instrument which is very agile on the pitch axis but slow on the roll axis. Only one azimuth is available for acquiring images. This problem is equivalent to the space mission problem [17] with transition times. The authors describe a partial enumeration algorithm for this problem and give preliminary results for randomly generated instances.

The work reported in [30] describes the so-called Window-Constrained Packing problem (WCP). It differs from the problems presented above in the fact that the evaluation function is a priority-weighted sum of observation durations under suitability functions. In other words, observations have non-fixed durations (ranging between a maximum and a minimum) and preference is given to higher priority observations (with longer durations) and to the observations which are well-placed inside their time window. There are no transition times between observations. This problem is similar to the space mission problem [17], with a particular gain function, depending on the starting times of the observations. The authors present some approximate algorithms for solving the WCP problem. A first one is a fast but not very accurate greedy algorithm. A second one is similar to the first, but it includes some look-ahead. It is a little better but has expensive computation times. These algorithms have been tested on randomly generated instances. Actually, the WCP problem is a simplified short-term version of the real scheduling problem. The paper investigates the associated mid-term and long-term scheduling problems and their connections, as well as different objective functions.

4   A Tabu Search Algorithm for AEOS Problem

In this section, we describe all the necessary component of a tabu resolution to solve the problem of managing the agile earth observing satellite.

4.1   Search Space Definition

Definition 1. *The unconstrained search space S consists of all the vectors of the pairs (x_i, t_i), $i = 1...2n$:*

$$S = \{((x_1, t_1), (x_2, t_2), ..., (x_{2n}, t_{2n}))/ \; \forall i \in [1, \; 2n]: \; x_i \in \{0, 1\} \; and \; t_i \in I\!R\}$$

In this formulation, i is a shot number among the $2 \times n$ possible ones issued from the n strips (2 shots per strip). t_i is the (unconstrained) beginning time acquisition of the shot i.

Definition 2. *The totally constrained search space X is a subset of vectors in S that satisfy all the imperative constraints (1) to (4):*

$$X = \{s \in S/ \; all \; the \; elements \; of \; s \; satisfy \; the \; constraints \; (1) \; to \; (4)\}$$

Each vector $s \in X$ is *a consistent configuration* (all the constraints are satisfied), which is evaluated by its corresponding gain $G(s) = \sum_{k=1}^{m} g_k \times f(\frac{1}{s(k)} \sum_{i \in p(k)} su(i).x_i)$ (see Section 3). We denote by $|s|$ the number of the shots that are selected in s ($|s| = \sum_{i=1\cdots 2n} x_i$). This constrained search space is used by our TS algorithm.

4.2 Neighborhood Definition

Now, we introduce the neighborhood function $\mathcal{N}$ over the totally constrained search space X. This function $\mathcal{N} : X \to (2^X - \emptyset)$ is defined as follows:

Let $s = ((x_1, t_1), (x_2, t_2), ..., (x_{2n}, t_{2n}))$ be a consistent configuration (all the constraints of the problem are satisfied), $s' = ((x_1', t_1'), (x_2', t_2'), ..., (x_{2n}', t_{2n}'))$ is a neighbor of s, i.e. $s' \in \mathcal{N}(s)$, if and only if the following conditions are checked:

1. There is exactly one shot i (and eventually its twin if $tw(i) \neq 0$) which is not selected in s ($x_i = 0$) and selected in s' ($x_i' = 0$). In other words, we try to insert exactly one shot in s, and its twin if it exists.
2. For any inserted new shot i in s', the equation constraining one shot acquisition by strip must be satisfied. Formally, if the shots i and j are issued from the same strip then $x_i = x_j = 0$ and $x_i' = 1$; $x_j' = 0$. Moreover, the stereo constraint imposing the acquisition of the stereo strips in the same direction must be checked.
3. For any inserted new shot , we have $t_i \in [es(i), ls(i)]$. Moreover, if $tw(i) \neq 0$ then $t_{tw(i)} \in [es(tw(i)), ls(tw(i))]$.
4. For each shot i that satisfies the condition 1, $\forall k \in [1, 2n]$ such that $(i, k) \notin C4$ and $x_k = 1$ we have $x_k' = 0$ and if $tw(k) \neq 0$ then $x_{tw(k)}' = 0$. Moreover, if $tw(i) \neq 0$ then $\forall l \in [1, 2n]$ such that $(tw(i), l) \notin C4$ and $x_l = 1$ then $x_l' = 0$. In addition, if $tw(l) \neq 0$ then $x_{tw(l)}' = 0$.
5. For any shot i that satisfies the condition 1, $-3 \leq |s'| - |s| \leq 1$. Also, if $tw(i) \neq 0$ then $-6 \leq |s'| - |s| \leq 2$.
6. For any shot i that satisfies the condition 1 and such that its insertion in s requires at most the removal of two shots $j, k \in [1, 2n]$ with their twin shots if they exist ($-3 \leq |s'| - |s| \leq -1$, $x_j = x_k = 1$, $x_j' = x_k' = 0$ and $j \neq tw(k)$) then even $y_{j \to k} = 1$ or $y_{k \to j} = 1$ in s (k and j are acquired one behind the other in s).

Thus, the neighborhood of s is obtained by adding a free shot i ($x_i = 0$) by flipping x_i from 0 to 1 (condition 1), then removing some shots k (by flipping x_k from 1 to 0) to repair the violated constraints (condition 4). The condition 5 enforces a maximum of 2 shot removals if all the dropped shots are mono and 4 if they are stereo (in order to maintain the consistency of the stereo constraint, if a stereo shot is removed then its twin is removed too).

Moreover, this choice heuristic is also restricted to the removal of successive shots as described in the condition 6.

Hence, each configuration reached from s and according to the condition 1 to 6 is also consistent. Consequently, the elaborated neighborhood $\mathcal{N}(s)$ is consistent.

Neighborhood Evaluation

We evaluate $\mathcal{N}(s)$ according to the gain criterion. For this purpose, consider a configuration $s = ((x_1, t_1), (x_2, t_2), ..., (x_{2n}, t_{2n}))$ where $|s|$ images are selected and an image j such that $x_j = 0$ (i.e. j is not yet selected). The shot j can be inserted in s through $|s| + 1$ positions: before the first shot, after the last shot, or between two successive shots on the schedule s. Hence, the insertion of the shot j in each of the $|s| + 1$ positions is tested by allowing successive image removals (as explained above). If a position is tested positively then a neighborhood configuration s' is reached by inserting j at this position is s (and dropping some others images, if necessary). We evaluate s' by computing its corresponding gain value $G(s')$ (see Section 3). Among all the feasible insertions of j, the one that maximizes the gain value is selected. Consequently, at each step of the TS algorithm, we solve the decision problem of finding the best insertion position for each free shot in s according to the gain function.

At each iteration, TS algorithm examines the value of $G(s')$ for each candidate neighbor $s' \in \mathcal{N}(s)$ and chooses the one with the highest gain. Those operations are very time consuming and to overcome this weakness incremental computing techniques are used [6]. The main idea is to use a specific data structure containing for each possible move the corresponding gain and the resulting configuration if the insertion is really performed. Each time a move is carried out, the elements of this data structure which are affected by the move are updated accordingly.

The neighborhood is evaluated in accordance to Algorithm 1. The used notations are as follows:

- $\mathcal{L}_{cand}$ is the list of the candidate moves among the neighbors of the current configuration.
- $best\_gain$ is the best gain associated to the configuration obtained by the insertion of a free shot in the current configuration s and contained in $\mathcal{L}_{cand}$.
- $Generate\text{-}sub\text{-}schedules(s)$ is the function which generates sub-schedules from s by removing 0, 1 or 2 successive images (and their twins if necessary).
- The set of the sub-schedules produced by the function $Generate\text{-}sub\text{-}schedules(s)$ is denoted by D_s. The first element of D_s is s (no removal). We denote by $|D_s|$ the number of sub-schedules in D_s and by s_l the lth sub-schedule in D_s.

- p_m denotes the position numbered by m and situated between two shots which are scheduled respectively at the orders m and $m + 1$ according to their shooting start date. In particular, p_0 is the position before the first acquired shot and $p_{|s_l|}$ is the position after the last acquired shot in s_l.
- g^* is the gain associated to the best schedule already reached s^* from the beginning of the resolution, $g^* = G(s^*)$.
- $Insert(s_l, i, p_m)$ is a function which returns $True$ if the insertion of the shot i at position p_m in a sub-schedule s_l is feasible regarding to the problem constraints, $False$ otherwise.
- $Gain(s_l, i, p_m)$ is the associated gain to the configuration resulted from the insertion of the shot i in s_l at the position p_m.

Algorithm 1. Evaluate–$\mathcal{N}(s)$

begin
> $\mathcal{L}_{cand} \leftarrow \emptyset$;
> $best\_gain \leftarrow -\infty$;
> $D_s \leftarrow Generate\text{-}sub\text{-}schedules(s)$; % $D_s = \{s, s_1, s_2, ...\}$;
> **for** all i, such that $x_i = 0$ **do**
> > **for** $l = 1$ **to** $|D_s|$ **do**
> > > **for** $m = 0$ **to** $|s_l|$ **do**
> > > > **if** $Insert(s_l, i, p_m) = True$ **then**
> > > > > **if** $(Gain(s_l, i, p_m) > g^*)$ or (s_l is not tabu) **then**
> > > > > > **if** $Gain(s_l, i, p_m) > best\_gain$ **then**
> > > > > > > $\mathcal{L}_{cand} \leftarrow \emptyset$;
> > > > > > > $best\_gain \leftarrow Gain(s_l, i, p_m)$;
> > > > > > > $\mathcal{L}_{cand} \leftarrow \{(s_l, i, p_m)\}$;
> > > > > > **else**
> > > > > > > **if** $Gain(s_l, i, p_m) = best\_gain$ **then**
> > > > > > > > $\mathcal{L}_{cand} \leftarrow \mathcal{L}_{cand} \cup \{(s_l, i, p_m)\}$;
> **return** ($\mathcal{L}_{cand}$, $best\_gain$)
end;

(1)

In Algorithm 1, we start by generating the set D_s of the sub-schedules obtained by removing some shots. Then we try to insert each free shot i in those sub-schedules. If an insertion is successful then we calculate its corresponding gain. Afterward, this insertion becomes candidate, the list of the best candidates is saved in $\mathcal{L}_{cand}$ and their associated gain value is $best\_gain$. If a candidate move improves the current best gain then $\mathcal{L}_{cand}$ will contain only this move. The list $\mathcal{L}_{cand}$ will be used in the move heuristic as it will be explained in the next sections.

4.3 Tabu List Management and Move Heuristic

We define a move by the insertion of a free shot i (flipping x_i from 0 to 1) followed by the removal of the conflicting shots that do not satisfy the problem constraints (for each conflicting shot j, we flip x_j from 1 to 0). Now, we explain both the management of the tabu list and the heuristic used to select one of the move candidates.

Tabu List Management

The role of a *tabu list* is to prevent short-term cycling. In this order, when a shot i is selected and scheduled (a move is carried out), this shot is classified tabu (forbidden for any change) for a certain time called the *tabu tenure*. In our TS algorithm, this tenure is dynamically formulated by:

$$tabu(i) = iter + \alpha \times freq(i), \text{ where:}$$

- *iter* is the number of the current iteration of the tabu algorithm.
- $freq(i)$ counts the number of times that the shot i has been selected by the tabu algorithm (note that a shot can be inserted at a given iteration and be dropped some iterations later, then reselected after and so on).
- α is a variable parameter used to weight $tabu(i)$ according to $freq(i)$.

Moreover, a sequence of shots is tabu if all its shots are tabu, and not tabu if it contains at least one non-tabu shot. Likewise, we state that a sub-schedule obtained from a current schedule by removing a tabu sequence is also tabu, otherwise it is not tabu. Additionally, the direct azimuth corresponds to the natural move direction of the satellite and changing azimuth between two shootings is very costly in the terms of transition time.

Consequently, the acquisitions in the direct azimuth are preferred over the indirect ones. This preference is expressed by setting $\alpha = 2 \times \beta$ for the shots acquired in a direct azimuth and $\alpha = \beta$ for the indirect ones (the value of β is fixed empirically). Hence:

begin
    if i *is odd* **then**
         $tabu(i) = iter + 2 \times \beta \times freq(i)$;
    else
         $tabu(i) = iter + \beta \times freq(i)$;
end;

Recall that in Section 3 we use odd and even numbers to differentiate the shots regarding to their acquisition directions.

Move Heuristic

Once the neighborhood is evaluated, the selected shot (move) is the one that maximizes the gain value and does not remove a sequence of tabu shots. However, if a move strictly improves the best gain then the *aspiration criterion* is employed to cancel the tabu status of a sub-schedule. Therefore in Algorithm

1, the condition labeled [1] corresponds to either the best gain g^* is strictly improved (aspiration criterion) or the sub-schedule s_l is not tabu.

To summarize, we start by constructing $\mathcal{L}_{cand}$, we select randomly one candidate move (s_l, i, p_m) from $\mathcal{L}_{cand}$, then we insert the shot i in s_l at the position p_m to obtain the neighbor configuration s' that replaces the current one, which completes the move.

The Enhanced Aspiration Criteria

As described above, if some moves lead to the same gain value then one of them is randomly chosen. However, this selection criterion is not the most effective one. For this reason and to tone down the random effect, we have introduced a second objective function which is the minimization of the sum of the transition durations in a configuration s with the respects of the constraints of the initial problem. We denote this sum by
$$TdT(s) = \sum_{i=1}^{2n} \sum_{\substack{j=1 \\ j \neq i}}^{2n} x_i.x_j.y_{i \to j}.d(i \to j) \ .$$

The aim of this minimization is to obtain a shortest schedule in terms of the sum of the transition durations between the strip shootings. Indeed, the reduction of this sum can generate visibility time windows sufficiently broad and usable by the satellite to acquire new shots and without removing those that are already selected. The dedicated algorithm to achieve this secondary optimization is not detailed here. It is simply based on two operations: the exchange of the order of the shots and the inversion of the acquisition direction of the strips.

According to this new secondary objectif, we improve the aspiration criterion as follows. The tabu status of a sub-schedule is canceled if one of the two conditions below is verified:

1. if a move leads to a configuration s' better than the best configuration s^* found so far, i.e. $G(s') > G(s^*) = g^*$,
2. if a move leads to a configuration s' with the same gain value of s^* but with a lower TdT value than $TdT(s^*)$, i.e. $G(s') = G(s^*)$ and $TdT(s') < TdT(s^*)$.

4.4   Intensification and Diversification Phases

The tabu mechanism may lead to a state where no move is admissible (all moves are tabu). This occurs when each possible move has been tried a large number of times without improving the best configuration already reached s^*. In this case, we launch an intensification phase: when the gain cannot be improved (all the shots are tabu and the aspiration criterion is not satisfied) the intensification phase attempts to overcome this situation by exploiting the best schedule s^* as follows:

First step: The minimization of the sum of the transition durations of s^* is treated. If it succeeds (i.e. the $TdT(s^*)$ value is decreased) then the tabu status of all the shots are set to 0, and the tabu exploration is restarted from the s^*. This step is important. In fact, we have observed that during the experimental process if this first step has been achieved successfully then we may insert a free shot without any removal and consequently improve the best gain g^*.

Second step: When the first step fails, we decrease the β value by dividing it by 2, $\beta \leftarrow \beta/2$. Hence, the tabu durations are reduced (recall that β is a variable parameter of the tabu tenure of a shot, see Section 4.3). Accordingly, if $\beta > 1$ then the tabu status of all the shots are set to 0 and the tabu exploration is restarted from the best solution s^*.

As described above, the intensification phase alternates between two exclusive steps. In fact, the execution of the fist step (respectively, the second step) inhibits the execution of the second one (respectively, the first one). Its aim is to focus the exploration around the elements of the best solution by either reordering its selected strips and inverting their acquisition directions ($TdT(s^*)$ minimization) or by decreasing the tabu tenure of the shots. However, if the intensification phase does not improve the best gain then a diversification process is applied.

The role of diversification is to escape from the attractive zones of the search space corresponding to the local minima. For this reason, when both the tabu exploration and the intensification phase fail, we generate a new starting point (first schedule) different from the last one used at the beginning of the tabu exploration, we set the tabu status of all the shots to 0 and the β parameter to a new value fixed empirically, then we restart the tabu search from this new point.

4.5 Global Tabu Resolution

The TS algorithm follows a general scheme consisting of three iterative phases: exploration, intensification and diversification. All these resolution steps are given in Algorithm 2. It starts by initializing the different structures that it uses, such as the wg, $freq$ and $tabu$ values. The greedy algorithm (denoted $greedy()$) returns a first feasible solution s_0 constructed by successive image insertions without any removal. Let us give more details about this greedy algorithm.

In fact, to produce quickly a first feasible solution, we use a simple algorithm based on the operation of inserting a shot that does not require the removal of the already selected shots. Also, each shot $i \in [1, 2n]$ is weighted according to the number of times that it was selected by the greedy algorithm. We denote by $wg(i)$ this weight which is initialized to 0.

When it is launched, the greedy algorithm starts by sorting the $2n$ shots in the increasing order of their weights wg and stores them is a list Q. The first element of Q is the least selected shot by $greedy()$ and the last one is the most

Algorithm 2. $Tabu–AEOS()$
begin

    for $i=1$ *to* $2n$ **do**
        $wg(i) \leftarrow 0$; % initialization of the wg weights

    repeat
        % initialization of the tabu and the frequency values of each shot
        for $i=1$ *to* $2n$ **do**
            $x_i \leftarrow 0$;
            $freq(i) \leftarrow 0$;
            $tabu(i) \leftarrow 0$;

(1) $s \leftarrow s_0 \leftarrow greedy()$; % the initial configuration
        $s^* \leftarrow s$;   % initialization of the best configuration
        $g^* \leftarrow G(s^*)$;   % initialization of the best gain
        $\beta \leftarrow |s_0|^+$;   % initialization of the β factor
        $iter \leftarrow 0$;   % initialization of the iteration number
        repeat
            $(\mathcal{L}_{cand}, best\_gain) \leftarrow$ Evaluate-$\mathcal{N}(s)$;
            if $\mathcal{L}_{cand} \neq \emptyset$ **then**
                %% Tabu exploration
                $(s_l, i, p_m) \leftarrow randSelect(\mathcal{L}_{cand})$;
                $s \leftarrow propagate\_move(s_l, i, p_m)$; % Inserts i in s_l at p_m and prop-
                agates all the constraints
                $freq(i) \leftarrow freq(i) + 1$;
                $iter \leftarrow iter + 1$;
(2)   **if** $(i \bmod 2) = 1$ **then**
                  $\alpha \leftarrow 2 \times \beta$;

              else $\alpha \leftarrow \beta$;
              $tabu(i) \leftarrow iter + \alpha \times freq(i)$;
              if $(best\_gain > g^*)$ *or* $(best\_gain = g^*$ *and* $TdT(s) < TdT(s^*))$
              then
                  $s^* \leftarrow s$; $g^* \leftarrow best\_gain$;
            else
(3)   % intensification: step 1
              $(z, TdT_z) \leftarrow Minimize–TdT(s^*)$;
              if $TdT_z < TdT(s^*)$ **then**
                  $s^* \leftarrow z$;

              else
                  % intensification: step 2
              $\beta \leftarrow \beta/2$;
              for $i=1$ *to* $2n$ **do** $freq(i) \leftarrow 0$; $tabu(i) \leftarrow 0$;
              $s \leftarrow s^*$;
        until $(\beta < 1$ *or* $|s^*| = n)$;
        %% If $(\beta < 1)$ then Diversification phase
    until *(stop criterion or* $|s^*| = n)$;
    return (s^*, g^*);
end;

selected one. Hence, acquiring the shots according to their weights wg aims to favor the acquisition of the less selected shots by the greedy algorithm. This one selects one by one the shots in Q (starting by the first one) and attempts to insert it without dropping the already selected shots. When it is selected by $greedy()$, the weight $wg(i)$ of the shot i is incremented. Regarding to the weights wg and to their corresponding sorted shots list Q, two executions of Algorithm $greedy()$ will return two different configurations in terms of the selected strips. This feature allows us to deal with the diversification step that needs different starting points (see last paragraph of Section 4.4).

$|s_0|^+$ is the number of the selected shots in s_0 and which are acquired in the direct azimuth. The instruction labeled [2] tunes the value of the parameter α according to the shooting direction of an image and as explained in Section 4.3: we recall that an odd (even) number indicates a shot that is acquirable on the direct (indirect) azimuth. The $Minimze$–$TdT(s^*)$ function (label [3]) corresponds to the tabu algorithm dedicated to the minimization of $TdT(s^*)$ value. Finally, the $stop\ criterion$ is a limited execution time.

The diversification phase is launched when the intensification step fails. Hence, we execute a new iteration of the most external loop **repeat ... until** by generating a new starting point s_0 by the means of the greedy algorithm and executing the tabu search, ...

The different phases (exploration, intensification and diversification) use the same tabu search engine. Each one is triggered and stopped automatically by the tabu list management, i.e. whenever no more move is admissible.

5 Computational Results

The AEOS management problem was the subject of the 3rd international challenge organized by the French Society of Operations Research and Decision Analysis (ROADEF'2003), and proposed by the CNES[1] and ONERA[2] French space agencies.

The provided instances are artificially generated, such as:

- the number of requests (m value) is ranging from 2 to 375, and
- the number of strips (n value)is varying from 2 to 534 with a maximum of 113 stereo strips (n_{stereo} value).

Table 1 gives the properties of each of the 20 used instances[3].

Table 2 gives the best known gains as published during the ROADEF' 2003 challenge, in a booklet of abstracts [26] and also available on the WEB site of the challenge[3], but using different test parameters: Sun-Blade-1000 750Mhz/ 512MB workstation and 300 seconds running time.

---

[1] Acronym of "Centre National d'Etudes Spatiales".

[2] Acronym of "Office National d'Etudes et de Recherches Aérospatiales".

[3] Available from the WEB site of the challenge:
www.prism.uvsq.fr/ vdc/ROADEF/CHALLENGES/2003

Table 1 Test Instance Characteristics

Instance	m	n	n_{Stereo}
2 9 36	2	2	0
2 9 66	4	7	0
2 13 111	68	106	12
2 15 170	218	295	39
2 26 96	336	483	63
2 27 22	375	534	67
2 9 170	12	25	4
3 25 22	150	342	113
3 8 155	12	28	10
4 17 186	77	147	48
2 21 140	284	420	58
2 21 155	311	472	55
2 21 170	294	450	71
2 21 22	306	455	54
2 21 37	315	477	62
2 21 7	289	410	49
2 21 81	297	436	59
2 21 96	291	437	49
3 21 155	135	295	105
3 21 81	135	283	88

Table 2 The best known gains

Instance	Best known gain	Instance	Best known gain
2 9 36	10,423,440	2 21 140	1,029,892,360
2 9 66	115,710,959	2 21 155	1,150,632,847
2 13 111	563,597,071	2 21 170	891,060,370
2 15 170	719,417,220	2 21 22	1,160,366,840
2 26 96	1,005,301,900	2 21 37	954,965,580
2 27 22	967,910,750	2 21 7	842,378,700
2 9 170	191,358,231	2 21 81	986,679,410
3 25 22	425,983,220	2 21 96	1,133,044,250
3 8 155	121,680,360	3 21 155	4,6019,6570
4 17 186	185,406,200	3 21 81	373,553,350

The TS algorithms is implemented in C/C++ language and compiled using Visual C++. The experiments were carried on a PIV 1.9 Ghz PC with 512 MB of RAM. The execution time, corresponding to the stop criterion of Algorithm 2, is fixed to 1800 seconds in order to evaluate the behavior of the algorithm over a relatively long computation time. Table 3 gives the obtained results. The TS algorithm was run 10 times per instance with different random seeds and the following information are collected:

Table 3 The results obtained after 1800 seconds running time

Instance	Best			Worst	Average	
	Gain	N_s	Time	Gain	Gain	Time
2 9 36	10,423,440	2	<1	10,423,440	10,423,440	<1
2 9 66	115,710,959	7	<1	115,710,959	115,710,959	<1
2 13 111	563,597,071	54	90	563,597,071	563,597,071	90
2 15 170	719,417,220	40	221	719,417,220	719,417,220	561
2 26 96	1,005,301,900	35	1725	985,763,300	989,671,020	443
2 27 22	966,643,460	30	57	966,643,460	966,643,460	57
2 9 170	191,358,231	17	<1	191,358,231	191,358,231	<1
3 25 22	425,983,220	28	95	425,983,220	425,983,220	95
3 8 155	121,680,360	12	<1	121,680,360	121,680,360	<1
4 17 186	185,406,200	37	<1	185,406,200	185,406,200	<1
2 21 140	1,030,060,860	32	191	1,027,543,540	1,029,288,814	467
2 21 155	1,150,632,847	35	1074	1,129,245,020	1,132,568,329	762
2 21 170	914,978,310	40	573	906,992,592	912,183,616	1151
2 21 22	1,160,594,340	32	115	1,160,366,840	1,160,571,590	831
2 21 37	954,965,580	37	190	954,605,760	954,821,652	602
2 21 7	842,378,700	33	115	842,378,700	842,378,700	116
2 21 81	986,679,410	30	897	98,424,5930	985,894,172	778
2 21 96	113,4461,030	38	843	1,125,880,120	1,130,197,543	932
3 21 155	460,196,570	36	3	460,196,570	460,196,570	3
3 21 81	37,3553,350	28	16	373,553,350	373,553,350	16

- Over the 10 runs, the best gain value, the time needed to reach this gain and the number of the selected strips in the corresponding solution (columns 2, 3 and 4 of Table 3).
- Over the 10 runs, the worst gain (column 5), and the average gain and time (columns 6 and 7).

For the ten first instances (from 2 9 36 to 4 17 186, which we call A instances), all the best known gains are reached except for instance 2 27 22. These values are obtained after a maximum of 221 sec, except for instance 2 26 96, which required 1725 sec. For the ten instances 2 21 140 to 3 21 81 (B instances), all the best values are reached after less than 200 sec computing time except for the instances 2 21 155 and 2 21 81, which require 1074 and 897 sec respectively. Furthermore, the gains which are obtained for the instances 2 21 140, 2 21 170, 2 21 22 and 2 21 96 are improved after a maximum of 843 sec.

Several comments can be made about these results. First, our TS algorithm is efficient and robust. In fact, 15 best gains are reached and 4 are improved, even if the best known gains were obtained on different experimentation conditions and the optimal gains are unknown.

Concerning the robustness of the algorithm, the gap value between the best gain found and the average gain, calculated using the formula $100 \times (\frac{Best}{Average} - 1)$, is equal to 0% for 12 instances (the best known value is always reached), and less then 1.60% for the rest of the benchmark, which is a poor value. Secondly, these instances seems highly constrained since only a small number of candidate strips are selected in the best solutions.

6 Conclusion

We have presented in this paper a recall on the main characteristics of a Tabu Search algorithm. Applying such algorithm to solve a given problem is not an easy task. It requires a very good knowledges on the treated problem and it calls for caution during the specification of its various elements, mainly the tabu list management, the diversification and the intensification.

Like any other local search method, TS may also require spending an important time in tuning some of its parameters. Such empirical step is very important in order to ensure its efficiency and robustness.

It seems also hard to generalize our resolution to other combinatorial optimization problems. In fact, each problem has its own characteristics that make it singular in regards to a resolution by a tabu-like method. However, it is clear that all the implementations of Tabu Search share the same components, including the search space and the neighborhood definition, the neighborhood evaluation, the move heuristic, tabu tenure, the diversification and the intensification.

Neverless, Tabu Search remains a powerful approach that has been applied with great success to many difficult combinatorial optimization problems. We have illustrated such case by solving successfully a NP-hard optimization problem, which is the managing of an agile earth observing satellite. This problem is intractable effectively by a complete method. Only approximate methods, based on neighborhood, were able to meet with some effectiveness.

References

1. Benoist, T., Rottembourg, B.: Upper Bounds of the Maximal Revenue of an Earth Observation Satellite. 4OR: A Quarterly Journal of Operations Research 2(3), 235–249 (2004)
2. Bensana, E., Verfaillie, G., Agnèse, J.G., Bataille, N., Blumstein, D.: Exact and Approximate Methods for the Daily Management of an Earth Observation Satellite. In: Proceeding of the 4th International Symposium od Space Mission Operations and Ground Data Systems (spaceOps 1996) (1996)
3. Bensana, E., Lemaître, M., Verfaillie, G.: Earth Observation Satellite Management. Constraints: An International Jounal 4(3), 293–299 (1999)
4. Cordeau, J.F., Laporte, G., Mercier, A.: A Unified Tabu Search Heuristic for Vehicle Routing Problems with Time Windows. Journal of Operational Research Society 59, 928–936 (2001)
5. de Werra, D., Hertz, A.: Tabu Search Techniques: A Tutorial and an Application to Neural Networks. OR Spektrum 11, 131–141 (1989)

 6. Fleurent, C., Ferland, J.A.: Genetic and Hybrid Algorithms for Graph Coloring. Annals of Operations Research 63, 437–461 (1996)
 7. Gabrel, V.: Improved Linear Programming Bounds via Column Generation for Daily Scheduling of Earth Observation Satellite. Technical report, LIPN (1999)
 8. Gabrel, V., Moulet, A., Murat, C., Paschos, V.T.: A New model and derived algorithms for the satellite shot planning problem using graph theory concepts. Annals of Operations Research 69, 115–134 (1997)
 9. Gendreau, M.: An Introduction to Tabu Search. Handbook of Metaheuristics 57, 37–54 (2003)
10. Gendreau, M., Guertin, F., Potvin, J., Taillard, E.: Parallel tabu search for real-time vehicle routing and dispatching. Transportation Science 33(4), 381–390 (1999)
11. Gendreau, M., Laporte, G., Potvin, J.Y.: Metaheuristics for the capacitated vrp. pp. 129–154 (2001)
12. Gendreau, M., Soriano, P., Salvail, L.: Solving the maximum clique problem using a tabu search approach. Annals of Operations Research 41(1-4), 385–403 (1993)
13. Glover, F.: Tabu Search – Part I. ORSA Journal on Computing 1, 190–206 (1989)
14. Glover, F.: Tabu Search – Part II. ORSA Journal on Computing 2, 4–32 (1990)
15. Glover, F., Kochenberger, G.A.: Critical Event Tabu Search for Multidimensional Knapsack Problems. In: Osman, I.H., Kelly, J.P. (eds.) Meta-Heurisitics: Theory and Applications, pp. 407–428. Kluwer Academic Publishers, Dordrecht (1996)
16. Glover, F., Laguna, M.: Tabu Search. Kluwer Academic Publishers, Dordrecht (1997)
17. Hall, N.G., Magazine, M.J.: Maximizing the Value of a Space Mission. European Journal of Operational Research (78), 224–241 (1994)
18. Hanafi, S., Freville, A.: An Efficient Tabu Search Approach for the 0-1 Multidimensional Knapsack Problem. European Journal of Operational Research, Special Tabu Search Issue 106(2–3), 663–697 (1998)
19. Harrison, S.A., Price, M.E.: Task Scheduling for Satellite Based Imagery. In: Proceedings of the 18th Workshop of UK Planning and Scheduling Special Interest Group, pp. 64–78 (1999)
20. Hertz, A.: Finding a feasible course schedule using tabu search. Discrete Appl. Math. 35(3), 255–270 (1992)
21. Hertz, A., de Werra, D.: The Tabu Search Metaheuristic: How We Used It. Annals of Mathematics and Artificial Intelligence 1, 111–121 (1991)
22. McKendall Jr., A.R.: Improved Tabu search heuristics for the dynamic space allocation problem. Comput. Oper. Res. 35(10), 3347–3359 (2008)
23. Laguna, M., Barnes, J.W., Glover, F.: Tabu search methodology for a single machine scheduling problem. Journal of Intelligent Manufacturing 2, 63–74 (1991)
24. Lemaître, M., Verfaillie, G., Jouhaud, F., Lachiver, J.M., Bataille, N.: Selecting and Scheduling Observations of Agile Satellites. Aerospace Science and Technology 6(5), 367–381 (2002)
25. Liu, Y., Yi, Z., Wu, H., Ye, M., Chen, K.: A tabu search approach for the minimum sum-of-squares clustering problem. Inf. Sci. 178(12), 2680–2704 (2008)
26. ROADEF 2003 Challenge. Booklet of Abstracts. ROADEF society, France (February 2003)

27. Schiex, T., Fargier, F., Verfaillie, G.: Valued Constrained Satisfaction Problems: Hard and Easy problems. In: Proceedings of IJCAI 1995, 14th International Joint Conference on Artificial Intelligence, pp. 631–639 (1995)
28. Vasquez, M., Hao, J.K.: A Logic-Constrained Knapsack Formulation and a Tabu Algorithm for the Daily Photograph Scheduling of an Earth Observation Satellite. Journal of Computational Optimization and Applications 20(2), 137–157 (2001)
29. Vasquez, M., Hao, J.K.: Upper Bounds for the SPOT5 Daily Photograph Scheduling Problem. Journal of Combinatorial Optimization 7, 87–103 (2003)
30. Wolf, W.J., Sorensen, S.E.: Three Scheduling Algorithms Applied to the Earth Observing Systems Domain. Management Science 46(1), 146–168 (2000)

Reformulations in Mathematical Programming: A Computational Approach

Leo Liberti, Sonia Cafieri, and Fabien Tarissan

Abstract. Mathematical programming is a language for describing optimization problems; it is based on parameters, decision variables, objective function(s) subject to various types of constraints. The present treatment is concerned with the case when objective(s) and constraints are algebraic mathematical expressions of the parameters and decision variables, and therefore excludes optimization of black-box functions. A reformulation of a mathematical program P is a mathematical program Q obtained from P via symbolic transformations applied to the sets of variables, objectives and constraints. We present a survey of existing reformulations interpreted along these lines, some example applications, and describe the implementation of a software framework for reformulation and optimization.

1 Introduction

Optimization and decision problems are usually defined by their input and a mathematical description of the required output: a mathematical entity with an associated value, or whether a given entity has a specified mathematical property or not. Mathematical programming is a language designed to express almost all practically interesting optimization and decision problems.

Mathematical programming formulations can be categorized according to various properties, and rather efficient solution algorithms exist for many of the categories. As in most languages, the same semantics can be conveyed by many different syntactical expressions. In other words, there are many equivalent formulations for each given problem (what the term "equivalent" means in this context will be defined later). Furthermore, solution algorithms for mathematical programming formulations often rely on solving a sequence

Leo Liberti, Sonia Cafieri, and Fabien Tarissan
LIX, École Polytechnique, Palaiseau, 91128 France
e-mail: {liberti,cafieri,tarissan}@lix.polytechnique.fr

A. Abraham et al. (Eds.): Foundations of Comput. Intel. Vol. 3, SCI 203, pp. 153–234.
springerlink.com   © Springer-Verlag Berlin Heidelberg 2009

of different problems (often termed *auxiliary problems*) related to the original one: although these are usually not fully equivalent to the original problem, they may be relaxations, projections, liftings, decompositions (among others). Auxiliary problems are *reformulations* of the original problem.

Consider for example the Kissing Number Problem (KNP) in D dimensions [60], i.e. the determination of the maximum number of unit D-dimensional spheres that can be arranged around a central unit D-dimensional sphere. As all optimization problems, this can be cast (by using a bisection argument) as a sequence of decision problems on the cardinality of the current spheres configuration. Namely, given the positive integers D (dimension of Euclidean space) and N, is there a configuration of N unit spheres around the central one? For any fixed D, the answer will be affirmative or negative depending on the value of N. The highest N such that the answer is affirmative is the kissing number for dimension D. The decision version of the KNP can be cast as a nonconvex Nonlinear Programming (NLP) feasibility problem as follows. For all $i \leq N$, let $x_i = (x_{i1}, \ldots, x_{iD}) \in \mathbb{R}^D$ be the center of the i-th sphere. We look for a set of vectors $\{x_i \mid i \leq N\}$ satisfying the following constraints:

$$\forall\, i \leq N \quad ||x_i|| = 2$$
$$\forall\, i < j \leq N \quad ||x_i - x_j|| \geq 2$$
$$\forall\, i \leq N \quad -2 \leq x_i \leq 2.$$

It turns out that this problem is numerically quite difficult to solve, as it is very unlikely that the local NLP solution algorithm will be able to compute a feasible starting solution. Failing to find an initial feasible solution means that the solver will immediately abort without having made any progress. Most researchers with some practical experience in NLP solvers (such as e.g. SNOPT [41]), however, will immediately reformulate this problem into a more computationally amenable form by squaring the norms to get rid of a potentially problematic square root, and treating the reverse convex constraints $||x_i - x_j|| \geq 2$ as soft constraints by multiplying the right hand sides by a non-negative scaling variable α, which is then maximized:

$$\max \alpha \tag{1}$$
$$\forall\, i \leq N \quad ||x_i||^2 = 4 \tag{2}$$
$$\forall\, i < j \leq N \quad ||x_i - x_j||^2 \geq 4\alpha. \tag{3}$$
$$\forall\, i \leq N \quad -2 \leq x_i \leq 2 \tag{4}$$
$$\alpha \geq 0. \tag{5}$$

In this form, finding an initial feasible solution is trivial; for example, $x_i = (2, 0, \ldots, 0)$ for all $i \leq N$ will do. Subsequent solver iterations will likely be able to provide a better solution. Should the computed value of α be ≥ 1, the solution would be feasible in the hard constraints, too. Currently, we are aware of no optimization language environment that is able to perform the described reformulation automatically. Whilst this is not a huge limitation for NLP experts, people who simply wish to model a problem and get its solution will fail to obtain one, and may even be led into thinking that the formulation itself is infeasible.

Another insightful example of the types of limitations we refer to can be drawn from the KNP. We might wish to impose ordering constraints on some of the spheres to reduce the number of symmetric solutions. Ordering spheres packed on a spherical surface is hard to do in Euclidean coordinates, but it can be done rather easily in spherical coordinates, by simply stating that the value of a spherical coordinate of the i-th sphere must be smaller than the corresponding value in the j-th sphere. We can transform a Euclidean coordinate vector $x = (x_1, \ldots, x_D)$ in D-spherical coordinates $(\rho, \vartheta_1, \ldots, \vartheta_{D-1})$ such that $\rho = ||x||$ and $\vartheta \in [0, 2\pi]^{D-1}$ by means of the following equations:

$$\rho = ||x|| \tag{6}$$

$$\forall k \leq D \quad x_k = \rho \sin \vartheta_{k-1} \prod_{h=k}^{D-1} \cos \vartheta_h \tag{7}$$

(this yields another NLP formulation of the KNP). Applying the D-spherical transformation is simply a matter of term rewriting and algebraic simplification, and yet no currently existing optimization language environment offers such capabilities. By pushing things further, we might wish to devise an algorithm that dynamically inserts or removes constraints expressed in either Euclidean or spherical coordinates depending on the status of the current solution, and re-solves the (automatically) reformulated problem at each iteration. This may currently be done (up to a point) by optimization language environments such as AMPL [39], provided all constraints are part of a pre-specified family of parametric constraints. Creating new constraints by term rewriting, however, is not a task currently addressed by current mathematical programming implementations.

The limitations emphasized in the KNP example illustrate a practical need for very sophisticated software including numerical as well as symbolic algorithms, both applied to the unique goal of solving optimization problems cast as mathematical programming formulations. The current state of affairs is that there are many numerical optimization solvers and many Computer Algebra Systems (CAS) — such as Maple or Mathematica — whose efficiency is severely hampered by the full generality of their capabilities. In short, we would ideally need (small) parts of the symbolic kernels driving the existing CASes to be combined with the existing optimization algorithms, plus a number of super-algorithms capable of making automated, dynamic decisions on the type of reformulations that are needed to improve the current search process.

Although the above paradigm might seem far-fetched, it does in fact already exist in the form of the hugely successful CPLEX [52] solver targeted at solving Mixed-Integer Linear Programming (MILP) problems. The initial formulation provided by the user is automatically simplified and improved with a sizable variety of pre-processing steps which attempt to reduce the number of variables and constraints. Thereafter, at each node of the Branch-and-Bound algorithm, the formulation may be tightened as needed by inserting and removing additional valid constraints, in the hope that the current relaxed solution of the (automatically obtained) linear relaxation is improved. Advanced users may of course decide to tune many parameters controlling this process, but practitioners needing a practical answer can simply use default parameters and to let CPLEX decide what is best. Naturally, the task carried out by CPLEX is greatly simplified by the assumption that both objective function and constraints are linear forms, which is obviously not the case in a general nonlinear setting.

In this chapter we attempt to move some steps in the direction of endowing general mathematical programming with the same degree of algorithmic automation enjoyed by linear programming. We propose: (a) a theoretical framework in which mathematical programming reformulations can be formalized in a unified way, and (b) a literature review of the most successful existing reformulation and relaxation techniques in mathematical programming. Since an all-comprehensive literature review in reformulation techniques would extend this chapter to possibly several hundreds (thousands?) pages, only a partial review has been provided. In this sense, this should be seen as "work in progress" towards laying the foundations to a computer software which is capable of reformulating mathematical programming formulations automatically. Note also that for this reason, the usual mathematical notations have been translated to a data structure framework that is designed to facilitate computer implementation. Most importantly, "functions" — which as mathematical entities are interpreted as maps between sets — are represented by expression trees: what is meant by the expression $x + y$, for example, is really a directed binary tree on the vertices $\{+, x, y\}$ with arcs $\{(+, x), (+, y)\}$. For clarity purposes, however, we also provide the usual mathematical languages.

One last (but not least) remark is that reformulations can be seen as a new way of expressing a known problem. Reformulations are syntactical operations that may add or remove variables or constraints, whilst keeping the fundamental structure of the problem optima invariant. When some new variables are added and some of the old ones are removed, we can usually try to re-interpret the reformulated problem and assign a meaning to the new variables, thus gaining new insights to the problem. One example of this is given in Sect. 3.5. One other area in mathematical programming that provides a similarly clear relationship between mathematical syntax and semantics is LP duality with the interpretation of reduced costs. This is important insofar as it offers alternative interpretations to known problems, which gains new and useful insights.

The rest of this chapter is organized as follows. In Section 2 we propose a general theoretical framework of definitions allowing a unified formalization of mathematical programming reformulations. The definitions allow a consistent treatment of the most common variable and constraint manipulations in mathematical programming formulations. In Section 3 we present a systematic study of a set of well known reformulations. Most reformulations are listed as symbolic algorithms acting on the problem structure, although the equivalent transformation in mathematical terms is given for clarity purposes. In Section 4 we present a systematic study of a set of well known relaxations. Again, relaxations are listed as symbolic algorithms acting on the problem structure whenever possible, the equivalent mathematical transformation being given for clarity. Section 5 describes the implementation of ROSE, a Reformulation/Optimization Software Engine.

2 General Framework

In Sect. 2.1 we formally define what a mathematical programming formulation is. In Sect. 2.2 we discuss the expression tree function representation. Sect. 2.3 lists the most common standard forms in mathematical programming.

2.1 A Data Structure for Mathematical Programming Formulations

In this chapter we give a formal definition of a mathematical programming formulation in such terms that can be easily implemented on a computer. We then give several examples to illustrate the generality of our definition. We refer to a mathematical programming problem in the most general form:

$$\left. \begin{array}{r} \min\ f(x) \\ g(x) \lessgtr b \\ x \in X, \end{array} \right\} \tag{8}$$

where f, g are function sequences of various sizes, b is an appropriately-sized real vector, and X is a cartesian product of continuous and discrete intervals.

The precise definition of a mathematical programming formulation lists the different formulation elements: parameters, variables having types and bounds, expressions depending on the parameters and variables, objective functions and constraints depending on the expressions. We let $\mathbb{P}$ be the set of all mathematical programming formulations, and $\mathbb{M}$ be the set of all matrices. This is used in Defn. 1 to define leaf nodes in mathematical expression trees, so that the concept of a formulation can also accommodate multilevel and semidefinite programming problems. Notationwise, in a digraph (V, A) for all $v \in V$ we indicate by $\delta^+(v)$ the set of vertices u for which $(v, u) \in A$ and by $\delta^-(v)$ the set of vertices u for which $(u, v) \in A$.

Definition 1. *Given an alphabet $\mathcal{L}$ consisting of countably many alphanumeric names $N_{\mathcal{L}}$ and operator symbols $O_{\mathcal{L}}$, a mathematical programming formulation P is a 7-tuple $(\mathcal{P}, \mathcal{V}, \mathcal{E}, \mathcal{O}, \mathcal{C}, \mathcal{B}, \mathcal{T})$, where:*

- *$\mathcal{P} \subseteq N_{\mathcal{L}}$ is the sequence of parameter symbols: each element $p \in \mathcal{P}$ is a parameter name;*
- *$\mathcal{V} \subseteq N_{\mathcal{L}}$ is the sequence of variable symbols: each element $v \in \mathcal{V}$ is a variable name;*
- *$\mathcal{E}$ is the set of expressions: each element $e \in \mathcal{E}$ is a Directed Acyclic Graph (DAG) $e = (V_e, A_e)$ such that:*

  (a) $V_e \subseteq \mathcal{L}$ is a finite set
  (b) there is a unique vertex $r_e \in V_e$ such that $\delta^-(r_e) = \emptyset$ (such a vertex is called the root vertex)
  (c) vertices $v \in V_e$ such that $\delta^+(v) = \emptyset$ are called leaf vertices and their set is denoted by $\lambda(e)$; all leaf vertices v are such that $v \in \mathcal{P} \cup \mathcal{V} \cup \mathbb{R} \cup \mathbb{P} \cup \mathbb{M}$
  (d) for all $v \in V_e$ such that $\delta^+(v) \neq \emptyset$, $v \in O_{\mathcal{L}}$
  (e) two weight functions $\chi, \xi : V_e \to \mathbb{R}$ are defined on V_e: $\chi(v)$ is the node coefficient and $\xi(v)$ is the node exponent of the node v; for any vertex $v \in V_e$, we let $\tau(v)$ be the symbolic term of v: namely, $v = \chi(v)\tau(v)^{\xi(v)}$.

  elements of $\mathcal{E}$ are sometimes called expression trees; nodes $v \in O_{\mathcal{L}}$ represent an operation on the nodes in $\delta^+(v)$, denoted by $v(\delta^+(v))$, with output in $\mathbb{R}$;
- *$\mathcal{O} \subseteq \{-1, 1\} \times \mathcal{E}$ is the sequence of objective functions; each objective function $o \in \mathcal{O}$ has the form (d_o, f_o) where $d_o \in \{-1, 1\}$ is the optimization direction (-1 stands for minimization, $+1$ for maximization) and $f_o \in \mathcal{E}$;*
- *$\mathcal{C} \subseteq \mathcal{E} \times \mathcal{S} \times \mathbb{R}$ (where $\mathcal{S} = \{-1, 0, 1\}$) is the sequence of constraints c of the form (e_c, s_c, b_c) with $e_c \in \mathcal{E}, s_c \in \mathcal{S}, b_c \in \mathbb{R}$:*

$$c \equiv \begin{cases} e_c \leq b_c \text{ if } s_c = -1 \\ e_c = b_c \text{ if } s_c = 0 \\ e_c \geq b_c \text{ if } s_c = 1; \end{cases}$$

- *$\mathcal{B} \subseteq \mathbb{R}^{|\mathcal{V}|} \times \mathbb{R}^{|\mathcal{V}|}$ is the sequence of variable bounds: for all $v \in \mathcal{V}$ let $\mathcal{B}(v) = [L_v, U_v]$ with $L_v, U_v \in \mathbb{R}$;*
- *$\mathcal{T} \subseteq \{0, 1, 2\}^{|\mathcal{V}|}$ is the sequence of variable types: for all $v \in \mathcal{V}$, v is called a continuous variable if $\mathcal{T}(v) = 0$, an integer variable if $\mathcal{T}(v) = 1$ and a binary variable if $\mathcal{T}(v) = 2$.*

We remark that for a sequence of variables $z \subseteq \mathcal{V}$ we write $\mathcal{T}(z)$ and respectively $\mathcal{B}(z)$ to mean the corresponding sequences of types and respectively bound intervals of the variables in z. Given a formulation $P = (\mathcal{P}, \mathcal{V}, \mathcal{E}, \mathcal{O}, \mathcal{C}, \mathcal{B}, \mathcal{T})$, the *cardinality* of P is $|P| = |\mathcal{V}|$. We sometimes refer to a formulation by calling it an *optimization problem* or simply a *problem*.

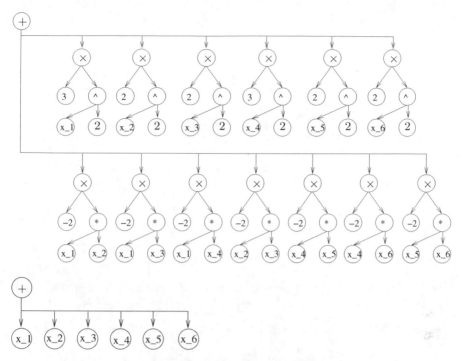

Fig. 1 The graphs e_1 (above) and e_2 (below) from Example 2.1

Definition 2. *Any formulation Q that can be obtained by P by a finite sequence of symbolic operations carried out on the data structure is called a problem transformation.*

Examples

In this section we provide some explicitly worked out examples that illustrate Defn. 1.

A quadratic problem

Consider the problem of minimizing the quadratic form $3x_1^2 + 2x_2^2 + 2x_3^2 + 3x_4^2 + 2x_5^2 + 2x_6^2 - 2x_1x_2 - 2x_1x_3 - 2x_1x_4 - 2x_2x_3 - 2x_4x_5 - 2x_4x_6 - 2x_5x_6$ subject to $x_1 + x_2 + x_3 + x_4 + x_5 + x_6 = 0$ and $x_i \in \{-1, 1\}$ for all $i \le 6$. For this problem,

- $\mathcal{P} = \emptyset$;
- $\mathcal{V} = (x_1, x_2, x_3, x_4, x_5, x_6)$;
- $\mathcal{E} = (e_1, e_2)$ where e_1, e_2 are the graphs shown in Fig. 1;
- $\mathcal{O} = (-1, e_1)$;
- $\mathcal{C} = ((e_2, 0, 0))$;
- $\mathcal{B} = ([-1, 1], [-1, 1], [-1, 1], [-1, 1], [-1, 1], [-1, 1])$;
- $\mathcal{T} = (2, 2, 2, 2, 2, 2)$.

Fig. 2 The BGBP instance in Example 2.1

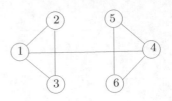

Balanced graph bisection

Example 2.1 is a (scaled) mathematical programming formulation of a balanced graph bisection problem instance. This problem is defined as follows.

BALANCED GRAPH BISECTION PROBLEM (BGBP). Given an undirected graph $G = (V, E)$ without loops or parallel edges such that $|V|$ is even, find a subset $U \subset V$ such that $|U| = \frac{|V|}{2}$ and the set of edges $C = \{\{u, v\} \in E \mid u \in U, v \notin U\}$ is as small as possible.

The problem instance considered in Example 2.1 is shown in Fig. 2. To all vertices $i \in V$ we associate variables $x_i = \left\{ \begin{array}{l} 1 \ i \in U \\ -1 \ i \notin U \end{array} \right.$. The number of edges in C is counted by $\frac{1}{4} \sum_{\{i,j\} \in E} (x_i - x_j)^2$. The fact that $|U| = \frac{|V|}{2}$ is expressed by requiring an equal number of variables at 1 and -1, i.e. $\sum_{i=1}^{6} x_i = 0$.

We can also express the problem in Example 2.1 as a particular case of the more general optimization problem:

$$\left.\begin{array}{rl} \min_x & x^\top L x \\ \text{s.t.} & x\mathbf{1} = 0 \\ & x \in \{-1, 1\}^6, \end{array}\right\}$$

where

$$L = \begin{pmatrix} 3 & -1 & -1 & -1 & 0 & 0 \\ -1 & 2 & -1 & 0 & 0 & 0 \\ -1 & -1 & 2 & 0 & 0 & 0 \\ -1 & 0 & 0 & 3 & -1 & -1 \\ 0 & 0 & 0 & -1 & 2 & -1 \\ 0 & 0 & 0 & -1 & -1 & 2 \end{pmatrix}$$

and $\mathbf{1} = (1, 1, 1, 1, 1, 1)^\top$. We represent this class of problems by the following mathematical programming formulation:

- $\mathcal{P} = (L_{ij} \mid 1 \le i, j \le 6)$;
- $\mathcal{V} = (x_1, x_2, x_3, x_4, x_5, x_6)$;
- $\mathcal{E} = (e_1', e_2)$ where e_1' is shown in Fig. 3 and e_2 is shown in Fig. 1 (below);
- $\mathcal{O} = (-1, e_1')$;
- $\mathcal{C} = ((e_2, 0, 0))$;

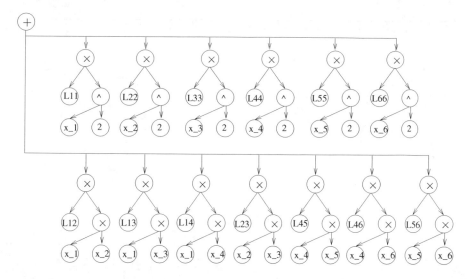

Fig. 3 The graph e'_1 from Example 2.1. $L'_{ij} = L_{ij} + L_{ji}$ for all i, j

- $\mathcal{B} = ([-1, 1], [-1, 1], [-1, 1], [-1, 1], [-1, 1], [-1, 1]);$
- $\mathcal{T} = (2, 2, 2, 2, 2, 2).$

The Kissing Number Problem

The kissing number problem formulation (1)-(5) is as follows:

- $\mathcal{P} = (N, D);$
- $\mathcal{V} = (x_{ik} \mid 1 \leq i \leq N \wedge 1 \leq k \leq D) \cup \{\alpha\};$
- $\mathcal{E} = (f, h_j, g_{ij} \mid 1 \leq i < j \leq N)$, where f is the expression tree for α, h_j is the expression tree for $||x_j||^2$ for all $j \leq N$, and g_{ij} is the expression tree for $||x_i - x_j||^2 - 4\alpha$ for all $i < j \leq N;$
- $\mathcal{O} = (1, f);$
- $\mathcal{C} = ((h_i, 0, 4) \mid i \leq N) \cup ((g_{ij}, 1, 0) \mid i < j \leq N);$
- $\mathcal{B} = [-2, 2]^{ND};$
- $\mathcal{T} = \{0\}^{ND}.$

As mentioned in Section 1, the kissing number problem is defined as follows.

KISSING NUMBER PROBLEM (KNP). Find the largest number N of non-overlapping unit spheres in $\mathbb{R}^D$ that are adjacent to a given unit sphere.

The formulation of Example 2.1 refers to the decision version of the problem: given integers N and D, is there an arrangement of N non-overlapping unit spheres in $\mathbb{R}^D$ adjacent to a given unit sphere?

Algorithm 1. The evaluation algorithm for expression trees

```
double Eval(node v) {
   double ρ;
   if (v ∈ O_L) {
      // v is an operator
      array α[|δ⁺(v)|];
      ∀ u ∈ δ⁺(v) {
         α(u) =Eval(u);
      }
      ρ = χ(v)v(α)^ξ(v);
   } else {
      // v is a constant value
      ρ = χ(v)v^ξ(v);
   }
   return ρ;
}
```

2.2 A Data Structure for Mathematical Expressions

Given an expression tree DAG $e = (V, A)$ with root node $r(e)$ and whose leaf nodes are elements of $\mathbb{R}$ or of $\mathbb{M}$ (the set of all matrices), the *evaluation* of e is the (numerical) output of the operation represented by the operator in node r applied to all the subnodes of r (i.e. the nodes adjacent to r); in symbols, we denote the output of this operation by $r(\delta^+(r))$, where the symbol r denotes both a function and a node. Naturally, the arguments of the operator must be consistent with the operator meaning. We remark that for leaf nodes belonging to $\mathbb{P}$ (the set of all formulations), the evaluation is not defined; the problem in the leaf node must first be solved and a relevant optimal value (e.g. an optimal variable value, as is the case with multilevel programming problems) must replace the leaf node.

For any $e \in E$, the *evaluation tree* of e is a DAG $\bar{e} = (\bar{V}, A)$ where $\bar{V} = \{v \in V \mid |\delta^+(v)| > 0 \vee v \in \mathbb{R}\} \cup \{x(v) \mid |\delta^+(v)| = 0 \wedge v \in \mathcal{V}\}$ (in short, the same as V with every variable leaf node replaced by the corresponding value $x(v)$). Evaluation trees are evaluated by Alg. 1. We can now naturally extend the definition of evaluation of e at a point x to expression trees whose leaf nodes are either in $\mathcal{V}$ or $\mathbb{R}$.

Definition 3. *Given an expression $e \in E$ with root node r and a point x, the evaluation $e(x)$ of e at x is the evaluation $r(\delta^+(r))$ of the evaluation tree $\bar{e}$.*

We consider a sufficiently rich operator set O_L including at least $+, \times$, power, exponential, logarithm, and trigonometric functions (for real arguments) and inner product (for matrix arguments). Note that since any term t is weighted by a multiplier coefficient $\chi(t)$ there is no need to employ a $-$ operator, for it suffices to multiply $\chi(t)$ by $-1 = \xi(v)$ in the appropriate term(s) t; a division u/v is expressed by multiplying u by v raised to the power -1. Depending on the problem form, it may sometimes be useful to enrich O_L with other (more complex) terms. In general, we view an operator in O_L as an atomic operation on a set of variables with cardinality at least 1.

A standard form for expressions

Since in general there is more than one way to write a mathematical expression, it is useful to adopt a standard form; whilst this does not resolve all ambiguities, it nonetheless facilitates the task of writing symbolic computation algorithms acting on the expression trees. For any expression node t in an expression tree $e = (V, A)$:

- if t is a sum:

  1. $|\delta^+(t)| \geq 2$
  2. no subnode of t may be a sum (sum associativity);
  3. no pair of subnodes $u, v \in \delta^+(t)$ must be such that $\tau(u) = \tau(v)$ (i.e. like terms must be collected); as a consequence, each sum only has one monomial term for each monomial type
  4. a natural (partial) order is defined on $\delta^+(t)$: for $u, v \in \delta^+(t)$, if u, v are monomials, u, v are ordered by degree and lexicographically

- if t is a product:

  1. $|\delta^+(t)| \geq 2$
  2. no subnode of t may be a product (product associativity);
  3. no pair of subnodes $u, v \in \delta^+(t)$ must be such that $\tau(u) = \tau(v)$ (i.e. like terms must be collected and expressed as a power)

- if t is a power:

  1. $|\delta^+(t)| = 2$
  2. the exponent may not be a constant (constant exponents are expressed by setting the exponent coefficient $\xi(t)$ of a term t)
  3. the natural order on $\delta^+(t)$ lists the base first and the exponent later.

The usual mathematical nomenclature (linear forms, polynomials, and so on) applies to expression trees.

2.3   Standard Forms in Mathematical Programming

Solution algorithms for mathematical programming problems read a formulation as input and attempt to compute an optimal feasible solution as output. Naturally, algorithms which exploit problem structure are usually more efficient than those that do not. In order to be able to exploit the structure of the problem, solution algorithms solve problems that are cast in a *standard form* that emphasizes the useful structure. In this section we review the most common standard forms.

Linear Programming

A mathematical programming problem P is a Linear Programming (LP) problem if (a) $|\mathcal{O}| = 1$ (i.e. the problem only has a single objective function); (b) e is a linear form for all $e \in \mathcal{E}$; and (c) $\mathcal{T}(v) = 0$ (i.e. v is a continuous variable) for all $v \in \mathcal{V}$.

An LP is in standard form if (a) $s_c = 0$ for all constraints $c \in \mathcal{C}$ (i.e. all constraints are equality constraints) and (b) $\mathcal{B}(v) = [0, +\infty]$ for all $v \in \mathcal{V}$. LPs are expressed in standard form whenever a solution is computed by means of the simplex method [27]. By contrast, if all constraints are inequality constraints, the LP is known to be in *canonical form*.

Mixed Integer Linear Programming

A mathematical programming problem P is a Mixed Integer Linear Programming (MILP) problem if (a) $|\mathcal{O}| = 1$; and (b) e is a linear form for all $e \in \mathcal{E}$.

A MILP is in standard form if $s_c = 0$ for all constraints $c \in \mathcal{C}$ and if $\mathcal{B}(v) = [0, +\infty]$ for all $v \in \mathcal{V}$. The most common solution algorithms employed for solving MILPs are Branch-and-Bound (BB) type algorithms [52]. These algorithms rely on recursively partitioning the search domain in a tree-like fashion, and evaluating lower and upper bounds at each search tree node to attempt to implicitly exclude some subdomains from consideration. BB algorithms usually employ the simplex method as a sub-algorithm acting on an auxiliary problem, so they enforce the same standard form on MILPs as for LPs. As for LPs, a MILP where all constraints are inequalities is in *canonical form*.

Nonlinear Programming

A mathematical programming problem P is a Nonlinear Programming (NLP) problem if (a) $|\mathcal{O}| = 1$ and (b) $\mathcal{T}(v) = 0$ for all $v \in \mathcal{V}$.

Many fundamentally different solution algorithms are available for locally solving NLPs, and most of them require different standard forms. One of the most widely used is Sequential Quadratic Programming (SQP) [41], which requires problem constraints $c \in \mathcal{C}$ to be expressed in the form $l_c \leq c \leq u_c$ with $l_c, u_c \in \mathbb{R} \cup \{-\infty, +\infty\}$. More precisely, an NLP is in SQP standard form if for all $c \in \mathcal{C}$ (a) $s_c \neq 0$ and (b) there is $c' \in \mathcal{C}$ such that $e_c = e_{c'}$ and $s_c = -s_{c'}$.

Mixed Integer Nonlinear Programming

A mathematical programming problem P is a Mixed Integer Nonlinear Programming (MINLP) problem if $|\mathcal{O}| = 1$. The situation as regards MINLP standard forms is generally the same as for NLPs, save that a few more works have appeared in the literature about standard forms for MINLPs

[113, 114, 96, 71]. In particular, the Smith standard form [114] is purpose-fully constructed so as to make symbolic manipulation algorithms easy to carry out on the formulation. A MINLP is in Smith standard form if:

- $\mathcal{O} = \{d_o, e_o\}$ where e_o is a linear form;
- $\mathcal{C}$ can be partitioned into two sets of constraints $\mathcal{C}_1, \mathcal{C}_2$ such that c is a linear form for all $c \in \mathcal{C}_1$ and $c = (e_c, 0, 0)$ for $c \in \mathcal{C}_2$ where e_c is as follows:

  1. $r(e_c)$ is the sum operator
  2. $\delta^+(r(e_c)) = \{\otimes, v\}$ where (a) $\otimes$ is a nonlinear operator where all subnodes are leaf nodes, (b) $\chi(v) = -1$ and (c) $\tau(v) \in \mathcal{V}$.

Essentially, the Smith standard form consists of a linear part comprising objective functions and a set of constraints; the rest of the constraints have a special form $\otimes(x_1, \ldots, x_p) - v = 0$ for some $p \in \mathbb{N}$, with $v, x_1, \ldots, x_p \in \mathcal{V}(P)$ and $\otimes$ a nonlinear operator in $O_{\mathcal{L}}$. By grouping all nonlinearities in a set of equality constraints of the form "variable = operator(variables)" (called *defining constraints*) the Smith standard form makes it easy to construct auxiliary problems. The Smith standard form can be constructed by recursing on the expression trees of a given MINLP [112] and is an opt-reformulation.

Solution algorithms for solving MINLPs are usually extensions of BB type algorithms [114, 71, 68, 124, 95].

Separable problems

A problem P is in separable form if (a) $\mathcal{O}(P) = \{(d_o, e_o)\}$, (b) $\mathcal{C}(P) = \emptyset$ and (c) e_o is such that:

- $r(e_o)$ is the sum operator
- for all distinct $u, v \in \delta^+(r(e_o))$, $\lambda(u) \cap \lambda(v) = \emptyset$,

where by slight abuse of notation $\lambda(u)$ is the set of leaf nodes of the subgraph of e_o whose root is u. The separable form is a standard form by itself. It is useful because it allows a very easy problem decomposition: for all $u \in \delta^+(r(e_o))$ it suffices to solve the smaller problems Q_u with $\mathcal{V}(Q_u) = \lambda(v) \cap \mathcal{V}(P)$, $\mathcal{O}(Q_u) = \{(d_o, u)\}$ and $\mathcal{B}(Q_u) = \{\mathcal{B}(P)(v) \mid v \in \mathcal{V}(Q_u)\}$. Then $\bigcup_{u \in \delta^+(r(e_o))} x(\mathcal{V}(Q_u))$ is a solution for P.

Factorable problems

A problem P is in factorable form [91, 130, 111, 124] if:

1. $\mathcal{O} = \{(d_o, e_o)\}$
2. $r(e_o) \in \mathcal{V}$ (consequently, the vertex set of e_o is simply $\{r(e_o)\}$)
3. for all $c \in \mathcal{C}$:

   - $s_c = 0$
   - $r(e_c)$ is the sum operator

- for all $t \in \delta^+(r(e_c))$, either (a) t is a unary operator and $\delta^+(t) \subseteq \lambda(e_c)$ (i.e. the only subnode of t is a leaf node) or (b) t is a product operator such that for all $v \in \delta^+(t)$, v is a unary operator with only one leaf subnode.

The factorable form is a standard form by itself. Factorable forms are useful because it is easy to construct many auxiliary problems (including convex relaxations, [91, 4, 111]) from problems cast in this form. In particular, factorable problems can be reformulated to emphasize separability [91, 124, 95].

D.C. problems

The acronym "d.c." stands for "difference of convex". Given a set $\Omega \subseteq \mathbb{R}^n$, a function $f : \Omega \to \mathbb{R}$ is a *d.c. function* if it is a difference of convex functions, i.e. there exist convex functions $g, h : \Omega \to \mathbb{R}$ such that, for all $x \in \Omega$, we have $f(x) = g(x) - h(x)$. Let C, D be convex sets; then the set $C \backslash D$ is a *d.c. set*. An optimization problem is *d.c.* if the objective function is d.c. and Ω is a d.c. set. In most of the d.c. literature, however [129, 116, 50], a mathematical programming problem is d.c. if:

- $\mathcal{O} = \{(d_o, e_o)\}$;
- e_o is a d.c. function;
- c is a linear form for all $c \in \mathcal{C}$.

D.C. programming problems have two fundamental properties. The first is that the space of all d.c. functions is dense in the space of all continuous functions. This implies that any continuous optimization problem can be approximated as closely as desired, in the uniform convergence topology, by a d.c. optimization problem [129, 50]. The second property is that it is possible to give explicit necessary and sufficient global optimality conditions for certain types of d.c. problems [129, 116]. Some formulations of these global optimality conditions [115] also exhibit a very useful algorithmic property: if at a feasible point x the optimality conditions do not hold, then the optimality conditions themselves can be used to construct an improved feasible point x'.

Linear Complementarity problems

Linear complementarity problems (LCP) are nonlinear feasibility problems with only one nonlinear constraint. An LCP is defined as follows [30], p. 50:

- $\mathcal{O} = \emptyset$;
- there is a constraint $c' = (e, 0, 0) \in \mathcal{C}$ such that (a) $t = r(e)$ is a sum operator; (b) for all $u \in \delta^+(t)$, u is a product of two terms v, f such that $v \in \mathcal{V}$ and $(f, 1, 0) \in \mathcal{C}$;
- for all $c \in \mathcal{C} \smallsetminus \{c'\}$, e_c is a linear form.

Essentially, an LCP is a feasibility problem of the form:

$$\left. \begin{array}{r} Ax \succeq b \\ x \geq 0 \\ x^\top (Ax - b) = 0, \end{array} \right\}$$

where $x \in \mathbb{R}^n$, A is an $m \times n$ matrix and $b \in \mathbb{R}^m$.

Many types of mathematical programming problems (including MILPs with binary variables [30, 53]) can be recast as LCPs or extensions of LCP problems [53]. Furthermore, some types of LCPs can be reformulated to LPs [86] and as separable bilinear programs [87]. Certain types of LCPs can be solved by an interior point method [58, 30].

Bilevel Programming problems

The bilevel programming (BLP) problem consists of two nested mathematical programming problems named the *leader* and the *follower* problem.

A mathematical programming problem P is a *bilevel programming problem* if there exist two programming problems L, F (the leader and follower problem) and a subset $\ell \neq \emptyset$ of all leaf nodes of $\mathcal{E}(L)$ such that any leaf node $v \in \ell$ has the form $(v, \mathcal{F})$ where $v \in \mathcal{V}(F)$.

The usual mathematical notation is as follows [32, 13]:

$$\left. \begin{array}{c} \min_y F(x(y), y) \\ \min_x f(x, y) \\ \text{s.t. } x \in X, \, y \in Y, \end{array} \right\} \tag{9}$$

where X, Y are arbitrary sets. This type of problem arises in economic applications. The leader knows the cost function of the follower, who may or may not know that of the leader; but the follower knows the optimal strategy selected by the leader (i.e. the optimal values of the decision variables of L) and takes this into account to compute his/her own optimal strategy.

BLPs can be reformulated exactly to MILPs with binary variables and vice-versa [13], where the reformulation is as in Defn. 6. Furthermore, two typical Branch-and-Bound (BB) algorithms for the considered MILPs and BLPs have the property that the the MILP BB can be "embedded" in the BLP BB (this roughly means that the BB tree of the MILP is a subtree of the BB tree of the BLP); however, the contrary does not hold. This seems to hint at a practical solution difficulty ranking in problems with the same degree of worst-case complexity (both MILPs and BLPs are **NP**-hard).

Semidefinite Programming problems

Consider known symmetric $n \times n$ matrices C, A_k for $k \leq m$, a vector $b \in \mathbb{R}^m$ and a symmetric $n \times n$ matrix $X = (x_{ij})$ where x_{ij} is a problem variable for all

$i, j \leq n$. The following is a *semidefinite programming problem* (SDP) in primal form:

$$\left.\begin{array}{r} \min_X \quad C \bullet X \\ \forall k \leq m \ A_k \bullet X = b_k \\ X \succeq 0, \end{array}\right\} \tag{10}$$

where $X \succeq 0$ is a constraint that indicates that X should be symmetric positive semidefinite, and $C \bullet X = \text{tr}(C^\top X) = \sum_{i,j} c_{ij} x_{ij}$. We also consider the SDP in dual form:

$$\left.\begin{array}{r} \max_{y,S} \quad b^\top y \\ \sum_{k \leq m} y_k A_k + S = C \\ S \succeq 0, \end{array}\right\} \tag{11}$$

where S is a symmetric $n \times n$ matrix and $y \in \mathbb{R}^m$. Both forms of the SDP problem are convex NLPs, so the duality gap is zero. Both forms can be solved by a particular type of polynomial-time interior point method (IPM), which means that solving SDPs is practically efficient [8, 125]. SDPs are important because they provide tight relaxations to (nonconvex) quadratically constrained quadratic programming problems (QCQP), i.e. problems with a quadratic objective and quadratic constraints (see Sect. 4.3).

SDPs can be easily modelled with the data structure described in Defn. 1, for their expression trees are linear forms where each leaf node contains a symmetric matrix. There is no need to explicitly write the semidefinite constraints $X \succeq 0, S \succeq 0$ because the solution IPM algorithms will automatically find optimal X, S matrices that are semidefinite.

3 Reformulations

In this section we define some types of reformulations and establish some links between them (Sect. 3.1) and we give a systematic study of various types of elementary reformulations (Sect. 3.2) and exact linearizations (Sect. 3.3). Sect. 3.5 provides a few worked out examples. In this summary, we tried to focus on two types of reformulations: those that are in the literature, but may not be known to every optimization practitioner, and those that represent the "tricks of the trade" of most optimization researchers but have never, or rarely, been formalized explicitly; so the main contributions of this section are systematic and didactic. Since the final aim of automatic reformulations is let the computer arrive at an alternative formulation which is easier to solve, we concentrated on those reformulations which simplified nonlinear terms into linear terms, or which reduced integer variables to continuous variables. By contrast, we did not cite important reformulations (such as the LP duality) which are fundamental in solution algorithms and alternative problem interpretation, but which do not significantly alter solution difficulty.

3.1 Reformulation Definitions

Consider a mathematical programming formulation $P = (\mathcal{P}, \mathcal{V}, \mathcal{E}, \mathcal{O}, \mathcal{C}, \mathcal{B}, \mathcal{T})$ and a function $x : \mathcal{V} \to \mathbb{R}^{|\mathcal{V}|}$ (called *point*) which assigns values to the variables.

Definition 4. *A point x is type feasible if:*

$$x(v) \in \begin{cases} \mathbb{R} & \text{if } \mathcal{T}(v) = 0 \\ \mathbb{Z} & \text{if } \mathcal{T}(v) = 1 \\ \{L_v, U_v\} & \text{if } \mathcal{T}(v) = 2 \end{cases}$$

for all $v \in \mathcal{V}$; x is bound feasible if $x(v) \in \mathcal{B}(v)$ for all $v \in \mathcal{V}$; x is constraint feasible if for all $c \in \mathcal{C}$ we have: $e_c(x) \leq b_c$ if $s_c = -1$, $e_c(x) = b_c$ if $s_c = 0$, and $e_c(x) \geq b_c$ if $s_c = 1$. A point x is feasible in P if it is type, bound and constraint feasible.

A point x feasible in P is also called a *feasible solution* of P. A point which is not feasible is called *infeasible*. Denote by $\mathcal{F}(P)$ the feasible points of P.

Definition 5. *A feasible point x is a local optimum of P with respect to the objective $o \in \mathcal{O}$ if there is a non-empty neighbourhood N of x such that for all feasible points $y \neq x$ in N we have $d_o f_o(x) \geq d_o f_o(y)$. A local optimum is strict if $d_o f_o(x) > d_o f_o(y)$. A feasible point x is a global optimum of P with respect to the objective $o \in \mathcal{O}$ if $d_o f_o(x) \geq d_o f_o(y)$ for all feasible points $y \neq x$. A global optimum is strict if $d_o f_o(x) > d_o f_o(y)$.*

Denote the set of local optima of P by $\mathcal{L}(P)$ and the set of global optima of P by $\mathcal{G}(P)$. If $\mathcal{O}(P) = \emptyset$, we define $\mathcal{L}(P) = \mathcal{G}(P) = \mathcal{F}(P)$.

Example 1. The point $x = (-1, -1, -1, 1, 1, 1)$ is a strict global minimum of the problem in Example 2.1 and $|\mathcal{G}| = 1$ as $U = \{1, 2, 3\}$ and $V \setminus U = \{4, 5, 6\}$ is the only balanced partition of V leading to a cutset size of 1.

It appears from the existing literature that the term "reformulation" is almost never formally defined in the context of mathematical programming. The general consensus seems to be that given a formulation of an optimization problem, a reformulation is a different formulation having the same set of optima. Various authors make use of this definition without actually making it explicit, among which [107, 114, 101, 81, 34, 44, 20, 98, 53, 30]. Many of the proposed reformulations, however, stretch this implicit definition somewhat. Liftings, for example (which consist in adding variables to the problem formulation), usually yield reformulations where an optimum in the original problem is mapped to a set of optima in the reformulated problem (see Sect. 3.2). Furthermore, it is sometimes noted how a reformulation in this sense is overkill because the reformulation only needs to hold at global optimality [1]. Furthermore, reformulations sometimes really refer to a change of variables, as is the case in [93]. Throughout the rest of this section we give various definitions for the concept of reformulation, and we explore the relations between them. We consider two problems

$$P = (\mathcal{P}(P), \mathcal{V}(P), \mathcal{E}(P), \mathcal{O}(P), \mathcal{C}(P), \mathcal{B}(P), \mathcal{T}(P))$$
$$Q = (\mathcal{P}(Q), \mathcal{V}(Q), \mathcal{E}(Q), \mathcal{O}(Q), \mathcal{C}(Q), \mathcal{B}(Q), \mathcal{T}(Q)).$$

Reformulations have been formally defined in the context of *optimization problems* (which are defined as decision problems with an added objective function). As was noted in Sect. 1, we see mathematical programming as a language used to describe and eventually solve optimization problems, so the difference is slim. The following definition is found in [13].

Definition 6. *Let P_A and P_B be two optimization problems. A reformulation $B(\cdot)$ of P_A as P_B is a mapping from P_A to P_B such that, given any instance A of P_A and an optimal solution of $B(A)$, an optimal solution of A can be obtained within a polynomial amount of time.*

This definition is directly inspired to complexity theory and **NP**-completeness proofs. In the more practical and implementation oriented context of this chapter, Defn. 6 has one weak point, namely that of polynomial time. In practice, depending on the problem and on the instance, a polynomial time reformulation may just be too slow; on the other hand, Defn. 6 may bar a non-polynomial time reformulation which might be actually carried out within a practically reasonable amount of time. Furthermore, a reformulation in the sense of Defn. 6 does not necessarily preserve local optimality or the number of global optima, which might in some cases be a desirable reformulation feature. It should be mentioned that Defn. 6 was proposed in a paper that was more theoretical in nature, using an algorithmic equivalence between problems in order to attempt to rank equivalent **NP**-hard problems by their Branch-and-Bound solution difficulty.

The following definition was proposed by H. Sherali [105].

Definition 7. *A problem Q is a reformulation of P if:*

- *there is a bijection $\sigma : \mathcal{F}(P) \to \mathcal{F}(Q)$;*
- *$|\mathcal{O}(P)| = |\mathcal{O}(Q)|$;*
- *for all $p = (e_p, d_p) \in \mathcal{O}(P)$, there is a $q = (e_q, d_q) \in \mathcal{O}(Q)$ such that $e_q = f(e_p)$ where f is a monotonic univariate function.*

Defn. 7 imposes a very strict condition, namely the bijection between feasible regions of the original and reformulated problems. Although this is too strict for many useful transformations to be classified as reformulations, under some regularity conditions on σ it presents some added benefits, such as e.g. allowing easy correspondences between partitioned subspaces of the feasible regions and mapping sensitivity analysis results from reformulated to original problem.

In the rest of the section we discuss alternative definitions which only make use of the concept of optimum (also see [73, 75]). These encompass a larger range of transformations as they do not require a bijection between the feasible regions, the way Defn. 7 does.

Definition 8. Q *is a local reformulation of* P *if there is a function* $\varphi :$ $\mathcal{F}(Q) \rightarrow \mathcal{F}(P)$ *such that (a)* $\varphi(y) \in \mathcal{L}(P)$ *for all* $y \in \mathcal{L}(Q)$, *(b)* φ *restricted to* $\mathcal{L}(Q)$ *is surjective. This relation is denoted by* $P \prec_\varphi Q$.

Informally, a local reformulation transforms all (local) optima of the original problem into optima of the reformulated problem, although more than one reformulated optimum may correspond to the same original optimum. A local reformulation does not lose any local optimality information and makes it possible to map reformulated optima back to the original ones; on the other hand, a local reformulation does not keep track of globality: some global optima in the original problem may be mapped to local optima in the reformulated problem, or vice-versa (see Example 2).

Example 2. Consider the problem $P \equiv \min\limits_{x \in [-2\pi, 2\pi]} x + \sin(x)$ and $Q \equiv$ $\min\limits_{x \in [-2\pi, 2\pi]} \sin(x)$. It is easy to verify that there is a bijection between the local optima of P and those of Q. However, although P has a unique global optimum, every local optimum in Q is global.

Definition 9. Q *is a global reformulation of* P *if there is a function* $\varphi :$ $\mathcal{F}(Q) \rightarrow \mathcal{F}(P)$ *such that (a)* $\varphi(y) \in \mathcal{G}(P)$ *for all* $y \in \mathcal{G}(Q)$, *(b)* φ *restricted to* $\mathcal{G}(Q)$ *is surjective. This relation is denoted by* $P \vartriangleleft_\varphi Q$.

Informally, a global reformulation transforms all global optima of the original problem into global optima of the reformulated problem, although more than one reformulated global optimum may correspond to the same original global optimum. Global reformulations are desirable, in the sense that they make it possible to retain the useful information about the global optima whilst ignoring local optimality. At best, given a difficult problem P with many local minima, we would like to find a global reformulation Q where $\mathcal{L}(Q) = \mathcal{G}(Q)$.

Example 3. Consider a problem P with $\mathcal{O}(P) = \{f\}$. Let Q be a problem such that $\mathcal{O}(Q) = \{\check{f}\}$ and $\mathcal{F}(Q) = \mathrm{conv}(\mathcal{F}(P))$, where $\mathrm{conv}(\mathcal{F}(P))$ is the convex hull of the points of $\mathcal{F}(P)$ and $\check{f}$ is the convex envelope of f over the convex hull of $\mathcal{F}(P)$ (in other words, $\check{f}$ is the greatest convex function underestimating f on $\mathcal{F}(P)$). Since the set of global optima of P is contained in the set of global optima of Q [49], the convex envelope is a global reformulation.

Unfortunately, finding convex envelopes in explicit form is not easy. A considerable amount of work exists in this area: e.g. for bilinear terms [91, 7], trilinear terms [92], fractional terms [122], monomials of odd degree [80, 66] the envelope is known in explicit form (this list is not exhaustive). See [119] for recent theoretical results and a rich bibliography.

Definition 10. Q *is an opt-reformulation (or exact reformulation) of* P *(denoted by* $P < Q$) *if there is a function* $\varphi : \mathcal{F}(Q) \rightarrow \mathcal{F}(P)$ *such that* $P \prec_\varphi Q$ *and* $P \vartriangleleft_\varphi Q$.

This type of reformulation preserves both local and global optimality information, which makes it very attractive. Even so, Defn. 10 fails to encompass those problem transformations that eliminate some global optima whilst ensuring that at least one global optimum is left. Such transformations are specially useful in Integer Programming problems having several symmetric optimal solutions: restricting the set of global optima in such cases may be beneficial. One such example is the pruning of Branch-and-Bound regions based on the symmetry group of the problem presented in [89]: the set of cuts generated by the procedure fails in general to be a global reformulation in the sense of Defn. 9 because the number of global optima in the reformulated problem is smaller than that of the original problem.

Lemma 1. *The relations* $\prec, \lhd, <$ *are reflexive and transitive, but in general not symmetric.*

Proof. For reflexivity, simply take φ as the identity. For transitivity, let $P \prec Q \prec R$ with functions $\varphi : \mathcal{F}(Q) \to \mathcal{F}(P)$ and $\psi : \mathcal{F}(R) \to \mathcal{F}(Q)$. Then $\vartheta = \varphi \circ \psi$ has the desired properties. In order to show that $\prec$ is not symmetric, consider a problem P with variables x and a unique minimum x^* and a problem Q which is exactly like P but has one added variable $w \in [0, 1]$. It is easy to show that $P \prec Q$ (take φ as the projection of (x, w) on x). However, since for all $w \in [0, 1]$ (x^*, w) is an optimum of Q, there is no function of a singleton to a continuously infinite set that is surjective, so $Q \nprec P$.

Given a pair of problems P, Q where $\prec, \lhd, <$ are symmetric on the pair, we call Q a *symmetric reformulation* of P. We remark also that by Lemma (1) we can compose elementary reformulations together to create chained reformulations (see Sect. 3.5 for examples).

Definition 11. *Any problem Q that is related to a given problem P by a formula $f(Q, P) = 0$ where f is a computable function is called an auxiliary problem with respect to P.*

Deriving the formulation of an auxiliary problem may be a hard task, depending on f. The most useful auxiliary problems are those whose formulation can be derived algorithmically in time polynomial in $|P|$.

We remark that casting a problem in a standard form is an opt-reformulation. A good reformulation framework should be aware of the available solution algorithms and attempt to reformulate given problems into the most appropriate standard form.

3.2 Elementary Reformulations

In this section we introduce some elementary reformulations in the proposed framework.

Objective function direction

Given an optimization problem P, the optimization direction d_o of any objective function $o \in \mathcal{O}(P)$ can be changed by simply setting $d_o \leftarrow -d_o$. This is an opt-reformulation where φ is the identity, and it rests on the identity $\min f(x) = -\max -f(x)$. We denote the effect of this reformulation carried out on each objective of a set $O \subseteq \mathcal{O}$ by $\mathrm{OBJDIR}(P, O)$.

Constraint sense

Changing constraint sense simply means to write a constraint c expressed as $e_c \leq b_c$ as $-e_c \geq -b_c$, or $e_c \geq b_c$ as $-e_c \leq -b_c$. This is sometimes useful to convert the problem formulation to a given standard form. This is an opt-reformulation where φ is the identity. It can be carried out on the formulation by setting $\chi(r(e_c)) \leftarrow -\chi(r(e_c))$, $s_c \leftarrow -s_c$ and $b_c = -b_c$. We denote the effect of this reformulation carried out for all constraints in a given set $C \subseteq \mathcal{C}$ by $\mathrm{CONSENSE}(P, C)$.

Liftings, restrictions and projections

We define here three important classes of auxiliary problems: liftings, restrictions and projections. Essentially, a lifting is the same as the original problem but with more variables. A restriction is the same as the original problem but with some of the variables replaced by either parameters or constants. A projection is the same as the original problem projected onto fewer variables. Whereas it is possible to give definitions of liftings and restrictions in terms of symbolic manipulations to the data structure given in Defn. 1, such a definition is in general not possible for projections. Projections and restrictions are in general not opt-reformulations nor reformulations in the sense of Defn. 7.

Lifting

A *lifting* Q of a problem P is a problem such that: $\mathcal{P}(Q) \supsetneq \mathcal{P}(P)$, $\mathcal{V}(Q) \supsetneq \mathcal{V}(P)$, $\mathcal{O}(Q) = \mathcal{O}(P)$, $\mathcal{E}(Q) \supsetneq \mathcal{E}(P)$, $\mathcal{C}(Q) = \mathcal{C}(P)$, $\mathcal{B}(Q) \supsetneq \mathcal{B}(P)$, $\mathcal{T}(Q) \supsetneq \mathcal{T}(P)$. This is an opt-reformulation where φ is a projection operator from $\mathcal{V}(Q)$ onto $\mathcal{V}(P)$: for $y \in \mathcal{F}(Q)$, let $\varphi(y) = (y(v) \mid v \in \mathcal{V}(P))$. We denote the lifting with respect to a new set of variables V by $\mathrm{LIFT}(P, V)$.

Essentially, a lifting is obtained by adding new variables to an optimization problem.

Restriction

A *restriction* Q of a problem P is such that:

- $\mathcal{P}(Q) \supseteq \mathcal{P}(P)$
- $\mathcal{V}(Q) \subsetneq \mathcal{V}(P)$

- $|\mathcal{O}(Q)| = |\mathcal{O}(P)|$
- $|\mathcal{C}(Q)| = |\mathcal{C}(P)|$
- for each $e \in \mathcal{E}(P)$ there is $e' \in \mathcal{E}(Q)$ such that e' is the same as e with any leaf node $v \in \mathcal{V}(P) \smallsetminus \mathcal{V}(Q)$ replaced by an element of $\mathcal{P}(Q) \cup \mathbb{R}$.

We denote the restriction with respect to a sequence of variable V with a corresponding sequence of values R by $\textsc{Restrict}(P, V, R)$.

Essentially, a restriction is obtained by fixing some variables at corresponding given values.

Projection

A *projection* Q of a problem P is such that:

- $\mathcal{P}(Q) \supseteq \mathcal{P}(P)$
- $\mathcal{V}(Q) \subsetneq \mathcal{V}(P)$
- $\mathcal{E}, \mathcal{O}, \mathcal{C}, \mathcal{B}, \mathcal{T}(Q)$ are so that for all $y \in \mathcal{F}(Q)$ there is $x \in \mathcal{F}(P)$ such that $x(v) = y(v)$ for all $v \in \mathcal{V}(Q)$.

In general, symbolic algorithms to derive projections depend largely on the structure of the expression trees in E. If E consists entirely of linear forms, this is not difficult (see e.g. [15], Thm. 1.1). We denote the projection onto a set of variables $V = \mathcal{V}(Q)$ as $\textsc{Proj}(P, V)$.

Essentially, $\mathcal{F}(Q) = \{y \mid \exists x \ (x, y) \in \mathcal{F}(P)\}$.

Equations to inequalities

Converting equality constraints to inequalities may be useful to conform to a given standard form. Suppose P has an equality constraint $c = (e_c, 0, b_c)$. This can be reformulated to a problem Q as follows:

- add two constraints $c_1 = (e_c, -1, b_c)$ and $c_2 = (e_c, 1, b_c)$ to $\mathcal{C}$;
- remove c from $\mathcal{C}$.

This is an opt-reformulation denoted by $\textsc{Eq2Ineq}(P, c)$.

Essentially, we replace the constraint $e_c = b_c$ by the two constraints $e_c \leq b_c, e_c \geq b_c$.

Inequalities to equations

Converting inequalities to equality constraints is useful to convert problems to a given standard form: a very well known case is the standard form of a Linear Programming problem for use with the simplex method. Given a constraint c expressed as $e_c \leq b_c$, we can transform it into an equality constraint by means of a lifting operation and a simple symbolic manipulation on the expression tree e_c, namely:

- add a variable v_c to $\mathcal{V}(P)$ with interval bounds $\mathcal{B}(v_c) = [0, +\infty]$ (added to $\mathcal{B}(P)$) and type $\mathcal{T}(v_c) = 0$ (added to $\mathcal{T}(P)$);

- add a new root node r_0 corresponding to the operator $+$ (sum) to $e_c = (V, A)$, two arcs $(r_0, r(e_c))$, (r_0, v) to A, and we then set $r(e_c) \leftarrow r_0$;
- set $s_c \leftarrow \cup$.

We denote this transformation carried out on the set of constraints C by SLACK(P, C). Naturally, for original equality constraints, this transformation is defined as the identity.

Performing this transformation on any number of inequality constraints results into an opt-reformulation.

Proposition 1. *Given a set of constraints $C \subseteq \mathcal{C}(P)$, the problem $Q = $ SLACK(P, C) is an opt-reformulation of P.*

Proof. We first remark that $\mathcal{V}(P) \subseteq \mathcal{V}(Q)$. Consider φ defined as follows: for each $y \in \mathcal{F}(Q)$ let $\varphi(y) = x = (y(v) \mid v \in \mathcal{V}(P))$. It is then easy to show that φ satisfies Defn. 10.

Absolute value terms

Consider a problem P involving a term $e = (V, A) \in \mathcal{E}$ where $r(e)$ is the absolute value operator $|\cdot|$ (which is continuous but not differentiable everywhere); since this operator is unary, there is a single expression node f such that $(r(e), f) \in A$. This term can be reformulated so that it is differentiable, as follows:

- add two continuous variables t^+, t^- with bounds $[0, +\infty]$;
- replace e by $t^+ + t^-$;
- add constraints $(f - t^+ - t^-, 0, 0)$ and $(t^+ t^-, 0, 0)$ to $\mathcal{C}$.

This is an opt-reformulation denoted by ABSDIFF(P, e).

Essentially, we replace all terms $|f|$ in the problem by a sum $t^+ + t^-$, and then add the constraints $f = t^+ + t^-$ and $t^+ t^- = 0$ to the problem.

Product of exponential terms

Consider a problem P involving a product $g = \prod_{i \leq k} h_i$ of exponential terms, where $h_i = e^{f_i}$ for all $i \leq k$. This term can be reformulated as follows:

- add a continuous variable w to $\mathcal{V}$ with $\mathcal{T}(w) = 0$ and bounds $\mathcal{B}(w) = [0, +\infty]$;
- add a constraint $c = (e_c, 0, 0)$ where $e_c = \sum_{i \leq k} f_i - \log(w)$ to $\mathcal{C}$;
- replace g with w.

This is an opt-reformulation denoted by PRODEXP(P, g). It is useful because many nonlinear terms (product and exponentials) have been the reduced to only one (the logarithm).

Essentially, we replace the product $\prod_i e^{f_i}$ by an added nonnegative continuous variable w and then add the constraint $\log(w) = \sum_i f_i$ to the problem.

Binary to continuous variables

Consider a problem P involving a binary variable $x \in \mathcal{V}$ with $(\mathcal{T}(x) = 2)$. This can be reformulated as follows:

- add a constraint $c = (e_c, 0, 0)$ to $\mathcal{C}$ where $e_c = x^2 - x$;
- set $\mathcal{T}(x) = 0$.

This is an opt-reformulation denoted by $\text{BIN2CONT}(P, x)$. Since a binary variable $x \in \mathcal{V}$ can only take values in $\{0, 1\}$, any univariate equation in x that has exactly $x = 0$ and $x = 1$ as solutions can replace the binary constraint $x \in \{0, 1\}$. The most commonly used is the quadratic constraint $x^2 = x$, sometimes also written as $x(x - 1) \geq 0 \wedge x \leq 1$ [118].

In principle, this would reduce all binary problems to nonconvex quadratically constrained problems, which can be solved by a global optimization (GO) solver for nonconvex NLPs. In practice, GO solvers rely on an NLP subsolver to do most of the computationally intensive work, and NLP solvers are generally not very good in handling nonconvex/nonlinear equality constraints such as $x^2 = x$. This reformulation, however, is often used in conjunction with the relaxation of binary linear and quadratic problems (see Sect. 4.4).

Integer to binary variables

It is sometimes useful, for different reasons, to convert general integer variables to binary (0-1) variables. One example where this yields a crucial step into a complex linearization is given in Sect. 3.5. There are two established ways of doing this: one entails introducing binary assignment variables for each integer values that the variable can take; the other involves the binary representation of the integer variable value. Supposing the integer variable value is n, the first way employs $O(n)$ added binary variables, whereas the second way only employs $O(\log_2(n))$. The first way is sometimes used to linearize complex nonlinear expressions of integer variables by transforming them into a set of constants to choose from (see example in Sect. 3.5). The second is often used in an indirect way to try and break symmetries in 0-1 problems: by computing the integer values of the binary representation of two 0-1 vectors x_1, x_2 as integer problem variables v_1, v_2, we can impose ordering constraints such as $v_1 \leq v_2 + 1$ to exclude permutations of x_1, x_2 from the feasible solutions.

Assignment variables

Consider a problem P involving an integer variable $v \in \mathcal{V}$ with type $\mathcal{T}(v) = 1$ and bounds $\mathcal{B}(v) = [L_v, U_v]$ such that $U_v - L_v > 1$. Let $V = \{L_v, \ldots, U_v\}$ be the variable domain. Then P can be reformulated as follows:

- for all $j \in V$ add a binary variable w_j to $\mathcal{V}$ with $\mathcal{T}(w_j) = 2$ and $\mathcal{B}(w_j) = [0, 1]$;
- add a constraint $c = (e_c, 0, 1)$ where $e_c = \sum_{j \in V} w_j$ to $\mathcal{C}$;

- add an expression $e = \sum_{j \in V} jw_j$ to $\mathcal{E}$;
- replace all occurrences of v in the leaf nodes of expressions in $\mathcal{E}$ with e.

This is an opt-reformulation denoted by $\text{INT2BIN}(P, v)$.

Essentially, we add assignment variables $w_j = 1$ if $v = j$ and 0 otherwise. We then add an assignment constraint $\sum_{j \in V} w_j = 1$ and replace v with $\sum_{j \in V} jw_j$ throughout the problem.

Binary representation

Consider a problem P involving an integer variable $v \in V$ with type $\mathcal{T}(v) = 1$ and bounds $\mathcal{B}(v) = [L_v, U_v]$ such that $U_v - L_v > 1$. Let $V = \{L_v, \ldots, U_v\}$ be the variable domain. Then P can be reformulated as follows:

- let b be the minimum exponent such that $|V| \leq 2^b$;
- add b binary variables $w_1, \ldots, w_b$ to $\mathcal{V}$ such that $\mathcal{T}(w_j) = 2$ and $\mathcal{B}(w_j) = [0, 1]$ for all $j \leq b$;
- add an expression $e = L_v + \sum_{j \leq b} w_j 2^j$
- replace all occurrences of v in the leaf nodes of expressions in $\mathcal{E}$ with e.

This is an opt-reformulation denoted by $\text{BINARYREP}(P, v)$.

Essentially, we write the binary representation of v as $L_v + \sum_{j \leq b} w_j 2^j$.

Feasibility to optimization problems

The difference between decision and optimization problems in computer science reflects in mathematical programming on the number of objective functions in the formulation. A formulation without objective functions models a feasibility problem; a formulation with one or more objective models an optimization problem. As was pointed out by the example in the introduction (see Sect. 1, p. 154), for computational reasons it is sometimes convenient to reformulate a feasibility problem in an optimization problem by introducing constraint tolerances. Given a feasibility problem P with $\mathcal{O} = \emptyset$, we can reformulate it to an optimization problem Q as follows:

- add a large enough constant M to $\mathcal{P}(Q)$;
- add a continuous nonnegative variable ε to $\mathcal{V}(Q)$ with $\mathcal{T}(\epsilon) = 0$ and $\mathcal{B}(\epsilon) = [0, M]$;
- for each equality constraint $c = (e_c, 0, b_c) \in \mathcal{C}$, apply $\text{EQ2INEQ}(P, c)$;
- add the expression ε to $\mathcal{E}(Q)$;
- add the objective function $o = (\varepsilon, -1)$ to $\mathcal{O}(Q)$;
- for each constraint $c = (e_c, s_c, b_c) \in \mathcal{C}$ (we now have $s_c \neq 0$), let $e_c' = e_c + s_c \varepsilon$ and $c' = (e_c', s_c, b_c)$; add c' to $\mathcal{C}(Q)$.

As the original problem has no objective function, the usual definitions of local and global optima do not hold. Instead, we define any point in $\mathcal{F}(P)$ to be both a local and a global optimum (see paragraph under Defn. 5). Provided the original problem is feasible, this is an opt-reformulation denoted by $\text{FEAS2OPT}(P)$.

Proposition 2. *Provided $\mathcal{F}(P) \neq \emptyset$, the reformulation* FEAS2OPT(P) *is an opt-reformulation.*

Proof. Let F be the projection of $\mathcal{F}(Q)$ on the space spanned by the variables of $\mathcal{P}$ (i.e. all variables of Q but ε, see Sect. 3.2), and let π be the projection map. We then have $\mathcal{F}(P) \subseteq F$ (this is because the constraints of Q essentially define a constraint relaxation of P, see Sect. 4.1 and Defn. 14). Let $x' \in \mathcal{F}(P)$. We define $\psi : F \to \mathcal{F}(P)$ to be the identity on $\mathcal{F}(P)$ and trivially extend it to $\mathcal{F}(Q) \smallsetminus F$ by setting $\psi(z) = x'$ for all $z \in \mathcal{F}(Q) \smallsetminus F$. The function $\phi = \psi \circ \pi$ maps $\mathcal{F}(Q)$ to $\mathcal{F}(P)$, and preserves local minimality by construction, as per Defn. 8. Since ε is bounded below by zero, and the restriction (see Sect. 3.2) of Q to $\varepsilon = 0$ is exactly P, any $x' \in \mathcal{G}(Q)$ is also in $\mathcal{F}(P)$. Moreover, by definition $\mathcal{G}(P) = \mathcal{F}(P)$ as $\mathcal{O}(P) = \emptyset$, showing that the identity (projected on F) preserves global minimality in the sense of Defn. 9.

3.3   Exact Linearizations

Definition 12. *An exact linearization of a problem P is an opt-reformulation Q of P where all expressions $e \in \mathcal{E}(P)$ are linear forms.*

Different nonlinear terms are linearized in different ways, so we sometimes speak of a linearization of a particular nonlinear term instead of a linearization of a given problem.

Piecewise linear objective functions

Consider a problem P having an objective function $o = (d_o, e_o) \in \mathcal{O}(P)$ and a finite set of expressions e_k for $k \in K$ such that $e_o = d_o \min_{k \in K} d_o e_k$ (this is a piecewise linear objective function of the form $\min \max_k e_k$ or $\max \min_k e_k$ depending on d_o). This can be linearized by adding one variable and $|K|$ constraints to the problem as follows:

- add a continuous variable t to $\mathcal{V}$ bounded in $[-\infty, +\infty]$;
- for all $k \in K$, add the constraint $c_k = (e_k - t, d_o, 0)$ to $\mathcal{C}$.

This is an opt-reformulation denoted by MINMAX(P).

Essentially, we can reformulate an objective function $\min \max_{k \in K} e_k$ as $\min t$ by adding a continuous variable t and the constraints $\forall k \in K \; t \geq e_k$ to the problem.

Product of binary variables

Consider a problem P where one of the expressions $e \in \mathcal{E}(P)$ is $\prod_{k \in \bar{K}} v_k$, where $v_k \in \mathcal{V}(P)$, $\mathcal{B}(v_k) = [0, 1]$ and $\mathcal{T}(v_k) = 2$ for all $k \in \bar{K}$ (i.e. v_k are binary 0-1 variables). This product can be linearized as follows:

- add a continuous variable w to $\mathcal{V}$ bounded in $[0, 1]$;

- add the constraint $(\sum_{k\in\bar{K}} v_k - w, -1, |\bar{K}| - 1)$ to $\mathcal{C}$;
- for all $k \in \bar{K}$ add the constraint $(w - v_k, -1, 0)$ to $\mathcal{C}$.

This is an opt-reformulation denoted by $\text{PRODBIN}(P, \bar{K})$.

Essentially, a product of binary variables $\prod_{k\in\bar{K}} v_k$ can be replaced by an added continuous variable $w \in [0, 1]$ and added constraints $\forall k \in \bar{K} \; w \le v_k$ and $w \ge \sum_{k\in\bar{K}} v_k - |\bar{K}| + 1$.

Proposition 3. *Given a problem P and a set $\bar{K} \subset \mathbb{N}$, the problem $Q = \text{PRODBIN}(P, \bar{K})$ is an opt-reformulation of P.*

Proof. Suppose first that $\forall k \in \bar{K}$, $v_k = 1$. We have to prove that $w = 1$ in that case. It comes from the hypothesis that $\sum_{k\in\bar{K}} v_k - |\bar{K}| + 1 = 1$ which implies by the last constraint that $w = 1$. The other constraints are all reduced to $w \le 1$ which are all verified.

Suppose now that at least one of the binary variable is equal to zero and call i the index of this variable. Since $\forall k \in \bar{K} \; w \le v_k$, we have in particular the constraint for $k = i$. This leads to $w = 0$ which is the expected value. Besides, it comes that $\sum_{k\in\bar{K}} v_k - |\bar{K}| \le -1$. We deduced from this inequality that the last constraint is verified by the value of w.

As products of binary variables model the very common AND operation, linearizations of binary products are used very often. Hammer and Rudeanu [46] cite [37] as the first published appearance of the above linearization for cases where $|\bar{K}| = 2$. For problems P with products $v_i v_j$ for a given set of pairs $\{i, j\} \in K$ where v_i, v_j are all binary variables, the linearization consists of $|Q|$ applications of $\text{Prodbin}(P, \{i, j\})$ for each $\{i, j\} \in K$. Furthermore, we replace each squared binary variable v_i^2 by simply v_i (as $v_i^2 = v_i$ for binary variables v_i). We denote this linearization by $\text{PRODSET}(P, K)$.

Product of binary and continuous variables

Consider a problem P involving products $v_i v_j$ for a given set K of ordered variable index pairs (i, j) where v_i is a binary 0-1 variable and v_j is a continuous variable with $\mathcal{B}(v_j) = [L_j, U_j]$. The problem can be linearized as follows:

- for all $(i, j) \in K$ add a continuous variable w_{ij} bounded by $[L_j, U_j]$ to $\mathcal{V}$;
- for all $(i, j) \in K$ replace the product terms $v_i v_j$ by the variable w_{ij};
- for all $(i, j) \in K$ add the constraints
  $(w_{ij}, -1, U_j v_i), (w_{ij}, 1, L_j v_i), (w_{ij}, -1, v_j - (1 - v_i)L_j), (w_{ij}, 1, v_j - (1 - v_i)U_j)$ to $\mathcal{C}$.

This is an opt-reformulation denoted by $\text{PRODBINCONT}(P, K)$.

Essentially, a product of a binary variable v_i and a continuous variable v_j bounded by $[L_j, U_j]$ can be replaced by an added variable w_{ij} and added constraints:

$$\begin{cases} w_{ij} \leq U_j v_i \\ w_{ij} \geq L_j v_i \\ w_{ij} \leq v_j - (1 - v_i) L_j \\ w_{ij} \geq v_j - (1 - v_i) U_j \end{cases}$$

Proposition 4. *Given a problem P and a set K of ordered variable index pairs (i, j), the problem $Q = \text{ProdBinCont}(P, K)$ is an opt-reformulation of P.*

Proof. We have to prove that the reformulation ensures $w_{ij} = v_i v_j$ for all possible values for v_i and v_j. We do it by cases on the binary variable v_i. Suppose first that $v_i = 0$. Then the two first constraints implies that $w_{ij} = 0$ which corresponds indeed to the product $v_i v_j$. It remains to see that the two other constraints don't interfere with this equality. In that case, the third constraint becomes $w_{ij} \leq v_j - L_j$. Since $v_j \geq L_j$ by definition, we have $v_j - L_j \geq 0$ implying that w_{ij} is less or equal to a positive term. With a similar reasoning, it comes from the fourth constraint that w_{ij} is greater or equal to a negative term. Thus, for the case $v_i = 0$, the constraints lead to $w_{ij} = 0$.

Suppose now that $v_i = 1$. The two first inequalities lead to $L_i \leq w_{ij} \leq U_j$ which corresponds indeed to the range of the variable. The two last constraints become $w_{ij} \geq v_j$ and $w_{ij} \leq v_j$. This implies $w_{ij} = v_j$ which is the correct result.

Complementarity constraints

Consider a problem P involving constraints of the form $c = (e_c, 0, 0)$ where (a) $r(e_c)$ is the sum operator, (b) for each node e outgoing from e_c, e is a product operator, (c) each of these product nodes e has two outgoing nodes f, g. We can linearize such a constraint as follows:

- for each product operator node e outgoing from $r(e_c)$ and with outgoing nodes f, g:

  1. add a (suitably large) constant parameter $M > 0$ to $\mathcal{P}$;
  2. add a binary variable w to $\mathcal{V}$ with $\mathcal{T}(v) = 2$ and $\mathcal{B} = [0, 1]$
  3. add the constraints $(f - Mw, -1, 0)$ and $(g + Mw, -1, M)$ to $\mathcal{C}$

- delete the constraint c.

Provided we set M as an upper bound to the maximum values attainable by f and g, this is an opt-reformulation which is also a linearization. We denote it by $\text{CCLin}(P)$.

Essentially, we linearize complementarity constraints $\sum_{k \in K} f_k g_k = 0$ by eliminating the constraint, adding 0-1 variables w_k for all $k \in K$ and the linearization constraints $f_k \leq Mw_k$ and $g_k \leq M(1 - w_k)$. This reformulation, together with ABSDIFF (see Sect. 3.2), provides an exact linearization (provided a suitably large but finite M exists) of absolute value terms.

Minimization of absolute values

Consider a problem P with a single objective function $o = (d_o, e_o) \in \mathcal{O}$ where $e_o = (-d_o) \sum_{k \in \bar{K}} e_k$ where the operator represented by the root node $r(e_k)$ of e_k is the absolute value $|\cdot|$ for all $k \in K \subseteq \bar{K}$. Since the absolute value operator is unary, $\delta^+(r(e_k))$ consists of the single element f_k. Provided f_k are linear forms, this problem can be linearized as follows. For each $k \in K$:

- add continuous variables t_k^+, t_k^- with bounds $[0, +\infty]$;
- replace e_k by $t_k^+ + t_k^-$;
- add constraints $(f_k - t_k^+ - t_k^-, 0, 0)$ to $\mathcal{C}$.

This is an opt-reformulation denoted by MINABS(P, K).

Essentially, we can reformulate an objective function $\min \sum_{k \in \bar{K}} |f_k|$ as $\min \sum_{k \in \bar{K}} (t_k^+ + t_k^-)$ whilst adding constraints $\forall k \in \bar{K}$ $f_k = t_k^+ + t_k^-$ to the problem. This reformulation is related to ABSDIFF(P, e) (see Sect. 3.2), however the complementarity constraints $t_k^+ t_k^- = 0$ are not needed because of the objective function direction: at a global optimum, because of the minimization of $t_k^+ + t_k^-$, at least one of the variables will have value zero, thus implying the complementarity.

Linear fractional terms

Consider a problem P where an expression in $\mathcal{E}$ has a sub-expression e with a product operator and two subnodes e_1, e_2 where $\xi(e_1) = 1$, $\xi(e_2) = -1$, and e_1, e_2 are affine forms such that $e_1 = \sum_{i \in V} a_i v_i + b$ and $e_2 = \sum_{i \in V} c_i v_i + d$, where $v \subseteq \mathcal{V}$ and $\mathcal{T}(v_i) = 0$ for all $i \in V$ (in other words e is a linear fractional term $\frac{a^\top v + b}{c^\top v + d}$ on continuous variables v). Assume also that the variables v only appear in some linear constraints of the problem $Av = q$ (A is a matrix and q is a vector in $\mathcal{P}$). Then the problem can be linearized as follows:

- add continuous variables α_i, β to $\mathcal{V}$ (for $i \in V$) with $\mathcal{T}(\alpha_i) = \mathcal{T}(\beta) = 0$;
- replace e by $\sum_{i \in V} a_i \alpha_i + b\beta$;
- replace the constraints in $Av = q$ by $A\alpha - q\beta = 0$;
- add the constraint $\sum_{i \in V} c_i \alpha_i + d\beta = 1$;
- remove the variables v from $\mathcal{V}$.

This is an opt-reformulation denoted by LINFRACT(P, e).

Essentially, α_i plays the role of $\frac{v_i}{c^\top v + d}$, and β that of $\frac{1}{c^\top v + d}$. It is then easy to show that e can be re-written in terms of α, β as $a^\top \alpha + b\beta$, $Av = q$ can be re-written as $A\alpha = q\beta$, and that $c^\top \alpha + d\beta = 1$. Although the original variables v are removed from the problem, their values can be obtained by α, β after the problem solution, by computing $v_i = \frac{\alpha_i}{\beta}$ for all $i \in V$.

3.4 Advanced Reformulations

In this section we review a few advanced reformulations in the literature.

Hansen's Fixing Criterion

This method applies to unconstrained quadratic 0-1 problems of the form $\min_{x \in \{0,1\}^n} x^\top Q x$ where Q is an $n \times n$ matrix [47], and relies on fixing some of the variables to values guaranteed to provide a global optimum.

Let P be a problem with $\mathcal{P} = \{n \in \mathbb{N}, \{q_{ij} \in \mathbb{R} \mid 1 \leq i, j \leq n\}\}$, $\mathcal{V} = \{x_i \mid 1 \leq i \leq n\}$, $\mathcal{E} = \{f = \sum_{i,j \leq n} q_{ij} x_i x_j\}$, $\mathcal{O} = \{(f, -1)\}$, $\mathcal{C} = \emptyset$, $\mathcal{B} = [0,1]^n$, $\mathcal{T} = 2$. This can be restricted (see Sect. 3.2) as follows:

- initialize two sequences $V = \emptyset, A = \emptyset$;
- for all $i \leq n$:

  1. if $q_{ii} + \sum_{j<i} \min(0, q_{ij}) + \sum_{j>i} \min(0, q_{ij}) > 0$ then append x_i to V and 0 to A;
  2. (else) if $q_{ii} + \sum_{j<i} \max(0, q_{ij}) + \sum_{j>i} \max(0, q_{ij}) < 0$ then append x_i to V and 1 to A;

- apply RESTRICT(P, V, A).

This opt-reformulation is denoted by FixQB(P).

Essentially, any time a binary variable consistently decreases the objective function value when fixed, independently of the values of other variables, it is fixed.

Compact linearization of binary quadratic problems

This reformulation concerns a problem P with the following properties:

- there is a subset of binary variables $x \subseteq \mathcal{V}$ with $|x| = n, \mathcal{T}(x) = 2, \mathcal{B}(x) = [0,1]^n$;
- there is a set $E = \{(i,j) \mid 1 \leq i \leq j \leq n\}$ in $\mathcal{P}$ such that the terms $x_i x_j$ appear as sub-expressions in the expressions $\mathcal{E}$ for all $(i,j) \in E$;
- there is an integer $K \leq n$ in $\mathcal{P}$ and a covering $\{I_k \mid k \leq K\}$ of $\{1, \ldots, n\}$ such that $(\sum_{i \in I_k} x_i, 0, 1)$ is in $\mathcal{C}$ for all $k \leq K$;
- there is a covering $\{J_k \mid k \leq K\}$ of $\{1, \ldots, n\}$ such that $I_k \subseteq J_k$ for all $k \leq K$ such that, letting $F = \{(i,j) \mid \exists k \leq K((i,j) \in I_k \times J_k \vee (i,j) \in J_k \times I_k)\}$, we have $E \subseteq F$.

Under these conditions, the problem P can be exactly linearized as follows:

- for all $(i,j) \in F$ add continuous variables w_{ij} with $\mathcal{T}(w_{ij}) = 0$ and $\mathcal{B}(w_{ij}) = [0,1]$;
- for all $(i,j) \in E$ replace sub-expression $x_i x_j$ with w_{ij} in the expressions $\mathcal{E}$;
- for all $k \leq K, j \in J_k$ add the constraint $(\sum_{i \in I_k} w_{ij} - x_j, 0, 0)$ to $\mathcal{C}$.
- for all $(i,j) \in F$ add the constraint $w_{ij} = w_{ji}$ to $\mathcal{C}$.

This opt-reformulation is denoted by $\text{RCLIN}(P, E)$. It was shown in [72] that this linearization is exact and has other desirable tightness properties. See [72] for examples.

Reduced RLT Constraints

This reformulation concerns a problem P with the following properties:

- there is a subset $x \subseteq V$ with $|x| = n$ and a set $E = \{(i, j) \mid 1 \leq i \leq j \leq n\}$ in P such that the terms $x_i x_j$ appear as sub-expressions in the expressions $\mathcal{E}$ for all $(i, j) \in E$;
- there is a number $m \leq n$, an $m \times n$ matrix $A = (a_{ij})$ and an m-vector b in P such that $(\sum_{j \leq n} a_{ij} x_j, 0, b_i) \in \mathcal{C}$ for all $i \leq m$.

Let $F = \{(i, j) \mid (i, j) \in E \vee \exists k \leq m (a_{kj} \neq 0)\}$. Under these conditions, P can be reformulated as follows:

- for all $(i, j) \in F$ add continuous variables w_{ij} with $\mathcal{T}(w_{ij}) = 0$ and $\mathcal{B}(w_{ij}) = [-\infty, +\infty]$;
- for all $(i, j) \in E$ replace sub-expression $x_i x_j$ with w_{ij} in the expressions $\mathcal{E}$;
- for all $i \leq n, k \leq m$ add the constraints $(\sum_{j \leq n} a_{kj} w_{ij} - b_k x_i, 0, 0)$ to $\mathcal{C}$: we call this linear system the *Reduced RLT Constraint System* (RCS) and $(\sum_{j \leq n} a_{kj} w_{ij}, 0, 0)$ the *companion* system;
- let $B = \{(i, j) \in F \mid w_{ij} \text{ is basic in the companion}\}$;
- let $N = \{(i, j) \in F \mid w_{ij} \text{ is non-basic in the companion}\}$;
- add the constraints $(w_{ij} - x_i x_j, 0, 0)$ for all $(i, j) \in N$.

This opt-reformulation is denoted by $\text{REDCON}(P)$, and its validity was shown in [70]. It is important because it effectively reduces the number of quadratic terms in the problem (only those corresponding to the set N are added). This reformulation can be extended to work with sparse sets E [81], namely sets E whose cardinality is small with respect to $\frac{1}{2}n(n + 1)$.

Essentially, the constraints $w_{ij} = x_i x_j$ for $(i, j) \in B$ are replaced by the RCS $\forall i \leq n \, (Aw_i = x_i)$, where $w_i = (w_{i1}, \ldots, w_{in})$.

3.5 Advanced Examples

We give in this section a few advanced examples that illustrate the power of the elementary reformulations given above.

The Hyperplane Clustering Problem

As an example of what can be attained by combining these simple reformulations presented in this chapter, we give a MINLP formulation to the

HYPERPLANE CLUSTERING PROBLEM (HCP) [29, 24]. Given a set of points $p = \{p_i \mid 1 \leq i \leq m\}$ in $\mathbb{R}^d$ we want to find a set of n hyperplanes $w = \{w_{j1}x_1 + \ldots + w_{jd} = w_j^0 \mid 1 \leq j \leq n\}$ in $\mathbb{R}^d$ and an assignment of points to hyperplanes such that the distances from the hyperplanes to their assigned points are minimized.

We then derive a MILP reformulation. For clarity, we employ the usual mathematical notation instead of the notation given Defn. 1.

The problem P can be modelled as follows:

- *Parameters.* The set of parameters is given by $p \in \mathbb{R}^{m \times d}, m, n, d \in \mathbb{N}$.
- *Variables.* We consider the hyperplane coefficient variables $w \in \mathbb{R}^{n \times d}$, the hyperplane constants $w^0 \in \mathbb{R}^n$, and the 0-1 assignment variables $x \in \{0, 1\}^{m \times n}$.
- *Objective function.* We minimize the total distance, weighted by the assignment variable:

$$\min \sum_{i \leq m} \sum_{j \leq n} |w_j p_i - w_j^0| x_{ij}.$$

- *Constraints.* We consider assignment constraints: each point must be assigned to exactly one hyperplane:

$$\forall i \leq m \quad \sum_{j \leq n} x_{ij} = 1,$$

and the hyperplanes must be nontrivial:

$$\forall j \leq n \quad \sum_{k \leq d} |w_{jk}| = 1,$$

for otherwise the trivial solution with $w = 0, w^0 = 0$ would be optimal.

This is a MINLP formulation because of the presence of the nonlinear terms (absolute values and products in the objective function) and of the binary assignment variables. We shall now apply several of the elementary reformulations presented in this chapter to obtain a MILP reformulation Q of P.

Let $K = \{(i, j) \mid i \leq m, j \leq n\}$.

1. Because x is nonnegative and because we are going to solve the reformulated MILP to global optimality, we can apply an reformulation similar to $\text{MINABS}(P, K)$ (see Sect. 3.3) to obtain an opt-reformulation P_1 as follows:

$$\min \sum_{i,j} (t_{ij}^+ x_{ij} + t_{ij}^- x_{ij})$$

$$\text{s.t.} \quad \forall i \sum_j x_{ij} = 1$$

$$\forall j \quad |w_j| \mathbf{1} = 1$$

$$\forall i, j \ t_{ij}^+ - t_{ij}^- = w_j p_i - w_j^0,$$

where $t_{ij}^+, t_{ij}^- \in [0, M]$ are continuous added variables bounded above by a (large and arbitrary) constant M which we add to the parameter set $\mathcal{P}$. We remark that this upper bound is enforced without loss of generality because w, w^0 can be scaled arbitrarily.

2. Apply PRODBINCONT(P_1, K) (see Sect. 3.3) to the products $t_{ij}^+ x_{ij}$ and $t_{ij}^- x_{ij}$ to obtain a opt-reformulation P_2 as follows:

$$\min \sum_{i,j} (y_{ij}^+ + y_{ij}^-)$$

$$\text{s.t.} \quad \forall i \ \sum_j x_{ij} = 1$$

$$\forall j \ |w_j| 1 = 1$$

$$\forall i, j \ t_{ij}^+ - t_{ij}^- = w_j p_i - w_j^0$$

$$\forall i, j \ y_{ij}^+ \leq \min(M x_{ij}, t_{ij}^+)$$

$$\forall i, j \ y_{ij}^+ \geq M x_{ij} + t_{ij}^+ - M$$

$$\forall i, j \ y_{ij}^- \leq \min(M x_{ij}, t_{ij}^-)$$

$$\forall i, j \ y_{ij}^- \geq M x_{ij} + t_{ij}^- - M,$$

where $y_{ij}^+, y_{ij}^- \in [0, M]$ are continuous added variables.

3. For each term $e_{jk} = |w_{jk}|$ apply ABSDIFF(P_2, e_{jk}) to obtain an opt-reformulation P_3 as follows:

$$\min \sum_{i,j} (y_{ij}^+ + y_{ij}^-)$$

$$\text{s.t.} \quad \forall i \ \sum_j x_{ij} = 1$$

$$\forall i, j \ t_{ij}^+ - t_{ij}^- = w_j p_i - w_j^0$$

$$\forall i, j \ y_{ij}^+ \leq \min(M x_{ij}, t_{ij}^+)$$

$$\forall i, j \ y_{ij}^+ \geq M x_{ij} + t_{ij}^+ - M$$

$$\forall i, j \ y_{ij}^- \leq \min(M x_{ij}, t_{ij}^-)$$

$$\forall i, j \ y_{ij}^- \geq M x_{ij} + t_{ij}^- - M$$

$$\forall j \ \sum_{k \leq d} (u_{jk}^+ + u_{jk}^-) = 1$$

$$\forall j, k \ u_{jk}^+ - u_{jk}^- = w_{jk}$$

$$\forall j, k \ u_{jk}^+ u_{jk}^- = 0,$$

where $u_{jk}^+, u_{jk}^- \in [0, M]$ are continuous variables for all j, k. Again, the upper bound does not enforce loss of generality. P_3 is an opt-reformulation of P: whereas P was not everywhere differentiable because of the absolute values, P_3 only involves differentiable terms.

4. We remark that the last constraints of P_3 are in fact complementarity constraints. We apply $\mathrm{CCLIN}(P_3)$ to obtain the reformulated problem Q:

$$\min \sum_{i,j} (y_{ij}^+ + y_{ij}^-)$$

$$\text{s.t.} \quad \forall i \ \sum_j x_{ij} = 1$$

$$\forall i,j \quad t_{ij}^+ - t_{ij}^- = w_j p_i - w_j^0$$

$$\forall i,j \quad y_{ij}^+ \le \min(Mx_{ij}, t_{ij}^+)$$

$$\forall i,j \quad y_{ij}^+ \ge Mx_{ij} + t_{ij}^+ - M$$

$$\forall i,j \quad y_{ij}^- \le \min(Mx_{ij}, t_{ij}^-)$$

$$\forall i,j \quad y_{ij}^- \ge Mx_{ij} + t_{ij}^- - M$$

$$\forall j \ \sum_{k \le d} (u_{jk}^+ + u_{jk}^-) = 1$$

$$\forall j,k \ u_{jk}^+ - u_{jk}^- = w_{jk}$$

$$\forall j,k \ u_{jk}^+ \le Mz_{jk}$$

$$\forall j,k \ u_{jk}^- \le M(1 - z_{jk}),$$

where $z_{jk} \in \{0,1\}$ are binary variables for all j,k. Q is a MILP reformulation of P (see Sect. 2.3).

This reformulation allows us to solve P by using a MILP solver — these have desirable properties with respect to MINLP solvers, such as numerical stability and robustness, as well as scalability and an optimality guarantee. A small instance consisting of 8 points and 2 planes in $\mathbb{R}^2$, with $p = \{(1,7), (1,1), (2,2), (4,3), (4,5), (8,3), (10,1), (10,5)\}$ is solved to optimality by the ILOG CPLEX solver [52] to produce the following output:

```
Normalized hyperplanes:
1: (0.452055) x_1 + (-1.20548) x_2 + (1.50685) = 0
2: (0.769231) x_1 + (1.15385) x_2 + (-8.84615) = 0
Assignment of points to hyperplanar clusters:
hyp_cluster 1 = { 2 3 4 8 }
hyp_cluster 2 = { 1 5 6 7 }.
```

Selection of software components

Large software systems consist of a complex architecture of interdependent, modular software components. These may either be built or bought off-the-shelf. The decision of whether to build or buy software components influences the cost, delivery time and reliability of the whole system, and should therefore be taken in an optimal way [26].

Consider a software architecture with n component slots. Let I_i be the set of off-the-shelf components and J_i the set of purpose-built components that can be plugged in the i-th component slot, and assume $I_i \cap J_i = \emptyset$. Let T be the maximum assembly time and R be the minimum reliability level. We want to select a sequence of n off-the-shelf or purpose-built components compatible with the software architecture requirements that minimize the total cost whilst satisfying delivery time and reliability constraints. This problem can be modelled as follows.

- *Parameters*:

  1. Let $N \in \mathbb{N}$;
  2. for all $i \leq n$, s_i is the expected number of invocations;
  3. for all $i \leq n, j \in I_i$, c_{ij} is the cost, d_{ij} is the delivery time, and μ_{ij} the probability of failure on demand of the j-th off-the-shelf component for slot i;
  4. for all $i \leq n, j \in J_i$, $\bar{c}_{ij}$ is the cost, t_{ij} is the estimated development time, τ_{ij} the average time required to perform a test case, p_{ij} is the probability that the instance is faulty, and b_{ij} the testability of the j-th purpose-built component for slot i.

- *Variables*:

  1. Let $x_{ij} = 1$ if component $j \in I_j \cup J_i$ is chosen for slot $i \leq n$, and 0 otherwise;
  2. Let $N_{ij} \in \mathbb{Z}$ be the (non-negative) number of tests to be performed on the purpose-built component $j \in J_i$ for $i \leq n$: we assume $N_{ij} \in \{0, \ldots, N\}$.

- *Objective function.* We minimize the total cost, i.e. the cost of the off-the-shelf components c_{ij} and the cost of the purpose-built components $\bar{c}_{ij}(t_{ij} + \tau_{ij}N_{ij})$:

$$\min \sum_{i \leq n} \left(\sum_{j \in I_i} c_{ij}x_{ij} + \sum_{j in J_i} \bar{c}_{ij}(t_{ij} + \tau_{ij}N_{ij})x_{ij} \right).$$

- *Constraints*:

  1. assignment constraints: each component slot in the architecture must be filled by exactly one software component

  $$\forall i \leq n \quad \sum_{j \in I_i \cup J_i} x_{ij} = 1;$$

  2. delivery time constraints: the delivery time for an off-the-shelf component is simply d_{ij}, whereas for purpose-built components it is $t_{ij} + \tau_{ij}N_{ij}$

  $$\forall i \leq n \quad \sum_{j \in I_i} d_{ij}x_{ij} + \sum_{j \in J_i}(t_{ij} + \tau_{ij}N_{ij})x_{ij} \leq T;$$

3. reliability constraints: the probability of failure on demand of off-the shelf components is μ_{ij}, whereas for purpose-built components it is given by

$$\vartheta_{ij} = \frac{p_{ij}b_{ij}(1-b_{ij})^{(1-b_{ij})N_{ij}}}{(1-p_{ij})+p_{ij}(1-b_{ij})^{(1-b_{ij})N_{ij}}},$$

so the probability that no failure occurs during the execution of the i-th component is

$$\varphi_i = e^{s_i\left(\sum\limits_{j\in I_i}\mu_{ij}x_{ij}+\sum\limits_{j\in J_i}\vartheta_{ij}x_{ij}\right)},$$

whence the constraint is

$$\prod_{i\leq n}\varphi_i \geq R;$$

notice we have three classes of reliability constraints involving two sets of added variables ϑ, φ.

This problem is a MINLP with no continuous variables. We shall now apply several reformulations to this problem (call it P).

1. Consider the term $g = \prod_{i\leq n}\varphi_i$ and apply PRODEXP(P, g) to P to obtain P_1 as follows:

$$\min\sum_{i\leq n}\left(\sum_{j\in I_i}c_{ij}x_{ij}+\sum_{j\in J_i}\bar{c}_{ij}(t_{ij}+\tau_{ij}N_{ij})x_{ij}\right)$$

$$\forall i\leq n \qquad \sum_{j\in I_i\cup J_i}x_{ij}=1$$

$$\forall i\leq n \quad \sum_{j\in I_i}d_{ij}x_{ij}+\sum_{j\in J_i}(t_{ij}+\tau_{ij}N_{ij})x_{ij}\leq T$$

$$\frac{p_{ij}b_{ij}(1-b_{ij})^{(1-b_{ij})N_{ij}}}{(1-p_{ij})+p_{ij}(1-b_{ij})^{(1-b_{ij})N_{ij}}}=\vartheta_{ij}$$

$$w\geq R$$

$$\sum_{i\leq n}s_i\left(\sum_{j\in I_i}\mu_{ij}x_{ij}+\sum_{j\in J_i}\vartheta_{ij}x_{ij}\right)=\log(w),$$

and observe that $w\geq R$ implies $\log(w)\geq\log(R)$ because the log function is monotonically increasing, so the last two constraints can be grouped into a simpler one not involving logarithms of problem variables:

$$\sum_{i\leq n}s_i\left(\sum_{j\in I_i}\mu_{ij}x_{ij}+\sum_{j\in J_i}\vartheta_{ij}x_{ij}\right)\geq\log(R).$$

2. We now make use of the fact that N_{ij} is an integer variable for all $i\leq n, j\in J_i$, and apply INT2BIN(P, N_{ij}). For $k\in\{0,\ldots,N\}$ we add assignment

variables ν_{ij}^k so that $\nu_{ij}^k = 1$ if $N_{ij} = k$ and 0 otherwise. Now for all $k \in [0, \ldots, N]$ we compute the constants $\vartheta^k = \frac{p_{ij} b_{ij} (1 - b_{ij})^{(1 - b_{ij})k}}{(1 - p_{ij}) + p_{ij}(1 - b_{ij})^{(1 - b_{ij})k}}$, which we add to the problem parameters. We remove the constraints defining ϑ_{ij} in function of N_{ij}: since the following constraints are valid:

$$\forall i \leq n, j \in J_i \quad \sum_{k \leq N} \nu_{ij}^k = 1 \tag{12}$$

$$\forall i \leq n, j \in J_i \quad \sum_{k \leq N} k \nu_{ij}^k = N_{ij} \tag{13}$$

$$\forall i \leq n, j \in J_i \sum_{k \leq N} \vartheta^k \nu_{ij}^k = \vartheta_{ij}, \tag{14}$$

the second constraints are used to replace N_{ij} and the third to replace ϑ_{ij}. The first constraints are added to the formulation. We obtain:

$$\min \sum_{i \leq n} \left(\sum_{j \in I_i} c_{ij} x_{ij} + \sum_{j \in J_i} \bar{c}_{ij} (t_{ij} + \tau_{ij} \sum_{k \leq N} k \nu_{ij}^k) x_{ij} \right)$$

$$\forall i \leq n \quad \sum_{j \in I_i \cup J_i} x_{ij} = 1$$

$$\forall i \leq n \quad \sum_{j \in I_i} d_{ij} x_{ij} + \sum_{j \in J_i} (t_{ij} + \tau_{ij} \sum_{k \leq N} k \nu_{ij}^k) x_{ij} \leq T$$

$$\sum_{i \leq n} s_i \left(\sum_{j \in I_i} \mu_{ij} x_{ij} + \sum_{j \in J_i} x_{ij} \sum_{k \leq N} \vartheta^k \nu_{ij}^k \right) \geq \log(R)$$

$$\forall i \leq n, j \in J_i \quad \sum_{k \leq N} \nu_{ij}^k = 1.$$

3. We distribute products over sums in the formulation to obtain the binary product sets $\{x_{ij} \nu_{ij}^k \mid k \leq N\}$ for all $i \leq n, j \in J_i$: by repeatedly applying the PRODBIN reformulation to all binary products of binary variables, we get a MILP opt-reformulation Q of P where all the variables are binary.

We remark that the MILP opt-reformulation Q derived above has a considerably higher cardinality than $|P|$. More compact reformulations are applicable in step 3 because of the presence of the assignment constraints (see Sect. 3.4).

Reformulation Q essentially rests on linearization variables w_{ij}^k which replace the quadratic terms $x_{ij} \nu_{ij}^k$ throughout the formulation. A semantic interpretation of step 3 is as follows. We notice that for $i \leq n, j \in J_i$, if $x_{ij} = 1$, then $x_{ij} = \sum_k \nu_{ij}^k$ (because only one value k will be selected), and if $x_{ij} = 0$, then $x_{ij} = \sum_k \nu_{ij}^k$ (because no value k will be selected). This means that

$$\forall i \leq n, j \in J_i \quad x_{ij} = \sum_{k \leq N} \nu_{ij}^k \tag{15}$$

is a valid problem constraint. We use it to replace x_{ij} everywhere in the formulation where it appears with $j \in I_i$, obtaining a opt-reformulation with x_{ij} for $j \in I_i$ and quadratic terms $\nu_{ij}^k \nu_{lp}^h$. Now, because of (12), these are zero when $(i,j) \neq (l,p)$ or $k \neq h$ and are equal to ν_{ij}^k when $(i,j) = (l,p)$ and $k = h$, so they can be linearized exactly by replacing them by either 0 or ν_{ij}^k according to their indices. What this really means is that the reformulation Q, obtained through a series of automatic reformulation steps, is a semantically different formulation defined in terms of the following decision variables:

$$\forall i \leq n, j \in I_i \quad x_{ij} = \begin{cases} 1 & \text{if } j \in I_i \text{ is assigned to } i \\ 0 & \text{otherwise.} \end{cases}$$

$$\forall i \leq n, j \in J_i, k \leq N \quad \nu_{ij}^k = \begin{cases} 1 \text{ if } j \in J_i \text{ is assigned to } i \text{ and there are } k \text{ tests to be performed} \\ 0 \quad \text{otherwise.} \end{cases}$$

This is an important hint to the importance of automatic reformulation in problem analysis: it is a syntactical operation, the result of which, when interpreted, can suggest a new meaning.

4 Relaxations

Loosely speaking, a relaxation of a problem P is an auxiliary problem of P whose feasible region is larger; often, relaxations are obtained by simply removing constraints from the formulation. Relaxations are useful because they often yield problems which are simpler to solve yet they provide a bound on the objective function value at the optimum.

Such bounds are mainly used in Branch-and-Bound type algorithms, which are the most common exact or ε-approximate (for a given $\varepsilon > 0$) solution algorithms for MILPs, nonconvex NLPs and MINLPs. Although the variants for solving MILPs, NLPs and MINLPs are rather different, they all conform to the same implicit enumeration search type. Lower and upper bounds are computed for the problem over the current variable domains. If the bounds are sufficiently close, a global optimum was found in the current domain: store it if it improves the incumbent (i.e. the current best optimum). Otherwise, partition the domain and recurse over each subdomain in the partition. Should a bound be worse off than the current incumbent during the search, discard the domain immediately without recursing on it. Under some regularity conditions, the recursion terminates. The Branch-and-Bound algorithm has been used on combinatorial optimization problems since the 1950s [6]. Its first application to nonconvex NLPs is [33]. More recently, Branch-and-Bound has evolved into Branch-and-Cut and Branch-and-Price for MILPs [94, 133, 52], which have been used to solve some practically difficult problems such as the Travelling Salesman Problem (TSP) [12]. Some recent MINLP-specific Branch-and-Bound approaches are [102, 10, 4, 5, 114, 124, 71].

A further use of bounds provided by mathematical programming formulations is to evaluate the performance of heuristic algorithms without an approximation guarantee [28]. Bounds are sometimes also used to guide heuristics [99].

In this section we define relaxations and review the most useful ones. In Sect. 4.1 we give some basic definitions. We then list elementary relaxations in Sect. 4.2 and more advanced ones in Sect. 4.3. We discuss relaxation strengthening in Sect. 4.4.

4.1 Definitions

Consider an optimization problem $P = (\mathcal{P}, \mathcal{V}, \mathcal{E}, \mathcal{O}, \mathcal{C}, \mathcal{B}, \mathcal{T})$ and let Q be such that: $\mathcal{P}(Q) \supseteq \mathcal{P}(P)$, $\mathcal{V}(Q) = \mathcal{V}(P)$, $\mathcal{E}(Q) \supseteq \mathcal{E}(P)$ and $\mathcal{O}(Q) = \mathcal{O}(P)$.

We first define relaxations in full generality.

Definition 13. *Q is a relaxation of P if (a)* $\mathcal{F}(P) \subseteq \mathcal{F}(Q)$; *(b) for all* $(f, d) \in \mathcal{O}(P)$, $(\bar{f}, \bar{d}) \in \mathcal{O}(Q)$ *and* $x \in \mathcal{F}(P)$, $\bar{d}\bar{f}(x) \geq df(x)$.

Defn. 13 is not used very often in practice because it does not say anything on how to construct Q. The following elementary relaxations are more useful.

Definition 14. *Q is a:*

- *constraint relaxation of P if* $\mathcal{C}(P) \subsetneq \mathcal{C}(Q)$;
- *bound relaxation of P if* $\mathcal{B}(P) \subsetneq \mathcal{B}(Q)$;
- *a continuous relaxation of P if* $\exists v \in \mathcal{V}(P)$ $(\mathcal{T}(v) > 0)$ *and* $\mathcal{T}(v) = 0$ *for all* $v \in \mathcal{V}(Q)$.

4.2 Elementary Relaxations

We shall consider two types of elementary relaxations: the continuous relaxation and the convex relaxation. The former is applicable to MILPs and MINLPs, and the latter to (nonconvex) NLPs and MINLPs. They are both based on the fact that whereas solving MILPs and MINLPs is considered difficult, there are efficient algorithms for solving LPs and convex NLPs. Since the continuous relaxation was already defined in Defn. 14 and trivially consists in considering integer/discrete variables as continuous ones, in the rest of this section we focus on convex relaxations.

Formally (and somewhat obviously), Q is a *convex relaxation* of a given problem P if Q is a relaxation of P and Q is convex. Associated to all sBB in the literature there is a (nonconvex) NLP or MINLP in standard form, which is then used as a starting point for the convex relaxation.

Outer approximation

Outer approximation (OA) is a technique for defining a polyhedral approximation of a convex nonlinear feasible region, based on computing tangents to the convex feasible set at suitable boundary points [31, 35, 57]. An outer approximation relaxation relaxes a convex NLP to an LP, (or a MINLP to a MILP) and is really a "relaxation scheme" rather than a relaxation: since the tangents to *all* boundary points of a convex set define the convex set itself, any choice of (finite) set of boundary points of the convex can be used to define a different outer approximation. OA-based optimization algorithms identify sets of boundary points that eventually guarantee that the outer approximation will be exact near the optimum. In [57], the following convex MINLP is considered:

$$\left.\begin{array}{c} \min L_0(x) + cy \\ \text{s.t. } L(x) + By \leq 0 \\ x^L \leq x \leq x^U \\ y \in \{0,1\}^q, \end{array}\right\} \tag{16}$$

where $L_0 : \mathbb{R}^n \to \mathbb{R}$, $L : \mathbb{R}^n \to \mathbb{R}^m$ are convex once-differentiable functions, $c \in \mathbb{R}^q$, B is an $m \times q$ matrix. For a given $y' \in \{0,1\}^q$, let $P(y')$ be (16) with y fixed at y'. Let $\{y^j\}$ be a sequence of binary q-vectors. Let $T = \{j \mid P(y^j)$ is feasible with solution $x^j\}$. Then the following is a MILP outer approximation for (16):

$$\left.\begin{array}{c} \min_{x,y,\eta} \qquad\qquad\qquad\qquad \eta \\ \forall j \in T \; L_0(x^j) + \nabla L_0(x^j)(x - x^j) + cy \leq \eta \\ \forall j \quad L(x^j) + \nabla L(x^j)(x - x^j) + By \leq 0 \\ x^L \leq x \leq x^U \\ y \in \{0,1\}^q, \end{array}\right\}$$

where x^j is the solution to $F(y^j)$ (defined in [35]) whenever $P(y^j)$ is infeasible. This relaxation is denoted by OUTERAPPROX(P, T).

αBB convex relaxation

The αBB algorithm [10, 4, 5, 36] targets single-objective NLPs where the expressions in the objective and constraints are twice-differentiable. The convex relaxation of the problem P:

$$\left.\begin{array}{c} \min_x \; f(x) \\ \text{s.t. } g(x) \leq 0 \\ h(x) = 0 \\ x^L \leq x \leq x^U \end{array}\right\} \tag{17}$$

is obtained as follows.

1. Apply the EQ2INEQ reformulation (see Sect. 3.2) to each nonlinear equality constraint in C, obtaining an opt-reformulation P_1 of P.
2. For every nonconvex inequality constraint $c = (e_c, s_c, b_c) \in C(P_1)$:

   a. if the root node r of the expression tree e_c is a sum operator, for every subnode $s \in \delta^+(r)$ replace s with a specialized convex underestimator if s is a bilinear, trilinear, linear fractional, fractional trilinear, univariate concave term. Otherwise replace with α-underestimator;
   b. otherwise, replace r with a specialized if s is a bilinear, trilinear, linear fractional, fractional trilinear, univariate concave term. Otherwise replace with α-underestimator.

The specialized underestimators are as follows: McCormick's envelopes for bilinear terms [91, 7], the second-level RLT bound factor linearized products [108, 107, 104] for trilinear terms, and a secant underestimator for univariate concave terms. Fractional terms are dealt with by extending the bilinear/trilinear underestimators to bilinear/trilinear products of univariate functions and then noting that $x/y = \phi_1(x)\phi_2(y)$ where ϕ_1 is the identity and $\phi_2(y) = 1/y$ [88]. Recently, the convex underestimator for trilinear terms have been replaced with the convex envelopes [92].

The general-purpose α-underestimator:

$$\alpha(x^L - x)^\top (x^U - x) \tag{18}$$

is a quadratic convex function that for suitable values of α is "convex enough" to overpower the generic nonconvex term. This occurs for

$$\alpha \geq \max\{0, -\frac{1}{2} \min_{x^L \leq x \leq x^U} \lambda(x)\},$$

where $\min \lambda(x)$ is the minimum eigenvalue of the Hessian of the generic nonconvex term in function of the problem variables.

The resulting αBB relaxation Q of P is a convex NLP. This relaxation is denoted by αBBRELAX(P).

Branch-and-Contract convex relaxation

The convex relaxation is used in the Branch-and-Contract algorithm [134], targeting nonconvex NLPs with twice-differentiable objective function and constraints. This relaxation is derived essentially in the same way as for the αBB convex relaxation. The differences are:

- the problem is assumed to only have inequality constraints of the form $c = (e_c, -1, 0)$;
- each function (in the objective and constraints) consists of a sum of nonlinear terms including: bilinear, linear fractional, univariate concave, and generic convex.

The convex relaxation is then constructed by replacing each nonconvex non-linear term in the objective and constraints by a corresponding envelope or relaxation. The convex relaxation for linear fractional term had not appeared in the literature before [134].

Symbolic reformulation based convex relaxation

This relaxation is used in the symbolic reformulation spatial Branch-and-Bound algorithm proposed in [113, 114]. It can be applied to all NLPs and MINLPs for which a convex underestimator and a concave overestimator are available. It consists in reformulating P to the Smith standard form (see Sect. 2.3) and then replacing every defining constraint with the convex and concave under/over-estimators. In his Ph.D. thesis [112], Smith had tried both NLP and LP convex relaxations, finding that LP relaxations were more reliable and faster to compute, although of course with slacker bounds. The second implementation of the sBB algorithm he proposed is described in [69, 71] and implemented in the $oo\mathcal{OPS}$ software framework [82]. Both versions of this algorithm consider under/overestimators for the following terms: bilinear, univariate concave, univariate convex (linear fractional being reformulated to bilinear). The second version also included estimators for piecewise convex/concave terms. One notable feature of this relaxation is that it can be adapted to deal with more terms. Some recent work in polyhedral envelopes, for example [119], gives conditions under which the sum of the envelopes is the envelope of the sum: this would yield a convex envelope for a sum of terms. It would then suffice to provide for a defining constraint in the Smith standard form linearizing the corresponding sum. The Smith relaxation is optionally strengthened via LP-based optimality and feasibility based range reduction techniques. After every range reduction step, the convex relaxation is updated with the new variable ranges in an iterative fashion until no further range tightening occurs [112, 69, 71].

This relaxation, denoted by SMITHRELAX(P) is at the basis of the sBB solver [71] in the $oo\mathcal{OPS}$ software framework [82], which was used to obtain solutions of many different problem classes: pooling and blending problems [48, 81], distance geometry problems [60, 62], and a quantum chemistry problem [63, 78].

BARON's convex relaxation

BARON (Branch And Reduce Optimization Navigator) is a commercial Branch-and-Bound based global optimization solver (packaged within the GAMS [23] modelling environment) which is often quoted as being the *de facto* standard solver for MINLPs [124, 123]. Its convex relaxation is derived essentially in the same way as for the symbolic reformulation based convex relaxation. The differences are:

- better handling of fractional terms [120, 121]

- advanced range reduction techniques (optimality, feasibility and duality based, plus a learning reduction heuristic)
- optionally, an LP relaxation is derived via outer approximation.

4.3 Advanced Relaxations

In this section we shall describe some more advanced relaxations, namely the Lagrangian relaxation, the semidefinite relaxation, the reformulation-linearization technique and the signomial relaxation.

Lagrangian relaxation

Consider a MINLP

$$\left. \begin{aligned} f^* = \min_x \; & f(x) \\ \text{s.t.} \; & g(x) \le 0 \\ & x \in X \subseteq \mathbb{R}^n, \end{aligned} \right\} \tag{19}$$

where $f : \mathbb{R}^n \to \mathbb{R}$ and $g : \mathbb{R}^n \to \mathbb{R}^m$ are continuous functions and X is an arbitrary set. The Lagrangian relaxation consists in "moving" the weighted constraints to the objective function, namely:

$$\left. \begin{aligned} L(\mu) = \inf_x \; & f(x) + \mu^\top g(x) \\ & x \in X \subseteq \mathbb{R}^n, \end{aligned} \right\}$$

for some nonnegative $\mu \in \mathbb{R}^m_+$. For all $x \in X$ with $g(x) \le 0$, we have $\mu^\top g(x) \le 0$, which implies $L(\mu) \le f^*$ for all $\mu \ge 0$. In other words, $L(\mu)$ provides a lower bound to (19) for all $\mu \ge 0$. Thus, we can improve the tightness of the relaxation by solving the Lagrangian problem

$$\max_{\mu \ge 0} L(\mu), \tag{20}$$

(namely, we attempt to find the largest possible lower bound). If (19) is an LP problem, it is easy to show that the Lagrangian problem (20) is the dual LP problem. In general, solving (20) is not a computationally easy task [95]. However, one of the nice features of Lagrangian relaxations is that they provide a lower bound for each value of $\mu \ge 0$, so (20) does not need to be solved at optimality. Another useful feature is that any subset of problem constraints can be relaxed, for X can be defined arbitrarily. This is useful for problems that are almost block-separable, i.e. those problems that can be decomposed in some independent subproblems bar a few constraints involving all the problem variables (also called complicating constraints). In these cases, one considers a Lagrangian relaxation of the complicating constraints and then solves a block-separable Lagrangian problem. This approach is called Lagrangian decomposition.

The Lagrangian relaxation has some interesting theoretical properties: (a) for convex NLPs it is a global reformulation [22]; (b) for MILPs, it is at least as tight as the continuous relaxation [133]; (c) for MINLPs, under some conditions (i.e. some constraint qualification and no equality constraints) it is at least as tight as any convex relaxation obtained by relaxing each nonconvex term or each constraint one by one [51], such as all those given in Sect. 4.2. Further material on the use of Lagrangian relaxation in NLPs and MINLPs can be found in [95, 51].

Consider a problem P such that $\mathcal{O}(P) = \{(e_o, d_o)\}$ and a subset of constraints $C \subseteq \mathcal{C}(P)$. A Lagrangian relaxation of C in P (denoted by $\text{LAGREL}(P, C)$) is a problem Q defined as follows.

- $\mathcal{V}(Q) = \mathcal{V}(P)$, $\mathcal{B}(Q) = \mathcal{B}(P)$, $\mathcal{T}(Q) = \mathcal{T}(P)$,
- $\mathcal{P}(Q) = \mathcal{P}(P) \cup \{\mu_c \mid c \in C\}$,
- $\mathcal{C}(Q) = \mathcal{C}(P) \smallsetminus C$,
- $\mathcal{O}(Q) = \{(e'_o, d'_o)\}$, where $e'_o = e_o + \sum_{c \in C} \mu_c c$.

The Lagrangian problem cannot itself be defined in the data structure of Defn. 1, for the max operator is only part of $O_{\mathcal{L}}$ as long as it has a finite number of arguments.

Semidefinite relaxation

As was pointed out in Sect. 2.3, SDPs provide very tight relaxations for quadratically constrained quadratic MINLPs (QCQP). A QCQP in general form is as follows [11]:

$$\left.\begin{array}{rl} \min_x & x^\top Q_0 x + a_0^\top x \\ \forall i \in I & x^\top Q_i x + a_i^\top x \leq b_i \\ \forall i \in E & x^\top Q_i x + a_i^\top x = b_i \\ & x^L \leq x \leq x^U \\ \forall j \in J & x_i \in \mathbb{Z}, \end{array}\right\} \tag{21}$$

where $I \cup E = \{1, \ldots, m\}$, $J \subseteq \{1, \ldots, n\}$, $x \in \mathbb{R}^n$, Q_i is an $n \times n$ symmetric matrix for all $i \leq m$. For general matrices Q_i and $J \neq \emptyset$, the QCQP is nonconvex. Optionally, the integer variables can be reformulated exactly to binary (see INT2BIN, Sect. 3.2) and subsequently to continuous (see BIN2CONT, Sect. 3.2) via the introduction of the constraints $x_i^2 - x_i = 0$ for all $i \in J$: since these constraints are quadratic, they can be accommodated in formulation (21) by suitably modifying the Q_i matrices. Many important applications can be modelled as QCQPs, including graph bisection (see Sect. 2.1) and graph partitioning [72], scheduling with communication delays [28], distance geometry problems such as the KNP (see Sect. 2.1) [60] and the Molecular Distance Geometry Problem (MDGP) [62, 77], pooling and blending problems from the oil industry [48, 81] and so on. The SDP relaxation of the QCQP, denoted by SDPRELAX(P) is constructed as follows:

- replace all quadratic products $x_i x_j$ in (21) with an added linearization variable X_{ij}
- form the matrix $X = (X_{ij})$ and the variable matrix

$$\bar{X} = \begin{pmatrix} 1 & x^\top \\ x & X \end{pmatrix}$$

- for all $0 \le i \le m$ form the matrices

$$\bar{Q}_i = \begin{pmatrix} -b_i & a_i^\top/2 \\ a_i/2 & Q_i \end{pmatrix}$$

- the following is an SDP relaxation for QCQP:

$$\left. \begin{array}{r} \min_X \bar{Q}_0 \bullet \bar{X} \\ \forall i \in I \;\; \bar{Q}_i \bullet \bar{X} \le 0 \\ \forall i \in E \;\; \bar{Q}_i \bullet \bar{X} = 0 \\ x^L \le x \le x^U \\ \bar{X} \succeq 0. \end{array} \right\} \tag{22}$$

As for the SDP standard form of Sect. 2.3, the SDP relaxation can be easily represented by the data structure described in Defn. 1.

Reformulation-Linearization Technique

The Reformulation-Linearization Technique (RLT) is a relaxation method for mathematical programming problems with quadratic terms. The RLT linearizes all quadratic terms in the problem and generates valid linear equation and inequality constraints by considering multiplications of bound factors (terms like $x_i - x_i^L$ and $x_i^U - x_i$) and constraint factors (the left hand side of a constraint such as $\sum_{j=1}^n a_j x_j - b \ge 0$ or $\sum_{j=1}^n a_j x_j - b = 0$). Since bound and constraint factors are always non-negative, so are their products: this way one can generate sets of valid problem constraints. In a sequence of papers published from the 1980s onwards (see e.g. [2, 108, 110, 107, 103, 111, 109]), RLT-based relaxations were derived for many different classes of problems, including IPs, NLPs, MINLPs in general formulation, and several real-life applications. It was shown that the RLT can be used in a lift-and-project fashion to generate the convex envelope of binary and general discrete problems [106, 3].

Basic RLT

The RLT consists of two symbolic manipulation steps: reformulation and linearization. The reformulation step is a reformulation in the sense of Defn. 10. Given a problem P, the reformulation step produces a reformulation Q' where:

- $\mathcal{P}(Q') = \mathcal{P}(P)$;
- $\mathcal{V}(Q') = \mathcal{V}(P)$;

- $\mathcal{E}(Q') \supseteq \mathcal{E}(P)$;
- $\mathcal{C}(Q') \supseteq \mathcal{C}(P)$;
- $\mathcal{O}(Q') = \mathcal{O}(P)$;
- $\mathcal{B}(Q') = \mathcal{B}(P)$;
- $\mathcal{T}(Q') = \mathcal{T}(P)$;
- $\forall x, y \in \mathcal{V}(P)$, add the following constraints to $\mathcal{C}(Q')$:

$$(x - L_x)(y - L_y) \geq 0 \tag{23}$$
$$(x - L_x)(U_y - y) \geq 0 \tag{24}$$
$$(U_x - x)(y - L_y) \geq 0 \tag{25}$$
$$(U_x - x)(U_y - y) \geq 0; \tag{26}$$

- $\forall x \in \mathcal{V}(P), c = (e_c, s_c, b_c) \in \mathcal{C}(P)$ such that e_c is an affine form, $s_c = 1$ and $b_c = 0$ (we remark that all linear inequality constraints can be easily reformulated to this form, see Sect. 3.2), add the following constraints to $\mathcal{C}(Q')$:

$$e_c(x - L_x) \geq 0 \tag{27}$$
$$e_c(U_x - x) \geq 0; \tag{28}$$

- $\forall x \in \mathcal{V}(P), c = (e_c, s_c, b_c) \in \mathcal{C}(P)$ such that e_c is an affine form, $s_c = 0$ and $b_c = 0$ (we remark that all linear equality constraints can be trivially reformulated to this form), add the following constraint to $\mathcal{C}(Q')$:

$$e_c x = 0. \tag{29}$$

Having obtained Q', we proceed to linearize all the quadratic products engendered by (23)-(29). We derive the auxiliary problem Q from Q' by reformulating Q' to Smith's standard form (see Sect. 2.3) and then performing a constraint relaxation with respect to all defining constraints. Smith's standard form is a reformulation of the lifting type, and the obtained constraint relaxation Q is a MILP whose optimal objective function value $\bar{f}$ is a bound to the optimal objective function value f^* of the original problem P. The bound obtained in this way is shown to dominate, or be equivalent to, several other bounds in the literature [3]. This relaxation is denoted by $\mathrm{RLTRELAX}(P)$.

We remark in passing that (23)-(26), when linearized by replacing the bilinear term xy with an added variable w, are also known in the literature as McCormick relaxation, as they were first proposed as a convex relaxation of the nonconvex constraint $w = xy$ [91], shown to be the convex envelope [7], and widely used in spatial Branch-and-Bound (sBB) algorithms for global optimization [114, 4, 5, 124, 71]. RLT constraints of type (29) have been the object of further research showing their reformulating power [67, 68, 70, 81, 72] (also see Sect. 3.4, where we discuss compact linearization of binary quadratic problems and reduced RLT constraints).

RLT Hierarchy

The basic RLT method can be extended to provide a hierarchy of relaxations, by noticing that we can form valid RLT constraints by multiplying sets of bound and constraint factors of cardinality higher than 2, and then projecting the obtained constraints back to the original variable space. In [106, 3] it is shown that this fact can be used to construct the convex hull of an arbitrary MILP P. For simplicity, we only report the procedure for MILP in standard canonical form (see Sect. 2.3) where all discrete variables are binary, i.e. $\mathcal{T}(v) = 2$ for all $v \in \mathcal{V}(P)$. Let $|\mathcal{V}(P)| = n$. For all integer $d \leq n$, let P_d be the relaxation of P obtained as follows:

- for all linear constraint $c = (e_c, 1, 0) \in \mathcal{C}(P)$, subset $V \subseteq \mathcal{V}(P)$ and finite binary sequence B with $|V| = |B| = d$ such that B_x is the x-th term of the sequence for $x \in V$, add the valid constraint:

$$e_c \left(\prod_{\substack{x \in V \\ B_x = 0}} x \right) \left(\prod_{\substack{x \in V \\ B_x = 1}} (1 - x) \right) \geq 0; \qquad (30)$$

  we remark that (30) is a multivariate polynomial inequality;
- for all monomials of the form

$$a \prod_{x \in J \subseteq \mathcal{V}(P)} x$$

  with $a \in \mathbb{R}$ in a constraint (30), replace $\prod_{x \in J} x$ with an added variable w_J (this is equivalent to relaxing a defining constraint $w_J = \prod_{x \in J}$ in the Smith's standard form restricted to (30)).

Now consider the projection X_d of P_d in the $\mathcal{V}(P)$ variable space (see Sect. 3.2). It can be shown that

$$\text{conv}(\mathcal{F}(P)) \subseteq \mathcal{F}(X_n) \subseteq \mathcal{F}(X_{n-1}) \ldots \subseteq \mathcal{F}(X_1) \subseteq \mathcal{F}(P).$$

We recall that for a set $Y \subseteq \mathbb{R}^n$, $\text{conv}(Y)$ is defined as the smallest convex subset of $\mathbb{R}^n$ containing Y.

A natural practical application of the RLT hierarchy is to generate relaxations for polynomial programming problems [103], where the various multivariate monomials generated by the RLT hierarchy might already be present in the problem formulation.

Signomial programming relaxations

A signomial programming problem is an optimization problem where every objective function is a signomial function and every constraint is of the form

$c = (g, s, 0)$ where g is a signomial function of the problem variables, and $s \neq 0$ (so signomial equality constraints must be reformulated to pairs of inequality constraints as per the Eq2Ineq reformulation of Sect. 3.2). A *signomial* is a term of the form:

$$a \prod_{k=1}^{K} x_k^{r_k}, \tag{31}$$

where $a, r_k \in \mathbb{R}$ for all $k \in K$, and the r_k exponents are assumed ordered so that $r_k > 0$ for all $k \leq m$ and $r_k < 0$ for $m \leq k \leq K$. Because the exponents of the variables are real constants, this is a generalization of a multivariate monomial term. A *signomial function* is a sum of signomial terms. In [19], a set of transformations of the form $x_k = f_k(z_k)$ are proposed, where x_k is a problem variable, z_k is a variable in the reformulated problem and f_k is suitable function that can be either exponential or power. This yields an opt-reformulation where all the inequality constraints are convex, and the variables z and the associated (inverse) defining constraints $x_k = f_k(z_k)$ are added to the reformulation for all $k \in K$ (over each signomial term of each signomial constraint).

We distinguish the following cases:

- If $a > 0$, the transformation functions f_k are exponential univariate, i.e. $x_k = e^{z_k}$. This reformulates (31) as follows:

$$\left. \begin{array}{c} a \dfrac{e^{\sum_{k \leq m} r_k z_k}}{\prod_{k=m+1}^{K} x_k^{|r_k|}} \\ \forall k \leq K \quad x_k = e^{z_k}. \end{array} \right\}$$

- If $a < 0$, the transformation functions are power univariate, i.e. $x_k = z_k^{\frac{1}{R}}$ for $k \leq m$ and $x_k = z_k^{-\frac{1}{R}}$ for $k > m$, where $R = \sum_{k \leq K} |r_k|$. This is also called a *potential transformation*. This reformulates (31) as follows:

$$\left. \begin{array}{c} a \prod_{k \leq K} z_k^{\frac{|r_k|}{R}} \\ \forall k \leq m \quad x_k = z_k^{\frac{1}{R}} \\ \forall k > m \quad x_k = z_k^{-\frac{1}{R}} \\ R = \sum_{k \leq K} |r_k|. \end{array} \right\}$$

This opt-reformulation isolates all nonconvexities in the inverse defining constraints. These are transformed as follows:

$$\forall k \leq K \ x_k = e^{z_k} \rightarrow \forall k \leq K \ z_k = \log x_k$$
$$\forall k \leq m \quad z_k = x_k^{R}$$
$$\forall k > m \quad z_k = x_k^{-R},$$

and then relaxed using a piecewise linear approximation as per Fig. 4. This requires the introduction of binary variables (one per turning point).

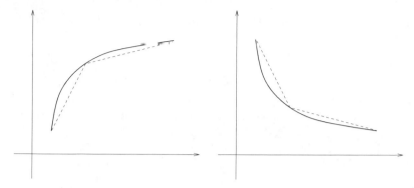

Fig. 4 Piecewise linear underestimating approximations for concave (left) and convex (right) univariate functions

The signomial relaxation is a convex MINLP; it can be further relaxed to a MILP by outer approximation of the convex terms, or to a convex NLP by continuous relaxation of the discrete variables. This relaxation is denoted by SIGNOMIALRELAX(P).

4.4 Valid Cuts

Once a relaxation has been derived, it should be strengthened (i.e. it should be modified so that the deriving bound becomes tighter). This is usually done by tightening the relaxation, i.e. by adding inequalities. These inequalities have the property that they are redundant with respect to the original (or reformulated) problem but they are not redundant with respect to the relaxation. Thus, they tighten the relaxation but do not change the original problem. In this section we discuss such inequalities for MILPs, NLPs and MINLPs.

Definition 15. *Given an optimization problem P and a relaxation Q, a valid inequality is a constraint $c = (e_c, s_c, b_c)$ such that the problem Q' obtained by Q from adding c to $\mathcal{C}(Q)$ has $\mathcal{F}(P) \subseteq \mathcal{F}(Q')$.*

Naturally, because Q can be seen as a constraint relaxation of Q', we also have $\mathcal{F}(Q') \subseteq \mathcal{F}(Q)$. Linear valid inequalities are very important as adding a linear inequality to an optimization problem usually does not significantly alter the solution time.

For any problem P and any $c \in \mathcal{C}(P)$, let $\mathcal{F}_c$ be the set of points in $\mathbb{R}^n$ that satisfy c. Let Q be a relaxation of P.

Definition 16. *A linear valid inequality c is a valid cut if there exists $y \in Q$ such that $y \notin \mathcal{F}_c$.*

Valid cuts are linear valid inequalities that "cut away" a part of the feasible region of the relaxation. They are used in two types of algorithms: cutting plane algorithms and Branch-and-Bound algorithms. The typical iteration of a cutting plane algorithm solves a problem relaxation Q (say with solution x'), derives a valid cut that cuts away x'; the cut is then added to the relaxation and the iteration is repeated. Convergence is attained when $x' \in \mathcal{F}(P)$. Cutting plane algorithms were proposed for MILPs [43] but then deemed to be too slow for practical purposes, and replaced by Branch-and-Bound. Cutting plane algorithms were also proposed for convex [56] and bilinear [59] NLPs, and pseudoconvex MINLPs [132, 131].

Valid cuts for MILPs

This is possibly the area of integer programming where the highest number of papers is published annually. It would be outside the scope of this chapter to relate on all valid cuts for MILPs, so we limit this section to a brief summary. The most effective cutting techniques usually rely on problem structure. See [94], Ch. II.2 for a good technical discussion on the most standard techniques, and [89, 90, 54] for recent interesting group-theoretical approaches which are applicable to large subclasses of IPs. Valid inequalities are generated by all relaxation hierarchies (like e.g. Chvátal-Gomory [133] or Sherali-Adams' [107]). The best known general-purpose valid cuts are the Gomory cuts [43], for they are simple to define and can be written in a form suitable for straightforward insertion in a simplex tableau; many strengthenings of Gomory cuts have been proposed (see e.g. [64]). Lift-and-project techniques are used to generate new cuts from existing inequalities [15]. Families of valid cuts for general Binary Integer Programming (BIP) problems have been derived, for example, in [16, 84], based on geometrical properties of the definition hypercube $\{0, 1\}^n$. In [16], inequalities defining the various faces of the unit hypercube are derived. The cuts proposed in [84] are defined by finding a suitable hyperplane separating a unit hypercube vertex $\bar{x}$ from its adjacent vertices. Intersection cuts [14] are defined as the hyperplane passing through the intersection points between the smallest hypersphere containing the unit hypercube and n half-lines of a cone rooted at the current relaxed solution of Q. Spherical cuts are similar to intersection cuts, but the considered sphere is centered at the current relaxed solution, with radius equal to the distance to the nearest integral point [74]. In [21], Fenchel duality arguments are used to find the maximum distance between the solution of Q and the convex hull of the $\mathcal{F}(P)$; this gives rise to provably deep cuts called *Fenchel cuts*. See [25] for a survey touching on the most important general-purpose MILP cuts, including Gomory cuts, Lift-and-project techniques, Mixed Integer Rounding (MIR) cuts, Intersection cuts and Reduce-and-split cuts.

Valid cuts for NLPs

Valid cuts for NLPs with a single objective function f subject to linear constraints are described in [50] (Ch. III) when an incumbent x^* with $f(x^*) = \gamma$

Fig. 5 A γ-valid cut

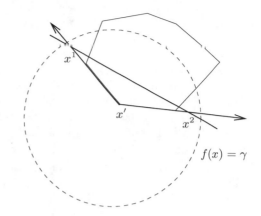

is known, in order to cut away feasible points x' with $f(x') > \gamma$. Such cuts are called γ-*valid cuts*. Given a nondegenerate vertex x' of the feasible polyhedron for which $f(x') > \gamma$, we consider the n polyhedron edges emanating from x'. For each $i \leq n$ we consider a point x^i on the i-th edge from x' such that $f(x^i) \geq \gamma$. The hyperplane passing through the intersection of the x^i is a γ-valid cut (see Fig. 5). More precisely, let Q be the matrix whose i-th column is $x^i - x'$ and e the unit n-vector. Then by [50] Thm. III.1 $eQ^{-1}(x - x') \geq 1$ defines a γ-valid cut. Under some conditions, we can find x^i such that $f(x) = x^i$ and define the strongest possible γ-valid cut, also called *concavity cut*.

The idea for defining γ-valid cuts was first proposed in [128]; this was applied to 0-1 linear programs by means of a simple reformulation in [100]. It is likely that this work influenced the inception of intersection cuts [14] (see Sect. 4.4), which was then used as the basis for current work on Reduce-and-Split cuts [9].

Some valid cuts for pseudoconvex optimization problems are proposed in [132]. An optimization problem is pseudoconvex if the objective function is a linear form and the constraints are in the form $c = (g, -1, 0)$ where $g(x)$ is a pseudoconvex function of the problem variable vector x. A function $g : S \subseteq \mathbb{R}^n \to \mathbb{R}$ is *pseudoconvex* if for all $x, y \in S$, $g(x) < g(y)$ implies $\nabla g(y)(x - y) < 0$. So it follows that for each $x, y \in S$ with $g(y) > 0$, there is a constant $\alpha \geq 1$ such that

$$g(y) + \alpha(\nabla g(y))(x - y) \leq g(x) \tag{32}$$

is a (linear) outer approximation to the feasible region of the problem. If g is convex, $\alpha = 1$ suffices.

In [95], Ch. 7 presents a non-exhaustive list of NLP cuts, applicable to a MINLP standard form ([95] Eq. (7.1): minimization of a linear objective subject to linear inequality constraints and nonlinear inequality constraints): linearization cuts (outer approximation, see Sect. 4.2), knapsack cuts (used for improving loose convex relaxations of given constraints), interval-gradient cuts (a linearization carried out on an interval where the gradient of a given constraint is defined), Lagrangian cuts (derived by solving Lagrangian subproblems), level cuts (defined for a given objective function upper bound), deeper cuts (used to tighten loose Lagrangian relaxation; they involve the solution of separation problems involving several variable blocks).

Another NLP cut based on the Lagrangian relaxation is proposed in [124]: consider a MINLP in the canonical form $\min_{g(x) \leq 0} f(x)$ and let $L(\cdot, \mu) = f(x) + \mu^\top g(x)$ be its Lagrangian relaxation. Let $\underline{f}$ be a lower bound obtained by solving L and $\bar{f}$ be an upper bound computed by evaluating f at a feasible point x'. From $\underline{f} \leq f(x) + \mu^\top g(x) \leq \bar{f} + \mu^\top g(x)$ one derives the valid cut $g_i(x) \geq -\frac{1}{\mu_i}(\bar{f} - \underline{f})$ for all $i \leq m$ (where $g : \mathbb{R}^n \to \mathbb{R}^m$).

Valid cuts for MINLPs

Naturally, both MILP and NLP cuts may apply to MINLPs. Some more specific MINLP cuts can be derived by reformulating integer variables to binary (see Sect. 3.2) and successively to continuous (see Sect. 3.2). The added quadratic constraints may then be relaxed in a Lagrangian (see Sect. 4.3) or SDP fashion (see Sect. 4.3) [98]: any of the NLP cuts described in Sect. 4.4 applied to such a reformulation is essentially a specific MINLP valid cut.

5 Reformulation/Optimization Software Engine

Although specific reformulations are carried out by most LP/MILP preprocessors [52, 45], and a few very simple reformulations are carried out by some mathematical programming language environments [39, 23], there is no software optimization framework that is able to carry out reformulations in a systematic way. In this section we describe the Reformulation/Optimization Software Engine (ROSE), a C++ software framework for optimization that can reformulate and solve mathematical programs of various types. ROSE is work in progress; currently, it is more focused on reformulation than optimization, but it has nonetheless a few native solvers (e.g. a Variable Neighbourhood Search (VNS) based algorithm for nonconvex NLPs [76]) and wrappers to various other external solvers (e.g. the LP solver GLPK [85] and the local NLP solver SNOPT [41]). In our research, we currently use ROSE's reformulation capabilities with AMPL's considerable set of numerical solvers in order to obtain solutions of complex MINLPs.

ROSE consists of a set of interfaces with external clients (currently, it has a direct user interface and an AMPL [39] interface), a problem class, a virtual solver class with many implementations, and an expression tree manipulation library called Ev3 (see Sect. 5.3). Reformulations may occur within the problem class, within the solvers, or within Ev3. Solvers may embed either a numerical solution algorithm or a symbolic reformulation algorithm, or both. The problem class builds the problem and simplifies it as much as possible; solvers are generally passed one or more problem together with a set of configuration parameters, and provide either a numerical solution or a reformulation. Reformulation solvers usually change the structure of their input problems; there is a special dedicated reformulation solver that makes an identical copy of the input problem. Most reformulation solvers acting on the mathematical expressions call specific methods within Ev3.

5.1 Development History

ROSE has a long history. Its "founding father" is the GLOP software ([71] Sect. 3.3), conceived and used by E. Smith to write his Ph.D. thesis [112] at CPSE, Imperial College, under the supervision of Prof. Pantelides. GLOP was never publically released, although test versions were used by CPSE students and faculty over a number of years. GLOP, however, was not so much a software framework rather than an implementation of the reformulation-based sBB algorithm described in [114]. The same algorithm (in a completely new implementation) as well as some other global optimization algorithms were put together in the $ooOPS$ (object-oriented OPtimization System) software framework ([71] Sect. 3.4), coded by the first author of this chapter during his Ph.D. thesis [69] at CPSE, Imperial College, and drawing a few software architecture ideas from its MILP predecessor, $ooMILP$ [127, 126]. The $ooOPS$ software framework [82] includes an sBB algorithm for MINLPs (which has a few glitches but works in a lot of instances), a VNS algorithm for nonconvex NLPs, a wrapper to the GO solver SobolOpt [61], and a wrapper to SNOPT. $ooOPS$ was used to compile the results of several research papers, but unfortunately Imperial College never granted the rights to distribute its source publically. Besides, $ooOPS$ used binary expression trees, which made it much more difficult to reformulate sums and products with more than two operands. The MINLP Object-oriented Reformulation/Optimization Navigator (MORON) was conceived to address these two limitations. MORON has an extensive API for dealing with both reformulation and optimization and includes: a prototypical Smith reformulator and convexifier ([71], Sect. 2.3 and 7.3); a preliminary version of the sBB algorithm; a wrapper to SNOPT. A lot of work was put into the development of Ev3, a separate expression tree library with reformulating capabilities [65]. Unfortunately, due to lack of time, development of MORON was discontinued. ROSE is MORON's direct descendant: it has a leaner API, almost the same software architecture (the main classes being `Problem` and `Solver`), and it uses Ev3 to handle expression trees. We expect to be able to publically distribute ROSE within the end of 2008; for using and/or contributing to its development, please contact the first author. We also remark that many of the ideas on which $ooOPS$'s and MORON's sBB solvers are built are also found in Couenne [18], a modern sBB implementation coded by P. Belotti within a CMU/IBM project, targeted at general MINLPs, and publically distributed within COIN-OR [83].

5.2　Software Architecture

The ROSE software relies on two main classes: `Problem` and `Solver`. The standard expected usage sequence is the following. The client (either the user or AMPL) constructs and configures a problem, selects and configures a solver, then solves a problem using the selected solver, and finally collects the output from the problem.

The `Problem` class has methods for reading in a problem, access/modify the problem description, perform various reformulations to do with adding/deleting variables and constraints, evaluate the problem expressions and their first and second derivatives at a given point, and test for feasibility of a given point in the problem. The `Solver` class is a virtual class that serves as interface for various solvers. Implementations of the solver class are passed a pointer to a `Problem` object and a set of user-defined configuration parameters. Solver implementations may either find numerical solutions and/or change the problem structure. Numerical solvers normally re-insert the numerical solution found within the `Problem` object. The output of a reformulation solver is simply the change carried out on the problem structure. Every action carried out on a mathematical expression, be it a function evaluation or a symbolic transformation, is delegated to the Ev3 library (see Sect. 5.3).

The `Problem` class

ROSE represents optimization problems in their *flat form* representation; i.e. variables, objective functions and constraints are arranged in simple linear lists rather than in jagged arrays of various dimensions. The reason for this choice is that languages such as AMPL and GAMS already do an excellent job of translating *structured form* problem formulations to their flat counterparts. Problems are defined in `problem.h` and `problem.cxx`.

This class rests on three `struct`s defining variables, objectives and constraints.

- `struct Variable`, storing the following information concerning decision variables.

  - `ID`, an integer (`int`) storing an ID associated to the variable. This ID does not change across reformulations, except in case of reformulations which delete variables. In this case, when a variable is deleted the IDs of the successive variables are shifted. The lists storing variable objects do not make any guarantee on the ordering of IDs across the list.
  - `Name`, a string (`std::string`) storing the variable name. This is only used for printing purposes.
  - `LB`, a floating point number (`double`) storing the variable lower bound.
  - `UB`, a floating point number (`double`) storing the variable upper bound.

- IsIntegral, a flag (bool) set to 1 if the variable is integer and 0 otherwise. Binary variables occur when IsIntegral is set to 1, LB to 0, and UB to 1.
- Persistent, a flag (bool) set to 1 if the variable cannot be deleted by reformulation algorithms, and 0 otherwise.
- Optimum, a floating point number (double) storing a value for the variable. Notwithstanding the name, this is not always the optimum value.

- struct Objective, storing the following information concerning objective functions.

  - ID, an integer (int) storing an ID associated to the objective. This ID does not change across reformulations, except in case of reformulations which delete objectives. In this case, when an objective is deleted the IDs of the successive objectives are shifted. The lists storing objective objects do not make any guarantee on the ordering of IDs across the list.
  - Name, a string(std::string) storing the objective name. This is not currently used.
  - Function, the expression tree (Expression) of the objective function.
  - FunctionFET, a fast evaluation expression tree (see Sect. 5.3 on p. 210) pointer (FastEvalTree*) corresponding to Function.
  - NonlinearPart, the expression tree (Expression) of the nonlinear part of Function. This may contain *copies* of subtrees of Function. The nonlinear part of an expression includes all subexpressions involving variables that appear nonlinearly at least once in the expression. For example, the nonlinear part of $x + y + z + yz$ is $y + z + yz$.
  - NonlinearPartFET, a fast evaluation expression tree pointer (FastEvalTree*) corresponding to NonlinearPart.
  - OptDir, a label (int) which is 0 if the objective is to be minimized and 1 if it is to be maximized.
  - Diff, the first-order partial derivatives (std::vector<Expression>) of Function.
  - DiffFET, the fast evaluation tree pointers (std::vector<FastEval Tree*>) corresponding to the first-order partial derivatives.
  - Diff2, the second-order partial derivatives (std::vector<std:: vector<Expression> >) of Function.
  - Diff2FET, the fast evaluation tree pointers (std::vector<std:: vector<FastEvalTree*> >) corresponding to second-order partial derivatives.

- struct Constraint, storing the following information concerning constraints.

  - ID, an integer (int) storing an ID associated to the constraint. This ID does not change across reformulations, except in case of reformulations

which delete constraints. In this case, when a constraint is deleted the
IDs of the successive constraints are shifted. The lists storing constraint
objects do not make any guarantee on the ordering of IDs across the list.

- Name, a string(`std::string`) storing the constraint name. This is not
  currently used.
- Function, the expression tree (`Expression`) of the constraint function.
- FunctionFET, a fast evaluation expression tree (see Sect. 5.3 on p. 210)
  pointer (`FastEvalTree*`) corresponding to Function.
- NonlinearPart, the expression tree (`Expression`) of the nonlinear part
  of Function. This may contain *copies* of subtrees of Function. The
  nonlinear part of an expression includes all subexpressions involving
  variables that appear nonlinearly at least once in the expression. For
  example, the nonlinear part of $x + y + z + yz$ is $y + z + yz$.
- NonlinearPartFET, a fast evaluation expression tree pointer
  (`FastEvalTree*`) corresponding to NonlinearPart.
- LB, a floating point number (`double`) storing the constraint lower
  bound.
- UB, a floating point number (`double`) storing the constraint upper
  bound.
- Diff, the first-order partial derivatives (`std::vector<Expression>`) of
  Function.
- DiffFET, the fast evaluation tree pointers (`std::vector<FastEval
  Tree*>`) corresponding to the first-order partial derivatives.
- Diff2, the second-order partial derivatives (`std::vector<std::
  vector<Expression> >`) of Function.
- Diff2FET, the fast evaluation tree pointers (`std::vector<std::
  vector<FastEvalTree*> >`) corresponding to second-order partial
  derivatives.

We remark that Constraint objects are expressed in the form LB $\leq$
Function $\leq$ UB; in order to deactivate one constraint side, use the defined constant MORONINFINITY (1×10^{30}).

Indexing of problem entities

Pointers to all variable, objective and constraint objects (also called *entities*)
in the problem are stored in STL `vectors`. Thus, on top of the entity indexing
given by the ID property, we also have the natural indexing associated to these
vectors, referred to as *local indexing*. Whereas ID-based indices are constant
throughout any sequence of reformulations, local indices refer to the current
problem structure. Direct and inverse mappings between indices and local
indices are given by the following Problem methods:

- `int GetVariableID(int localindex)`
- `int GetVarLocalIndex(int varID)`

- `int GetObjectiveID(int localindex)`
- `int GetObjLocalIndex(int objID)`
- `int GetConstraintID(int localindex)`
- `int GetConstrLocalIndex(int constrID)`.

Individual problem entities can be accessed/modified by their ID; a subset of the methods also exist in the "local index" version — such methods have the suffix `-LI` appended to their names. All indices in the API start from 1.

Parameters

The parameter passing mechanism is based on a **Parameters** class with the following methods.

class Parameters.

METHOD NAME	PURPOSE
int GetNumberOfParameters(void)	get number of parameters
string GetParameterName(int pID)	get name of parameter pID
int GetParameterType(int pID)	get type of parameter pID (0=int,1=bool,2=double,3=string)
int GetParameterIntValue(int pID)	get int value of parameter pID
bool GetParameterBoolValue(bool pID)	get bool value of parameter pID
double GetParameterDoubleValue(double pID)	get double value of parameter pID
string GetParameterStringValue(int pID)	get string value of parameter pID
void SetIntParameter(string parname, int)	set named parameter to int value
void SetBoolParameter(string parname, bool)	set named parameter to bool value
void SetDoubleParameter(string parname, double)	set named parameter to double value
void SetStringParameter(string parname, string)	set named parameter to string value
int GetIntParameter(string parname)	get int value of named parameter
int GetBoolParameter(string parname)	get bool value of named parameter
int GetDoubleParameter(string parname)	get double value of named parameter
int GetStringParameter(string parname)	get string value of named parameter

Problem *API*

The API of the **Problem** class is given in the tables on pages 211-212. Within the reformulation methods, the **Add-** methods automatically call a corresponding **New-** method to produce the next available ID. The **DeleteVariable** methods does not eliminate all occurrences of the variable from the problem (i.e. this is not a projection). The AMPL-based construction methods were made possible by an undocumented AMPL solver library feature that allows clients to access AMPL's internal binary trees [38, 40].

The Solver virtual class

Solver is a virtual class whose default implementation is an inactive (empty) solver. This is not a pure virtual class because it represent the union of all possible solver implementations, rather than the intersection; in other words, not all methods in **Solver** are implemented across all solvers (check the source files `solver*.h`, `solver*.cxx` to make sure).

class `UserCut`.

METHOD NAME	PURPOSE
UserCut(Expression e, double L, double U)	constructor
Expression Function	cut's body
FastEvalTree* FunctionFET	fast eval tree of Function
Expression NonlinearPart	nonlinear part of Function
FastEvalTree* NonlinearPartFET	fast eval tree of NonlinearPart
double LB	lower bound
double UB	upper bound
bool IsLinear	marks a linear cut
vector<Expression> Diff	derivatives
vector<FastEvalTree*> DiffFET	corresponding fast eval trees
vector<vector<Expression> > Diff2	2nd derivatives
vector<vector<FastEvalTree*> > Diff2FET	corresponding fast eval trees

Implementations of this class may be *numerical solvers*, working towards finding a solution, or *reformulation solvers*, working towards analysing or changing the problem structure. Normally, solvers are initialized and then activated. Problem bounds (both variable and constraint) can be changed dynamically by a solver without the original problem bounds being modified. Numerical solvers can add both linear and nonlinear cuts (see Sect. 4.4) to the formulation before solving it. Cuts are dealt with via two auxiliary classes `UserLinearCut` and `UserCut`.

Because of their simplicity, `UserLinearCut` and `UserCut` do not offer a full set/get interface, and all their properties are public. Cuts can only be added, never deleted; however, they can be enabled/disabled as needed.

Existing `Solver` implementations

Each solver implementation consists of a header and an implementation file. Currently, ROSE has three functional numerical solvers: VNS solver for nonconvex NLPs [76], a wrapper to SNOPT [42], a wrapper to GLPK [85]; and various reformulator solvers, among which: a problem analyser that returns problem information to AMPL, a problem copier that simply makes an identical copy of the current problem (for later reformulations), an outer approximation reformulator, a Smith standard form reformulator (see Sect. 2.3), a Smith convexifier (see Sect. 4.2), a PRODBINCONT reformulator (see Sect. 3.3), and various other partially developed solvers.

5.3   Ev3

Ev3 is a library providing expression tree functionality and symbolic transformations thereof (see Sect. 2.2). This library may also be used stand-alone, and the rest of this section actually refers to the stand-alone version. The only adaptation that was implemented for usage within ROSE was to provide additional structures for Fast Evaluation Trees (FETs). Ev3's native

class Problem. Basic methods.

METHOD NAME	PURPOSE
Problem(bool nosimplify)	constructor with optional nosimplify
void SetName(string)	set the problem name
string GetName(void)	get the problem name
bool IsProblemContinuous(void)	true if no integer variables
bool IsProblemLinear(void)	true if no nonlinear expressions
Problem* GetParent(void)	get parent problem in a tree of problems
int GetNumberOfChildren(void)	get number of children problems
Problem* GetChild(int pID)	get pID-th child in list of children problems
string GetFormulationName(void)	name of reform. assigned to this prob.
void SetOptimizationDirection(int oID, int minmax)	set opt. dir. of oID-th objective
void SetOptimizationDirectionLI(int li, int minmax)	local index version
int GetOptimizationDirection(int oID)	get opt. dir. of oID-th objective
int GetOptimizationDirectionLI(int li)	local index version
bool HasDeleted(void)	true if simplification deleted some entity
void SetSolved(bool s)	mark problem as solved/unsolved
bool IsSolved(void)	return solved/unsolved mark
void SetFeasible(int feas)	mark problem as feasible/infeasible
int IsFeasible(void)	return feasible/infeasible mark
Parameters GetParams(void)	returns a copy of the set of parameters
Parameters& GetParamsRef(void)	returns a reference to a set of parameters
void ReplaceParams(Parameters& prm)	replace the current set of parameters
int GetNumberOfVariables(void)	return number of variables
int GetNumberOfIntegerVariables(void)	return number of integer variables
int GetNumberOfObjectives(void)	return number of objectives
int GetNumberOfConstraints(void)	return number of constraints
Variable* GetVariable(int vID)	return pointer to variable entity
Variable* GetVariableLI(int li)	local index version
Variable* GetObjective(int vID)	return pointer to objective entity
Variable* GetObjectiveLI(int li)	local index version
Variable* GetConstraint(int vID)	return pointer to constraint entity
Variable* GetConstraintLI(int li)	local index version
void SetOptimalVariableValue(int vID, double val)	set optimal variable value
void SetOptimalVariableValueLI(int li, double val)	local index version
double GetOptimalVariableValue(int vID)	get optimal variable value
double GetOptimalVariableValueLI(int li)	local index version
void SetCurrentVariableValue(int vID, double val)	set optimal variable value
void SetCurrentVariableValueLI(int li, double val)	local index version
double GetCurrentVariableValue(int vID)	get optimal variable value
double GetCurrentVariableValueLI(int li)	local index version
bool TestConstraintsFeasibility(int cID, double tol, double& disc)	test feas. of current point w.r.t. a constraint
bool TestConstraintsFeasibility(double tol, double& disc)	test feasibility of current point in problem
bool TestVariablesFeasibility(double tol, double& disc)	test feasibility of current point in bounds
double GetStartingPoint(int vID)	get starting point embedded in the problem
double GetStartingPointLI(int localindex)	local index version
void SetOptimalObjectiveValue(int oID, double val)	set optimal obj. fun. value
double GetOptimalObjectiveValue(int oID)	get optimal obj. fun. value
void GetSolution(map<int,double>& ofval, map<int,double>& soln)	get solution
void GetSolutionLI(vector<double>& ofval, vector<double>& soln)	local index version
double GetObj1AdditiveConstant(void)	get additive constant of 1st objective

trees are very easy to change for reformulation needs, but unfortunately turn out to be slow to evaluate by Alg. 1. Since in most numerical algorithms for optimization the same expressions are evaluated many times, a specific data structure `fevaltree` with relative source files (`fastexpression.h`, `fastexpression.cxx`) have been added to Ev3. FETs are C-like n-ary (as opposed to binary) trees that have none of the reformulating facilities of

class Problem. Evaluation methods.

METHOD NAME	PURPOSE
double EvalObj(int oID)	evaluate an objective
double EvalNLObj(int oID)	evaluate the nonlinear part of an objective
double EvalObjDiff(int oID, int vID)	evaluate the derivative of an obj.
double EvalObjDiffNoConstant(int oID, int vID)	eval. non-const. part of a deriv.
double EvalObjDiff2(int oID, int vID1, int vID2)	evaluate 2nd derivative of an obj.
double EvalConstr(int cID)	evaluate a constraint
double EvalNLConstr(int cID)	evaluate nonlinear part of a constraint
double EvalConstrDiff(int cID, int vID)	evaluate a constr. derivative
double EvalConstrDiffNoConstant(int cID, int vID)	eval. non-const. part of constr. deriv.
double EvalConstrDiff2(int cID, int vID1, int vID2)	evaluate 2nd constr. derivative
bool IsObjConstant(int oID)	is the objective a constant?
bool IsObjDiffConstant(int oID, int vID)	is the obj. derivative a constant?
bool IsObjDiff2Constant(int oID, int vID1, int vID2)	is the 2nd obj. deriv. a const.?
bool IsConstrConstant(int cID)	is the constraint a constant?
bool IsConstrDiffConstant(int cID, int vID)	is the constr. deriv. a constant?
bool IsConstrDiff2(int cID, int vID1, int vID2)	is the 2nd constr. deriv. a const.?
bool IsConstrActive(int cID, double tol, int& LU)	is the constraint active L/U bound?

class Problem. Construction methods.

METHOD NAME	PURPOSE
void Parse(char* file)	parse a ROSE-formatted file
Ampl::ASL* ParseAMPL(char** argv, int argc)	parse AMPL-formatted .nl file

class Problem. Reformulation methods.

METHOD NAME	PURPOSE
int NewVariableID(void)	returns next available variable ID
int NewObjectiveID(void)	returns next available variable ID
int NewConstraintID(void)	returns next available variable ID
void AddVariable(string& n, bool i, bool pers, double L, double U, double v)	adds a new variable
void AddObjective(string& n, Expression e, int dir, double v)	adds a new objective
void AddConstraint(string& n, Expression e, double L, double U)	adds a new constraint
void DeleteVariable(int vID)	deletes a variable
void DeleteObjective(int oID)	deletes an objective
void DeleteConstraint(int cID)	deletes a constraint

class UserLinearCut.

METHOD NAME	PURPOSE
UserLinearCut(vector<pair<int,double> >&, double L, double U)	C++-style constructor
UserLinearCut(int* varIDs, double* coeffs, int size, double L, double U)	C-style constructor
double LB	lower bound
double UB	upper bound
int Nonzeroes	number of nonzeroes in linear form
int* Varindices	variable indices in row
double* Coeffs	coefficients of row

their Ev3 counterparts, but which are very fast to evaluate. Construction and evaluation of FETs is automatic and transparent to the user.

Architecture

The Ev3 software architecture is mainly based on 5 classes. Two of them, **Tree** and **Pointer**, are generic templates that provide the basic tree structure and a no-frills garbage collection based on reference count. Each object has a reference counter which increases every time a reference of that object is taken; the object destructor decreases the counter while it is positive,

class Solver. Basic and cut-related methods.

METHOD NAME	PURPOSE
string GetName(void)	get solver name
void SetProblem(Problem* p)	set the problem for the solver
Problem* GetProblem(void)	get the problem from the solver
bool CanSolve(int probtype)	can this solve a certain problem type?
	(0=LP,1=MILP,2=NLP,3=MINLP)
void Initialize(bool force)	initialize solver
bool IsProblemInitialized(void)	is solver initialized?
int Solve(void)	solve/reformulate the problem
Parameters GetParams(void)	get the parameter set
Parameters& GetParamsRef(void)	get a reference to the parameters
void ReplaceParams(Parameters& p)	replace the parameters
void SetOptimizationDirection(int maxmin)	set 1st objective opt. dir.
int GetOptimizationDirection(void)	get 1st objective opt. dir.
void GetSolution(map<int,double>& ofval, map<int,double>& soln)	get solution
void GetSolutionLI(vector<double>& ofval, vector<double>& soln)	get solution
void SetMaxNumberOfCuts(int)	set max number of cuts
int GetMaxNumberOfCuts(void)	get max number of cuts
int GetNumberOfCuts(void)	get number of cuts added till now
int AddCut(Expression e, double L, double U)	add a nonlienar cut
int AddCut(vector<pair<int,double> >&, double L, double U)	add a linear cut
double EvalCut(int cutID, double* xval)	evaluate a cut
double EvalNLCut(int cutID, double* xval)	evaluate the nonlinear part
double EvalCutDiff(int cutID, int vID, double* xval)	evaluate derivatives
double EvalCutDiffNoConstant(int cutID, int vID, double* xval)	as above, without constants
double EvalCutDiff2(int cutID, int vID1, int vID2, double* xval)	evaluated 2nd derivatives
bool IsCutLinear(int cutID)	is this cut linear?
void EnableCut(int cutID)	enables a cut
void DisableCut(int cutID)	disables a cut
void SetCutLB(int cutID, double L)	set lower bound
double GetCutLB(int cutID)	get lower bound
void SetCutUB(int cutID, double U)	set upper bound
double GetCutUB(int cutID)	get upper bound

only actually deleting the object when the counter reaches zero. This type of garbage collecting is due to Collins, 1960 (see [55]). Other two classes, Operand and BasicExpression, implement the actual semantics of an algebraic expression. The last class, ExpressionParser, implements a simple parser (based on the ideas given in [117]) which reads in a string containing a valid mathematical expression and produces the corresponding n-ary tree.

The Pointer class

This is a template class defined as

```
template<class NodeType> class Pointer {
  NodeType* node;
  int* ncount;
  // methods
};
```

The constructor of this class allocates a new integer for the reference counter ncount and a new NodeType object, and the copy constructor increases the counter. The destructor deletes the reference counter and invokes the delete method on the NodeType object. In order to access the data and methods

class `Solver`. Numerical problem information methods.

METHOD NAME	PURPOSE
void SetVariableLB(int vID, double LB)	set variable lower bound
double GetVariableLB(int vID)	get variable lower bound
void SetVariableUB(int vID, double UB)	set variable upper bound
double GetVariableUB(int vID)	get variable upper bound
void SetConstraintLB(int cID, double LB)	set constraint lower bound
double GetConstraintLB(int cID)	get constraint lower bound
void SetConstraintUB(int cID, double UB)	set constraint upper bound
double GetConstraintUB(int cID)	get constraint upper bound
void SetVariableLBLI(int li, double LB)	local index version
double GetVariableLBLI(int li)	local index version
void SetVariableUBLI(int li, double UB)	local index version
double GetVariableUBLI(int li)	local index version
void SetConstraintLBLI(int li, double LB)	local index version
double GetConstraintLBLI(int li)	local index version
void SetConstraintUBLI(int li, double UB)	local index version
double GetConstraintUBLI(int li)	local index version
void SetStartingPoint(int vID, double sp)	
void SetStartingPointLI(int li, double sp)	local index version
double GetStartingPoint(int vID)	get starting point
double GetStartingPointLI(int li)	local index version
bool IsBasic(int vID)	is variable basic?
bool IsBasicLI(int li)	local index version
double GetConstraintLagrangeMultiplier(int cID)	get Lagrange multiplier of constraint
double GetConstraintLagrangeMultiplierLI(int li)	local index version
double GetCutLagrangeMultiplier(int cutID)	get Lagrange multiplier of cut
double GetBoundLagrangeMultiplier(int varID)	get Lagrange multiplier of var. bound
double GetBoundLagrangeMultiplierLI(int li)	local index version
bool IsBasic(int varID)	is variable basic?
bool IsBasicLI(int li)	local index version

of the `NodeType` object pointed to by `node`, the `->` operator in the `Pointer` class is overloaded to return `node`.

A mathematical expression, in Ev3, is defined as a *pointer to a* `BasicExpression` *object* (see below for the definition of a `BasicExpression` object):

```
typedef Pointer<BasicExpression> Expression;
```

The `Tree` class

This is a template class defined as

```
template<class NodeType> class Tree {
  vector<Pointer<NodeType> > nodes;
  // methods
};
```

This is the class implementing the *n*-ary tree (subnodes are contained in the `nodes` vector). Notice that, being a template, the whole implementation is kept independent of the semantics of a `NodeType`. Notice also that because pointers to objects are pushed on the vector, algebraic substitution is very easy: just replace one pointer with another one. This differs from the implementation of GiNaC [17] where it appears that algebraic substitution is a more convoluted operation.

The `Operand` class

This class holds the information relative to each expression term, be they constants, variables or operators.

```
class Operand {
  int oplabel;          // operator label
  double value;         // if constant, value of constant
  long varindex;        // if variable, the variable index
  string varname;       // if variable, the variable name
  double coefficient;   // terms can be multiplied by a number
  double exponent;      // leaf terms can be raised to a number
  // methods
};
```

- `oplabel` can be one of the following labels (the meaning of which should be clear):

```
enum OperatorType {
  SUM, DIFFERENCE, PRODUCT, FRACTION, POWER, PLUS, MINUS, LOG,
  EXP, SIN, COS, TAN, COT, SINH, COSH, TANH, COTH, SQRT, VAR,
  CONST, ERROR
};
```

- `value`, the value of a constant numeric term, only has meaning if `oplabel` is CONST;
- `varindex`, the variable index, only has meaning if `oplabel` is VAR;
- every term, (variables, constants and operators), can be multiplied by a numeric coefficient. This makes it easy to perform symbolic manipulation on like terms (e.g. $x + 2x = 3x$).
- every leaf term (variables and constants) can be raised to a numeric power. This makes it easy to perform symbolic manipulation of polynomials.

Introducing numeric coefficients and exponents is a choice that has advantages as well as disadvantages. GiNaC, for example, does not explicitly account for numeric coefficients. The advantages are obvious: it makes symbolic manipulation very efficient for certain classes of basic operations (operations on like terms). The disadvantage is that the programmer has to explicitly account for the case where terms are assigned coefficients: whereas with a pure tree structure recursive algorithms can be formulated as "for each node, do something", this becomes more complex when numeric coefficients are introduced. Checks for non zero or non identity have to be performed prior to carrying out certain operations, as well as having to manually account for cases where coefficients have to be used. However, by setting both multiplicative and exponent coefficients to 1, the mechanism can to a certain extent be ignored and a pure tree structure can be recovered.

The BasicExpression class

This class is defined as follows:

```
class BasicExpression :
public Operand, public Tree<BasicExpression> {
  // methods
};
```

It includes no data of its own, but it inherits its semantic data from class `Operand` and its tree structure from template class `Tree` with itself (`BasicExpression`) as a base type. This gives `BasicExpression` an n-ary tree structure. Note that an object of class `BasicExpression` is *not* a `Pointer`, only its subnodes (if any) are stored as `Pointers` to other `BasicExpressions`. This is the reason why the client code should never explicitly use `BasicExpression`; instead, it should use objects `Expression`, which are defined as `Pointer<Basic-Expression>`. This allows the automatic garbage collector embedded in `Pointer` to work.

The ExpressionParser class

This parser originates from the example parser found in [117]. The original code has been extensively modified to support exponentiation, unary functions in the form $f(x)$, and creation of n-ary trees of type `Expression`. For an example of usage, see Section 5.3 below.

Application Programming Interface

The Ev3 API consists in a number of *internal methods* (i.e., methods belonging to classes) and *external methods* (functions whose declaration is outside the classes). Objects of type `class Expression` can be built from strings containing infix-format expressions (like, e.g. `"log(2*x*y)+ sin(z)"`) by using the built-in parser. However, they may also be built from scratch using the supplied construction methods (see Section 5.3 for examples). Since the fundamental type `Expression` is an alias for `Pointer<BasicExpression>`, and `BasicExpression` is in turn a mix of different classes (including a `Tree` with itself as a template type), calling internal methods of an `Expression` object may be confusing. Thus, for each class name involved in the definition of `Expression`, we have listed the calling procedure explicitly in the tables on pages 217-219.

Notes

- The lists given above only include the most important methods. For the complete lists, see the files `expression.h`, `tree.cxx`, `parser.h` in the source code distribution.

Class Operand. Call: `ret = (Expression e)->MethodName(args)`.

METHOD NAME	PURPOSE
int GetOpType(void)	returns the operator label
double GetValue(void)	returns the value of the constant leaf (takes multiplicative coefficient and exponent into account)
double GetSimpleValue(void)	returns the value (takes no notice of coefficient and exponent)
long GetVarIndex(void)	returns the variable index of the variable leaf
string GetVarName(void)	returns the name of the variable leaf
double GetCoeff(void)	returns the value of the multiplicative coefficient
double GetExponent(void)	returns the value of the exponent (for leaves)
void SetOpType(int)	sets the operator label
void SetValue(double)	sets the numeric value of the constant leaf
void SetVarIndex(long)	sets the variable index of the variable leaf
void SetVarName(string)	sets the name of the variable leaf
void SetExponent(double)	sets the exponent (for leaves)
void SetCoeff(double)	sets the multiplicative coefficient
bool IsConstant(void)	is the node a constant?
bool IsVariable(void)	is the node a variable?
bool IsLeaf(void)	is the node a leaf?
bool HasValue(double v)	is the node a constant with value v?
bool IsLessThan(double v)	is the node a constant with value $\leq v$?
void ConsolidateValue(void)	set value to `coeff*value*exponent` and set `coeff` to 1 and `exponent` to 1
void SubstituteVariableWithConstant (long int varindex, double c)	substitute a variable with a constant c

Template class Pointer<NodeType>. Call: `ret = (Expression e).MethodName(args)`.

METHOD NAME	PURPOSE
Pointer<NodeType> Copy(void)	returns a copy of this node
void SetTo(Pointer<NodeType>& t)	this is a reference of t
void SetToCopyOf(Pointer<NodeType>& t)	this is a copy of t
Pointer<NodeType> operator=(Pointer<NodeType> t)	assigns a reference of t to this
void Destroy(void)	destroys the node (collects garbage)

Template class Tree<NodeType>. Call: `ret = (Expression e)->MethodName(args)`.

METHOD NAME	PURPOSE
void AddNode(Pointer<NodeType>)	pushes a node at the end of the node vector
void AddCopyOfNode(Pointer<NodeType> n)	pushes a copy of node n at the end of the node vector
bool DeleteNode(long i)	deletes the i-th node, returns true if successful
void DeleteAllNodes(void)	empties the node vector
Pointer<NodeType> GetNode(long i)	returns a reference to the i-th subnode
Pointer<NodeType> * GetNodeRef(long i)	returns a pointer to the i-th subnode
Pointer<NodeType> GetCopyOfNode(long i)	returns a copy of the i-th subnode
long GetSize(void)	returns the length of the node vector

- There exist a considerable number of different constructors for Expression. See their purpose and syntax in files `expression.h`, `tree.cxx`. See examples of their usage in file `expression.cxx`.

Class BasicExpression (inherits from Operand, Tree<BasicExpression>).
Call: ret = (Expression e)->MethodName(args).

METHOD NAME	PURPOSE
string ToString(void)	returns infix notation expression in a string
void Zero(void)	sets this to zero
void One(void)	sets this to one
bool IsEqualTo(Expression&)	is this equal to the argument?
bool IsEqualToNoCoeff(Expression&)	[like above, ignoring multiplicative coefficient]
int NumberOfVariables(void)	number of variables in the expression
double Eval(double* v, long vsize)	evaluate; v[i] contains the value for variable with index i, v has length vsize
bool DependsOnVariable(long i)	does this depend on variable i?
int DependsLinearlyOnVariable(long i)	does this depend linearly on variable i? (0=nonlinearly, 1=linearly, 2=no dep.)
void ConsolidateProductCoeffs(void)	if node is a product, move product of all coefficients as coefficient of node
void DistributeCoeffOverSum(void)	if coeff. of a sum operand is not 1, distribute it over the summands
void VariableToConstant(long varindex, double c)	substitute a variable with a constant c
void ReplaceVariable(long vi1, long vi2, string vn2)	replace occurrences of variable vi1 with variable vi2 having name vn2
string FindVariableName(long vi)	find name of variable vi
bool IsLinear(void)	is this expression linear?
bool GetLinearInfo(...)	returns info about the linear part
Expression Get[Pure]LinearPart(void)	returns the linear part
Expression Get[Pure]NonlinearPart(void)	returns the nonlinear part
double RemoveAdditiveConstant(void)	returns any additive constant and removes it
void Interval(...)	performs interval arithmetics on the expression

Class ExpressionParser.

METHOD NAME	PURPOSE
void SetVariableID(string x, long i)	assign index i to variable x; var. indices start from 1 and increase by 1
long GetVariableID(string x)	return index of variable x
Expression Parse(char* buf, int& errors)	parse buf and return an Expression errors is the number of parsing errors occcurred

- Internal class methods usually return or set atomic information inside the object, or perform limited symbolic manipulation. Construction and extended manipulation of symbolic expressions have been confined to external methods. Furthermore, external methods may have any of the following characteristics:

  - they combine *references* of their arguments;
  - they may change their arguments;
  - they may change the order of the subnodes where the operations are commutative;
  - they may return one of the arguments.

  Thus, it is advisable to perform the operations on copies of the arguments when the expression being built is required to be independent of its subnodes. In particular, all the expression building functions (e.g. operator+(), ..., Log(), ...) do *not* change their arguments, whereas their -Link counterparts do.
- The built-in parser (ExpressionParser) uses linking and not copying (also see Section 5.3) of nodes when building up the expression.

Methods outside classes.

METHOD NAME	PURPOSE
Expression operator+(Expression a, Expression b)	returns symbolic sum of a, b
Expression operator-(Expression a, Expression b)	returns symbolic difference of a, b
Expression operator*(Expression a, Expression b)	returns symbolic product of a, b
Expression operator/(Expression a, Expression b)	returns symbolic fraction of a, b
Expression operator^(Expression a, Expression b)	returns symbolic power of a, b
Expression operator-(Expression a)	returns symbolic form of $-a$
Expression Log(Expression a)	returns symbolic $\log(a)$
Expression Exp(Expression a)	returns symbolic $\exp(a)$
Expression Sin(Expression a)	returns symbolic $\sin(a)$
Expression Cos(Expression a)	returns symbolic $\cos(a)$
Expression Tan(Expression a)	returns symbolic $\tan(a)$
Expression Sinh(Expression a)	returns symbolic $\sinh(a)$
Expression Cosh(Expression a)	returns symbolic $\cosh(a)$
Expression Tanh(Expression a)	returns symbolic $\tanh(a)$
Expression Coth(Expression a)	returns symbolic $\coth(a)$
Expression SumLink(Expression a, Expression b)	returns symbolic sum of a, b
Expression DifferenceLink(Expression a, Expression b)	returns symbolic difference of a, b
Expression ProductLink(Expression a, Expression b)	returns symbolic product of a, b
Expression FractionLink(Expression a, Expression b)	returns symbolic fraction of a, b
Expression PowerLink(Expression a, Expression b)	returns symbolic power of a, b
Expression MinusLink(Expression a)	returns symbolic form of $-a$
Expression LogLink(Expression a)	returns symbolic $\log(a)$
Expression ExpLink(Expression a)	returns symbolic $\exp(a)$
Expression SinLink(Expression a)	returns symbolic $\sin(a)$
Expression CosLink(Expression a)	returns symbolic $\cos(a)$
Expression TanLink(Expression a)	returns symbolic $\tan(a)$
Expression SinhLink(Expression a)	returns symbolic $\sinh(a)$
Expression CoshLink(Expression a)	returns symbolic $\cosh(a)$
Expression TanhLink(Expression a)	returns symbolic $\tanh(a)$
Expression CothLink(Expression a)	returns symbolic $\coth(a)$
Expression Diff(const Expression& a, long i)	returns derivative of a w.r.t variable i
Expression DiffNoSimplify(const Expression& a, long i)	returns unsimplified derivative of a w.r.t variable i
bool Simplify(Expression* a)	apply all simplification rules
Expression SimplifyCopy(Expression* a, bool& has_changed)	simplify a copy of the expression
void RecursiveDestroy(Expression* a)	destroys the whole tree and all nodes

- The symbolic derivative routine `Diff()` uses copying and not linking of nodes when building up the derivative.
- The method `BasicExpression::IsEqualToNoCoeff()` returns true if two expressions are equal apart from the multiplicative coefficient of the root node only. I.e., $2(x+y)$ would be deemed "equal" to $x+y$ (if 2 is a multiplicative coefficient, *not* an operand in a product) but $x + 2y$ would *not* be deemed "equal" to $x + y$.
- The `Simplify()` method applies all simplification rules known to Ev3 to the expression and puts it in standard form.
- The methods `GetLinearInfo()`, `GetLinearPart()`, `GetPureLinearPart()`, `GetNonlinearPart()`, `GetPureNonlinearPart()` return various types of linear and nonlinear information from the expression. Details concerning these methods can be found in the Ev3 source code files `expression.h`, `expression.cxx`.
- The method `Interval()` performs interval arithmetic on the expression. Details concerning this method can be found in the Ev3 source code files `expression.h`, `expression.cxx`.

- Variables are identified by a variable index, but they also know their variable name. Variable indices are usually assigned within the `ExpressionParser` object, with the `SetVariableID()` method. It is important that variable indices should start from 1 and increase monotonically by 1, as variable indices are used to index the array of values passed to the `Eval()` method.

Copying vs. Linking

One thing that is immediately noticeable is that this architecture gives a very fine-grained control over the construction of expressions. Subnodes can be copied or "linked" (i.e., a reference to the object is put in place, instead of a copy of the object — this automatically uses the garbage collection mechanism, so the client code does not need to worry about these details). Copying an expression tree entails a set of advantages/disadvantages compared to linking. When an expression is constructed by means of a copy to some other existing expression tree, the two expressions are thereafter completely independent. Manipulation one expression does not change the other. This is the required behaviour in many cases. The symbolic differentiation routine has been designed using copies because a derivative, in general, exists independently of its integral.

Linking, however, allows for facilities such as "propagated simplification", where some symbolic manipulation on an expression changes all the expressions having the manipulated expression tree as a subnode. This may be useful but calls for extra care. The built-in parser has been designed using linking because the "building blocks" of a parsed expression (i.e. its subnodes of all ranks) will not be used independently outside the parser.

Simplification Strategy

The routine for simplifying an expression repeatedly calls a set of simplification rules acting on the expression. These rules are applied to the expression as long as at least one of them manages to further simplify it.

Simplifications can be *horizontal*, meaning that they are carried out on the same list of subnodes (like e.g. $x + y + y = x + 2y$), or *vertical*, meaning that the simplification involves changing of node level (like e.g. application of associativity: $((x + y) + z) = (x + y + z)$).

The order of the simplification rules applied to an object `Expression e` is the following:

1. `e->ConsolidateProductCoeffs()`: in a product having n subnodes, collect all multiplicative coefficients, multiply them together, and set the result as the multiplicative coefficient of the whole product:

$$\prod_{i=1}^{n}(c_i f_i) = (\prod_{i=1}^{n} c_i)(\prod_{i=1}^{n} f_i).$$

2. `e->DistributeCoeffOverSum()`: in a sum with n subnodes and a non-unit multiplicative coefficient, distribute this coefficient over all subnodes in the sum:

$$c \sum_{i=1}^{n} f_i = \sum_{i=1}^{n} c f_i.$$

3. `DifferenceToSum(e)`: replace all differences and unary minus with sums, multiplying the coefficient of the operands by -1.

4. `SimplifyConstant(e)`: simplify operations on constant terms by replacing the `value` of the node with the result of the operation.

5. `CompactProducts(e)`: associate products; e.g. $((xy)z) = (xyz)$.

6. `CompactLinearPart(e)`: this is a composite simplification consisting of the following routines:

   a. `CompactLinearPartRecursive(e)`: recursively search all sums in the expression and perform horizontal and vertical simplifications on the coefficients of like terms.

   b. `ReorderNodes(e)`: puts each list of subnodes in an expression in *standard form* (also see Sect. 2.2):

   constant + monomials in rising degree + complicated operands

   (where *complicated operands* are sublists of subnodes).

7. `SimplifyRecursive(e)`: deals with the most common simplification rules, i.e.:

   - try to simplify like terms in fractions where numerator and denominator are both products;
   - $x \pm 0 = 0 + x = x$;
   - $x \times 1 = 1 \times x = x$;
   - $x \times 0 = 0 \times x = 0$;
   - $x^0 = 1$;
   - $x^1 = x$;
   - $0^x = 0$;
   - $1^x = 1$.

Differentiation

Derivative rules are the usual ones; the rule for multiplication is expressed in a way that allows for n-ary trees to be derived correctly:

$$\frac{\partial}{\partial x} \prod_{i=1}^{n} f_i = \sum_{i=1}^{n} \left(\frac{\partial f_i}{\partial x} \prod_{j \neq i} f_j \right).$$

Algorithms on n-ary Trees

We store mathematical expressions in a tree structure so that we can apply recursive algorithms to them. Most of these algorithms are based on the following model.

```
if expression is a leaf node
   do something
else
   recurse on all subnodes
   do something else
end if
```

In particular, when using Ev3, the most common methods used in the design of recursive algorithms are the following:

- `IsLeaf()`: is the node a leaf node (variable or constant)?
- `GetSize()`: find the number of subnodes of any given node.
- `GetOpType()`: return the type of operator node.
- `GetNode(int i)`: return the i-th subnode of this node (nodes are numbered starting from 0).
- `DeleteNode(int i)`: delete the i-th subnode of this node (care must be taken to deal with cases where all the subnodes have been deleted — Ev3 allows the creation of operators with 0 subnodes, although this is very likely to lead to subsequent errors, as it has no mathematical meaning).
- Use of the operators for manipulation of nodes: supposing *Expression e, f* contain valid mathematical expressions, the following are all valid expressions (the new expressions are created using copies of the old ones).

```
Expression e1 = e + f;
Expression e2 = e * Log(Sqrt(e^2 - f^2));
Expression e3 = e + f - f;//this is automatically by simplified to e
```

Ev3 usage example

The example in this section explains the usage of the methods which represent the core, high-level functionality of Ev3: fast evaluation, symbolic simplification and differentiation of mathematical expressions.

The following C++ code is a simple driver program that uses the Ev3 library. Its instructions should be self-explanatory. First, we create a "parser object" of type `ExpressionParser`. We then set the mapping variable names / variable indices, and we parse a string containing the mathematical expression $\log(2xy) + sin(z)$. We print the expression, evaluate it at the point $(2, 3, 1)$, and finally calculate its symbolic derivatives w.r.t. x, y, z, and print them.

```
#include "expression.h"
#include "parser.h"
int main(int argc, char** argv) {
  ExpressionParser p;     // create the parser object
  p.SetVariableID("x", 1) // map between symbols and variable indices
```

```
p.SetVariableID("y", 2) // x --> 0, y --> 1, z --> 2
p.SetVariableID("z", 3)
int parsererrors = 0;    // number of parser errors
/* call the parser's Parse method, which returns an Expression
   which is then used to initialize Expression e       */
Expression e(p.Parse("log(2*x*y)+sin(z)", parsererrors));
cout << "parsing errors: " << parsererrors << endl;
cout << "f = " << e->ToString() << endl; // print the expression
double val[3] = {2, 3, 1};
cout << "eval(2,3,1):    " << e->Eval(val, 3) << endl; // evaluate the expr.
cout << "numeric check: " << ::log(2*2*3)+::sin(1) << endl; // check result
// test diff
Expression de1 = Diff(e, 1);  // calculate derivative w.r.t. x
cout << "df/dx = " << de1->ToString() << endl; // print derivative
Expression de2 = Diff(e, 2);  // calculate derivative w.r.t. y
cout << "df/dy = " << de2->ToString() << endl; // print derivative
Expression de3 = Diff(e, 3);  // calculate derivative w.r.t. z
cout << "df/dz = " << de3->ToString() << endl; // print derivative
return 0;
}
```

The corresponding output is

```
parsing errors: 0
f = (log((2*x)*(y)))+(sin(z))
eval(2,3,1):    3.32638
numeric check: 3.32638
df/dx = (1)/(x)
df/dy = (1)/(y)
df/dz = cos(z)
```

Notes

- In order to evaluate a mathematical expression $f(x_1, x_2, \ldots, x_n)$, where x_i are the variables and i are the variable indices (starting from 1 and increasing by 1), we use the Eval() internal method, whose complete declaration is as follows:

  ```
  double Expression::Eval(double* varvalues, int size) const;
  ```

  The array of doubles varvalues contains size real constants, where size $>= n$. The variable indices are used to address this array (the value assigned to x_i during the evaluation is varvalues[i-1]), so it is important that the order of the constants in varvalues reflects the order of the variables. This method does not change the expression object being evaluated.

- The core simplification method is an external method with declaration

  ```
  bool Simplify(Expression* e);
  ```

  It consists of a number of different simplifications, as explained in Section 5.3. It takes a *pointer* to Expression as an argument, and it returns true if some simplification has taken place, and false otherwise. This method changes its input argument.

- The symbolic differentiation procedure is an external method:

  ```
  Expression Diff(const Expression& e, int varindex);
  ```

  It returns a simplified expression which is the derivative of the expression in the argument with respect to variable `varindex`. This method does not change its input arguments.
- External class methods take `Expressions` as their arguments. According as to whether they need to change their input argument or not, the `Expression` is passed by value, by reference, or as a pointer. This may be a little confusing at first, especially when using the overloaded `->` operator on `Expression` objects. Consider an `Expression` e object and a pointer `Expression* ePtr = &e`. The following calls are possibile:

  - `e->MethodName(args); (*ePtr)->MethodName(args);`
    Call a method in the `BasicExpression`, `Operand` or `Tree<>` classes.
  - `e.MethodName(args); (*ePtr).MethodName(args); ePtr->MethodName(args);`
    Call a method in the `Pointer<>` class.

  In particular, care must be taken between the two forms `e->MethodName()` and `ePtr->MethodName()` as they are syntactically very similar but semantically very different.

5.4   Validation Examples

As validation examples, we show ROSE's output on simple input problems by using two kind of reformulations. In order to ease the reading of the examples, we use an intuitive description format for MINLPs problems [71, pages 237–239]. It is worth noticing that the symbol '<' stands here for '$\leq$' and that we use an explicit boundary ($1e^{30}$) for dealing with infinity.

The first example performs the reformulation of products between continuous and binary variables.

Original Problem	ROSE Reformulation
```# ROSE problem:```	```# ROSE problem:```
```# Problem has 2 variables and 0 constraints```	```# Problem has 3 variables and 4 constraints```
```# Variables:```	```# Variables:```
```variables = 15 < x1 < 30 / Continuous,```	```variables = 15 < x1 < 30 / Continuous,```
```0 < x2 < 1 / Integer;```	```0 < x2 < 1 / Integer,```
	```15 < w3 < 30 / Continuous;```
```# Objective Function:```	```# Objective Function:```
```objfun = min [(x1)*(x2)];```	```objfun = min [w3];```
```# Constraints:```	```# Constraints:```
```constraints = 0;```	```constraints = [-1e+30 < (-30*x2)+(w3) < 0],```
	```[-1e+30 < (15*x2)+(-1*w3) < 0],```
	```[-1e+30 < (15)+(-1*x1)+(-15*x2)+(w3) < 0],```
	```[-1e+30 < (-30)+(x1)+(30*x2)+(-1*w3) < 0];```

As presented in Section 3.3, ROSE identifies all the terms involving a continuous and a binary variable (respectively `x1` and `x2` in the example) and add exactly one variable (w3 here) and four constraints. The reader might now check that both the objective function and the constraints are linear terms and that the computed values are similar in the two formulations of the problem.

The second example is an optimization problem whose objective function contains four nonlinear terms. We show how ROSE is able to find a convex relaxation for the problem using the convexifier reformulator (see Section 4.2).

Original Problem

```
# ROSE problem: convexifier
# Problem has 3 variables and 1 constraints
# Variables:

variables = -1 < x1 < 1 / Continuous,
-2 < y2 < 3 / Continuous,
1 < t3 < 2 / Continuous;
```

```
# Objective Function:
objfun = min [ (2*x1^2)+(y2^3)
               +((x1)*(y2))+((x1)/(t3)) ];
# Constraints:
constraints = [ 2 < (x1)+(y2) < 1e+30 ];
```

ROSE Reformulation

```
# ROSE problem: convexifier
# Problem has 9 variables and 18 constraints
# Variables:

variables = -1 < x1 < 1 / Continuous,
-2 < y2 < 3 / Continuous,
1 < t3 < 2 / Continuous,
0 < w4 < 2 / Continuous,
-8 < w5 < 27 / Continuous,
-3 < w6 < 3 / Continuous,
-1 < w7 < 1 / Continuous,
-12 < w8 < 33 / Continuous,
0.5 < z9 < 1 / Continuous;
```

```
# Objective Function:
objfun = min [ w8 ];

# Constraints:
constraints = [ 2 < (x1)+(y2) < 1e+30 ],
[ 0 < (w4)+(w5)+(w6)+(w7)+(-1*w8) < 0 ],
[ -2 < (4*x1)+(w4) < 1e+30 ],
[ -2 < (-4*x1)+(w4) < 1e+30 ],
[ -0.5 < (2*x1)+(w4) < 1e+30 ],
[ -0.5 < (-2*x1)+(w4) < 1e+30 ],
[ -2 < (-3*y2)+(w5) < 1e+30 ],
[ -54 < (-27*y2)+(w5) < 1e+30 ],
[ -1e+30 < (-6.75*y2)+(w5) < 6.75 ],
[ -1e+30 < (-12*y2)+(w5) < 16 ],
[ -2 < (2*x1)+(y2)+(w6) < 1e+30 ],
[ -3 < (-3*x1)+(-1*y2)+(w6) < 1e+30 ],
[ -1e+30 < (-3*x1)+(y2)+(w6) < 3 ],
[ -1e+30 < (2*x1)+(-1*y2)+(w6) < 2 ],
[ 0.5 < (-0.5*x1)+(w7)+(z9) < 1e+30 ],
[ -1 < (-1*x1)+(w7)+(-1*z9) < 1e+30 ],
[ -1e+30 < (-1*x1)+(w7)+(z9) < 1 ],
[ -1e+30 < (-0.5*x1)+(w7)+(-1*z9) < -0.5 ];
```

The reformulation process is performed in various steps. In order to explain how the reformulator/convexifier works, we show in the following how the original problem is modified during the main steps.

The first step consists in reformulating the problem to the Smith standard form. Each nonconvex term in the objective function is replaced by an added variable w and defining constraints of the form $w = nonconvex\,term$ are added to the problem. The objective function of the reformulated problem is one linearizing variable only, that is the sum of all the added variables, and a constraint for this equation is also added to the problem. We remark that the obtained reformulation is a lifting reformulation, since a new variable is added for each nonconvex term. This first-stage reformulation is the following:

```
# ROSE problem: convexifier
# Problem has 8 variables and 6 constraints
# Variables:

variables = -1 < x1 < 1 / Continuous,
-2 < y2 < 3 / Continuous,
1 < t3 < 2 / Continuous,
0 < w4 < 2 / Continuous,
-8 < w5 < 27 / Continuous,
-3 < w6 < 3 / Continuous,
-1 < w7 < 1 / Continuous,
-12 < w8 < 33 / Continuous;

# Objective Function:
objfun = min [ w8 ];

# Constraints:
constraints = [ 2 < (x1)+(y2) < 1e+30 ],
[ 0 < (-1*w4)+(2*x1^2) < 0 ],
[ 0 < (-1*w5)+(y2^3) < 0 ],
[ 0 < (-1*w6)+((x1)*(y2)) < 0 ],
[ 0 < (-1*w7)+((x1)/(t3)) < 0 ],
[ 0 < (w4)+(w5)+(w6)+(w7)+(-1*w8) < 0 ];
```

Then, each defining constraint is replaced by a convex under-estimator and concave over-estimator of the corresponding nonlinear term. In particular, the term 2*x1^2 is treated as a convex univariate function f(x) and a linear under-estimator is obtained by considering five tangents to f at various given points, an over-estimator is obtained by considering the secant through the points (x1^L,f(x1^L)),(x1^U,f(x1^U)), where x1^L, and x1^U are the bounds on x1. For the term y2^3, where the range of y2 includes zero, the linear relaxation given in [80] is used. McCormick's envelopes are considered for the bilinear term x1*y2. The fractional term is reformulated as bilinear by considering z=1/t3 and McCormick's envelopes are exploited again. We obtain the following relaxation:

```
# ROSE problem: convexifier
# Problem has 9 variables and 22 constraints
# Variables:

variables = -1 < x1 < 1 / Continuous,
-2 < y2 < 3 / Continuous,
1 < t3 < 2 / Continuous,
0 < w4 < 2 / Continuous,
-8 < w5 < 27 / Continuous,
-3 < w6 < 3 / Continuous,
-1 < w7 < 1 / Continuous,
```

```
-12 < w8 < 33 / Continuous,
0.5 < z9 < 1 / Continuous;

# Objective Function:
objfun = min [ w8 ];

# Constraints:
constraints = [ 2 < (x1)+(y2) < 1e+30 ],
[ 0 < (-1*w4)+(2*x1^2) < 0 ],
[ 0 < (-1*w5)+(y2^3) < 0 ],
[ 0 < (-1*w6)+((x1)*(y2)) < 0 ],
[ 0 < (-1*w7)+((x1)/(t3)) < 0 ],
[ 0 < (w4)+(w5)+(w6)+(w7)+(-1*w8) < 0 ],
[ -2 < (4*x1)+(w4) < 1e+30 ],
[ -2 < (-4*x1)+(w4) < 1e+30 ],
[ -0.5 < (2*x1)+(w4) < 1e+30 ],
[ -0.5 < (-2*x1)+(w4) < 1e+30 ],
[ -2 < (-3*y2)+(w5) < 1e+30 ],
[ -54 < (-27*y2)+(w55) < 1e+30 ],
[ -1e+30 < (-6.75*y2)+(w5) < 6.75 ],
[ -1e+30 < (-12*y2)+(w5) < 16 ],
[ -2 < (2*x1)+(y2)+(w6) < 1e+30 ],
[ -3 < (-3*x1)+(-1*y2)+(w6) < 1e+30 ],
[ -1e+30 < (-3*x1)+(y2)+(w6) < 3 ],
[ -1e+30 < (2*x1)+(-1*y2)+(w6) < 2 ],
[ 0.5 < (-0.5*x1)+(w7)+(z9) < 1e+30 ],
[ -1 < (-1*x1)+(w7)+(-1*z9) < 1e+30 ],
[ -1e+30 < (-1*x1)+(w7)+(z9) < 1 ],
[ -1e+30 < (-0.5*x1)+(w7)+(-1*z9) < -0.5 ];
```

Finally, the Smith defining constraints are removed, obtaining the final reformulation (of the relaxation type).

6 Conclusion

This chapter contains a study of mathematical programming reformulation and relaxation techniques. Section 1 presents some motivations towards such a study, the main being that Mixed Integer Nonlinear Programming solvers need to be endowed with automatic reformulation capabilities before they can be as reliable, functional and efficient as their industrial-strength Mixed Integer Linear Programming solvers are. Section 2 presents a general framework for representing and manipulating mathematical programming formulations, as well as some definitions of the concept of reformulation together with some theoretical results; the section is concluded by listing some of the most common standard forms in mathematical programming. In Section 3 we present a partial systematic study of existing reformulations. Each reformulation is presented both in symbolic algorithmic terms (i.e. a prototype for carrying out the reformulation automatically in terms of the provided data structures is always supplied) and in the more usual mathematical terms. This should be seen as the starting point for a more exhaustive study: eventually, all known useful reformulations might find their place in an automatic reformulation preprocessing software for Mixed Integer Nonlinear Programming. In Section 4, we attempt a similar work with respect to relaxations. Section 5 describes the implementation of ROSE, a reformulation/optimization software engine.

Acknowledgements. Financial support by ANR grant 07-JCJC-0151 and by the EU NEST "Morphex" project grant is gratefully acknowledged. We also wish to thank: Claudia D'Ambrosio and David Savourey for help on the ROSE implementation; Pierre Hansen, Nenad Mladenović, Frank Plastria, Hanif Sherali and Tapio Westerlund for many useful discussions and ideas; Kanika Dhyani and Fabrizio Marinelli for providing interesting application examples.

References

1. Adams, W., Forrester, R., Glover, F.: Comparisons and enhancement strategies for linearizing mixed 0-1 quadratic programs. Discrete Optimization 1, 99–120 (2004)
2. Adams, W., Sherali, H.: A tight linearization and an algorithm for 0-1 quadratic programming problems. Management Science 32(10), 1274–1290 (1986)
3. Adams, W., Sherali, H.: A hierarchy of relaxations leading to the convex hull representation for general discrete optimization problems. Annals of Operations Research 140, 21–47 (2005)
4. Adjiman, C., Dallwig, S., Floudas, C., Neumaier, A.: A global optimization method, αBB, for general twice-differentiable constrained NLPs: I. Theoretical advances. Computers & Chemical Engineering 22(9), 1137–1158 (1998)
5. Adjiman, C.S., Androulakis, I.P., Floudas, C.A.: A global optimization method, αBB, for general twice-differentiable constrained NLPs: II. Implementation and computational results. Computers & Chemical Engineering 22(9), 1159–1179 (1998)
6. Aho, A., Hopcroft, J., Ullman, J.: Data Structures and Algorithms. Addison-Wesley, Reading (1983)
7. Al-Khayyal, F., Falk, J.: Jointly constrained biconvex programming. Mathematics of Operations Research 8(2), 273–286 (1983)
8. Alizadeh, F.: Interior point methods in semidefinite programming with applications to combinatorial optimization. SIAM Journal on Optimization 5(1), 13–51 (1995)
9. Andersen, K., Cornuéjols, G., Li, Y.: Reduce-and-split cuts: Improving the performance of mixed-integer Gomory cuts. Management Science 51(11), 1720–1732 (2005)
10. Androulakis, I.P., Maranas, C.D., Floudas, C.A.: *alpha*BB: A global optimization method for general constrained nonconvex problems. Journal of Global Optimization 7(4), 337–363 (1995)

11. Anstreicher, K.: SDP versus RLT for nonconvex QCQPs. In: Floudas, C., Pardalos, P. (eds.) Proceedings of Advances in Global Optimization: Methods and Applications, Mykonos, Greece (2007)
12. Applegate, D., Bixby, R., Chvátal, V., Cook, W.: The Travelling Salesman Problem: a Computational Study. Princeton University Press, Princeton (2007)
13. Audet, C., Hansen, P., Jaumard, B., Savard, G.: Links between linear bilevel and mixed 0-1 programming problems. Journal of Optimization Theory and Applications 93(2), 273–300 (1997)
14. Balas, E.: Intersection cuts — a new type of cutting planes for integer programming. Operations Research 19(1), 19–39 (1971)
15. Balas, E.: Projection, lifting and extended formulation in integer and combinatorial optimization. Annals of Operations Research 140, 125–161 (2005)
16. Balas, E., Jeroslow, R.: Canonical cuts on the unit hypercube. SIAM Journal on Applied Mathematics 23(1), 61–69 (1972)
17. Bauer, C., Frink, A., Kreckel, R.: Introduction to the ginac framework for symbolic computation within the C++ programming language. Journal of Symbolic Computation 33(1), 1–12 (2002)
18. Belotti, P., Lee, J., Liberti, L., Margot, F., Wächter, A.: Branching and bound reduction techniques for non-convex MINLP. Optimization Methods and Software (submitted)
19. Björk, K.M., Lindberg, P., Westerlund, T.: Some convexifications in global optimization of problems containing signomial terms. Computers & Chemical Engineering 27, 669–679 (2003)
20. Bjorkqvist, J., Westerlund, T.: Automated reformulation of disjunctive constraints in MINLP optimization. Computers & Chemical Engineering 23, S11–S14 (1999)
21. Boyd, E.: Fenchel cutting planes for integer programs. Operations Research 42(1), 53–64 (1994)
22. Boyd, S., Vandenberghe, L.: Convex Optimization. Cambridge University Press, Cambridge (2004)
23. Brook, A., Kendrick, D., Meeraus, A.: GAMS, a user's guide. ACM SIGNUM Newsletter 23(3-4), 10–11 (1988)
24. Caporossi, G., Alamargot, D., Chesnet, D.: Using the computer to study the dyamics of the handwriting processes. In: Suzuki, E., Arikawa, S. (eds.) DS 2004. LNCS (LNAI), vol. 3245, pp. 242–254. Springer, Heidelberg (2004)
25. Cornuéjols, G.: Valid inequalities for mixed integer linear programs. Mathematical Programming B 112(1), 3–44 (2008)
26. Cortellessa, V., Marinelli, F., Potena, P.: Automated selection of software components based on cost/reliability tradeoff. In: Gruhn, V., Oquendo, F. (eds.) EWSA 2006. LNCS, vol. 4344, pp. 66–81. Springer, Heidelberg (2006)
27. Dantzig, G.: Linear Programming and Extensions. Princeton University Press, Princeton (1963)
28. Davidović, T., Liberti, L., Maculan, N., Mladenović, N.: Towards the optimal solution of the multiprocessor scheduling problem with communication delays. In: MISTA Proceedings (2007)
29. Dhyani, K.: Personal communication (2007)
30. Di Giacomo, L.: Mathematical programming methods in dynamical nonlinear stochastic supply chain management. Ph.D. thesis, DSPSA, Università di Roma "La Sapienza" (2007)

31. Duran, M., Grossmann, I.: An outer-approximation algorithm for a class of mixed-integer nonlinear programs. Mathematical Programming 36, 307–339 (1986)
32. Falk, J., Liu, J.: On bilevel programming, part I: General nonlinear cases. Mathematical Programming 70, 47–72 (1995)
33. Falk, J., Soland, R.: An algorithm for separable nonconvex programming problems. Management Science 15, 550–569 (1969)
34. Fischer, A.: New constrained optimization reformulation of complementarity problems. Journal of Optimization Theory and Applications 99(2), 481–507 (1998)
35. Fletcher, R., Leyffer, S.: Solving mixed integer nonlinear programs by outer approximation. Mathematical Programming 66, 327–349 (1994)
36. Floudas, C.: Deterministic Global Optimization. Kluwer Academic Publishers, Dordrecht (2000)
37. Fortet, R.: Applications de l'algèbre de Boole en recherche opérationelle. Revue Française de Recherche Opérationelle 4, 17–26 (1960)
38. Fourer, R.: Personal communication (2004)
39. Fourer, R., Gay, D.: The AMPL Book. Duxbury Press, Pacific Grove (2002)
40. Galli, S.: Parsing AMPL internal format for linear and non-linear expressions, B.Sc. dissertation, DEI, Politecnico di Milano, Italy (2004)
41. Gill, P.: User's Guide for SNOPT 5.3. Systems Optimization Laboratory, Department of EESOR, Stanford University, California (1999)
42. Gill, P.: User's guide for SNOPT version 7. In: Systems Optimization Laboratory. Stanford University, California (2006)
43. Gomory, R.: Essentials of an algorithm for integer solutions to linear programs. Bulletin of the American Mathematical Society 64(5), 256 (1958)
44. Grant, M., Boyd, S., Ye, Y.: Disciplined convex programming. In: Liberti and Maculan [79], pp. 155–210
45. Guéret, C., Prins, C., Sevaux, M.: Applications of optimization with Xpress-MP. Dash Optimization, Bilsworth (2000)
46. Hammer, P., Rudeanu, S.: Boolean Methods in Operations Research and Related Areas. Springer, Berlin (1968)
47. Hansen, P.: Method of non-linear 0-1 programming. Annals of Discrete Mathematics 5, 53–70 (1979)
48. Haverly, C.: Studies of the behaviour of recursion for the pooling problem. ACM SIGMAP Bulletin 25, 19–28 (1978)
49. Horst, R.: On the convexification of nonlinear programming problems: an applications-oriented approach. European Journal of Operations Research 15, 382–392 (1984)
50. Horst, R., Tuy, H.: Global Optimization: Deterministic Approaches, 3rd edn. Springer, Berlin (1996)
51. Horst, R., Van Thoai, N.: Duality bound methods in global optimization. In: Audet, C., Hansen, P., Savard, G. (eds.) Essays and Surveys in Global Optimization, pp. 79–105. Springer, Berlin (2005)
52. ILOG: ILOG CPLEX 11.0 User's Manual. ILOG S.A., Gentilly, France (2008)
53. Judice, J., Mitra, G.: Reformulation of mathematical programming problems as linear complementarity problems and investigation of their solution methods. Journal of Optimization Theory and Applications 57(1), 123–149 (1988)
54. Kaibel, V., Pfetsch, M.: Packing and partitioning orbitopes. Mathematical Programming 114(1), 1–36 (2008)

55. Kaltofen, E.: Challenges of symbolic computation: My favorite open problems. Journal of Symbolic Computation 29, 891–919 (2000), citeseer.nj.nec.com/article/kaltofen00challenge.html
56. Kelley, J.: The cutting plane method for solving convex programs. Journal of SIAM VIII(6), 703–712 (1960)
57. Kesavan, P., Allgor, R., Gatzke, E., Barton, P.: Outer-approximation algorithms for nonconvex mixed-integer nonlinear programs. Mathematical Programming 100(3), 517–535 (2004)
58. Kojima, M., Megiddo, N., Ye, Y.: An interior point potential reduction algorithm for the linear complementarity problem. Mathematical Programming 54, 267–279 (1992)
59. Konno, H.: A cutting plane algorithm for solving bilinear programs. Mathematical Programming 11, 14–27 (1976)
60. Kucherenko, S., Belotti, P., Liberti, L., Maculan, N.: New formulations for the kissing number problem. Discrete Applied Mathematics 155(14), 1837–1841 (2007)
61. Kucherenko, S., Sytsko, Y.: Application of deterministic low-discrepancy sequences in global optimization. Computational Optimization and Applications 30(3), 297–318 (2004)
62. Lavor, C., Liberti, L., Maculan, N.: Computational experience with the molecular distance geometry problem. In: Pintér, J. (ed.) Global Optimization: Scientific and Engineering Case Studies, pp. 213–225. Springer, Berlin (2006)
63. Lavor, C., Liberti, L., Maculan, N., Chaer Nascimento, M.: Solving Hartree-Fock systems with global optimization metohds. Europhysics Letters 5(77), 50,006p1–50,006p5 (2007)
64. Letchford, A., Lodi, A.: Strengthening Chvátal-Gomory cuts and Gomory fractional cuts. Operations Research Letters 30, 74–82 (2002)
65. Liberti, L.: Framework for symbolic computation in C++ using n-ary trees. Tech. rep., CPSE, Imperial College London (2001)
66. Liberti, L.: Comparison of convex relaxations for monomials of odd degree. In: Tseveendorj, I., Pardalos, P., Enkhbat, R. (eds.) Optimization and Optimal Control. World Scientific, Singapore (2003)
67. Liberti, L.: Reduction constraints for the global optimization of NLPs. International Transactions in Operational Research 11(1), 34–41 (2004)
68. Liberti, L.: Reformulation and convex relaxation techniques for global optimization. 4OR 2, 255–258 (2004)
69. Liberti, L.: Reformulation and convex relaxation techniques for global optimization. Ph.D. thesis, Imperial College London, UK (2004)
70. Liberti, L.: Linearity embedded in nonconvex programs. Journal of Global Optimization 33(2), 157–196 (2005)
71. Liberti, L.: Writing global optimization software. In: Liberti and Maculan [79], pp. 211–262
72. Liberti, L.: Compact linearization of binary quadratic problems. 4OR 5(3), 231–245 (2007)
73. Liberti, L.: Reformulations in mathematical programming: Definitions. In: Aringhieri, R., Cordone, R., Righini, G. (eds.) Proceedings of the 7th Cologne-Twente Workshop on Graphs and Combinatorial Optimization, pp. 66–70. Università Statale di Milano, Crema (2008)
74. Liberti, L.: Spherical cuts for integer programming problems. International Transactions in Operational Research 15, 283–294 (2008)

75. Liberti, L.: Reformulations in mathematical programming: Definitions and systematics. RAIRO-RO (accepted for publication)
76. Liberti, L., Dražic, M.: Variable neighbourhood search for the global optimization of constrained NLPs. In: Proceedings of GO Workshop, Almeria, Spain (2005)
77. Liberti, L., Lavor, C., Maculan, N.: Double VNS for the molecular distance geometry problem. In: Proc. of Mini Euro Conference on Variable Neighbourhood Search, Tenerife, Spain (2005)
78. Liberti, L., Lavor, C., Nascimento, M.C., Maculan, N.: Reformulation in mathematical programming: an application to quantum chemistry. Discrete Applied Mathematics (accepted for publication)
79. Liberti, L., Maculan, N. (eds.): Global Optimization: from Theory to Implementation. Springer, Berlin (2006)
80. Liberti, L., Pantelides, C.: Convex envelopes of monomials of odd degree. Journal of Global Optimization 25, 157–168 (2003)
81. Liberti, L., Pantelides, C.: An exact reformulation algorithm for large nonconvex NLPs involving bilinear terms. Journal of Global Optimization 36, 161–189 (2006)
82. Liberti, L., Tsiakis, P., Keeping, B., Pantelides, C.: ooOPS. Centre for Process Systems Engineering, Chemical Engineering Department, Imperial College, London, UK (2001)
83. Lougee-Heimer, R.: The common optimization interface for operations research: Promoting open-source software in the operations research community. IBM Journal of Research and Development 47(1), 57–66 (2003)
84. Maculan, N., Macambira, E., de Souza, C.: Geometrical cuts for 0-1 integer programming. Tech. Rep. IC-02-006, Instituto de Computação, Universidade Estadual de Campinas (2002)
85. Makhorin, A.: GNU Linear Programming Kit. Free Software Foundation (2003), http://www.gnu.org/software/glpk/
86. Mangasarian, O.: Linear complementarity problems solvable by a single linear program. Mathematical Programming 10, 263–270 (1976)
87. Mangasarian, O.: The linear complementarity problem as a separable bilinear program. Journal of Global Optimization 6, 153–161 (1995)
88. Maranas, C.D., Floudas, C.A.: Finding all solutions to nonlinearly constrained systems of equations. Journal of Global Optimization 7(2), 143–182 (1995)
89. Margot, F.: Pruning by isomorphism in branch-and-cut. Mathematical Programming 94, 71–90 (2002)
90. Margot, F.: Exploiting orbits in symmetric ILP. Mathematical Programming B 98, 3–21 (2003)
91. McCormick, G.: Computability of global solutions to factorable nonconvex programs: Part I — Convex underestimating problems. Mathematical Programming 10, 146–175 (1976)
92. Meyer, C., Floudas, C.: Trilinear monomials with mixed sign domains: Facets of the convex and concave envelopes. Journal of Global Optimization 29, 125–155 (2004)
93. Mladenović, N., Plastria, F., Urošević, D.: Reformulation descent applied to circle packing problems. Computers and Operations Research 32(9), 2419–2434 (2005)
94. Nemhauser, G., Wolsey, L.: Integer and Combinatorial Optimization. Wiley, New York (1988)

95. Nowak, I.: Relaxation and Decomposition Methods for Mixed Integer Nonlinear Programming. Birkhäuser, Basel (2005)
96. Pantelides, C., Liberti, L., Tsiakis, P., Crombie, T.: Mixed integer linear/nonlinear programming interface specification. Global Cape-Open Deliverable WP2.3-04 (2002)
97. Pardalos, P., Romeijn, H. (eds.): Handbook of Global Optimization, vol. 2. Kluwer Academic Publishers, Dordrecht (2002)
98. Plateau, M.C.: Reformulations quadratiques convexes pour la programmation quadratique en variables 0-1. Ph.D. thesis, Conservatoire National d'Arts et Métiers (2006)
99. Puchinger, J., Raidl, G.: Relaxation guided variable neighbourhood search. In: Proc. of Mini Euro Conference on Variable Neighbourhood Search, Tenerife, Spain (2005)
100. Raghavachari, M.: On connections between zero-one integer programming and concave programming under linear constraints. Operations Research 17(4), 680–684 (1969)
101. van Roy, T., Wolsey, L.: Solving mixed integer programming problems using automatic reformulation. Operations Research 35(1), 45–57 (1987)
102. Ryoo, H., Sahinidis, N.: Global optimization of nonconvex NLPs and MINLPs with applications in process design. Computers & Chemical Engineering 19(5), 551–566 (1995)
103. Sherali, H.: Global optimization of nonconvex polynomial programming problems having rational exponents. Journal of Global Optimization 12, 267–283 (1998)
104. Sherali, H.: Tight relaxations for nonconvex optimization problems using the reformulation-linearization/convexification technique (RLT). In: Pardalos and Romeijn [97], pp. 1–63
105. Sherali, H.: Personal communication (2007)
106. Sherali, H., Adams, W.: A hierarchy of relaxations between the continuous and convex hull representations for zero-one programming problems. SIAM Journal of Discrete Mathematics 3, 411–430 (1990)
107. Sherali, H., Adams, W.: A Reformulation-Linearization Technique for Solving Discrete and Continuous Nonconvex Problems. Kluwer Academic Publishers, Dodrecht (1999)
108. Sherali, H., Alameddine, A.: A new reformulation-linearization technique for bilinear programming problems. Journal of Global Optimization 2, 379–410 (1992)
109. Sherali, H., Liberti, L.: Reformulation-linearization technique for global optimization. In: Floudas, C., Pardalos, P. (eds.) Encyclopedia of Optimization, 2nd edn., pp. 3263–3268. Springer, New York (2008)
110. Sherali, H., Tuncbilek, C.: New reformulation linearization/convexification relaxations for univariate and multivariate polynomial programming problems. Operations Research Letters 21, 1–9 (1997)
111. Sherali, H., Wang, H.: Global optimization of nonconvex factorable programming problems. Mathematical Programming 89, 459–478 (2001)
112. Smith, E.: On the optimal design of continuous processes. Ph.D. thesis, Imperial College of Science, Technology and Medicine, University of London (1996)
113. Smith, E., Pantelides, C.: Global optimisation of nonconvex MINLPs. Computers & Chemical Engineering 21, S791–S796 (1997)

114. Smith, E., Pantelides, C.: A symbolic reformulation/spatial branch-and-bound algorithm for the global optimisation of nonconvex MINLPs. Computers & Chemical Engineering 23, 457–478 (1999)
115. Strekalovsky, A.: On global optimality conditions for d.c. programming problems. Technical Paper, Irkutsk State University (1997)
116. Strekalovsky, A.: Extremal problems with d.c. constraints. Computational Mathematics and Mathematical Physics 41(12), 1742–1751 (2001)
117. Stroustrup, B.: The C++ Programming Language, 3rd edn. Addison-Wesley, Reading (1999)
118. Sutou, A., Dai, Y.: Global optimization approach to unequal sphere packing problems in 3d. Journal of Optimization Theory and Applications 114(3), 671–694 (2002)
119. Tardella, F.: Existence and sum decomposition of vertex polyhedral convex envelopes. Tech. rep., Facoltà di Economia e Commercio, Università di Roma "La Sapienza" (2007)
120. Tawarmalani, M., Ahmed, S., Sahinidis, N.: Global optimization of 0-1 hyperbolic programs. Journal of Global Optimization 24, 385–416 (2002)
121. Tawarmalani, M., Sahinidis, N.: Semidefinite relaxations of fractional programming via novel techniques for constructing convex envelopes of nonlinear functions. Journal of Global Optimization 20(2), 137–158 (2001)
122. Tawarmalani, M., Sahinidis, N.: Convex extensions and envelopes of semicontinuous functions. Mathematical Programming 93(2), 247–263 (2002)
123. Tawarmalani, M., Sahinidis, N.: Exact algorithms for global optimization of mixed-integer nonlinear programs. In: Pardalos and Romeijn [97], pp. 65–86
124. Tawarmalani, M., Sahinidis, N.: Global optimization of mixed integer nonlinear programs: A theoretical and computational study. Mathematical Programming 99, 563–591 (2004)
125. Todd, M.: Semidefinite optimization. Acta Numerica 10, 515–560 (2001)
126. Tsiakis, P., Keeping, B.: $oo\mathcal{MILP}$ – a C++ callable object-oriented library and the implementation of its parallel version using corba. In: Liberti and Maculan [79], pp. 155–210
127. Tsiakis, P., Keeping, B., Pantelides, C.: $oo\mathcal{MILP}$. Centre for Process Systems Engineering, Chemical Engineering Department, Imperial College, London, UK, 0.7 edn (2000)
128. Tuy, H.: Concave programming under linear constraints. Soviet Mathematics, 1437–1440 (1964)
129. Tuy, H.: D.c. optimization: Theory, methods and algorithms. In: Horst, R., Pardalos, P. (eds.) Handbook of Global Optimization, vol. 1, pp. 149–216. Kluwer Academic Publishers, Dordrecht (1995)
130. Wang, X., Change, T.: A multivariate global optimization using linear bounding functions. Journal of Global Optimization 12, 383–404 (1998)
131. Westerlund, T.: Some transformation techniques in global optimization. In: Liberti and Maculan [79], pp. 45–74
132. Westerlund, T., Skrifvars, H., Harjunkoski, I., Pörn, R.: An extended cutting plane method for a class of non-convex MINLP problems. Computers & Chemical Engineering 22(3), 357–365 (1998)
133. Wolsey, L.: Integer Programming. Wiley, New York (1998)
134. Zamora, J.M., Grossmann, I.E.: A branch and contract algorithm for problems with concave univariate, bilinear and linear fractional terms. Journal of Global Optimization 14, 217–249 (1999)

Graph-Based Local Elimination Algorithms in Discrete Optimization[*]

Oleg Shcherbina

Abstract. The aim of this chapter is to provide a review of structural decomposition methods in discrete optimization and to give a unified framework in the form of local elimination algorithms (LEA). This chapter is organized as follows. Local elimination algorithms for discrete optimization (DO) problems (DOPs) with constraints are considered; a classification of dynamic programming computational procedures is given. We introduce Elimination Game and Elimination tree. Application of bucket elimination algorithm from constraint satisfaction (CS) to solving DOPs is done. We consider different local elimination schemes and related notions. Clustering that merges several variables into single meta-variable defines a promising approach to solve DOPs. This allows to create a quotient (condensed) graph and apply a local block elimination algorithm. In order to describe a block elimination process, we introduce Block Elimination Game. We discuss the connection of aforementioned local elimination algorithmic schemes and a way of transforming the directed acyclic graph (DAG) of computational LEA procedure to the tree decomposition.

1 Introduction

The use of discrete optimization (DO) models and algorithms makes it possible to solve many practical problems in scheduling theory, network

Oleg Shcherbina
Faculty of Mathematics,
University of Vienna
Nordbergstrasse 15, A-1090 Vienna,
Austria
e-mail: `oleg.shcherbina@univie.ac.at`

[*] Research supported by FWF (Austrian Science Funds) under the project P17948-N13.

optimization, routing in communication networks, facility location, optimization in enterprise resource planning, and logistics (in particular, in supply chain management [36]). The field of artificial intelligence includes aspects like theorem proving, SAT in propositional logic (see [23], [50]), robotics problems, inference calculation in Bayesian networks [66], scheduling, and others.

Many real-life DO problems contain a huge number of variables and/or constraints that make the models intractable for currently available DO solvers. NP-hardness refers to the worst-case complexity of problems. Recognizing problem instances that are better (and easier for solving) than these "worst cases" is a rewarding task given that better algorithms can be used for these easy cases.

Complexity theory has proved that universality and effectiveness are contradictory requirements to algorithm complexity. But the complexity of some class of problems decreases if the class may be divided into subsets and the special structure of these subsets can be used in the algorithm design.

To meet the challenge of solving large scale DO problems (DOPs) in reasonable time, there is an urgent need to develop new decomposition approaches [22], [82], [75]. Large-scale DOPs are characterized not only by huge size but also by special or sparse structure. The block form of many DO problems is usually caused by the weak connectedness of subsystems of real systems. One of the first examples of large sparse linear programming (LP) problems which DANTZIG started to study was a class of staircase LP problems for dynamic planning [27], [29], [28]. Further examples of staircase linear programs (see FOURER [42]) for multiperiod planning, scheduling, and assignment, and for multistage structural design, are included in a set of staircase test problems collected by HO & LOUTE [57]. Staircase linear programs have also been derived in connection with linearly constrained optimal control and stochastic programming [103]. Problems of optimal hotel apartments assignment, linear dynamic programming, labor resources allocation, control on hierarchic structures (usually having tree-like structure), multistage integer stochastic programming, network problems may be considered as examples of DO problems which have staircase structure (see [89], [90]). The well known SAT problem stems from classical investigations by logicians of propositional satisfiability and has over 90 years of history. It is possible to represent a SAT problem as a sparse DO problem [58]. Some applied facility location problems can be formulated as set covering problems, set packing problems, node packing problems [73]. Another class of sparse DO problems is a production lot-sizing problem [73]. The frequency assignment problem (FAP) [65] in mobile telephone systems communication is a hard problem as it is closely related to the graph coloring problem. One of the well known decomposition approaches to solving DOPs is Lagrangean decomposition that consists of isolating sets of constraints to obtain separate and easy to solve DO problems. Lagrangean decomposition removes the complicating constraints from the constraint set and inserts them into the objective function. Most Lagrangean decomposition methods deal with special row structures. Block angular structures with complicating variables and with complicating variables and constraints can be decomposed using Benders decomposition [13] and cross decomposition [99]. The Dantzig-Wolfe decomposition principle of LP has its equivalent in integer programming [98]. This approach uses the reformulation that gives rise to an integer master problem, whose typically large number of variables is dealt with implicitly by using an integer programming column generation procedure, also known as branch-and-price algorithm [9] that allows solving large-scale DOPs in recent years. NEMHAUSER ([74], p. 9) mentioned, however, that

> ... the overall idea of using branch and bound with linear programming relaxation has not changed.

Usually, DOPs from applications have a special structure, and the matrices of constraints for large-scale problems have a lot of zero elements (sparse matrices). Among decomposition approaches appropriate for solving such problems we mention poorly known local decomposition algorithms using the special block matrix structure of constraints and half-forgotten nonserial dynamic programming algorithms (NSDP) (BERTELE & BRIOSCHI [14], [15], [16], DECHTER [31], [32], [33], [34], HOOKER [58]) which can exploit sparsity in the dependency graph of a DOP and allow to compute a solution in stages such that each of them uses results from previous stages.

Recently, there has been growing interest in graph-based approaches to decomposition [19]; one of them is tree decomposition (TD). COURCELLE [25] and ARNBORG et al. [6] showed that several NP-hard problems posed in monadic second-order logic can be solved in polynomial time using dynamic programming techniques on input graphs with bounded treewidth. Thus graph-based decomposition approaches have gained importance. Graph-based structural decomposition techniques, e.g., nonserial dynamic programming (NSDP) (BERTELE, BRIOSCHI [16], ESOGBUE & MARKS [37], HOOKER [58], Martelli & Montanari [68], Mitten & Nemhauser [69], NEUMAIER & SHCHERBINA [76], ROSENTHAL [86], SHCHERBINA [91]), WILDE & BEIGHTLER [101] and its modifications (bucket elimination [32], Seidel's invasion method [87]), tree decomposition combined with dynamic programming [35], [21] and its variants [77], hypertree [47] and hinge decomposition [60], [49] are promising decomposition approaches that allow exploiting the structure of discrete problems in constraint satisfaction (CS) [43] and DO.

It is important that aforementioned methods use just the local information (i.e., information about elements of given element's neighborhood) in a process of solving discrete problems. It is possible to propose a class of local elimination algorithms as a general framework that allows to calculate some **global information** about a solution of the entire problem using **local computations** [62], [66], [95]. Note that a main feature in aforementioned problems is the locality of information, a definition of elements' neighborhoods and studying them.

The use of local information (see [104], [105], [39], [94], [97]) is very important in studying complex discrete systems and in the development of decomposition methods for solving large sparse discrete problems; these problems
simultaneously belong to the fields of discrete optimization [73], [40], [78], [79],
[88], artificial intelligence [32], [48], [72], [81], and databases [10]. In linear algebra, multifrontal techniques for solving sparse systems of linear equations
were developed (see [85]); these methods are also of the decomposition nature.
In [104], local algorithms for computing information are introduced. A local
algorithm A examines the elements in the order specified by an ordering algorithm A_π, calculates the function ϕ whose value at each step determines the
form of the information marks, and labels the element using local information
about the elements in its neighborhood. The function ϕ that induces the algorithm depends on two variables: the first ranges over the set of all elements
and the second ranges over the set of neighborhoods. Local decomposition
algorithms (see [89], [90]) in DO problems have a specific feature. Namely,
rather than calculating predicates, they use Bellman's optimality principle
[12] to find optimal solutions of the subproblems corresponding to blocks of
the DO problem. A step of the local algorithm $\mathfrak{A}$ changes the neighborhood
and replaces the index p by $p + 1$ (however, one can increment the index by
an arbitrary number replacing S_p by $S_{p+\rho}$; at each step of the algorithm, for
every fixed set of variables of the boundary ring, the values of the variables of
the corresponding neighborhood are stored, which is an important difference
of the local algorithm $\mathfrak{A}$ from A: information about variables in the solutions
of the subproblems is stored rather than information about the predicates.
ZHURAVLEV proposed to call it **indicator information**.

Tree and branch decomposition algorithms have been shown to be effective for DO problems like the traveling salesman problem [24], frequency
assignment [65] etc. (see a survey paper [55]). A paper [4] surveys algorithms
that use tree decompositions. Most of works based on tree decomposition
approach only present theoretical results [61], see the recent surveys [55],
[92]. Thus these methods are not yet recognized tools of operations research
practitioners.

Some implementations of NSDP are known [16], [38], however, generally,
it remains some "obscure" tool for operations research modellers. Usually,
tree decomposition approaches and NSDP are considered in the literature
separately, without reference to the close relation between these methods.
We try to indicate a close relation between these methods.

A need to solve large-scale discrete problems with special structure using
graph-based structural decomposition methods provides the main motivation
for this chapter. Here we try to answer a number of questions about tree
decomposition and NSDP in solving DO problems. What are they? How and
where can they be applied? What consists a connection between different
structural decomposition methods, such as tree decomposition and nonserial
dynamic programming?

The aim of this chapter is to provide a review of structural decomposition methods and to give a unified framework in the form of **local elimination algorithms** [94]. We propose here the general approach which consists of viewing a decomposition of some DO problem as being represented by a DAG whose nodes represent subproblems that only contain local information. The nodes are connected by arcs that represent the dependency of the local information in the subproblems. A subproblem that is higher in the hierarchy may use the information (or knowledge) obtained in the dependent subproblems.

This chapter is organized as follows: In section 2 we introduce local elimination algorithms for solving discrete problems. In Section 3 we survey necessary terminology and notions for discrete optimization problems and their graph representations. In Section 4 we consider local variable elimination schemes for solving DO problems with constraints and discuss a classification of dynamic programming (DP) computational procedure. Elimination Game is introduced. Application of the bucket elimination algorithm from CS to solving DO problems is done. Then, in Section 5, we consider a local block elimination scheme and related notions. As a promising abstraction approach of solving DOPs we define clustering that merges several variables into a single meta-variable. This allows us to create a quotient (condensed) graph and apply a local block elimination algorithm. In Section 6 a tree decomposition scheme is introduced. Connection of of the local elimination algorithmic schemes with tree decomposition and a way of transforming the DAG of computational local elimination procedure to tree decomposition are discussed.

2   Local Elimination Algorithms for Solving Discrete Problems

The structure of discrete optimization problems is determined either by the original elements (e.g., variables) with a system of **neighborhoods** specified for them and with the order of searching through those elements using a **local elimination algorithm** or by various derived structures (e.g., block or treeblock structures). Both original and derived structures can be specified by the so called **structural graph**. The **structural graph** can be the interaction graph of the original elements (for example, between the variables of the problem) or the **quotient [45] (condensed [51]) graph**. The quotient graph can be obtained by merging a set of original elements (for example, a subgraph) into a condensed element. The original subset (subgraph) that formed the condensed element is called the **detailed graph** of this element. A local elimination algorithm (LEA) [94] eliminates local elements of the problem's structure defined by the structural graph by computing and storing local information about these elements in the form of new dependencies added to the problem. Thus, the local elimination procedure consists of two parts:

A. The **forward part** eliminates elements, computes and stores local solutions, and finally computes the value of the objective function;

B. The **backward part** finds the global solution of the whole problem using the tables of local solutions; the global solution gives the optimal value of the objective function found while performing the forward part of the procedure.

The LEA analyzes a **neighborhood** $Nb(x)$ of the current element x in the structural graph of the problem, applies an **elimination operator** (which depends on the particular problem) to that element, calculates the function $h(Nb(x))$ that contains **local information** about x, and finds the local solution $x^*(Nb(x))$. Next, the element x is eliminated, and a clique is created from the elements of $Nb(x)$. The elimination of elements and the creation of cliques changes the structural graph and the neighborhoods of elements. The backward part of the local elimination algorithm reconstructs the solution of the whole problem based on the local solutions $x^*(Nb(x))$.

The algorithmic scheme of the LEA is a **DAG** in which the vertices correspond to the local subproblems and the edges reflect the informational dependence of the subproblems on each other.

3 Discrete Optimization Problems and Their Graph Representations

3.1 *Notions and Definitions*

Consider a sparse DOP in the following form

$$F(x_1, x_2, \ldots, x_n) = \sum_{k \in K} f_k(X^k) \to \max \tag{1}$$

subject to the constraints

$$g_i(X_{S_i})\, R_i\, 0, \quad i \in M = \{1, 2, \ldots, m\}, \tag{2}$$

$$x_j \in D_j, \quad j \in N = \{1, \ldots, n\}, \tag{3}$$

where
$X = \{x_1, \ldots, x_n\}$ is a set of discrete variables, $X^k \subseteq \{x_1, x_2, \ldots, x_n\}, k \in K = \{1, 2, \ldots, t\}$, t – number of components in the objective function, $S_i \subseteq \{1, 2, \ldots, n\}$, $R_i \in \{\leq, =, \geq\}, i \in M$; D_j is a finite set of admissible values of variable x_j, $j \in N$. Functions $f_k(X^k)$, $k \in K$ are called components of the objective function and can be defined in tabular form. We use here the notation: if $S = \{j_1, \ldots, j_q\}$ then $X_S = \{x_{j_1}, \ldots, x_{j_q}\}$.

In order to avoid complex notation, without loss of generality, we consider further a DOP with linear constraints and binary variables:

$$\max_X f(X) = \max_X \sum_{k \in K} f_k(X^k), \tag{4}$$

subject to

$$A_{iS_i} X_{S_i} \leq b_i, \ i \in M = \{1, 2, \ldots, m\}, \tag{5}$$

$$x_j = 0, 1, \ j \in N = \{1, \ldots, n\}. \tag{6}$$

We shall consider further a linear objective function (7):

$$f(x_1, \ldots, x_n) = f(X) = C_N X_N = \sum_{j=1}^{n} c_j x_j \to \max \tag{7}$$

Definition 1. *[16]. Variables $x \in X$ and $y \in X$ interact in DOP with constraints (we denote $x \sim y$) if they both appear either in the same component of the objective function, or in the same constraint (in other words, if variables are both either in a set X^k, or in a set X_{S_i}).*

Introduce a graph representation of the DOP. Description of the DOP structure may be done with various detailization. The structural graph of the DOP defines which variables are in which constraints. Structure of a DOP can be defined either by interaction graph of initial elements (variables in the DOP) or by various derived structures, e.g., block structures, block-tree structures defined by so called **quotient** (condensed or compressed [7], [8], [54]) graph.

Concrete choice of a structural graph of the DOP defines different local elimination schemes: nonserial dynamic programming, block decomposition, tree decomposition etc.

If the DOP is divided into blocks corresponding to subsets of variables (meta-variables) or to subsets of constraints (meta-constraints), then block structure can be described by a structural quotient (condensed) graph, whose meta-nodes correspond to subsets of the variables of blocks and meta-edges correspond to adjacent blocks (see below, in section 5.1).

An **interaction graph** [16] (**dependency graph** by HOOKER [58]) represents a structure of the DOP in a natural way.

Definition 2. *[16]. **Interaction graph** of the DOP is an undirected graph $G = (X, E)$, such that*

1. Vertices X of G correspond to variables of the DOP;
2. Two vertices of G are adjacent iff corresponding variables interact.

Further, we shall use the notion of vertices that correspond one-to-one to variables.

Definition 3. *Set of variables interacting with a variable $x \in X$ is denoted by $Nb(x)$ and called the **neighborhood** of the variable x. For corresponding vertices a neighborhood of a vertex x is a set of vertices of interaction graph that are linked by edges with x. Denote the latter neighborhood as $Nb_G(x)$.*

Introduce the following notions:

1. Neighborhood of a set $S \subseteq X$, $Nb_G(S) = \bigcup_{x \in S} Nb_G(x) - S$.
2. Closed neighborhood of a set $S \subseteq X$, $Nb_G[S] = Nb_G(S) \cup S$.

4 Local Variable Elimination Algorithms in Discrete Optimization

4.1 Nonserial Dynamic Programming and Classification of DP Formulations

NSDP exploits only **local computations** to solve global discrete optimization problems and is, therefore, a particular instance of local elimination algorithm. It appeared in 1961 with ARIS [3] (see [11], [14], [15], [69]) but is poorly known to the optimization community. This approach is used in Artificial Intelligence under the names "Variable Elimination" or "Bucket Elimination" [32]. NSDP being a natural and general decomposition approach to sparse problems solving, considers a set of constraints and an objective function as recursively computable function [58]. This allows to compute a solution in stages such that each of them uses results from previous stages. This requires a reduced effort to find the solution.

Thus, the DP algorithm can be applied to find the optimum of the entire problem by using the connected optimizations of the smaller DO subproblems with the aid of existing optimization solvers.

It is worth noting that NSDP is implicit in Hammer and Rudeanu's "basic method" for pseudoboolean optimization [52]. CRAMA, HANSEN, AND JAUMARD [26] discovered that the basic method can exploit the structure of a DOP with the usage of so-called **co-occurrence graph** (interaction graph). It was found that the complexity of the algorithm depends on **induced width** of this graph, which is defined for a given ordering of the variables. Consideration of the variables in the right order may result in a smaller induced width and faster solution [59].

In [16] mostly DO problems without constraints were considered. Here, we consider an application of NSDP variable elimination algorithm to solving DO problems with constraints.

One of the most useful graph-based interpretations is a representation of computational DP procedure as a direct acyclic graph (DAG) [93] whose vertices are associated with subproblems and whose edges express information interdependence between subproblems.

Every DP algorithm has an underlying DAG structure that usually is implicit [30]: the dependencies between subproblems in a DP formulation can be represented by a DAG. Each node in the DAG represents a subproblem. A directed edge from node A to node B indicates that the solution to the subproblem represented by node A is used to compute the solution to the subproblem represented by node B (Fig.1). The DAG is explicit only when

Fig. 1 Precedence of
subproblems A and B

Fig. 2 Underlying DAG
of subproblems

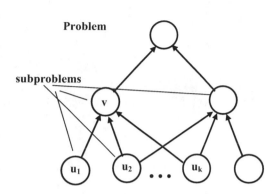

we have a graph optimization problem (say, a shortest path problem). Having
nodes $u_1, \ldots, u_k$ point to v means "subproblem v can only be solved once the
solutions to $u_1, \ldots, u_k$ are known" (Fig. 2). Thus, the DP formulation can
be described by the DAG of the computational procedure of a DP algorithm
(underlying DAG [30]). LI & WAH [100] proposed to classify various DP computational procedures or DP formulations on the basis of the dependencies
between subproblems from the underlying DAG.

The nodes of the DAG can be organized into levels such that subproblems at
a particular level depend only on subproblems at previous levels. In this case,
the DP procedure (formulation) can be categorized as follows. If subproblems
at all levels depend only on the results of subproblems at the immediately preceding levels, the procedure (formulation) is called a **serial** DP procedure (formulation), otherwise, it is called a **nonserial** DP procedure (formulation).

Example 1. The simplest optimization problem is the **serial** unconstrained
discrete optimization problem [16]

$$\max_X f(X) = \max_X \sum_{i \in K} f_i(X^i),$$

where $X = \{x_1, \ldots, x_n\}$ is a set of discrete variables.

$$K = \{1, 2, \ldots, n-1\}; \quad X^i = \{x_i, x_{i+1}\}.$$

In fig. 3 it is shown an interaction graph of the serial DO problem.

Fig. 3 Interaction graph for the serial formulation of unconstrained DOP

4.2 Discrete Optimization Problem with Constraints

Consider the DOP (7), (5), (6) and suppose without loss of generality that variables are eliminated in the order $x_1, \ldots, x_n$. Using the local variable elimination scheme eliminate the first variable x_1. x_1 is in a set of constraints with the indices in U_1:

$$U_1 = \{i \mid x_1 \in S_i\}$$

Together with x_1, constraints in U_1 contain variables from $Nb(x_1)$.

The following subproblem P_1 corresponds to the variable x_1 of the DOP:

$$h_{x_1}(Nb(x_1)) = \max_{x_1}\{c_1 x_1 | A_{iS_i} X_{S_i} \leq b_i,\ i \in U_1,\ x_j = 0, 1,\ x_j \in Nb[x_1]\}$$

Then the initial DOP can be transformed in the following way:

$$\max_{x_1,\ldots,x_n} \left\{\sum C_N X_N | A_{iS_i} X_{S_i} \leq b_i,\ i \in M,\ x_j = 0, 1,\ j \in N\right\} =$$

$$\max_{x_2,\ldots,x_n} \{C_{N-\{1\}} X_{N-\{1\}} + h_{x_1}(Nb(x_1) | A_{iS_i} X_{S_i} \leq b_i,\ i \in M - U_1,\ x_j = 0, 1,\ j = 2, \ldots, n\}$$

The last problem has $n - 1$ variables; from the initial DOP were excluded constraints with the indices in U_1 and from the objective function the term $c_1 x_1$; there appeared a new objective function term $h_{x_1}(Nb(x_1))$. Due to this fact the interaction graph associated with the new problem is changed: a vertex x_1 is eliminated and its neighbors have become connected (due to the appearance a new term $h_{x_1}(Nb(x_1))$ in the objective). It can be noted that a graph induced by vertices of $Nb(x_1)$ is complete, i.e. is a clique. Denote the new interaction graph G^1 and find all neighborhoods of variables in G^1. NSDP eliminates the remaining variables one by one in an analogous manner. We have to store tables with optimal solutions at each stage of this process. At the stage n of the described process we eliminate a variable x_n and find an optimal value of the objective function. Then a backward step of the local elimination procedure is performed using the tables with solutions.

4.3 Elimination Game, Combinatorial Elimination Process, and Underlying DAG of the LAE Computational Procedure

Consider a sparse discrete optimization problem (1) — (3) whose structure is described by an undirected interaction graph $G = (X, E)$. Solve this problem with a local elimination algorithm (LEA). LEA uses an ordering α of X [84]: Given a graph $G = (X, E)$ an **ordering** α of X is a bijection $\alpha : X \leftrightarrow \{1, 2, \ldots, n\}$ where $n = |X|$.

G_α and X_α are correspondingly an ordered graph and an ordered vertex set. Sometimes the ordering will be denoted as $x_1, \ldots, x_n$, i.e. $\alpha(x_i) = i$ and i will be considered as an index of the vertex x_i.

In G_α, a **monotone neighborhood** $\overline{Nb}_G^\alpha(x_i)$ ([18], [84]) of $x_i \in X$ is a set of vertices **monotonely adjacent** to a vertex x_i, i.e.

$$\overline{Nb}_G^\alpha(x_i) = \{x_j \in Nb_G(x_i) | j > i\}.$$

The graph G_x [85] obtained from $G = (X, E)$ by

(i) adding edges so that all vertices in $Nb_G(x)$ are pairwise adjacent, and
(ii) deleting x and its incident edges

is the x–**elimination** graph of G. This process is called the **elimination** of the vertex x.

Given an ordering $x_1, x_2, \ldots, x_n$, the LEA proceeds in the following way: it subsequently eliminates $x_1, x_2, \ldots, x_n$ in the current graph and computes an associated local information about vertices from $h_{x_i}(Nb(x_i))$ [94]. This can be described by the **combinatorial elimination process** [85]:

$$G_0 = G, G_1, \ldots, G_{j-1}, G_j, \ldots, G_n$$

where G_j is the x_j–elimination graph of G_{j-1} and $G_n = \emptyset$.

The process of interaction graph transformation corresponding to the LEA scheme is known as **Elimination Game** which was first introduced by PARTER [80] as a graph analogy of Gaussian elimination. The input of the elimination game is a graph G and an ordering α of G (i.e. $\alpha(x) = i$ if x is i-th vertex in the ordering α). Elimination Game according to [53] consists in the following. At each step i, the neighborhood of vertex x_i is turned into a clique, and x_i is deleted from the graph. This is referred to as eliminating vertex x_i. We obtain a graph $G_{x_i}^{(i)}$. The filled graph $G_\alpha^+ = (X, E_\alpha^+)$ is obtained by adding to G all the edges added by the algorithm. The resulting filled graph G_α^+ is a triangulation of G (FULKERSON & GROSS [44]), i.e., a chordal graph.

Let us introduce the notion for the **elimination tree** (etree) [67]. Given a graph $G = (X, E)$ and an ordering α, the **elimination tree** is a directed tree $\overrightarrow{T}_\alpha$ that has the same vertices X as G and its edges are determined by a parent relation defined as follows: the parent x is the **first vertex** (according to the ordering α) of the monotone neighborhood $\overline{Nb}_{G_\alpha^+}^\alpha(x)$ of x in the filled graph G_α^+.

Using the parent relation introduced above we can define a directed filled graph $\overrightarrow{G}_\alpha^+$.

The underlying DAG of a local variable elimination scheme can be constructed using Elimination Game. At step i, we represent the computation of the function $h_{x_i}(Nb_{G_{x_{i-1}}}(x_i))$ as a node of the DAG (corresponding to the vertex x_i). Then, this node containing variables $(x_i, Nb_{G_{x_{i-1}}}^{(i-1)}(x_i))$ is linked with a first x_j (accordingly to the ordering α) which is in $Nb_{G_{x_{i-1}}^{(i-1)}}(x_i)$.

It is easy to see that the elimination tree is the DAG of the computational procedure of the LEA.

Example 2. Consider a DOP (P) with binary variables:

$$
\begin{aligned}
2x_1 + 3x_2 + \; x_3 + 5x_4 + 4x_5 + 6x_6 + \; x_7 \quad &\to \max \\
3x_1 + 4x_2 + \; x_3 \qquad\qquad\qquad\qquad\qquad &\le 6, \quad (C_1) \\
2x_2 + 3x_3 + 3x_4 \qquad\qquad\qquad\quad\; &\le 5, \quad (C_2) \\
2x_2 \qquad\quad + 3x_5 \qquad\qquad\qquad &\le 4, \quad (C_3) \\
2x_3 \qquad\qquad + 3x_6 + 2x_7 \quad &\le 5, \quad (C_4) \\
x_j = 0, 1, \; j = 1, \ldots, 7. \qquad\qquad\qquad\qquad &
\end{aligned}
$$

The interaction graph is shown in Fig. 4 (a). Elimination Game results and graphs $G_{x_i}^{(i)}$ are in Fig. 5. Associated underlying DAG of NSDP procedure for the variable ordering $\{x_5, x_2, x_1, x_4, x_3, x_6, x_7\}$ is shown in Fig. 4 (b).

4.4 Bucket Elimination

Bucket elimination (BE) is proposed in [32] as a version of NSDP for solving CSPs. Now, we consider a modification of the BE algorithm for solving DOPs. The BE algorithm works as follows: Assume we are given an order $x_1, \ldots, x_n$ of the variables of the DOP. BE starts by creating n "buckets", one for each variable x_j. BE algorithm uses as input ordered set of variables and a set of constraints. To each variable x_j is corresponded a bucket $\Sigma^{(x_j)}$, i.e., a set of constraints and components of objective function built as follows: In the bucket $\Sigma^{(x_j)}$ of variable x_j we put all constraints that contain x_j but do not contain any variable having a higher index. We now iterate on j from n to 1, eliminating one bucket at a time. Algorithm finds new components of the objective applying so called "elimination operator" (in our case the latter consists on solving associated DO subproblems) to all constraints and components of the objective function of the bucket under consideration. New components of the objective function reflecting an impact of variable x_j on the rest part of the DO problem, are located in corresponding lower buckets.

Consider an application of BE to solving the DOP with constraints from Example 2. We use an elimination ordering $\alpha : \{x_5, x_2, (x_1, x_4), x_3, (x_6, x_7)\}$. Variables (x_1, x_4) shall be eliminated in block since they are indistinguishable. Build buckets (subsets of constraints) beginning from last (due order α) block (x_6, x_7). A bucket $\Sigma^{(x_6, x_7)}$ includes all constraints of the DOP containing the variables $x_6, \; x_7$, i.e., the bucket $\Sigma^{(x_6, x_7)}$ consists of constraint C_4: $\Sigma^{(x_6, x_7)} = \{C_4\}$. Similarly: $\Sigma^{(x_3)} = \{C_1, C_2\}$, $\Sigma^{(x_1, x_4)} = \emptyset$, $\Sigma^{(x_2)} = \{C_3\}$, $\Sigma^{(x_5)} = \emptyset$.

We solve a DO subproblem associated with the bucket $\Sigma^{(x_6, x_7)}$:
For each binary assignment x_3, we compute values x_6, x_7 such that

$$
h_{x_6, x_7}(x_3) = \max_{x_6, x_7} \{6x_6 + x_7 \mid 2x_3 + 3x_6 + 2x_7 \le 5, x_j \in \{0, 1\}\}.
$$

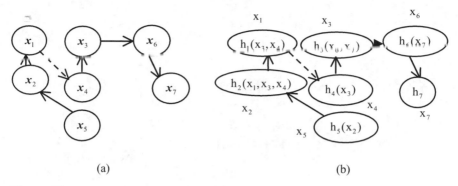

Fig. 4 Elimination tree of the DOP (a) Computing the information while eliminating variables in the LEA computational procedure (b) (example 2)

The function $h_{x_6,x_7}(x_3)$ is placed in the bucket $\Sigma^{(x_3)}$. Consider the DO subproblem associated with this bucket

$$h_{x_3}(x_1, x_2, x_4) = \max_{x_3}\left[\, x_3 + h_{x_6,x_7}(x_3)\right]$$

$$3x_1 + 4x_2 + \ x_3 \qquad\qquad \leq 6,$$
$$2x_2 + 3x_3 + 3x_4 \ \leq 5,$$
$$x_j = 0, 1, \ j = 1, 2, 3, 4.$$

We place the function $h_{x_3}(x_1, x_2, x_4)$ in the bucket $\Sigma^{(x_1,x_4)}$ and solve the problem

$$h_{x_1,x_4}(x_2) = \max_{x_1,x_4}\{2x_1 + 5x_4 + h_{x_3}(x_1, x_2, x_4) \mid x_j \in \{0,1\}\}.$$

Build the corresponding table 3.

Function $h_{x_1,x_4}(x_2)$ is placed in the bucket $\Sigma^{(x_2)}$. A new DO subproblem left to be solved

Table 1 Calculation of $h_{x_6,x_7}(x_3)$

x_3	h_{x_6,x_7}	x_6^*	x_7^*
0	7	1	1
1	6	1	0

Table 2 Calculation of $h_{x_3}(x_1,x_2,x_4)$

x_1	x_2	x_4	h_{x_3}	x_3^*
0	0	0	7	1
0	0	1	7	0
0	1	0	7	1
0	1	1	7	0
1	0	0	7	1
1	0	1	7	0
1	1	0	-	-
1	1	1	-	-

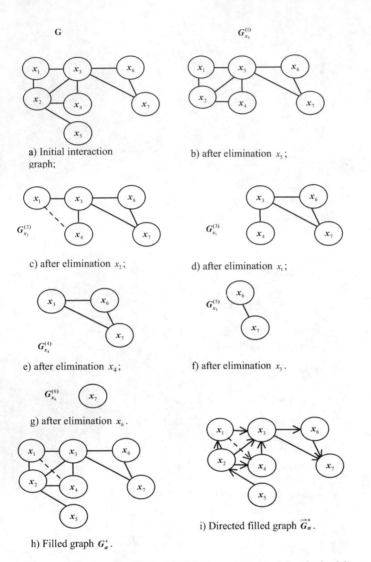

a) Initial interaction graph;

b) after elimination x_5;

c) after elimination x_2;

d) after elimination x_1;

e) after elimination x_4;

f) after elimination x_3.

g) after elimination x_6.

h) Filled graph G_a^+.

i) Directed filled graph $\overrightarrow{G_a}^+$.

Fig. 5 Elimination Game. Fill-in is represented by dashed lines

$$h_{x_2}(x_5) = \max_{x_2}\{3x_2 + h_{x_1,x_4}(x_2) \mid 2x_2 + 3x_5 \leq 4, x_j \in \{0,1\}\}$$

Place $h_{x_2}(x_5)$ in the last bucket $\Sigma^{(x_5)}$. The new subproblem is:

$$h_{x_5} = \max_{x_5}\{4x_5 + h_{x_2}(x_5) \mid x_j \in \{0,1\}\},$$

its solution is $h_5 = 18$, $x_5^* = 1$ and the maximal objective value is 18.

Table 3 Calculation of $h_{x_1,x_4}(x_2)$

x_2	h_{x_1,x_4}	x_1^*	x_4^*
0	14	1	1
1	12	0	1

Table 4 Calculation of $h_{x_2}(x_5)$

r_5	h_{x_2}	r_2^*
0	15	1
1	14	0

To find the optimal values of the variables, it is necessary to do backward step of the BE procedure: from the last table 4 using $x_5 = 1$ we have $x_2^* = 0$. Considering the table 3 we have for $x_2 = 0 : x_1^* = 1$, $x_4^* = 1$. From the table 2: $x_1 = 1$, $x_2 = 0$, $x_4 = 1 \Rightarrow x_3^* = 0$. Table 1: $x_3 = 0 \Rightarrow x_6^* = 1$, $x_7^* = 1$.

The solution is $(1, \ 0, \ 0, \ 1, \ 1, \ 1, \ 1)$, optimal objective value is 18.

5 Block Local Elimination Scheme

5.1 Partitions, Clustering, and Quotient Graphs

The local elimination procedure can be applied to elimination of not only separate variables but also to sets of variables and can use the so called "elimination of variables in blocks" ([16], [90]), which allows to eliminate several variables in block. Local decomposition algorithm [90] actually implements the local block elimination algorithm. If the DOP is divided into blocks corresponding to subsets of variables (meta-variables), then block structure can be described with the aid of a structural condensed graph whose meta-nodes correspond to subsets of the variables or blocks and meta-edges correspond to adjacent blocks.

Applying the method of merging variables into meta-variables allows to obtain **condensed** or meta-DOPs which have a simpler structure. If the resulting meta-DOP has a nice structure (e.g., a tree structure) then it can be solved efficiently.

The structural graph of the meta-DOP is obtained by collapsing merged nodes into a single meta-node and connecting the meta-node with all nodes that were adjacent with some of the merged nodes. Such a graph usually is called a quotient graph.

An **ordered** partition of a set X is a decomposition of X into ordered sequence of pairwise disjoint nonempty subsets whose union is all of X.

Partitioning is a fundamental operation on graphs. One variant of it is to partition the vertex set X to three sets $X = U \cup S \cup W$, such that U and W are balanced, meaning that neither of them is too small, and S is small. Removing S along with all edges incident on it separates the graph into two connected components. S is called a **separator**. In general, graph partitioning is NP-hard. Since graph partitioning is difficult in general, there is a need for approximation algorithms. A popular algorithm in this respect is MeTiS [70], which has a good implementation available in the public domain.

Taking advantage of **indistinguishable** variables (two variables are indistinguishable if they have the same closed neighborhood [1], [7], [54], [8]) it is possible to compute a **quotient (condensed) graph** which is formed by merging all vertices with the same neighborhoods into a single meta-node. Let $\mathbf{x}$ be a **block** of a graph G [5], i.e., a maximal set of indistinguishable with v vertices. Clearly, the blocks of G partition X since indistinguishability is an **equivalence relation** defined on the original vertices.

An equivalence relation on a set induces a partition on it, and also any partition induces an **equivalence relation**. Given a graph $\Gamma = (X, E)$, let $\mathbf{X}$ be a partition on the vertex set X:

$$\mathbf{X} = \{\mathbf{x_1}, \mathbf{x_2}, \ldots, \mathbf{x_m}\}.$$

That is, $\cup_{i=1}^{m} \mathbf{x_i} = X$ and $\mathbf{x_i} \cap \mathbf{x_k} = \emptyset$ for $i \neq k$. We define the **quotient graph** of G with respect to the partition $\mathbf{X}$ to be the graph

$$G/\mathbf{X} = (\mathbf{X}, \mathcal{E}),$$

where $(\mathbf{x_i}, \mathbf{x_k}) \in \mathcal{E}$ if and only if $Nb_G(\mathbf{x_i}) \cap \mathbf{x_k} \neq \emptyset$.

The quotient graph $\mathbf{G}(\mathbf{X}, \mathcal{E})$ is an equivalent representation of the interaction graph $G(X, E)$, where $\mathbf{X}$ is a set of blocks (or indistinguishable sets of vertices), and $\mathcal{E} \subseteq \mathbf{X} \times \mathbf{X}$ be the edges defined on $\mathbf{X}$. A **local block elimination** scheme is one in which the vertices of each block are eliminated contiguously [5]. As an application of a clustering technique we consider below a block local elimination procedure [16] where the elimination of the block (i.e., a subset of variables) can be seen as the merging of its variables into a meta-variable.

The merges done define a so called **synthesis tree** [102] on the variables.

Definition 4. *A synthesis tree of an initial DOP P is a tree whose leaves correspond to the variables of the initial DOP P, and where each intermediate node is a meta-variable corresponding to the combination of its children nodes.*

Using the synthesis tree it is possible to "decode" meta-variables and find the solution of the initial DOP.

Consider an ordered partition $\mathbf{X}$ of the set X of the variables into blocks:

$$\mathbf{X} = (\mathbf{x_1}, \ldots, \mathbf{x_p}), \quad p \leq n,$$

where $\mathbf{x_l} = X_{K_l}$ (K_l is a set of indices corresponding to $\mathbf{x_l}$, $l = 1, \ldots, p$). For this ordered partition $\mathbf{X}$, the DOP P: (7), (5), (6) can be solved by the LEA using **quotient interaction graph G**.

A. Forward part
Consider first the block $\mathbf{x_1}$. Then

$$\max_X \{C_N X_N | A_{iS_i} X_{S_i} \le b_i, \ i \in M, \ x_j = 0, 1, \ j \in N\} =$$

$$\max_{X_{K_2},\dots,X_{K_p}} \{C_{N-K_1} X_{N-K_1} + h_1(Nb(X_{K_1}) | A_{iS_i} X_{S_i} \le b_i, \ i \in M - U_1,$$

$$x_j = 0, 1, \ j \in N - K_1\}$$

where $U_1 = \{i : S_i \cap K_1 \ne \emptyset\}$ and

$$h_1(Nb(X_{K_1})) = \max_{X_{K_1}} \{C_{K_1} X_{K_1} | A_{iS_i} X_{S_i} \le b_i, \ i \in U_1, \ x_j = 0, 1, \ x_j \in Nb[\mathbf{x_1}]\}.$$

The first step of the local block elimination procedure consists of solving, using complete enumeration of X_{K_1}, the following optimization problem

$$h_1(Nb(X_{K_1})) = \max_{X_{K_1}} \{C_{K_1} X_{K_1} | A_{iS_i} X_{S_i} \le b_i, \ i \in U_1, \ x_j = 0, 1, \ x_j \in Nb[\mathbf{x_1}]\}, \tag{8}$$

and storing the optimal local solutions X_{K_1} as a function of the neighborhood of X_{K_1}, i.e., $X_{K_1}^*(Nb(X_{K_1}))$.

The maximization of $f(X)$ over all feasible assignments $Nb(X_{K_1})$, is called the **elimination of the block** (or meta-variable) X_{K_1}. The optimization problem left after the elimination of X_{K_1} is:

$$\max_{X-X_{K_1}} \{C_{N-K_1} X_{N-K_1} + h_1(Nb(X_{K_1})) | A_{iS_i} X_{S_i} \le b_i, \ i \in M - U_1,$$

$$x_j = 0, 1, \ j \in N - K_1\}.$$

Note that it has the same form as the original problem, and the tabular function $h_1(Nb(X_{K_1}))$ may be considered as a new component of the modified objective function. Subsequently, the same procedure may be applied to the elimination of the blocks – meta-variables $\mathbf{x_2} = X_{K_2}, \dots, \mathbf{x_p} = X_{K_p}$, in turn. At each step j the new component $h_{\mathbf{x_j}}$ and optimal local solutions $X_{K_j}^*$ are stored as functions of $Nb(X_{K_j} | X_{K_1}, \dots, X_{K_{j-1}})$, i.e., the set of variables interacting with at least one variable of X_{K_j} in the current problem, obtained from the original problem by the elimination of $X_{K_1}, \dots, X_{K_{j-1}}$. Since the set $Nb(X_{K_p} | X_{K_1}, \dots, X_{K_{p-1}})$ is empty, the elimination of X_{K_p} yields the optimal value of objective $f(X)$.

B. Backward part
This part of the procedure consists of the consecutive choice of $X_{K_p}^*$, $X_{K_{p-1}}^*, \dots, X_{K_1}^*$, i.e., the optimal local solutions from the stored tables $X_{K_1}^*(Nb(X_{K_1})), X_{K_2}^*(Nb(X_{K_2} | X_{K_1})), \dots, X_{K_p}^* | X_{K_{p-1}}, \dots, X_{K_1}$.

Block elimination game and underlying DAG
It is possible to extend EG to the case of the block elimination. The input of extended EG is an initial interaction graph G and a partition

$\mathbf{X} = \{\mathbf{x_1}, \ldots, \mathbf{x_p}\}$ of vertices of G. At each step ν $(1 \leq \nu \leq p)$ of EG, the neighborhood $Nb(\mathbf{x}_\nu)$ of $\mathbf{x}_\nu$ is turned into a clique, and $\mathbf{x}_\nu$ is deleted from the graph G. The filled graph $G_{\mathbf{X}}^+ = (X, E^+)$ is obtained by adding to G all the edges added by the algorithm. The resulting filled graph $G_{\mathbf{X}}^+$ is a triangulation of G, i.e., a chordal graph [6].

Underlying DAG of the local block elimination procedure contains nodes corresponding to computing of functions $h_{\mathbf{x}_i}(Nb_{G_{\mathbf{X}}^{(i-1)}}(\mathbf{x}_i))$ and is a **generalized elimination tree**.

Example 3. **Local block elimination for unconstrained DOP.**
Consider an unconstrained DOP

$$\max_X [f_1(x_1, x_2, x_3) + f_2(x_2, x_3, x_4) + f_3(x_2, x_5) + f_4(x_3, x_6, x_7)],$$

where $$X = (x_1, x_2, x_3, x_4, x_5, x_6, x_7)$$

and functions f_1, f_2, f_3, f_4 are given in the following tables.

Consider an ordered partition of the variables of the set into blocks:

$$\mathbf{x_1} = \{x_5\}, \ \mathbf{x_2} = \{x_1, x_2, x_4\}, \ \mathbf{x_3} = \{x_6, x_7\}, \ \mathbf{x_4} = \{x_3\}.$$

Interaction graph for this problem is the same as in Fig. 4 (a).

For the ordered partition $\mathbf{X} = \{\mathbf{x_1}, \mathbf{x_2}, \mathbf{x_3}, \mathbf{x_4}\}$, this unconstrained DO problem may be solved by the LEA. Initial interaction graph with partition presented by dashed lines is shown in Fig. 6 (a), quotient interaction graph is in Fig. 6 (b), and the DAG of the block local elimination computational procedure is shown in Fig. 7.

A. Forward part
Consider first the block $\mathbf{x_1} = \{x_5\}$. Then $Nb(\mathbf{x_1}) = \{x_2\}$. Solve using complete enumeration the following optimization problem

$$h_{\mathbf{x_1}}(Nb(\mathbf{x_1})) = \max_{x_5} f_3(x_2, x_5),$$

Table 5 f_1

x_1	x_2	x_3	f_1
0	0	0	2
0	0	1	3
0	1	0	4
0	1	1	0
1	0	0	5
1	0	1	2
1	1	0	4
1	1	1	1

Table 6 f_2

x_1	x_2	x_3	f_2
0	0	0	3
0	0	1	1
0	1	0	5
0	1	1	2
1	0	0	4
1	0	1	1
1	1	0	3
1	1	1	0

Table 7 f_3

x_2	x_5	f_3
0	0	6
0	1	2
1	0	4
1	1	5

Table 8 f_4

x_3	x_6	x_7	f_4
0	0	0	5
0	0	1	2
0	1	0	3
0	1	1	4
1	0	0	2
1	0	1	1
1	1	0	3
1	1	1	6

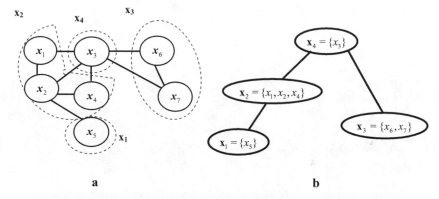

a   b

Fig. 6 Interaction graph of the DOP with partition (dashed) (a) and quotient interaction graph (b) (example 3)

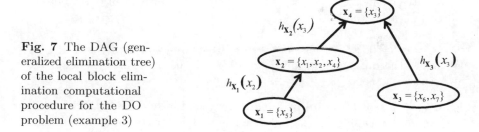

Fig. 7 The DAG (generalized elimination tree) of the local block elimination computational procedure for the DO problem (example 3)

and store the optimal local solutions $\mathbf{x_1}$ as a function of a neighborhood, i.e., $\mathbf{x_1}^*(Nb(\mathbf{x_1}))$.

Eliminate the block $\mathbf{x_1}$ and consider the block $\mathbf{x_2} = \{x_1, x_2, x_4\}$. $Nb(\mathbf{x_2}) = \{x_3\}$. Now the problem to be solved is

$$h_{\mathbf{x_2}}(x_3) = \max_{x_1, x_2, x_4} \{h_{\mathbf{x_1}}(x_2) + f_1(x_1, x_2, x_3) + f_2(x_2, x_3, x_4)\}.$$

Build the corresponding table 10.

Table 9 Calculation of $h_{\mathbf{x_1}}(x_2)$ **Table 10** Calculation of $h_{\mathbf{x_2}}(x_3)$

x_2	$h_{\mathbf{x_1}}(x_2)$	x_5^*
0	6	0
1	5	1

x_3	$h_{\mathbf{x_2}}(x_3)$	x_1^*	x_2^*	x_4^*
0	14	1	0	0
1	14	0	0	0

Eliminate the block $\mathbf{x_2}$ and consider the block $\mathbf{x_3} = \{x_6, x_7\}$. The neighbor of $\mathbf{x_3}$ is x_3: $Nb(\mathbf{x_3}) = \{x_3\}$. Solve the DOP containing x_3:

$$h_{\mathbf{x_3}}(x_3) = \max_{x_6,x_7}\{f_4(x_3, x_6, x_7), x_j \in \{0,1\}\}$$

and build the table 11.

Table 11 Calculation of $h_{\mathbf{x_3}}(x_3)$

x_3	$h_{\mathbf{x_3}}(x_3)$	x_6^*	x_7^*
0	5	0	0
1	6	0	1

Eliminate the block $\mathbf{x_3}$ and consider the block $\mathbf{x_4} = \{x_3\}$. $Nb(\mathbf{x_4}) = \emptyset$. Solve the DOP:

$$h_{\mathbf{x_4}} = \max_{x_3}\{h_{\mathbf{x_2}}(x_3) + h_{\mathbf{x_3}}(x_3), x_j \in \{0,1\}\} = 20,$$

where $x_3^* = 1$.

B. Backward part
Consecutively find $\mathbf{x_3}^*, \mathbf{x_2}^*, \mathbf{x_1}^*$, i.e., the optimal local solutions from the stored tables 11, 10, 9:
$x_3^* = 1 \Rightarrow x_6^* = 1$, $x_7^* = 1$ (table 11);
$x_3^* = 1 \Rightarrow x_1^* = 0$, $x_2^* = 0$, $x_4^* = 0$ (table 10); $x_2^* = 0 \Rightarrow x_5^* = 0$ (table 9).
We found the optimal solution to be $(0, 0, 1, 0, 0, 1, 1)$, the maximum objective value is 20.

Example 4. **Local block elimination for constrained DOP**
Consider the DOP from example 2 and an ordered partition of the variables of the set into blocks:

$$\mathbf{x_1} = \{x_5\}, \ \mathbf{x_2} = \{x_1, x_2, x_4\}, \ \mathbf{x_3} = \{x_6, x_7\}, \ \mathbf{x_4} = \{x_3\}.$$

For the ordered partition $\{\mathbf{x_1}, \mathbf{x_2}, \mathbf{x_3}, \mathbf{x_4}\}$, this constrained optimization problem may be solved by the LEA.

A. Forward part
Consider first the block $\mathbf{x_1} = \{x_5\}$. Then $Nb(\mathbf{x_1}) = \{x_2\}$. Solve the following problem containing x_5 in the objective and the constraints:

$$h_{\mathbf{x_1}}(Nb(\mathbf{x_1})) = \max_{x_5}\{4x_5 \mid 2x_2 + 3x_5 \le 4, x_j \in \{0,1\}\}$$

and store the optimal local solutions $\mathbf{x_1}$ as a function of a neighborhood, i.e., $\mathbf{x_1}^*(Nb(\mathbf{x_1}))$. Eliminate the block $\mathbf{x_1}$. and consider the block $\mathbf{x_2} = \{x_1, x_2, x_4\}$. $Nb(\mathbf{x_2}) = \{x_3\}$. Now the problem to be solved is

$$h_{\mathbf{x_2}}(x_3) = \max_{x_1, x_2, x_4}\{h_{\mathbf{x_1}}(x_2) + 2x_1 + 3x_2 + 5x_4\}$$

subject to

$$3x_1 + 4x_2 + \quad x_3 \qquad \le 6,$$
$$2x_2 + 3x_3 + 3x_4 \le 5,$$
$$x_j = 0, 1, \ j = 1, 2, 3, 4.$$

Build the corresponding table 13. Eliminate the block $\mathbf{x_2}$ and consider the

Table 12 Calculation of $h_{\mathbf{x_1}}(x_2)$ **Table 13** Calculation of $h_{\mathbf{x_2}}(x_3)$

x_2	$h_{\mathbf{x_1}}(x_2)$	x_5^*
0	4	1
1	0	0

x_3	$h_{\mathbf{x_2}}(x_3)$	x_1^*	x_2^*	x_4^*
0	11	1	0	1
1	6	1	0	0

block $\mathbf{x_3} = \{x_6, x_7\}$. The neighbor of $\mathbf{x_3}$ is x_3: $Nb(\mathbf{x_3}) = \{x_3\}$. Solve the DOP containing x_3:
$$h_{\mathbf{x_3}}(x_3) = \max_{x_6, x_7}\{h_{\mathbf{x_2}} + x_3 + 6x_6 + x_7 \mid 2x_3 + 3x_6 + 2x_7 \le 5, x_j \in \{0,1\}\}$$
and build the table 14.

Table 14 Calculation of $h_{\mathbf{x_3}}(x_3)$

x_3	$h_{\mathbf{x_3}}(x_3)$	x_6^*	x_7^*
0	18	1	1
1	12	1	0

Eliminate the block $\mathbf{x_3}$ and consider the block $\mathbf{x_4} = \{x_3\}$. $Nb(\mathbf{x_4}) = \emptyset$. Solve the DOP:

$$h_{\mathbf{x_4}} = \max_{x_3}\{h_{\mathbf{x_3}}(x_3), x_j \in \{0,1\}\} = 18,$$

where $x_3^* = 0$.

B. Backward part

Consecutively find $\mathbf{x_3}^*, \mathbf{x_2}^*, \mathbf{x_1}^*$, i.e., the optimal local solutions from the stored tables 14, 13, 12. $x_3^* = 0 \Rightarrow x_6^* = 1$, $x_7^* = 1$ (table 14); $x_3^* = 0 \Rightarrow x_1^* = 1$, $x_2^* = 0$, $x_4^* = 1$ (table 13); $x_3^* = 0 \Rightarrow x_5^* = 1$ (table 12). We found the optimal global solution to be $(1, 0, 0, 1, 1, 1, 1)$, the maximum objective value is 18.

6 Tree Structural Decompositions in Discrete Optimization

Tree structural decomposition methods use partitioning of constraints and use as a meta-tree a structural graph . Dynamic programming algorithm starts at the leaves of the meta-tree and proceeds from the smaller to the larger subproblems (corresponding to the subtrees) that is to say, bottom-up in the rooted tree.

6.1 Tree Decomposition and Methods of Its Computing

Aforementioned facts and an observation that many optimization problems which are hard to solve on general graphs are easy on trees make detection of tree structures in a graph a very promising solution. It can be done with such powerful tool of the algorithmic graph theory as a **tree decomposition** and the treewidth as a measure for the "tree-likeness" of the graph [83]. It is worth noting that in [56] is discussed a number of other useful parameters like branch-width, rank-width (clique-width) or hypertree-width.

Definition 5. *Let* $G = (X, E)$ *be a graph. A **tree decomposition** of G is a pair $(T; \mathbf{Y})$ with $T = (I; F)$ a tree and $\mathbf{Y} = \{\mathbf{y_i} \mid I \in I\}$ a family of subsets of X, one for each node of T, such that*

- *(i)* $\bigcup_{i \in I} \mathbf{y_i} = X$,
- *(ii) for every edge* $(x, y) \in X$ *there is an* $i \in I$ *with* $x \in \mathbf{y_i}$, $y \in \mathbf{y_i}$,
- *(iii) (intersection property) for all* $i, j, l \in I$, *if* $i < j < l$, *then* $\mathbf{y_i} \cap \mathbf{y_l} \subseteq \mathbf{y_j}$.

Note that tree decomposition uses partition of constraints, i.e., it can be considered as a dual structural decomposition method. The best known complexity bounds are given by the "treewidth" tw (ROBERTSON, SEYMOUR [83]) of an interaction graph associated with a DOP. This parameter is related to some topological properties of the interaction graph. Tree decomposition and the treewidth (ROBERTSON, SEYMOUR [83]) play a very important role in algorithms, for many NP-complete problems on graphs that are otherwise intractable become polynomial time solvable when these graphs have a tree decomposition with restricted maximal size of cliques (or have a bounded treewidth [6], [20], [21]). It leads to a time complexity in $O(n \cdot 2^{tw+1})$. Tree decomposition methods aim to merge variables such that the meta-graph is a tree of meta-vertices.

The procedure to solve a DO problem with bounded treewidth involves two steps: (1) computation of a good tree decomposition, and (2) application of a dynamic programming algorithm that solves instances of bounded treewidth in polynomial time.

Thus, a tree decomposition algorithm can be applied to solving DOPs using the following steps:

(i) generate the interaction graph for a DOP (P);
(ii) using an ordering for Elimination Game add edges in the interaction graph to produce a (chordal) filled graph,
(iii) build the elimination tree and information flows (see Fig 4(b));
(iv) identify the maximum cliques, apply an absorption and build subproblems;
(v) produce a tree decomposition;
(vi) solve the DO subproblems for each meta-node and combine the results using LEA.

As finding an optimal tree decomposition is NP-hard, approximate optimal tree decompositions using triangulation of a given graph are often exploited. Let us list existing methods of computing tree decomposition using a survey of them in [61]. **Optimal triangulations** algorithms have an exponential time complexity. Unfortunately, their implementations do not have much interest from a practical viewpoint. For example, the algorithm described in [41] has time complexity $O(n^4 \cdot (1.9601^n))$ [61]. A paper [46] has shown that the algorithm proposed in [96] cannot solve small graphs (50 vertices and 100 edges). The approach of [46] using a branch and bound algorithm, seems promising for computing optimal triangulations. **Approximation algorithms** approximate the optimum by a constant factor. Although their complexity is often polynomial in the treewidth [2], this approach seems unusable due to a big hidden constant. **Minimal triangulation** computes a set C' such that, for every subset $C'' \subset C'$, the graph $G' = (X, C \cup C'')$ is not triangulated. The algorithms LEX-M [84] and LB [17] have a polynomial time complexity of $O(ne')$ with e' the number of edges in the triangulated graph. **Heuristic triangulation** methods build a perfect elimination order by adding some edges to the initial graph. They can be easily implemented and often do this work in polynomial time without providing any minimality warranty. In practice, these heuristics compute triangulations reasonably close to the optimum [64].

Experimental comparative study of four triangulation algorithms, LEX-M, LB, min-fill and MCS was done in [61]. Min-fill orders the vertices from 1 to n by choosing the vertex which leads to add a minimum number of edges when completing the subgraph induced by its unnumbered neighbors. Paper [61] claims that LB and min-fill obtain the best results.

6.2 Computing Tree Decompositions for NSDP Schemes

Given a triangulated (or chordal) graph, the set of its maximal cliques corresponds to the family of subsets associated with a tree decomposition (so called **clique tree** [18]). When we exploit tree decomposition, we only

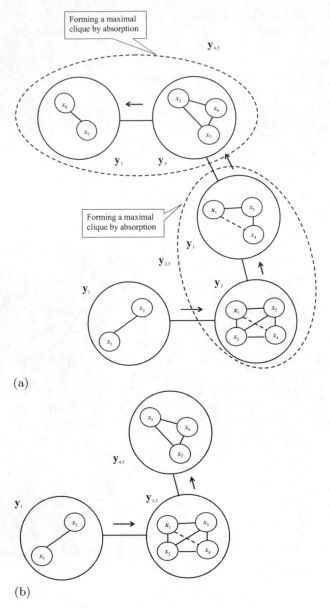

(a)

(b)

Fig. 8 Tree decomposition for the NSDP procedure (example 2) before (a) and after absorption (b)

consider approximations of optimal triangulations by clique trees. Hence, the time complexity is then $O(n \cdot 2^{w^+ +1})$ $(w + 1 \leq w^+ + 1 \leq n)$. The space complexity is $O(n \cdot s \cdot 2^s)$ with s the size of the largest minimal separator [61].

Usually, tree decomposition is considered in the literature separately from NSDP issues. But there is a close connection between these two structural decomposition approaches. Moreover, it is easy to see that a tree decomposition can be obtained from the DAG of the computational NSDP procedure (this fact was noted in [63]).

Consider example 2 and build a tree decomposition associated with the corresponding NSDP procedure. Associated underlying DAG of NSDP procedure for the variable ordering $\{x_5, x_2, x_1, x_4, x_3, x_6, x_7\}$ is shown in Fig. 4 (b). As was mentioned above, this underlying DAG of local variable elimination (the elimination tree) is constructed using Elimination Game. A node i of the DAG is containing variables $(\alpha_i, Nb_{G_{x_{i-1}}}^{(i-1)}(x_i))$ is linked with the first x_j (accordingly to the ordering α) which is in $Nb_{G_{x_{i-1}}^{(i-1)}}(x_i)$. Nodes and edges of desired tree decomposition correspond one-by-one to nodes and edges of the underlying DAG. Each node of the tree decomposition is indeed a meta-node containing a subset of vertices of the interaction graph G. This subset induces a subgraph in G that was condensed to generate the meta-node. Restore these subgraphs for each meta-node of the tree decomposition.

Proposition 1. *Graph structure obtained by this construction from the underlying DAG of the NSDP procedure is a tree decomposition.*

Proof is in [63].

In our example 2, we observe that the first (accordingly to ordering α) meta-node corresponds to the variable x_5 and contains variables (vertices) x_2, x_5 (i.e., $x_5 \cup Nb(x_5)$). Subgraph induced by these vertices can be constructed using the interaction graph G (Fig. 4 a). This subgraph is shown in Fig. 8 (a) — the meta-node $\mathbf{y_1}$. Next meta-node $\mathbf{y_2}$ of the tree decomposition corresponds to the variable x_2 and contains variables x_1, x_2, x_3, x_4. The corresponding induced subgraph (clique) is shown inside the meta-node $\mathbf{y_2}$ in Fig. 9 (a). Continuing in analogous way we obtain the tree decomposition as shown in Fig. 8 (a).

It is easy to see that some cliques in this tree decomposition are not maximal and can be absorbed by other cliques. In the case, when one clique contains another clique, the second clique can be absorbed into the first one. Thus, the clique corresponding to the meta-node $\mathbf{y_2}$ is absorbed by clique $\mathbf{y_3}$ (we denote a result of absorption as a clique $\mathbf{y_{2,3}}$. The clique $\mathbf{y_5}$ is absorbed by clique $\mathbf{y_4}$ forming a clique $\mathbf{y_{4,5}}$. After absorptions done we obtain a clique tree (Fig. 8 (b)) containing only maximal cliques. These maximal cliques correspond to constraints of the DOP. In Fig. 8 (b) maximal cliques and links between them are shown. Local decomposition algorithm [90] that uses a dynamic programming paradigm can be applied to this clique tree. Other possible way of finding the clique tree is using maximal spanning tree in the dual graph.

6.3 Applying the Local Decomposition Algorithm to Solving Do Problem

To describe how tree decompositions are used to solve problems with the local decomposition algorithm, let us assume we find a tree decomposition of a graph G. Since this tree decomposition is represented as a rooted tree T, the ancestor/descendant relation is well-defined. We can associate to each meta-node $\mathbf{y}$ the subgraph of G made up by the vertices in $\mathbf{y}$ and all its descendants, and all the edges between those vertices. Starting at the leaves of the tree T, one computes information typically stored in a table, in a bottom-up manner for each bag until we reach the root. This information is sufficient to solve the subproblem for the corresponding subgraph. To compute the table for a node of the tree decomposition, we only need the information stored in the tables of the children (i.e. direct descendants) of this node. The DO problem for the entire graph can then be solved with the information stored in the table of the root of T. Consider example 2 and exploit the tree decomposition (clique tree) shown in Fig. 8 (b). Let us solve the subproblem corresponding to the block $\mathbf{y}_1$. Since this block is adjacent to the block $\mathbf{y}_{2,3}$, we have to solve a DOP with variables $\mathbf{y}_1 - \mathbf{y}_{2,3}$ for all possible assignments $\mathbf{y}_1 \cap \mathbf{y}_{2,3}$. Thus, since $\mathbf{y}_1 - \mathbf{y}_{2,3} = \{x_5\}$ and $\mathbf{y}_1 \cap \mathbf{y}_{2,3} = \{x_2\}$, then induced subproblem has the form:

$$h_{\mathbf{y}_1}(x_2) = \max_{x_5} \{4x_5\}$$

subject to

$$2x_2 + 3x_5 \le 4, \quad x_j = 0,1, \ j \in \{2,5\}$$

Solution of the problem can be written in a tabular form (see table 15).

Since $\mathbf{y}_{2,3} - \mathbf{y}_4 = \{x_1, x_2, x_3, x_4\} - \{x_3, x_6, x_7\} = \{x_1, x_2, x_4\}$ and $\mathbf{y}_{2,3} \cap \mathbf{y}_4 = \{x_3\}$, next subproblem corresponding to the leaf (or meta-node) $\mathbf{y}_{2,3}$ of the clique tree is

$$h_{\mathbf{y}_{2,3}}(x_3) = \max_{x_1, x_2, x_4} \{h_{\mathbf{y}_1} + 2x_1 + 3x_2 + 5x_4\}$$

subject to

$$3x_1 + 4x_2 + x_3 \le 6, \quad 2x_2 + 3x_3 + 3x_4 \le 5, \quad x_j = 0,1, \ j \in \{1,2,3,4\}$$

Solution of this subproblem is in table 16. The last problem corresponding to the block $\mathbf{y}_{4,5}$ left to be solved is:

$$h_{\mathbf{y}_{4,5}} = \max_{x_3, x_6, x_7} \{h_{\mathbf{y}_{2,3}}(x_3) + x_3 + 6x_6 + x_7\}$$

s.t.

$$2x_3 + 3x_6 + 2x_7 \le 5, \quad x_j = 0,1, \ j \in \{3,6,7\}$$

Table 15 Calculation of $h_{\mathbf{y_1}}(x_2)$ **Table 16** Calculation of $h_{\mathbf{y_{2,3}}}(x_3)$

x_2	$h_{\mathbf{y_1}}$	$x_5^*(x_2)$
0	4	1
1	0	0

x_3	$h_{\mathbf{y_{2,3}}}$	$x_1^*(x_3)$	$x_2^*(x_3)$	$x_4^*(x_3)$
0	11	1	0	1
1	6	1	0	0

Table 17 Calculation of $h_{\mathbf{y_{4,5}}}$

$h_{\mathbf{y_{4,5}}}$	x_3^*	x_6^*	x_7^*
18	0	1	1

The maximal objective value is 18. To find the optimal values of the variables, it is necessary to do a backward step of the dynamic programming procedure: from table 17 we have $x_3^* = 0$, $x_6^* = 1$, $x_7^* = 1$. From the table 16 using the information $x_3^* = 0$ we find $x_1^* = 1$, $x_2^* = 0$, $x_4^* = 1$. Considering table 15 we have for $x_2^*=0$: $x_5^* = 1$. The solution is $(1, 0, 0, 1, 1, 1, 1)$; the maximal objective value is 18.

7 Conclusion

This paper reviews the main graph-based local elimination algorithms for solving DO problems. The main aim of this paper is to unify and clarify the notation and algorithms of various structural DO decomposition approaches. We hope that this will allow us to apply the aforementioned decomposition techniques to develop competitive algorithms which will be able to solve practical real-life discrete optimization problems.

References

1. Amestoy, P.R., Davis, T.A., Duff, I.S.: An approximate minimum degree ordering algorithm. SIAM J. on Matrix Analysis and Applications 17, 886–905 (1996)
2. Amir, E.: Efficient approximation for triangulation of minimum treewidth. In: Proceedings of UAI (2001)
3. Aris, R.: The optimal design of chemical reactors. Academic Press, New York (1961)
4. Arnborg, S.: Efficient algorithms for combinatorial problems on graphs with bounded decomposability — A survey. BIT 25, 2–23 (1985)
5. Arnborg, S., Corneil, D.G., Proskurowski, A.: Complexity of finding embeddings in a k-tree. SIAM J. Alg. Disc. Meth. 8(2), 277–284 (1987)
6. Arnborg, S., Lagergren, J., Seese, D.: Easy problems for tree-decomposable graphs. J. of Algorithms 12, 308–340 (1991)
7. Ashcraft, C.: Compressed graphs and the minimum degree algorithm. SIAM J. Sci. Comput. 16(6), 1404–1411 (1995)

8. Ashcraft, C., Liu, J.W.H.: Robust ordering of sparse matrices using multisection. SIAM J. Matrix Anal. Appl. 19(3), 816–832 (1995)
9. Barnhart, C., Johnson, E.L., Nemhauser, G.L., Savelsbergh, M.W.P., Vance, P.H.: Branch and price: Column generation for solving huge integer programs. Operations Research 46, 316–329 (1998)
10. Beeri, C., Fagin, R., Maier, D., Yannakakis, M.: On the desirability of acyclic database schemes. Journal ACM 30, 479–513 (1983)
11. Beightler, C.S., Johnson, D.B.: Superposition in branching allocation problems. Journal of Mathematical Analysis and Applications 12, 65–70 (1965)
12. Bellman, R., Dreyfus, S.: Applied Dynamic Programming. Princeton University Press, Princeton (1962)
13. Benders, J.F.: Partitioning procedures for solving mixed-variables programming problems. Numerische Mathematik 4, 238–252 (1962)
14. Bertele, U., Brioschi, F.: A new algorithm for the solution of the secondary optimization problem in nonserial dynamic programming. Journal of Mathematical Analysis and Applications 27, 565–574 (1969)
15. Bertele, U., Brioschi, F.: Contribution to nonserial dynamic programming. Journal of Mathematical Analysis and Applications 28, 313–325 (1969)
16. Bertele, U., Brioschi, F.: Nonserial Dynamic Programming. Academic Press, New York (1972)
17. Berry, A.: A wide-range efficient algorithm for minimal triangulation. In: Proceedings of SODA (1999)
18. Blair, J.R.S., Peyton, B.: An introduction to chordal graphs and clique trees. In: Graph theory and sparse matrix computation. Springer, New York (1993)
19. Bodlaender, H.L. (ed.): WG 2003. LNCS, vol. 2880. Springer, Heidelberg (2003)
20. Bodlaender, H.L.: Treewidth: Algorithmic techniques and results. In: Privara, I., Ružička, P. (eds.) MFCS 1997. LNCS, vol. 1295. Springer, Heidelberg (1997)
21. Bodlaender, H., Koster, A.M.C.A.: Combinatorial optimization on graphs of bounded treewidth. Computer Journal 51, 255–269 (2008)
22. Burkard, R.E., Hamacher, H.W., Tind, J.: On General Decomposition Schemes in Mathematical Programming. Mathematical Programming Studies 24: Festschrift on the occasion of the 70 th birthday of George B. Dantzig, 238–252 (1985)
23. Cook, S.A.: The complexity of theorem-proving procedures. In: Proc. 3rd Ann. ACM Symp. on Theory of Computing Machinery, New York (1971)
24. Cook, W., Seymour, P.D.: Tour merging via branch-decomposition. INFORMS Journal on Computing 15, 233–248 (2003)
25. Courcelle, B.: The monadic second-order logic of graphs I: Recognizable sets of finite graphs. Information and Computation 85, 12–75 (1990)
26. Crama, Y., Hansen, P., Jaumard, B.: The basic algorithm for pseudo-boolean programming revisited. Discrete Applied Mathematics 29, 171–185 (1990)
27. Dantzig, G.B.: Programming of interdependent activities II: Mathematical model. Econometrica 17, 200–211 (1949)
28. Dantzig, G.B.: Time-staged methods in linear programming. Comments and early history. In: Dantzig, G.B., et al. (eds.) Large-Scale Linear Programming, IIASA, Laxenburg, Austria, pp. 3–16 (1981)
29. Dantzig, G.B.: Solving staircase linear programs by a nested block-angular method. Technical Report 73-1. Stanford Univ., Dept. of Operations Research, Stanford (1973)

30. Dasgupta, S., Papadimitriou, C.H., Vazirani, U.V.: Algorithms. McGraw-Hill, New York (2006)
31. Dechter, R.: Constraint networks. In: Encyclopedia of Artificial Intelligence, 2nd edn. Wiley, New York (1992)
32. Dechter, R.: Bucket elimination: A unifying framework for reasoning. Artificial Intelligence 113, 41–85 (1999)
33. Dechter, R., El Fattah, Y.: Topological parameters for time-space tradeoff. Artificial Intelligence 125, 93–118 (2001)
34. Dechter, R.: Constraint processing. Morgan Kaufmann, San Francisco (2003)
35. Dechter, R., Pearl, J.: Tree clustering for constraint networks. Artificial Intelligence 38, 353–366 (1989)
36. Dolgui, A., Soldek, J., Zaikin, O. (eds.): Supply chain optimisation: product/process design, facilities location and flow control. Series: Applied Optimization, vol. 94, XVI. Springer, Heidelberg (2005)
37. Esogbue, A.O., Marks, B.: Non-serial dynamic programming – A survey. Operational Research Quarterly 25, 253–265 (1974)
38. Fernandez-Baca, D.: Nonserial dynamic programming formulations of satisfiability. Information Processing Letters 27, 323–326 (1988)
39. YuYu, F.: On solving discrete programming problems of special form. Economics and Mathematical Methods 1, 262–270 (1965) (Russian)
40. Floudas, C.A.: Nonlinear and mixed-integer optimization: fundamentals and applications. Oxford University Press, Oxford (1995)
41. Fomin, F., Kratsch, D., Todinca, I.: Exact (exponential) algorithms for treewidth and minimum fill-in. In: Díaz, J., Karhumäki, J., Lepistö, A., Sannella, D. (eds.) ICALP 2004. LNCS, vol. 3142, pp. 568–580. Springer, Heidelberg (2004)
42. Fourer, R.: Staircase matrices and systems. SIAM Review 26(1), 1–70 (1984)
43. Freuder, E.: Constraint solving techniques. In: Tyngu, E., Mayoh, B., Penjaen, J. (eds.) Constraint Programming of series F: Computer and System Sciences. NATO ASI Series, pp. 51–74 (1992)
44. Fulkerson, D.R., Gross, O.A.: Incidence matrices and interval graphs. Pacific J. of Mathematics 15, 835–855 (1965)
45. George, J.A., Liu, J.W.H.: Computer Solution of Large Sparse Positive Definite Systems. Prentice-Hall Inc., Englewood Cliffs (1981)
46. Gogate, V., Dechter, R.: A complete anytime algorithm for treewidth. In: Proceedings of UAI (2004)
47. Gottlob, G., Leone, N., Scarcello, F.: A comparison of structural CSP decomposition methods. Artificial Intelligence 124, 243–282 (2000)
48. Gottlob, G., Szeider, S.: Fixed-parameter algorithms for artificial intelligence, constraint satisfaction and database problems. The Computer Journal 51, 303–325 (2008)
49. Gyssens, M., Jeavons, P.G., Cohen, D.A.: Decomposing constraint satisfaction problems using database techniques. Artificial Intelligence 66, 57–89 (1994)
50. Gu, J., Purdom, P.W., Franco, J., Wah, B.W.: Algorithms for the satisfiability (SAT) problem: A survey. Satisfiability Problem Theory and Applications (1997)
51. Harary, F., Norman, R.Z., Cartwright, D.: Structural Models: An Introduction to the Theory of Directed Graphs. John Wiley & Sons, Chichester (1965)
52. Hammer, P.L., Rudeanu, S.: Boolean Methods in Operations Research and Related Areas. Springer, Heidelberg (1968)

53. Heggernes, P., Eisenstat, S.C., Kumfert, G., Pothen, A.: The Computational Complexity of the Minimum Degree Algorithm. Techn. report UCRL-ID-148375. Lawrence Livermore National Laboratory (2001), http://www.llnl.gov/tid/lof/documents/pdf/241278.pdf

54. Hendrickson, B., Rothberg, E.: Improving the run time and quality of nested dissection ordering. SIAM J. Sci. Comput. 20(2), 468–489 (1998)

55. Hicks, I.V., Koster, A.M.C.A., Kolotoglu, E.: Branch and tree decomposition techniques for discrete optimization. In: Tutorials in Operations Research. INFORMS, New Orleans (2005),
http://ie.tamu.edu/People/faculty/Hicks/bwtw.pdf

56. Hliněný, P., Oum, S., Seese, D., Gottlob, G.: Width parameters beyond treewidth and their applications. The Computer Journal 51, 326–362 (2008)

57. Ho, J.K., Loute, E.: A set of staircase linear programming test problems. Mathematical Programming 20, 245–250 (1981)

58. Hooker, J.N.: Logic-based Methods for Optimization: Combining Optimization and Constraint Satisfaction. John Wiley & Sons, Chichester (2000)

59. Hooker, J.N.: Logic, optimization and constraint programming. INFORMS Journal on Computing 14, 295–321 (2002)

60. Jeavons, P.G., Gyssens, M., Cohen, D.A.: Decomposing constraint satisfaction problems using database techniques. Artificial Intelligence 66, 57–89 (1994)

61. Jégou, P., Ndiaye, S.N., Terrioux, C.: Computing and exploiting treedecompositions for (Max-)CSP. In: van Beek, P. (ed.) CP 2005. LNCS, vol. 3709, pp. 777–781. Springer, Heidelberg (2005)

62. Jensen, F.V., Lauritzen, S.L., Olesen, K.G.: Bayesian updating in causal probabilistic networks by local computations. Computat. Statist. Quart. 4, 269–282 (1990)

63. Kask, K., Dechter, R., Larrosa, J., Dechter, A.: Unifying cluster-tree decompositions for reasoning in graphical models. Artificial Intelligence 160, 165–193 (2005)

64. Kjaerulff, U.: Triangulation of graphs – algorithms giving small total state space. Techn.report. Aalborg, Denmark (1990)

65. Koster, A.M.C.A., van Hoesel, C.P.M., Kolen, A.W.J.: Solving frequency assignment problems via tree-decomposition. In: Broersma, H.J., et al. (eds.) 6th Twente workshop on graphs and combinatorial optimization. Univ. of Twente, Enschede, Netherlands (1999)

66. Lauritzen, S.L., Spiegelhalter, D.J.: Local computation with probabilities on graphical structures and their application to expert systems. J. Roy. Statist. Soc. Ser. B 50, 157–224 (1988)

67. Liu, J.W.H.: The role of elimination trees in sparse factorization. SIAM Journal on Matrix Analysis and Applications 11, 134–172 (1990)

68. Martelli, A., Montanari, U.: Nonserial Dynamic Programming: On the Optimal Strategy of Variable Elimination for the Rectangular Lattice. Journal of Mathematical Analysis and Applications 40, 226–242 (1972)

69. Mitten, L.G., Nemhauser, G.L.: Multistage optimization. Chemical Engineering Progress 54, 52–60 (1963)

70. Karypis, G., Kumar, V.: MeTiS - a software package for partitioning unstructured graphs, partitioning meshes, and computing fill-reducing orderings of sparse matrices. Version 4, University of Minnesota (1998),
http://www-users.cs.umn.edu/~karypis/metis

71. Mitten, L.G., Nemhauser, G.L.: Multistage optimization. Chemical Engineering Progress 54, 52–60 (1963)
72. Neapolitan, R.E.: Probabilistic Reasoning in Expert Systems. Wiley, New York (1990)
73. Nemhauser, G.L., Wolsey, L.A.: Integer and Combinatorial Optimization. John Wiley & Sons, Chichester (1988)
74. Nemhauser, G.L.: The age of optimization: solving large-scale real-world problems. Operations Research 42, 5–13 (1994)
75. Nowak, I.: Lagrangian decomposition of block-separable mixed-integer all-quadratic programs. Mathematical Programming 102, 295–312 (2005)
76. Neumaier, A., Shcherbina, O.: Nonserial dynamic programming and local decomposition algorithms in discrete programming (submitted, 2008), http://www.optimization-online.org/DB_HTML/2006/03/1351.html
77. Pang, W., Goodwin, S.D.: A new synthesis algorithm for solving CSPs. In: Proc. of the 2nd Int. Workshop on Constraint-Based Reasoning. Key West (1996)
78. Pardalos, P.M., Du, D.Z. (eds.): Handbook of combinatorial optimization, vol. 1, 2, and 3. Kluwer Academic Publishers, Dordrecht (1998)
79. Pardalos, P.M., Wolkowicz, H. (eds.): Novel approaches to hard discrete optimization. Fields Institute, American Mathematical Society (2003)
80. Parter, S.: The use of linear graphs in Gauss elimination. SIAM Review 3, 119–130 (1961)
81. Pearl, J.: Probabilistic reasoning in intelligent systems. Morgan Kaufmann, San Mateo (1988)
82. Ralphs, T.K., Galati, M.V.: Decomposition in integer linear programming. In: Karlof, J. (ed.) Integer Programming: Theory and Practice (2005)
83. Robertson, N., Seymour, P.D.: Graph minors. II. Algorithmic aspects of tree width. J. of Algorithms 7, 309–322 (1986)
84. Rose, D., Tarjan, R., Lueker, G.: Algorithmic aspects of vertex elimination on graphs. SIAM J. on Computing 5, 266–283 (1976)
85. Rose, D.J.: A graph-theoretic study of the numerical solution of sparse positive definite systems of linear equations. In: Read, R.C. (ed.) Graph Theory and Computing, pp. 183–217. Academic Press, New York (1972)
86. Rosenthal, A.: Dynamic programming is optimal for nonserial optimization problems. SIAM J. Comput. 11, 47–59 (1982)
87. Seidel, P.: A new method for solving constraint satisfaction problems. In: Proc. of the 7th IJCAI, Vancouver, Canada, pp. 338–342 (1981)
88. Sergienko, I.V., Shylo, V.P.: Discrete Optimization: Problems, Methods, Studies, Naukova Dumka, Kiev (2003)
89. Shcherbina, O.: A local algorithm for integer optimization problems. USSR Comput. Math. Phys. 20, 276–279 (1980)
90. Shcherbina, O.A.: On local algorithms of solving discrete optimization problems. Problems of Cybernetics (Moscow) 40, 171–200 (1983)
91. Shcherbina, O.: Nonserial dynamic programming and tree decomposition in discrete optimization. In: Proc. of Int. Conference on Operations Research Operations Research 2006, Karlsruhe, September 6-8, pp. 155–160. Springer, Berlin (2006)
92. Shcherbina, O.A.: Tree decomposition and discrete optimization problems: A survey. Cybernetics and Systems Analysis 43, 549–562 (2007)

93. Shcherbina, O.A.: Methodological issues of dynamic programming. Dynamich Sistemy 22, 21–36 (2007) (in Russian)
94. Shcherbina, O.A.: Local elimination algorithms for solving sparse discrete problems. Comput. Math. and Math. Phys. 48, 152–167 (2008)
95. Shenoy, P.P., Shafer, G.: Propagating belief functions using local computations. IEEE Expert 1, 43–52 (1986)
96. Shoikhet, K., Geiger, D.: A practical algorithm for finding optimal triangulation. In: Proceedings of AAAI (1997)
97. Urrutia, J.: Local solutions for global problems in wireless networks. J. of Discrete Algorithms 5, 395–407 (2007)
98. Vanderbeck, F., Savelsbergh, M.: A generic view at the Dantzig-Wolfe decomposition approach in mixed integer programming. Operations Research Letters 34, 296–306 (2006)
99. Van Roy, T.J.: Cross decomposition for mixed integer programming. Mathematical Programming 25, 46–63 (1983)
100. Wah, B.W., Li, G.-J.: Systolic processing for dynamic programming problems. Circuits Systems Signal Process 7, 119–149 (1988)
101. Wilde, D., Beightler, C.: Foundations of Optimization. Prentice-Hall, Englewood Cliffs (1967)
102. Weigel, R., Faltings, B.: Compiling constraint satisfaction problems. Artificial Intelligence 115, 257–287 (1999)
103. Wets, R.J.B.: Programming under uncertainty: The equivalent convex program. SIAM J. Appl. Math. 14, 89–105 (1966)
104. Yu I, Z.: Selected Works. Magistr, Moscow (1998) (in Russian)
105. Yu I, Z., Losev, G.: Neighborhoods in problems of discrete mathematics. Cybern. Syst. Anal. 31, 183–189 (1995)

Evolutionary Approach to Solving Non-stationary Dynamic Multi-Objective Problems

Zikrija Avdagić, Samim Konjicija, and Samir Omanović

This chapter aims at presenting the general problem of decision making in unknown, complex or changing environment by an extension of static multi-objective optimization problem. General optimization problem is defined, which encompasses not just dynamics, but also change in the optimization problem itself, with focus on changing number of objectives used to evaluate potential solutions.

In order to solve the defined problem, a variant of multi-objective genetic algorithm was used. Since the chapter doesn't focus on the performance of the algorithm used for solving the problem, but tends to demonstrate the approach, experimental results produced by tests with MOGA are presented. These experimental results clearly demonstrate, that MOGA successfully led the population of potential solutions to the problem for different test cases, such as homogenous, non-homogenous, and the problem with changing number of objectives. Decision-making based on ranking of the potential solutions has also been demonstrated.

1 Introduction

Modern technical systems are capable of developing a very complex interaction with their environment. But one of the basic preconditions for sucess of this interaction is a good information on that environment.

Zikrija Avdagić
Faculty of Electrical Engineering, University of Sarajevo, Bosnia and Herzegovina
e-mail: zikrija.avdagic@etf.unsa.ba

Samim Konjicija
Faculty of Electrical Engineering, University of Sarajevo, Bosnia and Herzegovina
e-mail: samim.konjicija@etf.unsa.ba

Samir Omanović
Faculty of Electrical Engineering, University of Sarajevo, Bosnia and Herzegovina
e-mail: samir.omanovic@etf.unsa.ba

A. Abraham et al. (Eds.): Foundations of Comput. Intel. Vol. 3, SCI 203, pp. 267–289.
springerlink.com © Springer-Verlag Berlin Heidelberg 2009

If a simple industrial controller is analyzed, its output results from adjustment of parameters chosen in accordance to the control goals, based on known features of the system [1]. The performance of such controller is very good, but only in circumstances taken into account during its synthesis. Using various approaches, such as estimation of behavior of the system with statistically known disturbances, introduction of adaptation of controller parameters etc. [2], a wider problem domain can be encompassed, nevertheless it is still very clearly defined in advance. The question is how to lead the interaction of the system with its environment, when the description of that environment becomes too complex for explicit treatment or deficiant. In other words, how to act in conditions of unsuficiently known or changing environment? On the other side, interaction of living organisms with their environment happens very successfully all the time in exactly such conditions of uncertainty. Without an exact description of interaction with his environment, behavior of a human being is always dynamic and adaptable to a concrete circumstances in much wider scope, which excels capabilities of any existing technical adaptive system.

Imitation of some aspects of human's behavior in conditions of uncertainty has inspired various approaches and techniques of artificial intelligence [3]. Yet, the problem domain still stays highly constrained and predefined, so that any radical change in model of interaction with environment requires direct intervention, which goes back into system's structure.

This chapter describes a part of the research in which the problem of inference and decision making in conditions of uncertainty was presented by a multi-objective problem whith dynamic and changing nature. Therefore, it was necessary to extend the definition of multi-objective problem. It was necessary, as well, to signify the effects, which changing dimensionality of the space of objective values has on the definition of the precedence in this space.

Evolutionary algorithms, especially genetic algorithm as their typical representative have been successfully used to solve dynamic problems [5], as well as various multi-objective problems [6] [7]. An approach to apply a multi-objective evolutionary algorithm to solving the defined dynamic multi-objective problem of search for solution was demonstrated in this chapter.

The application of evolutionary algorithms to solving dynamic multi-objective problem will be demonstrated on problems defined on the basis of a class of test functions with known features.

Dynamic multi-objective problem defined based on a class of test functions with known features has been chosen in order to evaluate the application of the proposed approach, in conditions when the features of the problem are known (features of test functions, Pareto front etc.). In other words, although the nature and features of the problem are known, it will be modelled as the problem of decision making in unknown environment, and after its solution it will be possible to have insight into the features of the solution acquired. This problem will also be used to demonstrate the common approach to decision making on choice of a unique point for solution to the problem, since due to known features of the problem, it will be possible to get clear insight into the effects of different decisions on the subjective quality of the solution. It is clear that such insight into features of the solution to practical problems is not available.

2 General Optimization Problem

Definition of multi-objective optimization problem was given in [4] [7]. Although it clearly defines basic elements of multi-objective optimization problem, it is limited to static (without the existence of independent variables) and stationary problems (problems with unchanging elements). This definition will now be modified into the definition of general optimization problem, in order to encompass non-stationary and dynamic problems as well:

Definition 1. *General optimization problem includes the set of n problem variables*

$$\mathbf{x}(t) = [x_1(t), x_2(t), ..., x_n(t)]^T, t \in t_0, t_f \tag{1}$$

where t_0 represents initial, and t_f final value of independent variable t, the set of $m(\mathbf{x}(t), t)$ objectives (criteria)

$$\mathbf{Q}(\mathbf{x}(t), t) = [q_1(\mathbf{x}(t), t), q_2(\mathbf{x}(t), t), ..., q_m(\mathbf{x}(t), t)]^T \tag{2}$$

and the set of $\mathbf{r}(t)$ constraints on values of problem variables

$$\mathbf{G}(\mathbf{x}(t), t) = [g_1(\mathbf{x}(t), t), g_2(\mathbf{x}(t), t), ..., g_r(\mathbf{x}(t), t)]^T \tag{3}$$

The problem variables are functions of independent variable. The criteria and constraints are functions of problem variables and independent variable. The goal of optimization is to determine optimum of vector function $\mathbf{Q}(\mathbf{x}(t), t)$, satisfying the constraints, i.e. to determine:

$$opt\mathbf{Q}(\mathbf{x}(t), t), g_k(\mathbf{x}(t), t) \leqslant 0, k = 1, 2, ..., r(t). \tag{4}$$

The following can be noticed from this definition:

- the dimensionality of the problem space is constant, since this work deals only with the systems with fixed structure, so the dimensionality of the problem space, as well as of the state vector is always constant
- the trajectories of problem variables $\mathbf{x}(t)$ in certain range of value of independent variable t are analyzed,
- the alternatives considered are the trajectories of problem variables $\mathbf{x}(t)$ from the set of feasible values $\Omega(t)$, defined by the system of constraints in form $g_k(\mathbf{x}(t), t)$, whose shape and total number $r(t)$ depends on independent variable,
- the value of each feasible trajectory of a problem variable is being measured using criteria in form $q_i(\mathbf{x}(t), t)$, whose shape and total number $m(\mathbf{x}, t)$ also depends on independent variable, and the alternative $\mathbf{x}(t)$.

The general optimization problem defined in this way encompasses all the classes of problems mentioned in the beginning of this section. On the other hand, the general optimization problem reflects the essence of the problem in the focus of this work, which is search for solution of certain problem when the subject is in changing, unknown or complex environment. The general optimization problem enables to, as the environment changes (what can be a consequence of update of the model of environment based on newly obtained information), modify problem being solved even during its solving. A question of choice of appropriate method for solving the problem formulated in such way remains still open, since practically all classical approaches fall off, due to their inherent limitations on the nature of a problem being solved. Besides, the most practical problems that are intended to be solved using the described approach are NP-hard.

On the other hand, very small sensitivity of performance of an algorithm for various domains of problems is one of basic features of evolutionary algorithms, whose the best known representative is genetic algorithms (GA). Genetic algorithms are search algorithms based on the mechanics of natural selection, which combine survival of the fittest among string structures, that efficiently exploit information present in the population [8]. Contrary to most other algorithms, which pose strict requirements on preconditions of their application and on solving the problem where they achieve the highest performance, GA is usually applied to black-box type of problems, without sufficient knowledge on their structure [9]. Also, GA is applied to the problems of optimization in presence of disturbances. In order for this to be possible, the preconditions for application of GA for solving such problems are very rare. Observing in average, the performance of GA is far better from the performance of any problem-specific algorithm, alghough GA possesses definitively worse performance compared to any problem-specific algorithm, when comparison is done just for the problems from domain of this problem-specific algorithm.

A lot of literature can be found, which treats in depth various topics of genetic algorithms [8] [9] [10] [11], so they will not be addressed in this chapter.

3 Dynamic Multi-Objective Problem Defined on a Class of Test Functions

The sets of test functions play very important role in research of features and behaviour, as well as the efficiency of search algorithms in general, as well as the evolutionary algorithms. There exist sets of test functions for single-criterion algorithms, which contain functions with various features, what enables testing various aspects of search algorithms [12]. A large number of test functions can be found in literature, which are proposed for exploring the behavior and effects of application of multi-objective evolutionary algorithms as well [7], [6], [13]. These test functions, being already defacto standard, are usually limited to two or three criteria. The main reason for that lies in the fact that, although various numerical performance measures for an algorithm exist, visualization still gives the best insight into real features of the solution found by certain algorithm. Since, with increase in number of criteria, there exist no efficient way for visualization of the set of values, it is quite difficult to synthesize test functions for larger number of criteria. As a good approach, various scalable problems have been proposed. They enable to determine the number of criteria freely, still maintaining the knowledge on the features of Pareto front [7], [6].

Meanwhile, all these test functions are static, and just in a couple of recent years, the proposals on how to modify these static functions in order to evaluate behaviour of algoriths when applied to solving dynamic problems can be found in literature [14] [15]. The choice of test functions should be done carefully, since it was stated many times in literature, that the measured performance of certain algorithm greatly depends on the chosen set of test functions, and that inadequate choice of test functions generates conclusions that don't hold generally (see No-free-lunch theorems) [16].

The test cases have been designed so that the application of a variant of multi-objective genetic algorithm to different subclasses of general optimization problem could have been tested:

- dynamic multi-objective test problem,
- homogenous non-stationary multi-objective test problem,
- non-homogenous non-stationary multi-objective test problem,
- non-homogenous non-stationary multi-objective test problem with changing number of criteria during a single run of the algorithm.

Dynamic multi-objective test problem demonstrates finding the optimal trajectory in the search space, based on the pre-defined objectives. Such test problem is stationary, which means that the problem elements don't change.

Non-stationary test problems are designed so that certain element or elements of the problem change with independent variable.

Homogenous non-stationary test problem possesses such feature, that the objectives change with independent variable, but the Pareto form moves coherently [5]. Non-homogenous non-stationary test problem doesn't possess this feature.

Non-homogenous non-stationary test problem with changing number of criteria enables us to test how the population behaves, whent even the number of objectives changes with independent variable [17].

These test cases encompass the basic types of general optimization problem that can be derived as models of real-life applications.

MOGA was used for solving the test cases. It is a modification of GA with Pareto ranking proposed by Fonseca and Flemming, as the first multi-objective GA, which explicitly takes care of non-dominated solutions maintaining the population diversity at the same time [18]. Although many other modifications of multi-objective GA exist nowadays, MOGA has been chosen due to its simplicity, and having in mind that the focus of the research was not on the performance of the algorithm itself.

3.1 Dynamic Multi-Objective Test Problem

In order to illustrate application of multi-objective GA for solving dynamic multi-objective problem, it was necessary to choose criteria in form:

$$q_i(\mathbf{x}(k \cdot T), k) = H_i(\mathbf{x}(N \cdot T), N) + \sum_{k=0}^{N-1} F_i(\mathbf{x}(k \cdot T), k) \tag{5}$$

as generalized form of the common single-objective dynamic problem found in the literature [19] [20]. The function $H_i(\mathbf{x}(N \cdot T), N)$ determines the influence of final state on the value of criterion q_i , whereas the function $F_i(\mathbf{x}(k \cdot T), k)$ determines contribution of states $\mathbf{x}(k \cdot T)$ on the value of the same criterion. In other words, the function $H_i(\mathbf{x}(N \cdot T), N)$ where the trajectory ends, whereas the function $F_i(\mathbf{x}(k \cdot T), k)$ determines the way on which the trajectory reaches the final state.

As an example, let's consider the problem of determining the trajectory in two-dimensional space from the initial point $\mathbf{x}^0$ to final point $\mathbf{x}^N$, with:

$$\mathbf{x}^0 = \begin{bmatrix} 0 \\ 0 \end{bmatrix}, \mathbf{x}^N = \begin{bmatrix} 1 \\ 0.5 \end{bmatrix} \tag{6}$$

where $N = 5$. So, the trajectory consists of six points in two-dimensional space (including the initial point), and each potential trajectory will be evaluated by two criteria:

$$
\begin{aligned}
q_1(\mathbf{x}(k \cdot T), k) &= 100 - 50 \cdot d_1(x_1^N, x_1(5 \cdot T)) \\
&\quad - 5 \cdot \sum_{k=0}^{4} d(\mathbf{x}(k \cdot T), \mathbf{x}((k+1) \cdot T)) \\
q_2(\mathbf{x}(k \cdot T), k) &= 100 - 50 \cdot d_1(x_2^N, x_2(5 \cdot T)) \\
&\quad - 5 \cdot \sum_{k=0}^{4} ln(1 + |x_1(k \cdot T) - x_2((k+1) \cdot T)|)
\end{aligned}
\tag{7}
$$

where $d(\mathbf{x}^i, \mathbf{x}^j)$ defines the distance between the points $\mathbf{x}^i$ and $\mathbf{x}^j$:

$$d(\mathbf{x}^i, \mathbf{x}^j) = \sqrt{(x_1^i - x_1^j)^2 + (x_2^i - x_2^j)^2} \tag{8}$$

and $d_1(\mathbf{x}^i, \mathbf{x}^j)$ is defined as:

$$d_1(x_r^i, x_r^j) = ln(1 + |x_r^i - x_r^j|) \tag{9}$$

It is necessary to determine the maximal value for each criterion. It can be noticed that the first criterion has larger value for shorter trajectories, and when the final point lies closer to the goal regarding the first variable. The second criterion will have larger value when both components of point **x** have similar values, and when the final point lies closer to the goal regarding the second variable. The criteria used for this test problem demonstrate also the freedom of choice of criteria, when an evolutionary algorithm is used for solving the problem. Namely, both criteria are non-linear and the second one is non-smooth.

Determining non-dominated solutions

MOGA was used for solving the problem, with an individual represented by a vector of 10 real numbers, with the first five representing increment for the variable x_1, and the second five representing increment for the variable x_2:

$$\Delta \mathbf{x} = [\Delta x_1(0), \Delta x_1(T), \Delta x_1(2 \cdot T), ...,$$
$$\Delta x_1(4 \cdot T), \Delta x_2(0), \Delta x_2(T), \Delta x_2(2 \cdot T), ..., \Delta x_2(4 \cdot T)] \tag{10}$$

The increment vectors have been evaluated by using MOGA, and the trajectories have been defined by following equations:

$$x_1((k+1) \cdot T) = x_1(k \cdot T) + \Delta x_1(k \cdot T)$$
$$x_2((k+1) \cdot T) = x_2(k \cdot T) + \Delta x_2(k \cdot T) \tag{11}$$

The increments have been limited to $\Delta x_i(k \cdot T) \in [-0.3, 0.3]$. Parameter settings for MOGA are given in Table 1. The problem of determining proper parameters for GA in order to achieve good performance is a specific problem [12]. Meanwhile, standard suggested values for parameters have been used in this work, since the purpose of test problem was to illustrate the approach of application of evolutionary algorithm for solving dynamic multi-objective problem, and not to achieve optimal performance of this algorithm.

Figure 1 represents values of individuals of the final population. This figure shows also non- dominated solutions from the final solution. Out of 200 individuals in the final population, 143 are non-dominated, which means that 71,5% individuals of the final population lies in vicinity of the Pareto front. In order to demonstrate that random search can't produce even similarly qualitative results, as well as in order to get insight into shape of the set of values, 100,000 random trajectories have been generated, and their values have been determined. It can be seen that randomly generated trajectories lie far from the Pareto front.

Figure 2 illustrates the development of population in the space of values each 100 generations. Figure 3 shows some of non-dominated resulting trajectories. It can be noticed, that the shown trajectories represent varius compromises between the two criteria. So, the trajectory a) prefers the criterion q_1 which depends on the length of trajectory, to the criterion q_2 which

Table 1 Parameter settings for MOGA used for solving dynamic test problem

Parameter	Value
Discretization step	10^{-2}
Population size (Chromosomes)	200
Crossover operator	two-point crossover
Crossover probability	0.75
Mutation operator	binary mutation
Mutation probability	0.005
Number of generations	200

Fig. 1 Values for the final population (•), non-dominated solutions (+) and random trajectories (x)

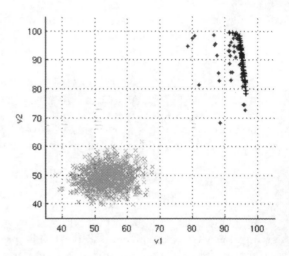

depends on the distance of the final point from the goal. The trajectory b) preferes the criterionj q_2 to the criterion q_1, so the length of the trajectory is increased, but the final point is closer to the goal. The trajectory c) represents the best compromise regarding both criteria, and this point is the closest to the ideal $v_0 = [100, 100]^T$ which lies out of the set of values. It can be noticed, that trajectories b) and c) lead almost linearly to the point $x(4 \cdot T)$, and then bend towards the goal. It is the consequence of the fact that the criterion q_2 has greater value when $x_1(k \cdot T)$ and $x_2(k \cdot T)$ are similar.

Decision making on choice of solution from the set of non-dominated solutions

When the set V_{sol} consisting of 143 non-dominated solutions was extracted from the final population, it was necessary to choose one trajectory as a solution to the problem. Therefore, an example of decision making on choice of the solution to the problem will be demonstrated, by using ranking of solutions regarding the target point [21]. This approach to decision making ranks all the potential solutions to the problem depending on the number

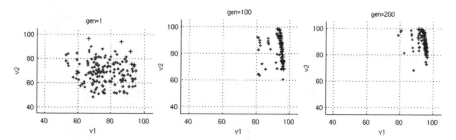

Fig. 2 Development of the population each 100 generations

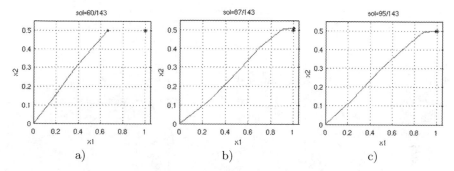

Fig. 3 Trajectories for different compromises between criteria q_1 and q_2: a) Preferring the criterion q_1, b) Preferring the criterion q_2, c) Compromise solution

of the values of objectives for each potential solution, which outperform the pre-chosen target values (Figure 4). Several variants of this approach exist, some of which introduce priority of objectives.

Ranking the non-dominated solutions. Let criteria q_1 and q_2 have identical priority. In order to perform ranking of trajectories $\mathbf{x}(k \cdot T) \in \Omega_{sol}$, $k \in [0, 5]$, it is necessary to choose target point in the space of values. Let's take this

Fig. 4 Values for the final population ($\bullet$), non-dominated solutions ($+$) and random trajectories ($\times$)

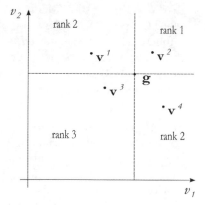

target point in vicinity of Pareto front, as close as possible to the ideal $\mathbf{v}_0 = [100, 100]^T$ but inside the set of values. When the target point lies out of the set of values, none of solutions has rank 1, and depending on the choice of this target point, more or less of solutions get rank 1. So, if a point $\mathbf{g} = [94, 98]^T$ was chosen for target, we get a unique solution shown in Fig. 3.c) having rank 1. But if a point $\mathbf{g} = [93, 98]^T$ was chosen for target, total of 6 points have rank 1, and in this case the choice of any of these points is equal.

3.2   Non-stationary Multi-objective Test Problem

During the choice of test functions for demonstrating the approach to solving non-stationary multi-objective problem of search for solution, it is necessary to take into account the basic features of such problem:

- it should be easy to extend the number of criteria,
- it should be possible to dynamically modify criteria,

Having in mind the mentioned features, test functions generated according to the principle DTLZ (Deb-Thiele-Laumanns-Zitzler) [16] have been chosen, with modification:

$$q_1(\mathbf{x}) = a_1 \cdot x_1^{c_1} \cdot x_2^{c_1} \cdot \ldots \cdot x_{m-1}^{c_1} \cdot (1 - x_m)^{c_1} \cdot g_1(\mathbf{x}) + b_1$$
$$q_2(\mathbf{x}) = a_2 \cdot x_1^{c_2} \cdot x_2^{c_2} \cdot \ldots \cdot (1 - x_{m-1})^{c_2} \cdot (1 - x_m)^{c_2} \cdot g_2(\mathbf{x}) + b_2$$

$$\vdots$$

$$q_{m-1}(\mathbf{x}) = a_{m-1} \cdot x_1^{c_{m-1}} \cdot (1 - x_2)^{c_{m-1}} \cdot \ldots \cdot (1 - x_{m-1})^{c_{m-1}} \cdot \tag{12}$$
$$\cdot (1 - x_m)^{c_{m-1}} \cdot g_{m-1}(\mathbf{x}) + b_{m-1}$$
$$q_m(\mathbf{x}) = a_m \cdot (1 - x_1)^{c_m} \cdot (1 - x_2)^{c_m} \cdot \ldots \cdot (1 - x_{m-1})^{c_m} \cdot$$
$$\cdot (1 - x_m)^{c_m} \cdot g_m(\mathbf{x}) + b_m$$

where $g_i = 1 - d_i \cdot cos(20 \cdot \pi \cdot x_i)$, and a_i, b_i, c_i and d_i represent real numbers. These real numbers can be used as parameters for adjustment of features of the set of values. Additionally, when functions of independent variable are used for these parameters instead of constant values, a class of non-stationary test functions can be generated, whose features are still known. It is obvious that for such case it is necessary to have number of criteria equal to the number of variables.

Three cases have been tested:

- Homogenous non-stationary problem, where the set of values continually changes with generation of GA(generation of GA is directly proportional to the independent variable),
- Non-homogenous non-stationary problem, with continually changing set of values, together with changes in Pareto front, as consequence of modification of criteria,
- Non-homogenous non-stationary problem with changing number of criteria.

Fig. 5 Random trajec-
tory of minimal point

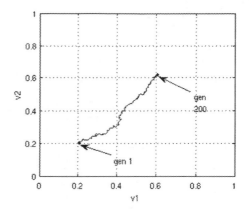

Table 2 Parameter settings for MOGA used for solving dynamic test problem

Parameter	Value
Discretization step	10^{-2}
Population size (Chromosomes)	100
Crossover operator	uniform crossover
Crossover probability	0.75
Mutation operator	binary mutation
Mutation probability	0.01
Number of generations	200

Homogenous non-stationary problem with continual change of set of values

In this test case, the set of values has coherently been changing with each generation of GA, each time for random value. Two-criterial case has been analyzed, and the trajectory of minimal point of the set of values is shown in Fig. 5.

Values of parameters of test functions were $a_i = 1$, $d_i = 0$, $b_i = b_i(k)$, where k designates the generation of GA, which now represents the independent variable. Such parameter values have been chosen in order to get simpler form of Pareto front, as well as dependence of the set of values on the generation of MOGA, as independent variable. Two cases have been taken for value c_i: $c_i = 1$, which results in the set of values with linear borders, and $c_i = 2$, which results in the set of values with non-linear borders.

MOGA has been used for solving the problem, where the vector of 2 real numbers represented the individual, and these real numbers defined the values of problem variables x_j in E^2. Parameters settings for MOGA are given in Table 2.

In order to follow the performance of MOGA while tracking the changes
in the set of values, and since the real Pareto front was known for used test
functions, two measures have been used. The first one was the measure of
covering the Pareto front [22], which has been calculated for the population
of each generation according to the expression:

$$cpf(PF_{true}, PF_{sol}) = \frac{card\{\mathbf{b} \in PF_{sol} | \exists \mathbf{a} \in PF_{true} : \mathbf{a}\rho\mathbf{b}\}}{card\{PF_{sol}\}} \tag{13}$$

where PF_{true} represents the true Pareto front, and PF_{sol} the set of non-
dominated solutions in the current population. The second measure used
was the hyper-volume formed by all the solutions from PF_{sol} in reference to
the minimal point of the set of values [6].

The measure of covering the Pareto front reflects how close are the
solutions of population to the points at the Pareto front, whereas the hyper-
volume reflects both vicinity to the Pareto front and the distribution of solutions along the Pareto front.

Figure 6 represents the set of values, and the development of population
each 50 generations for the set of values with linear borders, whereas the
figures 7 and 8 represent the measure of covering the Pareto front and the
hyper-volume through generations.

It can be noticed in Fig. 6 that the population very rapidly converges towards the set of Pareto optimal points, which reflects in the space of values
in grouping the points representing the values of solutions in the vicinity
of Pareto front. After that, the population very closely tracks the movement of Pareto front, and the population consists mosty of non-dominated
points. Additionally, the distribution of points along the Paret front is quit
satisfactory, which can be noticed also in Fig. 7, where the measure of
covering the Pareto front very rapidly decreases. At the same time, the
hyper-volume shown in Fig. 8 promptly reaches the values, which is kept
sucessfuly thorugh generations. The average value of the measure of covering
the Pareto front for all generations is 0.8232, and the average hyper-volume
is 16.301.

Similar results are collected for the set of values with non-linear borders.
So, the Figure 9 shows the set of alternatives, the set of values and the
development of population each 50 generations, whereas the figures 10 and 11
show the measure of covering the Pareto front and the hyper-voluem through
generations. The average value of the measure of convering the Pareto front
is now 0.9084, and the average hyper-volume is 75.0147.

It can be concluded similar as in the previous test case. The population
has again rapidly converged towards the set of Pareto optimal points, i.e. the
Pareto front, and it quite satisfactory tracked the changes in the set of values
all the time.

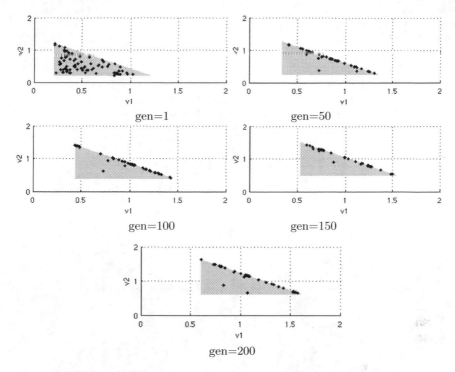

gen=1   gen=50

gen=100   gen=150

gen=200

Fig. 6 Development of solutions with change in criteria for the set of values with linear borders

Fig. 7 Measure of covering the Pareto front through generations for the set of values with linear borders

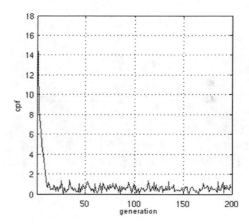

Non-homogenous non-stationary problem with continual change of set of values

In this test case, the set of values has moved with each generation of GA, each time for random value as in the previous case, but additionally the

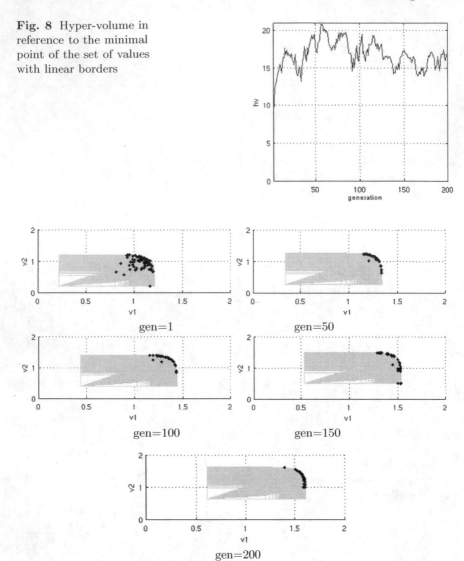

Fig. 8 Hyper-volume in reference to the minimal point of the set of values with linear borders

Fig. 9 Development of solutions with change in criteria for the set of values with non-linear borders

shape of the set of values was continually changing. The parameter values for test functions were again $a_i = 1$, $d_i = 0$, $b_i = b_i(k)$, where k represents the number of generation of GA, which represents the independent variable. Nevertheless, the function based on random value used also for the parameter bi was used to produce the value of the parameter c_i, so its value now was $c_i(k) = 5 \cdot b_i(k) - 1$. This produced change of shape of the set of values, from convex non-linear, through linear, to non-convex non- linear. In fact

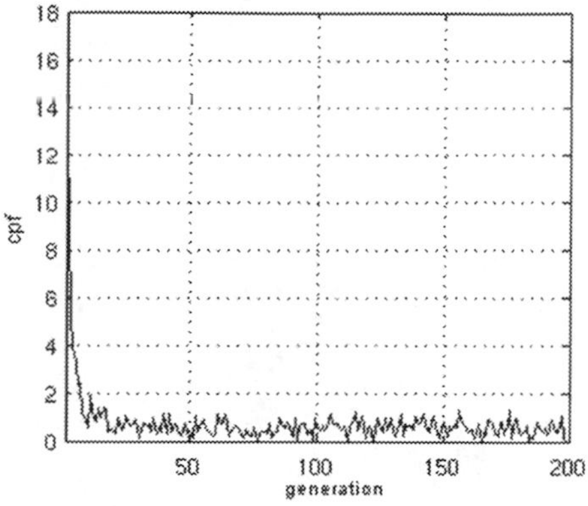

Fig. 10 Measure of covering the Pareto front through generations for the set of values with non-linear borders

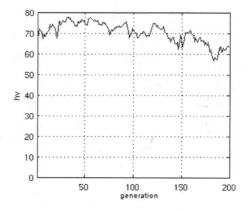

Fig. 11 Hyper-volume in reference to the minimal point of the set of values with non-linear borders

only Pareto front was changing in the mentioned way, whereas all the other borders were without change. For solving this problem, MOGA was used again, with identical coding scheme as in the previous cases, and with the parameter values given in the table 2.

The true Pareto front has been determined during the algorithm run, as well as the measures of covering the Pareto front and hyper-volume in reference to the minimal point of the set of values.

Figure 12 shows the set of alternatives, the set of values and the development of population each 50 generations, whereas the figures 13 and 14 show the measure of covering the Pareto front and the hyper-volume through generations.

The population still very quickly converges towards the set of Pareto optimal points, andtracks very good its changes. The distribution of points along the Pareto front is also quite good, what can be seen in the figure 12, where the measure of covering the Pareto front decreases very quickly. But in the figure 14 it can be noticed that the value of hyper-volume constantly decreases, although the population of points tracks very good the movement of Pareto front. This is the consequence of the fact that the hyper-volume of the

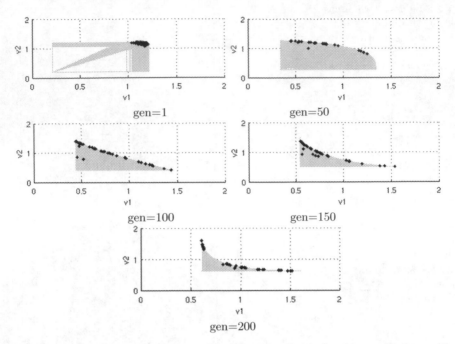

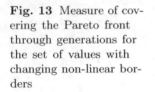

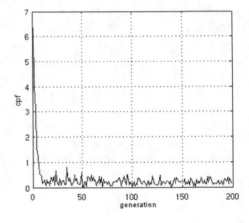

Fig. 12 Development of solutions with change in criteria for the set of values with changing non-linear borders

Fig. 13 Measure of covering the Pareto front through generations for the set of values with changing non-linear borders

whole set of values diminishes as the shape of the Pareto front changes from convex to non-convex. Therefore, it can be concluded that the hyper-volume measure defined in such way as used in this work can't be used as a qualitative performance measure for non-stationary problems with non-homogenous change of the set of values. Contrary, the measure of covering the Pareto front still provides a good insight into the quality of solutions, with very important notice that determining the true Pareto fron by complete search of the set

Fig. 14 Hyper-volume
in reference to the min-
imal point of the set
of values with changing
non-linear borders

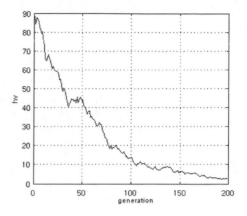

of values represents highly complicated and numerically complex problem,
what is not realizable for more complex problems.

The average performance measures for the whole run of MOGA are 0.3118
for average measure of covering the Pareto front, and 24.3927 for average
hyper-volume.

Non-homogenous non-stationary problem with changing number of criteria

The purpose of this test case is to demonstrate how the population adapts
when the number of criteria used for evaluating the value of individuals
changes. Firstly, the case with non-changing criteria, but with changing number of criteria is presented. The criteria with parameters $a_i = 1$, $b_i = 0$,
$c_i = 0.2$, $d_i = 0$ have been used. After that, the general case is presented, with
non- homogenously changing set of values including the Pareto front, what
was achieved using the parameter values $b_i = b_i(k)$ and $c_i(k) = 5 \cdot b_i(k) - 1$,
as in the previous case. In the presented test case, two criteria were used
from generation 1 to generation 40, from generation 81 to generation 120,
and from generation 161 to generation 200, whereas three criteria were used
from generation 41 to generation 80 and from generation 121 to generation
160. MOGA has been used again for solving the problem, with vector of three
real numbers (now the problem space represents the subset of E^3) coded into
chromosome, and with the parameter values of MOGA identical to the ones
given in the Table 2.

Figure 15 presents the Pareto front and the population for the first gener-
ation, one generation before (40th generation, 80th generation, 120th gener-
ation) and after (41st generation, 81st generation) the change in number of
criteria has happened. In this way, the effect can be noticed how the change in
number of criteria influences the qulaity of solutions, which up to the change
has hapened were in the set of non-dominated solution, in the vicinity of or
at the Pareto front. It can be noticed very clearly, that the solutions have
converged to the Pareto front (points of Pareto front are designated by "+").

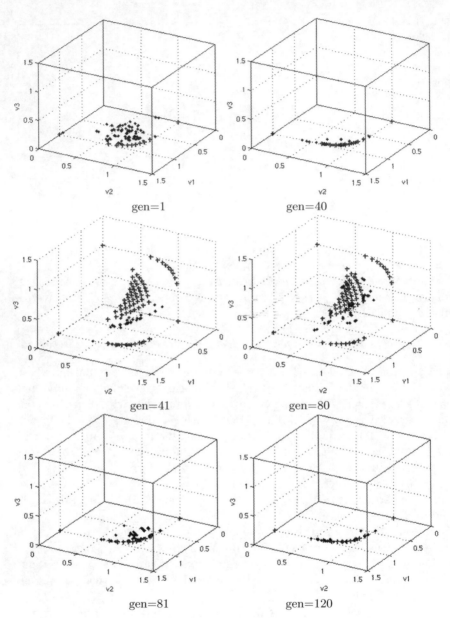

Fig. 15 Development of solutions with change in number of criteria

But after the increase in number of criteria to three (generation 41), the solutions cover just the part of the Pareto front, for which $v_3 = 0$.

The population gradually develops in concordance with the new model of the problem, which now has three criteria, and after certain number of generations (5-10 generations were sufficient in the test case) the solutions

Fig. 16 Measure of covering the Pareto front for the set of values with changing number of criteria

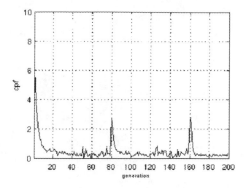

distribute along the Pareto front, which now represents a subset of the space E^3. By generation 80 the solutions are good distributed over the Pareto front. But the decrease in number of criteria causes now that quite a large nuber of solutions lies out of the Pareto front, which is now again a subset of the space E^2, and a certain number of generations is necessary for these solutions to converge again to the Pareto front. By the generation 120 all the solutions have converged very good to the Pareto front.

Figure 15 shows that the population adapts very good to the changes in the number of criteria. The same can be concluded by taking insight into the change in the measure of covering the Pareto front through generations (Figure 16).

It can be clearly noticed that the measure of covering Pareto front increases instantly for generations 81 and 161. In both of these cases, the number of criteria dropped from three to two. By this change, a part of the solutions, whish were non-dominated in the space E^3 became dominated, and 5-10 generations were necessary for the measure of covering the Pareto front to drop again to small value. Change in the number of criteria from two to three (generation 41 and generation 121) didn't cause substantial increase in the value of this measure. This is the consequence of the fact, that the solutions non-dominated in the space E^2 retain their non- dominance after the change has occured, with the notice that their distribution over the Pareto front was not satisfactory (Figure 15). Besides, it can be also noticed that the value of the measure of covering the Pareto front changes more intensively for the cases when three criteria were used. This was the consequence of the fact that the problem being solved becomes more complex with increase in the number of criteria, and it was more difficult for the algorithm to find the population, which was good distributed over the Pareto front in the space E^3.

Figure 17 shows the Pareto front and the population for the generation 1, one generation before (40th generation, 80th generation, 120th generation) and after (41st generation, 81st generation) the increase i.e. decrease in number of criteria, for the case when both the set of values and the Pareto front were changing non-homogenously. Despite the additional changes in the set of values and the Pareto front, the identical conclusions regarding the ability

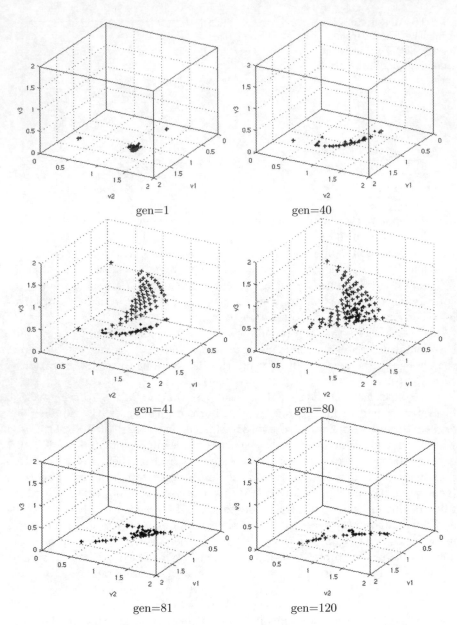

Fig. 17 Development of solutions with change in the number of criteria and non-homogenous change of the set of values

of population to find the solutions converging to the Pareto front after each change in the number of criteria as in the previous case still hold. Figure 18 shows the change in the measure of covering the Pareto front through generations. It can be noticed that the measure was very changable from generation

Fig. 18 Measure of covering the Pareto front for the set of values with changing number of criteria and non-homogenous change of the set of values and the Pareto front

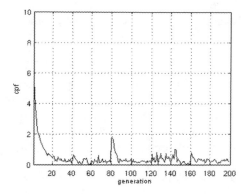

121 to generation 150, when three criteria were used, and when the shape of the Pareto front was unsuitalbe, so the algorithm needed more generations for finding the solutions, which adequately cover the Pareto front.

As a general conclusion, it can be said that the population of chromosomes of the multi-objective GA, since it has not converget to a unique point in the space of values but the solutions remain distributed over the Pareto front, retains sufficient diversity to efficiently adapt both to the changes of the set of values and to the changes in the number of criteria. Practically, contrary to the GA with single criterion, there was no need for introduction of additional mechanisms for retaining the population diversity. During the research of the effects which changing environment has on GA, it was shown that the single criterion GA without additional mechanisms for retaining the population diversity is not capable of tracking the changes in the environment [23].

4 Conclusion and Future Works

The basic thesis of this work consisted of an idea to model a problem faced during decision making in unknown or changing environment by a non-stationary dynamic multicriteral problem of search for solution.

A definition of general optimization problem has been introduced, which encompasses the non-stationary problems of search for solution This created a basis for modeling a problem of decision making in unknown or changing environment by a non-stationary dynamic multi-objective problem of search for solution.

By defining the previously mentioned model, it was necessary to choose an algorithm to solve it. It was implied that there exist two basic goals, which ought to be fulfilled by a set of non-dominated solutions, which are as good covering the Pareto front in the sense of vicinity as possible, and as good distribution over the Pareto front as possible.

Several examples of application of the model of non-stationary dynamic multi-objective problem of search for solution have been given. These examples included solving the problem of optimization of test functions in form of the problems discussed in this work. It was demonstrated that evolutionary algorithm successfully solved these problems.

The results of test conducted with MOGA have been presented in this chapter. Since there exist much more efficient variants of multi-objective genetic algorithm nowadays, they should also be tested in the described approach.

The authors of this chapter have already applied the presented approach to the problem of determining the trajectory of a mobile robot in unknown and changing environment [17], but it should be tested in some other fields of application as well.

References

1. Ziegler, J.G., Nichols, N.B.: Optimum Settings for Automatic Controllers. Journal of Dynamic Systems, Measurement, and Control 115(2B), 220–222 (1993)
2. Sastry, S., Bodson, M.: Adaptive Control: Stability, Convergence, and Robustness. Prentice-Hall, New York (1994)
3. Angelov, P.P.: Evolving Rule-Based Models, A Tool for Design of Flexible Adaptive Systems. Springer, New York (2002)
4. Cohon, J.L.: Multiobjective Programming and Planning. Dover Publications Inc., New York (1978/2003)
5. Branke, J.: Evolutionary Optimization in Dynamic Environments. Kluwer Academic Publishers, New York (2002)
6. Deb, K.: Multi-Objective Optimization using Evolutionary Algorithms. John Wiley & Sons, Ltd., West Sussex (2002)
7. Coello, C.A., Van Veldhuizen, D.A., Lamont, G.B.: Evolutionary Algorithms for Solving Multi-Objective Problems. Kluwer Academic Publishers, New York (2002)
8. Goldberg, D.E.: Genetic Algorithms in Search, Optimization, and Machine Learning. Addison-Wesley Longman Inc., New York (1989)
9. Michalewicz, Z.: Genetic Algorithms+Data Structures=Evolution Programs. Springer, New York (1996)
10. Bäck, T., Fogel, D.B., Michalewicz, Z.: Evolutionary Computation 1: Basic Algorithms and Operators. Institute of Physics Publishing, Bristol (2000)
11. Bäck, T., Fogel, D.B., Michalewicz, Z.: Evolutionary Computation 2: Advanced Algorithms and Operators. Institute of Physics Publishing, Bristol (2000)
12. Konjicija, S.: Improvement of Performance of Feed-forward Artificial Neural Network Using Genetic Algorithm with Adaptive Operators. MA Thesis, University of Sarajevo, Sarajevo (2003)
13. Coello, C.A.: An Updated Survey of GA-Based Multiobjective Optimization Techniques. ACM Computing Surveys 32(2) (2000)
14. Farina, M., Deb, K., Amato, P.: Dynamic Multiobjective Optimization Problems: Test Cases, Approximations, and Applications. IEEE Transactions on Evolutionary Computation 8(5) (October 2004)
15. Mehnen, J., Wagner, T., Rudolph, G.: Evolutionary Optimization of Dynamic Mulit-objective Test Functions. Technical Report, University of Dortmund, Dortmund (2006)

16. Deb, K., Thiele, L., Laumanns, M., Zitzler, E.: Scalable Test Problems for Evolutionary Multi-Objective Optimization. Technical Report 112, Computer Engineering and Networks Laboratory (TIK), Swiss Federal Institute of Technology (ETH), Zürich (2001)
17. Konjicija, S.: Evolutionary Approach to Finding a Solution of Dynamic Multi-criterial Optimization Problem in Processes of Inference. PhD thesis, University of Sarajevo, Sarajevo (2007)
18. Fonseca, C.M., Fleming, P.J.: Genetic Algorithms for Multiobjective Optimization: Formulation, Discussion and Generalization. In: Forest, S. (ed.) Genetic Algorithms: Proceedings of the Fifth International Conference. Morgan Kauffman, San Mateo (1993)
19. Syrmos, V., Lewis, F.: Optimal Control, 2nd edn. Willey Interscience, New York (2007)
20. Stengel, R.: Optimal Control and Estimation. Dover Publications Inc., New York (1994)
21. MacCrimmon, K.R.: An Overview of Multiple Objective Decision Making. In: Cochrane, J.L., Zeleny, M. (eds.) Multiople Criteria Decision Making, pp. 18–44. University of South Carolina Press (1973)
22. Zitzler, E.: Evolutionary Algorithms for Multiobjective Optimization: Methods and Applications. PhD thesis, ETH-Swiss Federal Institute of Technology, Zürich (1999)
23. Konjicija, S., Lacevic, B., Avdagic, Z.: Performance of Genetic Algorithm with Adaptive Mutation Probability Dependant on Fitness in Dynamic Environments. In: Trappl, R. (ed.) Proceedings of the 18th European Meeting on Cybernetics and Systems Research, Vienna (2006)

Turbulent Particle Swarm Optimization Using Fuzzy Parameter Tuning

Ajith Abraham and Hongbo Liu

Abstract. Particle Swarm Optimization (PSO) algorithm is a stochastic search technique, which has exhibited good performance across a wide range of applications. However, very often for multi-modal problems involving high dimensions the algorithm tends to suffer from premature convergence. Premature convergence could make the PSO algorithm very difficult to arrive at the global optimum or even a local optimum. Analysis of the behavior of the particle swarm model reveals that such premature convergence is mainly due to the decrease of velocity of particles in the search space that leads to a total implosion and ultimately fitness stagnation of the swarm. This paper introduces Turbulence in the Particle Swarm Optimization (TPSO) algorithm to overcome the problem of stagnation. The algorithm uses a minimum velocity threshold to control the velocity of particles. TPSO mechanism is similar to a turbulence pump, which supplies some power to the swarm system to explore new neighborhoods for better solutions. The algorithm also avoids clustering of particles and at the same time attempts to maintain diversity of population. We attempt to theoretically analyze that the algorithm converges with a probability of 1 towards the global optimal. The parameter, the minimum velocity threshold of the particles is tuned adaptively by a fuzzy logic controller embedded in the TPSO algorithm, which is further called as Fuzzy Adaptive

Ajith Abraham
Centre for Quantifiable Quality of Service in Communication Systems,
Norwegian University of Science and Technology, NO-7491 Trondheim, Norway
e-mail: ajith.abraham@ieee.org
http://www.softcomputing.net

Ajith Abraham and Hongbo Liu
School of Computer Science and Engineering, Dalian Maritime University,
116026 Dalian, China

Hongbo Liu
Department of Computer, Dalian University of Technology, 116023 Dalian, China
e-mail: lhb@dlut.edu.cn

A. Abraham et al. (Eds.): Foundations of Comput. Intel. Vol. 3, SCI 203, pp. 291–312.
springerlink.com   © Springer-Verlag Berlin Heidelberg 2009

TPSO (FATPSO). We evaluated the performance of FATPSO and compared it with the Standard PSO (SPSO), Genetic Algorithm (GA) and Simulated Annealing (SA). The comparison was performed on a suite of 20 widely used benchmark problems. Empirical results illustrate that the FATPSO could prevent premature convergence very effectively. It clearly outperforms the considered methods, especially for high dimension multi-modal optimization problems.

1 Introduction

Particle Swarm Optimization (PSO) algorithm is mainly inspired by social behaviour patterns of organisms that live and interact within large groups. In particular, PSO incorporates swarming behaviours observed in flocks of birds, schools of fish, or swarms of bees, and even human social behavior, from which the idea of swarm intelligence is emerged ([14]). It could be applied to solve various function optimization problems, or the problems that can be transformed to function optimization problems. PSO has exhibited good performance across a wide range of applications ([19, 15, 25, 26, 1, 21, 5, 22]). However, its performance deteriorates as the dimensionality of the search space increases, especially for multi-modal optimization problems ([13, 20]). PSO algorithm often demonstrates faster convergence speed in the first phase of the search, and then slows down or even stops as the number of generations is increased. Once the algorithm slows down, it is difficult to achieve better fitness values. This state is called as stagnation or premature convergence. The trajectory of particles was given a lot of importance rather than their velocities. In this paper, we attempt to discuss the relation between the algorithm convergence and the velocities of the particles. It is found that the stagnation state is mainly due to a decrease of velocity of particles in the search space which leads to a total implosion and ultimately fitness stagnation of the swarm. We introduce Turbulent Particle Swarm Optimization (TPSO) algorithm to improve the optimization performance and overcome the premature convergence problem. The basic idea is to drive those lazy particles and get them to explore new search spaces. TPSO uses a minimum velocity threshold to control the velocity of particles and also avoids clustering of particles and maintains diversity of population in the search space. The minimum velocity threshold of the particles is tuned adaptively by using a fuzzy logic controller in the algorithm, which is further called as Fuzzy Adaptive TPSO (FATPSO).

The Chapter is organized as follows. Particle swarm optimization is reviewed briefly and the effects on the change of the velocities of particles are analyzed in Section 2. In Section 3, we describe the TPSO model and the fuzzy adaptive processing method. Experiment settings, results and discussions are given in Section 4 followed by some conclusions in the last Section.

2   Particle Swarm Optimization

Particle swarm optimization refers to a relatively new family of algorithms
that may be used to find optimal (or near optimal) solutions to numerical
and qualitative problems. Some researchers have done much work on its study
and development during the recent years ([29, 20, 16, 12]). We review briefly
the standard particle swarm model, and then analyze the various effects in
the change in the velocities of particles.

2.1   Standard Particle Swarm Model

The particle swarm model consists of a swarm of particles, which are initialized with a population of random candidate solutions. They move iteratively
through the d-dimension problem space to search the new solutions, where
the fitness f can be calculated as the certain qualities measure. Each particle has a position represented by a position-vector $\mathbf{p}_i$ (i is the index of the
particle), and a velocity represented by a velocity-vector $\mathbf{v}_i$. Each particle re-
members its own best position so far in a vector $\mathbf{p}_i^{\#}$, and its j-th dimensional
value is $p_{ij}^{\#}$. The best position-vector among the swarm so far is then stored
in a vector $\mathbf{p}^*$, and its j-th dimensional value is p_j^*. During the iteration time
t, the update of the velocity from the previous velocity to the new velocity
is determined by Eq.(1). The new position is then determined by the sum of
the previous position and the new velocity by Eq.(2).

$$
\begin{aligned}
v_{ij}(t) = & w v_{ij}(t-1) + c_1 r_1 (p_{ij}^{\#}(t-1) - p_{ij}(t-1)) \\
& + c_2 r_2 (p_j^*(t-1) - p_{ij}(t-1))
\end{aligned}
\tag{1}
$$

$$
p_{ij}(t) = p_{ij}(t-1) + v_{ij}(t)
\tag{2}
$$

where r_1 and r_2 are the random numbers, uniformly distributed within the
interval [0,1] for the j-th dimension of i-th particle. c_1 is a positive constant,
called as coefficient of the self-recognition component, c_2 is a positive constant, called as coefficient of the social component. The variable w is called
as the inertia factor, which value is typically setup to vary linearly from 1 to
near 0 during the iterated processing. From Eq.(1), a particle decides where
to move next, considering its own experience, which is the memory of its best
past position, and the experience of its most successful particle in the swarm.

In the particle swarm model, the particle searches the solutions in the
problem space within a range $[-s, s]$ (If the range is not symmetrical, it can
be translated to the corresponding symmetrical range). In order to guide the
particles effectively in the search space, the maximum moving distance during
one iteration is clamped in between the maximum velocity $[-v_{max}, v_{max}]$
given in Eq.(3), and similarly for its moving range given in Eq.(4):

$$v_{i,j} = sign(v_{i,j})min(|v_{i,j}|, v_{max}) \tag{3}$$

$$p_{i,j} = sign(p_{i,j})min(|p_{i,j}|, p_{max}) \tag{4}$$

The value of v_{max} is $\alpha \times s$, with $0.1 \leq \alpha \leq 1.0$ and is usually chosen to be s, i.e. $\alpha = 1$.

2.2 Velocities Analysis in Particle Swarm

Some previous studies have discussed the trajectory of particles and the convergence of the algorithm ([3, 29, 27]). It has been shown that the trajectories of the particles oscillate as different sinusoidal waves and converge quickly, sometimes prematurely. We analyze the effects of the change in the velocities of particles.

The gradual change of the particle's velocity can be explained geometrically. During each iteration, the particle is attracted towards the location of the best fitness achieved so far by the particle itself and by the location of the best fitness achieved so far across the whole swarm. From Eq.(1), $v_{i,j}$ can attain a smaller value, but if the second term and the third term in RHS of Eq.(1) are both small, it cannot resume a larger value and could eventually loose the exploration capabilities in the future iterations. Such situations could occur even in the early stages of the search. When the second term and the third term in RHS of Eq.(1) are zero, $v_{i,j}$ will be damped quickly with the ratio of w. In other words, if a particle's current position coincides with the global best position/particle, the particle will only move away from this point if its previous velocity and w are non-zero. If their previous velocities are very close to zero, then all the particles will stop moving once they catch up with the global best particle, which many lead to premature convergence. In fact, this does not even guarantee that the algorithm has converged to a local minimum and it merely means that all the particles have converged to the best position discovered so far by the swarm. This state owes to the second term and the third term in the RHS of Eq.(1), the cognitive components of the PSO. But if the cognitive components of the PSO algorithm are invalidated, all particles always search the solutions using the initial velocities. Then the algorithm is merely a degenerative stochastic search without the characteristics of PSO.

3 Turbulent Swarm Optimization

We introduce a new velocity update approach for the particles in PSO, and analyze its effect on the particle's behavior. We also illustrate a Fuzzy Logic Controller (FLC) scheme to adaptively control the parameters ([11, 30, 17]).

3.1 Velocity Update of the Particles

As discussed in the previous Section, one of the main reason for premature convergence of PSO is due to the stagnation of the particles exploration of a new search space. We introduce a strategy to drive those lazy particles and let them explore better solutions. If a particle's velocity decreases to a threshold v_c, a new velocity is assigned using Eq.(6). Thus, we present the turbulent particle swarm optimization using new velocity update equations:

$$
\begin{aligned}
v_{ij}(t) = w\hat{v} &+ c_1 r_1(x_{ij}^{\#}(t-1) - x_{ij}(t-1)) \\
&+ c_2 r_2(x_j^*(t-1) - x_{ij}(t-1))
\end{aligned}
\tag{5}
$$

$$
\hat{v} =
\begin{cases}
v_{ij} & \text{if } |v_{ij}| \geq v_c \\
u(-1,1)v_{max}/\rho & \text{if } |v_{ij}| < v_c
\end{cases}
\tag{6}
$$

where $u(-1,1)$ is the random number, uniformly distributed with the interval [-1,1], and ρ is the scaling factor to control the domain of the particle's oscillation according to v_{max}. v_c is the minimum velocity threshold, a tunable threshold parameter to limit the minimum of the particles' velocity. Fig. 1 illustrates the trajectory of a single particle in standard particle swarm optimization (SPSO) and turbulent particle swarm optimization (TPSO) respectively.

The change of the particle's situation is directly correlated to two parameter values, v_c and ρ. A large v_c shortens the oscillation period, and it provides a great probability for the particles to leap over local minima using the same number of iterations. But a large v_c compels particles in the quick "flying" state, which leads them not to search the solution and forcing them not to refine the search. In other words, a large v_c facilitates a global search while a

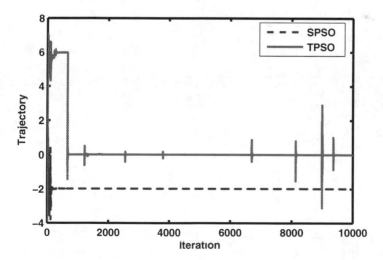

Fig. 1 Trajectory of a single particle

smaller value facilitates a local search. By changing it dynamically, the search ability is dynamically adjusted. The value of ρ changes directly the particle oscillation domain. It is possible for particles not to jump over the local minima if there would be a large local minimum available in the objective search space. But the particle trajectory would more prone to oscillate because of a smaller value of ρ. For the desired exploration-exploitation trade-off, we divide the particle search into three stages. In the first stage the values for v_c and ρ are set at large and small values respectively. In the second stage, v_c and ρ are set at medium values and in the last stage, v_c is set at a small value and ρ is set at a large value. This enable the particles to take very large steps to explore solutions in the early stages, by scanning the whole solution space for good local minima and then in the final stages particles perform a fine grain search. The use of fuzzy logic would be suitable for dynamically tuning the velocity threshold, since it starts a run with an initial value which is changed during the run. By using the fuzzy control approach, the parameters can be adaptively regulated according to the problem environment.

3.2   Fuzzy Parameter Control

A Fuzzy Logic Controller (FLC) is composed of a knowledge base, that includes the information given by the expert in the form of linguistic control rules, a fuzzification interface, which has the effect of transforming crisp data into fuzzy sets, an inference system, that uses them together with the knowledge base to make inference by means of a reasoning method, and a defuzzification interface, that translates the fuzzy control action thus obtained to a real control action using a defuzzification method [4]. The generic structure of an FLC is shown in Figure 2.

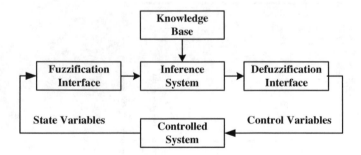

Fig. 2 Generic structure of an FLC

In the proposed algorithm, two variables are selected as inputs to the fuzzy system: the Current Best Performance Evaluation ($CBPE$) ([24]) and the Current Velocity (CV) of the particle. For adapting to a wide range of optimization problems, $CBPE$ is normalized as Eq.(7):

$$NCBPE = \frac{CBPE - CBPE_{min}}{CBPE_{max} - CBPE_{min}} \qquad (7)$$

where $CBPE_{min}$ is the estimated (or real) minimum, $CBPE_{max}$ is the worst solution to the minimization problem, which usually is the $CBPE$ at half the number of iterations. If we do not have any prior information about the objective function and if it is difficult to estimate $CBPE_{min}$ and $CBPE_{max}$, we can do some preliminary experiments by decreasing linearly from 1 to 0 during the run. One of the output variables is ρ, the scaling factor to control the domain of the particle's oscillation. Another is Vck, which controls the change of the velocity threshold according to Eq.(8):

$$v_c = e - [10(1 + Vck)] \qquad (8)$$

The fuzzy inference system is listed briefly as follows:
[System]
Name=$'FATPSO'$

[Input1] Name=$'NCBPE'$
Range=[0 1]
NumMFs=3
MF1='Low':'gaussmf', [0.005 0]
MF2='Medium':'gaussmf', [0.03 0.1]
MF3='High':'gaussmf', [0.25 1]

[Input2]
Name=$'CV'$
Range=[0 1e-006]
NumMFs=2
MF1='Low':'trapmf', [0 0 1e-030 1e-020]
MF2='High':'trapmf', [1e-010 1e-008 1e-006 1e-006]

[Output1]
Name=$'Vck'$
Range=[-1 2.2]
NumMFs=3
MF1='Low':'trimf', [-1 -0.8 -0.5]
MF2='Medium':'trimf', [-0.6 0 0.2]
MF3='High':'trimf', [0.1 1.1 2.2]

[Output2]
Name=$'\rho'$
Range=[1 120]
NumMFs=3
MF1='Small':'trimf', [1 1 4]
MF2='Medium':'trimf', [2.214 10.71 59.29]
MF3='Large':'trimf', [47.15 120 120]

[Rules]
1 1, 3 0 (1) : 1
2 0, 2 0 (1) : 1
3 2, 1 0 (1) : 1
1 1, 0 3 (1) : 2
2 0, 0 2 (1) : 2
3 2, 0 1 (1) : 2

In the above mentioned list, there are three parts: the first part is the configuration of the fuzzy system, the second one is the definition of the membership functions, and the third one is the rule base. There are two inputs and two outputs based on six rules. In the rule base, the first two columns correspond to the input variables, the second two columns correspond to the output variables, the fifth column displays the weight applied to each rule, and the sixth column is short form that indicates whether this is an AND (1) rule or an OR (2) rule. The numbers in the first four columns refer to the index number of the membership function, in which the number 1 encodes fuzzy set 'Low', 2 encodes 'Medium', and 3 encodes 'High'. For example, the first rule is "If ($NCBPE$ is Low) and (CV is Low) then (Vck is High) with the weight 1". The general structure of the FATPSO is illustrated in Algorithm 1.

Algorithm 1. FATPSO

---
01. Initialize parameters and the particles
02. While (the end criterion is not met) do
03. $t = t + 1$
04.   Calculate the fitness value of each particle
05. $x^* = argmin_{i=1}^{n}(f(x^*(t-1)), f(x_1(t)),$
06. $f(x_2(t)), \cdots, f(x_i(t)), \cdots, f(x_n(t)))$
07.   For $i = 1$ to n
08. $x_i^{\#}(t) = argmin_{i=1}^{n}(f(x_i^{\#}(t-1)), f(x_i(t))$
09.   For $j = 1$ to d
10.   If $abs(v_{ij}) < 1e - 6$
11.   Obtain the velocity threshold
12.   {
13.   fismat = readfis('$FATPSO.fis$')
14. FO = evalfis([$NCBPE$ CV], fismat)
15.   }
16.   Endif
17.   Update the j-th dimension value of $\mathbf{x}_i$
18.   and $\mathbf{v}_i$ according to Eqs. (1), (2) and (3)
19.   Next j
20.   Next i
21. End While

---

4   Convergence Analysis of TPSO

For analyzing the convergence of the proposed algorithm, we first introduce the definitions and lemmas [8, 9, 10], and then theoretically prove that the proposed variable neighborhood particle swarm algorithm converges with a probability 1 or strongly towards the global optimal.

Consider the problem (P) as

$$(P) = \begin{cases} min f(\mathbf{x}) \\ \mathbf{x} \in \Omega = [-s, s]^n \end{cases} \tag{9}$$

where $\mathbf{x} = (x_1, x_2, \cdots, x_n)^T$. $\mathbf{x}^*$ is the global optimal solution to the problem (P), let $f^* = f(\mathbf{x}^*)$. Let

$$D_0 = \{\mathbf{x} \in \Omega | f(\mathbf{x}) - f^* < \varepsilon\} \tag{10}$$
$$D_1 = \Omega \setminus D_0$$

for every $\varepsilon > 0$.

Assume that the i-th dimensional value of the particle's velocity decreases to a threshold v_c, then the shaking strategy is activated, and a turbulent velocity is generated by Eq.(6). In $u(-1, 1)v_{max}/\rho$, $u(-1, 1)$ is a normal distributed random number within the interval [-1,1], and the scaling factor ρ is a positive constant to control the domain of the particle's oscillation according to v_{max}. Therefore the turbulent velocity $\hat{v}$ belongs to the normal distribution. If $v_{max} = s$, then $\hat{v} \sim [-\frac{s}{\rho}, \frac{s}{\rho}]$. During the iterated procedure from the time t to $t+1$, let q_{ij} denote that $\mathbf{x}(t) \in D_i$ and $\mathbf{x}(t+1) \in D_j$. Accordingly the particles' positions in the swarm could be classified into four states: q_{00}, q_{01}, q_{10} and q_{01}. Obviously $q_{00} + q_{01} = 1$, $q_{10} + q_{11} = 1$.

Definition 1 (Convergence in terms of probability). *Let ξ_n a sequence of random variables, and ξ a random variable, and all of them are defined on the same probability space. The sequence ξ_n converges with a probability of ξ if*

$$\lim_{n \to \infty} P(|\xi_n - \xi| < \varepsilon) = 1 \tag{11}$$

for every $\varepsilon > 0$.

Definition 2 (Convergence with a probability of 1). *Let ξ_n a sequence of random variables, and ξ a random variable, and all of them are defined on the same probability space. The sequence ξ_n converges almost surely or almost everywhere or with probability of 1 or strongly towards ξ if*

$$P\left(\lim_{n \to \infty} \xi_n = \xi\right) = 1; \tag{12}$$

or

$$P\left(\bigcap_{n=1}^{\infty} \bigcup_{k \geq n} [|\xi_n - \xi| \geq \varepsilon]\right) = 0 \tag{13}$$

for every $\varepsilon > 0$.

Lemma 1 (Borel-Cantelli Lemma). *Let $\{A_k\}_{k=1}^{\infty}$ be a sequence of events occurring with a certain probability distribution, and let A be the event consisting of the occurrences of a finite number of events A_k for $k = 1, 2, \cdots$. Then*

$$P\left(\bigcap_{n=1}^{\infty} \bigcup_{k \geq n} A_k\right) = 0 \tag{14}$$

if

$$\sum_{n=1}^{\infty} P(A_n) < \infty; \tag{15}$$

$$P\left(\bigcap_{n=1}^{\infty} \bigcup_{k \geq n} A_k\right) = 1 \tag{16}$$

if the events are totally independent and

$$\sum_{n=1}^{\infty} P(A_n) = \infty. \tag{17}$$

Lemma 2 (Particle State Transference). $q_{01} = 0$; $q_{00} = 1$; $q_{11} \leq c \in (0,1)$ *and* $q_{10} \geq 1 - c \in (0,1)$.

Proof. In the proposed algorithm, the best solution is updated and saved during the whole iterated procedure. So $q_{01} = 0$ and $q_{00} = 1$.

Let $\hat{\mathbf{x}}$ is the position with the best fitness among the swarm so far as the time t, i.e. $\hat{\mathbf{x}} = \mathbf{p}^*$. As the definition in Eq. (10), $\exists r > 0$, when $\|\mathbf{x} - \hat{\mathbf{x}}\|_{\infty} \leq r$, we have $|f(\mathbf{x}) - f^*| < \varepsilon$. Denote $Q_{\hat{\mathbf{x}},r} = \{x \in \Omega | \|\mathbf{x} - \hat{\mathbf{x}}\|_{\infty} \leq r\}$. Accordingly

$$Q_{\hat{\mathbf{x}},r} \subset D_0 \tag{18}$$

Then,

$$P\{(\mathbf{x} + \Delta\mathbf{x}) \in Q_{\hat{\mathbf{x}},r}\} = \prod_{i=1}^{n} P\{|x_i + \Delta x_i - \hat{x}_i| \leq r\} \tag{19}$$

$$= \prod_{i=1}^{n} P\{\hat{x}_i - x_i - r \leq \Delta x_i \leq \hat{x}_i - x_i + r\}$$

where x_i, Δx_i and $\hat{x}_i$ are the i-th dimensional values of $\mathbf{x}$, $\Delta\mathbf{x}$ and $\hat{\mathbf{x}}$, respectively. Moreover, $\hat{v} \sim [-\frac{s}{\rho}, \frac{s}{\rho}]$, so that

$$P((\mathbf{x} + \Delta\mathbf{x}) \in Q_{\hat{\mathbf{x}},r}) = \prod_{i=1}^{n} \int_{\hat{x}_i - x_i - r}^{\hat{x}_i - x_i + r} \frac{\rho}{2\sqrt{2\pi}s} e^{-\frac{\rho^2 y^2}{2s^2}} dy \qquad (20)$$

Denote $P_1(\mathbf{x}) = P\{(\mathbf{x} + \Delta\mathbf{x}) \in Q_{\hat{\mathbf{x}},r}\}$ and $\mathbb{C}$ is the convex closure of level set for the initial particle swarm. According to Eq. (20), $0 < P_1(\mathbf{x}) < 1$ ($\mathbf{x} \in \mathbb{C}$). Again, since $\mathbb{C}$ is a bounded closed set, so $\exists \hat{\mathbf{y}} \in \mathbb{C}$,

$$P_1(\hat{\mathbf{y}}) = \min_{\mathbf{x} \in \mathbb{C}} P_1(\mathbf{x}), \quad 0 < P_1(\hat{\mathbf{y}}) < 1. \qquad (21)$$

Considering synthetically Eqs. (18) and (21), so that

$$q_{10} \geq P_1(\mathbf{x}) \geq P_1(\hat{\mathbf{y}}) \qquad (22)$$

Let $c = 1 - P_1(\hat{\mathbf{y}})$, thus,

$$q_{11} = 1 - q_{10} \leq 1 - P_1(\hat{\mathbf{y}}) = c \quad (0 < c < 1) \qquad (23)$$

and

$$q_{10} \geq 1 - c \in (0, 1) \qquad (24)$$

$\square$

Theorem 1. *Assume that the TPSO algorithm provides position series* $\mathbf{p}_i(t)(i = 1, 2, \cdots, n)$ *at time t by the iterated procedure.* $\mathbf{p}^*$ *is the best position among the swarm explored so far, i.e.*

$$\mathbf{p}^*(t) = \arg \min_{1 \leq i \leq n} (f(\mathbf{p}^*(t-1)), f(\mathbf{p}_i(t))) \qquad (25)$$

Then,

$$P\left(\lim_{t \to \infty} f(\mathbf{p}^*(t)) = f^*\right) = 1 \qquad (26)$$

Proof. For $\forall \varepsilon > 0$, let $p_k = P\{|f(\mathbf{p}^*(k)) - f^*| \geq \varepsilon\}$, then

$$p_k = \begin{cases} 0 & \text{if } \exists T \in \{1, 2, \cdots, k\}, \mathbf{p}^*(T) \in D_0 \\ \bar{p}_k & \text{if } \mathbf{p}^*(t) \notin D_0, t = 1, 2, \cdots, k \end{cases} \qquad (27)$$

According to Lemma 2,

$$\bar{p}_k = P\{\mathbf{p}^*(t) \notin D_0, t = 1, 2, \cdots, k\} = q_{11}^k \leq c^k. \qquad (28)$$

Hence,

$$\sum_{k=1}^{\infty} p_k \leq \sum_{k=1}^{\infty} c^k = \frac{c}{1-c} < \infty. \qquad (29)$$

According to Lemma 1,

$$P\left(\bigcap_{t=1}^{\infty} \bigcup_{k \geq t} |f(\mathbf{p}^*(k)) - f^*| \geq \varepsilon\right) = 0 \qquad (30)$$

As defined in Definition 2, the sequence $f(\mathbf{p}^*(t))$ converges almost surely or almost everywhere or with probability 1 or strongly towards f^*. The theorem is proven.   □

5   Experiments and Discussions

In our experiments the algorithms used for comparison were mainly SPSO (standard PSO) ([6]), FATPSO (fuzzy adaptive turbulent PSO), Genetic Algorithm(GA) ([2]) and Simulated Annealing (SA) ([18, 28]). The four algorithms share many similarities. GA and SA are powerful stochastic global search and optimization methods, which are also inspired from the nature like the PSO.

Genetic algorithms mimic an evolutionary natural selection process. Generations of solutions are evaluated according to a fitness value and only those candidates with high fitness values are used to create further solutions via crossover and mutation procedures.

Simulated annealing is based on the manner in which liquids freeze or metals re-crystalize in the process of annealing. In an annealing process, a

Table 1 Parameter settings for the algorithms

SPSO		
	Swarm size	20
	Self-recognition coefficient c_1	1.49
	Social coefficient c_2	1.49
	Inertia weight w	$0.9 \rightarrow 0.1$
FATPSO		
	Swarm size	20
	Self-recognition coefficient c_1	1.49
	Social coefficient c_2	1.49
	Inertia weight w	0.7
GA		
	Size of the population	20
	Probability of crossover	0.8
	Probability of mutation	0.02
SA		
	Number operations before temperature adjustment	20
	Number of cycles	10
	Temperature reduction factor	0.85
	Vector for control step of length adjustment	2

Table 2 Numerical benchmark functions

Rosenbrock (f_1):

$$f_1 = \sum_{i=1}^{n}(100(x_{i+1} - x_i^2)^2 + (x_i - 1)^2);$$
$$\mathbf{x} \in [-2.048, 2.048]^n,$$
$$min(f_1(\mathbf{x}^*)) = f_1(\mathbf{1}) = 0.$$

Quadric (f_2):

$$f_2 = \sum_{i=1}^{n}(\sum_{j=1}^{i} x_j)^2;$$
$$\mathbf{x} \in [-100, 100]^n,$$
$$min(f_2(\mathbf{x}^*)) = f_2(\mathbf{0}) = 0.$$

Schwefel 2.22 (f_3):

$$f_3 = \sum_{i=1}^{n}|x_i| + \prod_{i=1}^{n}|x_i|;$$
$$\mathbf{x} \in [-10, 10]^n,$$
$$min(f_3(\mathbf{x}^*)) = f_3(\mathbf{0}) = 0.$$

Schwefel 2.26 (f_4):

$$f_4 = 418.9829n - \sum_{i=1}^{n}(x_i sin(\sqrt{|x_i|}));$$
$$\mathbf{x} \in [-500, 500]^n,$$
$$min(f_4(\mathbf{x}^*)) = f_4(\mathbf{420.9687}) \approx 0.$$

Levy (f_5):

$$f_5(\mathbf{x}) = \frac{\pi}{n}\Big(ksin^2(\pi y_1) + \sum_{i=1}^{n-1}((y_i - a)^2$$
$$(1 + ksin^2(\pi y_{i+1}))) + (y_n - a)^2\Big),$$
$$y_i = 1 + \frac{1}{4}(x_i - 1), \quad k = 10, \quad a = 1;$$
$$\mathbf{x} \in [-10, 10]^n,$$
$$min(f_5(\mathbf{x}^*)) = f_5(\mathbf{1}) = 0.$$

Generalized Shubert (f_6):

$$f_6 = \prod_{i=1}^{n}\sum_{j=1}^{5}(jcos((j + 1)x_i + j));$$
$$\mathbf{x} \in [-10, 10]^n,$$
$$min(f_6(\mathbf{x}^*)) \text{ is unknown.}$$

Rastrigin (f_7):

$$f_7 = \sum_{i=1}^{n}(x_i^2 - 10cos(2\pi x_i) + 10)$$
$$\mathbf{x} \in [-5.12, 5.12]^n,$$
$$min(f_7(\mathbf{x}^*)) = f_7(\mathbf{0}) = 0.$$

Griewank (f_8):

$$f_8 = \frac{1}{4000}\sum_{i=1}^{n}x_i^2 - \prod_{i=1}^{n}cos(\frac{x_i}{\sqrt{i}}) + 1;$$
$$\mathbf{x} \in [-300, 300]^n,$$
$$min(f_8(\mathbf{x}^*)) = f_8(\mathbf{0}) = 0.$$

Ackley (f_9):

$$f_9 = -20exp(-0.2\sqrt{\frac{1}{n}\sum_{i}^{n}x_i^2})$$
$$-exp(\frac{1}{n}\sum_{i=1}^{n}cos(2\pi x_i)) + 20 + e;$$
$$\mathbf{x} \in [-32, 32]^n,$$
$$min(f_9(\mathbf{x}^*)) = f_9(\mathbf{0}) = 0.$$

Zakharov (f_{10}):

$$f_{10} = \sum_{i}^{n}x_i^2 + (\sum_{i}^{n}\frac{1}{2}ix_i)^2 + (\sum_{i}^{n}\frac{1}{2}ix_i)^4;$$
$$\mathbf{x} \in [-10, 10]^n,$$
$$min(f_{10}(\mathbf{x}^*)) = f_{10}(\mathbf{0}) = 0.$$

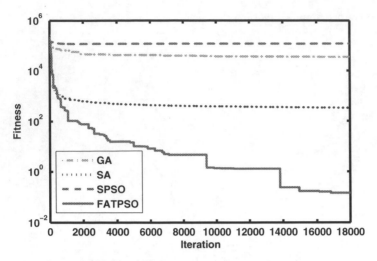

Fig. 3 30-D Quadric (f_2) function performance

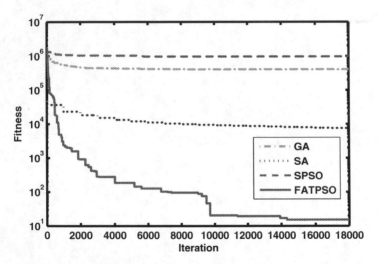

Fig. 4 100-D Quadric (f_2) function performance

melt, initially at high temperature and disordered, is slowly cooled so that
the system at any time is approximately in thermodynamic equilibrium. In
terms of computational simulation, a global minimum would correspond to
such a "frozen"(steady) ground state at the temperature $T = 0$.

Both methods are valid and efficient methods in numeric programming
and have been employed in various fields due to their strong convergence
properties. In the experiments, the specific parameter settings for each of the
considered algorithms are described in Table 1. Each algorithm was tested
with all the numerical functions shown in Table 2. The first two functions,

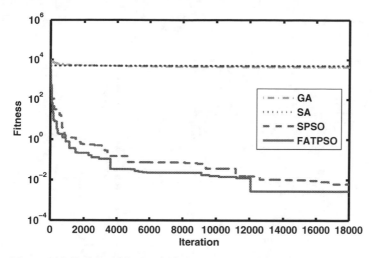

Fig. 5 30-D Schwefel 2.26 (f_4) function performance

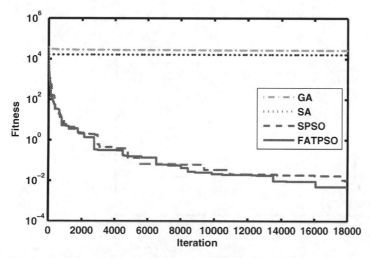

Fig. 6 100-D Schwefel 2.26 (f_4) function performance

namely Rosenbrock's and Quadric function, have a single minimum, while the other functions are highly multimodal with multiple local minima. A new function, Generalized Shubert was constructed temporarily for which global minimum function is unknown for us. It is also useful for us to validate the algorithms without knowing the optimal value. Some of the functions have the sum of their variables, some of them have the product (multiplying), some of them have dimensional effect (ix_i). We tested the algorithms on the different functions in 30 and 100 dimensions, yielding a total of 20 numerical benchmarks. For each of these functions, the goal was to find the global

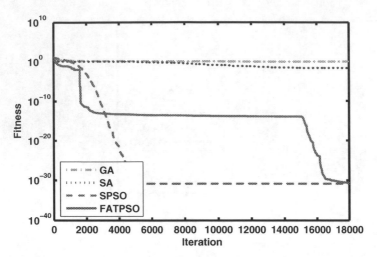

Fig. 7 30-D Levy (f_5) function performance

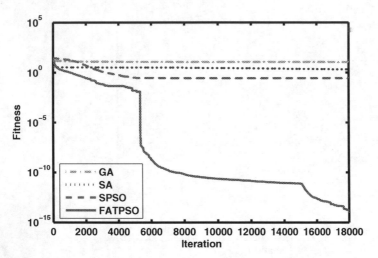

Fig. 8 100-D Levy (f_5) function performance

minima. Each algorithm (for each benchmark) was repeated 10 times with different random seeds. Each trial used a fixed number of 18,000 iterations. The objective functions were evaluated 360,000 times in each trial. Since the swarm size in all PSOs was 20, the size of the population in GA was 20 and the number operations before temperature adjustment (SA) were 20. The average fitness values of the best solutions throughout the optimization run were recorded and the averages and the standard deviations were calculated from the 10 different trials. The standard deviation indicates the differences in the results during the 10 different trials.

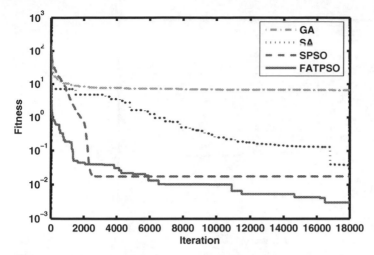

Fig. 9 30-D Griewank (f_8) function performance

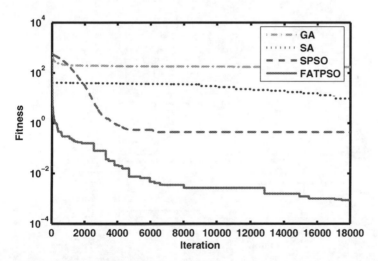

Fig. 10 100-D Griewank (f_8) function performance

Figures 3 to 12 illustrate the mean best function values for the ten functions with two different dimensions (i.e. 30-D and 100-D) using the four algorithms. Each algorithm for different dimensions of the same objective function has similar performance. But in general, the higher the dimension is, the higher the fitness values are. It is observed that for almost all algorithms, the solutions get trapped in a local minimum within the first 2000 iterations except for FATPSO. For the low dimensional problems, SA is usually a cost-efficient choice. For example, SA for 30-D f_8 has a good performance than that in other situations. It is interesting that even if other algorithms are very close

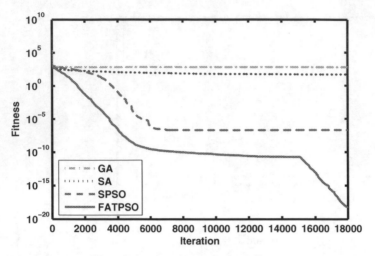

Fig. 11 30-D Zakharov (f_{10}) function performance

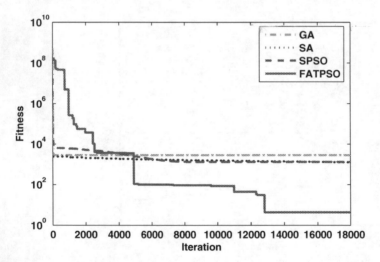

Fig. 12 100-D Zakharov (f_{10}) function performance

to or better than FATPSO in 30-D benchmarks, but a very large difference emerges in the case of 100-D benchmark problems. FATPSO becomes much better than other algorithms in general besides for f_4. The averages and the standard deviations for 10 trials are showed in Table 3. The larger the averages are, wider the standard deviations are usually. There is not too large difference of the standard deviations between the different algorithms for the same benchmark functions. Referring to the empirical results depicted in Table 3, for most of considered functions, FATPSO demonstrated a consistent performance pattern among all the considered algorithms. FATPSO

Table 3 Performance comparison for the function optimization problems

f	D	SPSO	FAPSO	GA	SA
f_1	30	25.4594 ±16.5424	1.1048e-004 ±0.0017	222.9510 ±26.4874	29.0552 ±4.8291
	100	228.6963 ±675.7348	6.9026e-004 ±0.0080	7.2730e+003 ±459.1044	138.3233 ±38.1029
f_2	30	1.1927e+005 ±41.3785	2.9699 ±24.9744	3.7843e+004 ±4.4308e+003	382.7578 ±103.9384
	100	9.6398e+005 ±3.7652e+004	54.0376 ±482.4480	4.0615e+005 ±2.2613e+004	9.5252e+003 ±4.8500+003
f_3	30	2.3732e-008 ±0.3763	5.9520e-006 ±1.3009e-005	20.2291 ±1.4324	0.4991 ±1.8212
	100	55.5606 ±2.3719e-007	9.2702e-004 ±2.6465	1.2391e+013 ±1.2269e+017	23.4349 ±5.0520
f_4	30	0.0501 ±0.2215	0.0279 ±0.1086	4.5094e+003 ±294.7204	4.9754e+003 ±4.2394
	100	0.0481 ±0.7209	0.0220 ±0.6902	2.7101e+004 ±528.3332	1.6131e+004 ±51.7519
f_5	30	1.4685e-031 ±0.0021	1.5535e-030 ±2.6040e-012	1.0734 ±0.1996	0.1617 ±0.4583
	100	0.2806 ±2.1761	2.6011e-011 ±0.1219	11.4534 ±0.4760	2.8817 ±0.4526
f_6	30	-7.4305e+033 ±2.3497e+033	-4.0465e+034 ±1.2176e+034	-5.1931e+020 ±6.9217e+020	-1.5457e+032 ±1.2010e+016
	100	-2.9776e+096 ±1.2330e+096	-3.2111e+114 ±2.4430e+114	-1.5347e+055 ±9.4580e+054	-3.0040e+104 ±4.2442e+101
f_7	30	33.7291 ±17.7719	8.4007e-010 ±9.3676	204.0560 ±6.8450	32.7997 ±6.9936
	100	391.0421 ±176.3618	19.9035 ±115.9034	1.2070e+003 ±23.8156	177.8810 ±37.7808
f_8	30	0.0177 ±0.3157	0.0102 ±0.0149	6.8463 ±0.6060	0.3193 ±1.7880
	100	0.4400 ±14.4633	0.0720 ±0.6945	179.5966 ±7.3908	31.4270 ±11.4656
f_9	30	0.6206 ±0.2996	5.4819e-004 ±0.0086	1.7437 ±0.0427	0.6606 ±0.0657
	100	1.0666 ±0.3921	0.0011 ±0.0059	2.3570 ±0.0079	1.0167 ±0.0532
f_{10}	30	2.0098e-007 ±52.8218	5.911e-011 ±0.0626	659.0997 ±12.0276	62.2253 ±46.5389
	100	1.3223e+003 ±1.4259e+003	90.1373 ±1.7697e+004	2.8632e+003 ±4.7935e-013	1.5625e+003 ±294.7468

performed extremely well with the exception of 30-D f_4, 100-D f_4, 30-D f_5, 30-D f_{10}, in which the results have little difference between the considered algorithms. It is to be noted that FATPSO could be an ideal choice for solving complex problems (example f_2) when all other algorithms failed to give a better solution.

6   Conclusions

We introduced the Turbulent Particle Swarm Optimization (TPSO) as an alternative method to overcome the problem of premature convergence in the conventional PSO algorithm. TPSO uses a minimum velocity threshold to control the velocity of particles. TPSO mechanism is similar to a turbulence pump, which supply some power to the swarm system. The basic idea is to control the velocity the particles to get out of possible local optima and continue exploring optimal search spaces. The minimum velocity threshold can make the particle continue moving and maintain the diversity of the population until the algorithm converges. We proposed a fuzzy logic based system to tune adaptively the velocity threshold, which is further called as Fuzzy adaptive TPSO (FATPSO). We evaluated and compared the performance of SPSO, FATPSO, GA and SA algorithms on a suite of 20 widely used benchmark problems. The results from our research demonstrated that FATPSO generally outperforms most of the other considered algorithms, especially for high dimensional, multimodal functions.

Acknowledgements. The authors would like to thank Prof. Xiukun Wang, Drs. Ran He and Bo Li for their scientific collaboration in this research work. This work was partially supported by NSFC (60873054).

References

1. Boeringer, D.W., Werner, D.H.: Particle swarm optimization versus genetic algorithms for phased array synthesis. IEEE Transactions on Antennas and Propagation 52(3), 771–779 (2004)
2. Cantu-Paz, E.: Efficient and Accurate Parallel Genetic Algorithms. Kluwer Academic Publishers, Dordrecht (2000)
3. Clerc, M., Kennedy, J.: The particle swarm-explosion, stability, and convergence in a multidimensional complex space. IEEE Transactions on Evolutionary Computation 6, 58–73 (2002)
4. Cordón, O., Herrera, F., Peregrin, A.: Applicability of the fuzzy operators in the design of fuzzy logic controllers. Fuzzy Sets and Systems 86, 15–41 (1997)
5. Du, F., Shi, W.K., Chen, L.Z., Deng, Y., Zhu, Z.F.: Infrared image segmentation with 2-D maximum entropy method based on particle swarm optimization. Pattern Recognition Letters 26, 597–603 (2005)
6. Eberhart, R.C., Shi, Y.H.: Comparison between genetic algorithms and particle swarm optimization. In: Proceedings of IEEE International Conference on Evolutionary Computation, pp. 611–616 (1998)
7. Eiben, A.E., Hinterding, R., Michalewicz, Z.: Parameter control in evolutionary algorithms. IEEE Transations on Evolutionary Computation 3(2), 124–141 (1999)

8. Feller, W.: An Introduction to Probability Theory and Its Application, 3rd edn. John Wiley & Sons, Chichester (1968)
9. Guo, C., Tang, H.: Global convergence properties of evolution stragtegics. Mathematica Numerica Sinica 23(1), 105–110 (2001)
10. He, R., Wang, Y., Wang, Q., Zhou, J., Hu, C.: An improved particle swarm optimization based on self-adaptive escape velocity. Journal of Software 16(12), 2036–2044 (2005)
11. Herrera, F., Lozano, M.: Fuzzy adaptive genetic algorithms: design, taxonomy, and future directions. Soft Computing 7, 545–562 (2003)
12. Jiang, C.W., Etorre, B.: A hybrid method of chaotic particle swarm optimization and linear interior for reactive power optimisation. Mathematics and Computers in Simulation 68, 57–65 (2005)
13. Kennedy, J., Spears, W. M.: Matching algorithms to problems: an experimental test of the particle swarm and some genetic algorithms on the multimodal problem generator. In: Proceedings of the IEEE International Conference on Evolutionary Computation, pp. 78–83 (1998)
14. Kennedy, J., Eberhart, R.: Swarm intelligence. Morgan Kaufmann Publishers, Inc., San Francisco (2001)
15. Lu, W.Z., Fan, H.Y., Lo, S.M.: Application of evolutionary neural network method in predicting pollutant levels in downtown area of Hong Kong. Neurocomputing 51, 387–400 (2003)
16. Mahfouf, M., Chen, M.Y., Linkens, D.A.: Adaptive weighted swarm optimization for multiobjective optimal design of alloy steels. In: Yao, X., Burke, E.K., Lozano, J.A., Smith, J., Merelo-Guervós, J.J., Bullinaria, J.A., Rowe, J.E., Tiño, P., Kabán, A., Schwefel, H.-P. (eds.) PPSN 2004. LNCS, vol. 3242, pp. 762–771. Springer, Heidelberg (2004)
17. Mark, L., Shay, E.: A fuzzy-based lifetime extension of genetic algorithms. Fuzzy Sets and Systems 149, 131–147 (2005)
18. Orosz, J.E., Jacobson, S.H.: Analysis of static simulated annealing algorithms. Journal of Optimzation theory and Applications 115(1), 165–182 (2002)
19. Parsopoulos, K.E., Vrahatis, M.N.: Recent approaches to global optimization problems through particle swarm optimization. Natural Computing 1, 235–306 (2002)
20. Parsopoulos, K.E., Vrahatis, M.N.: On the computation of all global minimizers through particle swarm optimization. IEEE Transactions on Evolutionary Computation 8(3), 211–224 (2004)
21. Phan, H.V., Lech, M., Nguyen, T.D.: Registration of 3D range images using particle swarm optimization. In: Maher, M.J. (ed.) ASIAN 2004. LNCS, vol. 3321, pp. 223–235. Springer, Heidelberg (2004)
22. Schute, J.F., Groenwold, A.A.: A study of global optimization using particle swarms. Journal of Global Optimization 31, 93–108 (2005)
23. Shi, Y.H., Eberhart, R.C., Chen, Y.: Implementation of evolutionary fuzzy systems. IEEE Transactions on Fuzzy System 7(2), 109–119 (1999)
24. Shi, Y.H., Eberhart, R.C.: Fuzzy adaptive particle swarm optimization. In: Proceedings of IEEE International Conference on Evolutionary Computation, pp. 101–106 (2001)
25. Sousa, T., Silva, A., Neves, A.: Particle swarm based data mining algorithms for classification tasks. Parallel Computing 30, 767–783 (2004)
26. Ting, T., Rao, M., Loo, C.K., Ngu, S.S.: Solving unit commitment problem using hybrid particle swarm optimization. Journal of Heuristics 9, 507–520 (2003)

27. Trelea, I.C.: The particle swarm optimization algorithm: convergence analysis and parameter selection. Information Processing Letters 85(6), 317–325 (2003)
28. Triki, E., Collette, Y., Siarry, P.: A theoretical study on the behavior of simulated annealing leading to a new cooling schedule. European Journal of Operational Research 166, 77–92 (2005)
29. van den Bergh, F.: An analysis of particle swarm optimizers, PhD thesis, University of Pretoria, South Africa (2002)
30. Yun, Y.S., Gen, M.: Performance analysis of adaptive genetic algorithms with fuzzy logic and heuristics. Fuzzy Optimization and Decision Making 2, 161–175 (2003)

Part II
Global Optimization Algorithms:
Applications

An Evolutionary Approximation for the Coefficients of Decision Functions within a Support Vector Machine Learning Strategy

Ruxandra Stoean, Mike Preuss, Catalin Stoean, Elia El-Darzi, and D. Dumitrescu

Abstract. Support vector machines represent a state-of-the-art paradigm, which has nevertheless been tackled by a number of other approaches in view of the development of a superior hybridized technique. It is also the proposal of present chapter to bring support vector machines together with evolutionary computation, with the aim to offer a simplified solving version for the central optimization problem of determining the equation of the hyperplane deriving from support vector learning. The evolutionary approach suggested in this chapter resolves the complexity of the optimizer, opens the 'black-box' of support vector training and breaks the limits of the canonical solving component.

1 Introduction

This chapter puts forward a hybrid approach which embraces the geometrical consideration of learning within support vector machines (SVMs) while it

Ruxandra Stoean and Catalin Stoean
Department of Computer Science, University of Craiova, Romania
e-mail: {ruxandra.stoean,catalin.stoean}@inf.ucv.ro

Mike Preuss
Department of Computer Science, University of Dortmund, Germany
e-mail: mike.preuss@cs.uni-dortmund.de

Elia El-Darzi
Department of Computer Science, University of Westminster, UK
e-mail: eldarze@westminster.ac.uk

D. Dumitrescu
Department of Computer Science, University of Cluj-Napoca, Romania
e-mail: ddumitr@cs.ubbcluj.ro

A. Abraham et al. (Eds.): Foundations of Comput. Intel. Vol. 3, SCI 203, pp. 315–346.
springerlink.com   © Springer-Verlag Berlin Heidelberg 2009

considers the estimation for the coefficients of the decision surface through the direct search capabilities of evolutionary algorithms (EAs).

The SVM framework views artificial learning from an interesting perception: A hyperplane geometrically discriminates between training samples and the coefficients of its equation have to be determined, with respect to both the particular prediction ability and the generalization capacity. On the other hand, EAs are universal optimizers that generate solutions based on abstract principles of evolution and heredity. The aim of this work thus becomes to approximate the coefficients of the decision hyperplane through a canonical EA.

The motivation for the emergence of this combined technique resulted from several findings. SVMs are a top performing tool for data mining, however, the inner-workings of the optimization component are rather constrained and very complex. On the other hand, the adaptable EAs achieve learning relatively difficult from a standalone perspective. Taking advantage of the original interpretation of learning within SVMs and the flexible optimization nature of EAs, hybridization aims to accomplish an improved methodology. The novel approach augments support vector learning to become more 'white-box' and to be able to converge independent of the properties of the underlying kernel for a potential decision function. Apart from the straightforward evolution of hyperplane coefficients, an additional aim of the chapter is to investigate the treatment of several other variables involved in the learning process. Furthermore, it is demonstrated that the transparent evolutionary alternative is performed at no additional effort as regards the parametrization of the EA. Last but not least, on a different level from the theoretical reasons, the hybridized technique offers a simple and efficient tool for solving practical problems. Several real-world test cases served not only as benchmark, but also for application of the proposed architecture, and results bear out the initial assumption that an evolutionary approach is useful in terms of deriving the coefficients of such learning structures.

The research objectives and aims of this chapter will be carried out through the following original aspects:

- The hybrid technique will consider the learning task as in SVMs but use an EA to solve the optimization problem of determining the decision function.
- Classification and regression particularities will be treated separately. The optimization problem will be tackled through two possible EAs: One will allow for a more relaxed, adaptive evolutionary learning condition, while the second will be more similar to support vector training.
- Validation will be achieved by considering five diverse real-world learning tasks.
- Besides comparing results, the potential of the utilized, simplistic EA through parametrization is to be investigated.
- To enable handling large data sets, the first adaptive EA approach will be enhanced by the use of a chunking technique, with the purpose of resulting in a more versatile approach.

- The behavior of a crowding-based EA on preserving the performance of the technique will be examined with the purpose of a future employment for the coevolution of nonstandard kernels.
- The second methodology, which is more straightforward, will be generalized through the additional evolution of internal parameters within SVMs; a very general method of practical importance is therefore desired to be achieved.

The chapter contributes some key elements to both EAs and SVMs:

- The hybrid approach combines the strong characteristics of the two important artificial intelligence fields, namely: The original learning concept of SVMs and the flexibility of the direct search and optimization power of EAs.
- The novel alternative approach simplifies the support vector training.
- The proposed hybridization offers the possibility of a general evolutionary solution to all SVM components.
- The novel technique opens the direction towards the evolution and employment of nonstandard kernels.

The remainder of this chapter is organized as follows: Section 2 outlines the primary concepts and mechanisms underlying SVMs. Section 3 illustrates the means to achieve the evolution of the coefficients for the learning hyperplane. Section 4 describes the insides of the technique and its application to real-world problems. The chapter closes with several conclusions and ideas for future enhancement.

2   The SVM Learning Scheme

SVMs are a powerful approach to data mining tasks. Their originality and performance emerge as a result of the inner learning methodology, which is based on the geometrical relative position of training samples.

2.1   A Viewpoint on Learning

Given $\{(x_i, y_i)\}_{i=1,2,...,m}$, a training set where every $x_i \in R^n$ represents a data sample and each y_i corresponds to a target, a learning task is concerned with the discovery of the optimal function that minimizes the discrepancy between the given targets of data samples and the predicted ones; the outcome of previously "unknown" samples, $\{(x_i', y_i')\}_{i=1,2,...,p}$, is then tested.

The SVM technique is equally suited for classification and regression problems. The task for classification is to achieve an optimal separation of given samples into classes. SVMs assume the existence of a separating surface between every two classes labelled as -1 and 1. The aim then becomes the discovery of the appropriate decision hyperplane.

The standard assignment of SVMs for regression is to find the optimal function to be fitted to the data such that it achieves at most ϵ deviation from the actual targets of samples; the aim is thus to estimate the optimal regression coefficients of such a function.

2.2 SVM Separating Hyperplanes

If training examples are known to be linearly separable, then there exists a linear hyperplane of equation (1), which separates the samples according to classes. In (1), w and b are the coefficients of the hyperplane and $\langle \rangle$ denotes the scalar product.

$$\langle w, x_i \rangle - b = 0, w \in \Re^n, b \in \Re, x_i \in R^n, i = 1, 2, ..., m . \tag{1}$$

The positive data samples lie on the corresponding side of the hyperplane and their negative counterparts on the opposite side. As a stronger statement for linear separability [1], each of the positive and negative samples lies on the corresponding side of a matching supporting hyperplane for the respective class (denoted by y_i) (2).

$$y_i(\langle w, x_i \rangle - b) > 1, i = 1, 2, ..., m . \tag{2}$$

SVMs must determine the optimal values for the coefficients of the decision hyperplane that separates the training data with as few exceptions as possible. In addition, according to the principle of Structural Risk Minimization [2], separation must be performed with a maximal margin between classes. This high generalization ability implies a minimal $\|w\|$. In summary, the SVM classification of linear separable data with a linear hyperplane leads to the optimization problem (3).

$$\begin{cases} \min_{w,b} \|w\|^2 \\ \text{subject to } y_i(\langle w, x_i \rangle - b) \geq 1, i = 1, 2, ..., m . \end{cases} \tag{3}$$

Generally, the training samples are not linearly separable. In the nonseparable case, it is obvious that a linear separating hyperplane is not able to build a partition without any errors. However, a linear separation that minimizes training error can be applied to derive a solution to the classification problem [3]. The idea is to relax the separability statement through the introduction of slack variables, denoted by ξ_i for every training example. This relaxation can be achieved by observing the deviations of data samples from the corresponding supporting hyperplane, i.e. from the ideal condition of data separability. Such a deviation corresponds to a value of $\frac{\pm \xi_i}{\|w\|}$, $\xi_i \geq 0$ [4]. These values may indicate different nuanced digressions, but only a ξ_i higher than unity signals a classification error (Fig. 1). Minimization of training error is achieved by adding the indicator of error for every data sample into the separability statement while minimizing their sum. Hence, the SVM classification of linear nonseparable data with a linear hyperplane leads to the primal optimization

Fig. 1 The separating
and supporting linear hy-
perplanes for the nonsep-
arable training subsets
(*squares* denote positive
samples, while *circles*
stand for the negative
ones). The support vec-
tors are *circled* and the
misclassified data point
is *highlighted*

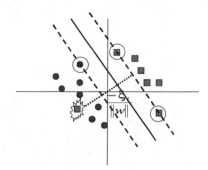

problem (4), where C represents the penalty for errors and is what is called a hyperparameter (free parameter) of the SVM method.

$$\begin{cases} \min_{w,b} \|w\|^2 + C \sum_{i=1}^m \xi_i, C > 0 \\ \text{subject to } y_i(\langle w, x_i \rangle - b) \geq 1 - \xi_i, \xi_i \geq 0, i = 1, 2, ..., m \ . \end{cases} \quad (4)$$

If a linear hyperplane does not provide satisfactory results for the classifi-cation task, then a nonlinear decision surface can be formulated. The initial space of training data samples can be nonlinearly mapped into a higher di-mensional one, called the feature space and further denoted by H, where a linear decision hyperplane can be subsequently built. The separating hy-perplane can achieve an accurate classification in the feature space which corresponds to a nonlinear decision function in the initial space (Fig. 2). The procedure therefore leads to the creation of a linear separating hyperplane that would, as before, minimize training error; however, in this case, it will perform in the feature space. Accordingly, a nonlinear map $\Phi : R^n \to H$ is considered and data samples from the initial space are mapped into H.

As in the classical SVM solving procedure, vectors appear only as part of scalar products, the issue can be further simplified by substituting the scalar product by what is referred to as kernel.

A kernel is defined as a function with the property given by (5).

$$K(x_i, x_j) = \langle \Phi(x_i), \Phi(x_j) \rangle, x_i, x_j \in R^n \ . \quad (5)$$

The kernel can be perceived as to express the similarity between samples. SVMs require the kernel to be a positive (semi-)definite function in order for the standard approach to find a solution to the optimization problem [5]. Such a kernel satisfies Mercer's theorem below and is, therefore, a scalar product in some space [6].

Theorem 1. *(Mercer) [3], [7], [8], [9]*
Let $K(x,y)$ be a continuous symmetric kernel that is defined in the closed interval $a \leq x \leq b$ and likewise for y. The kernel $K(x,y)$ can be expanded in the series

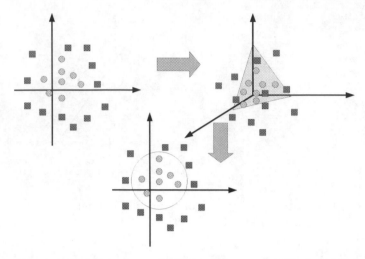

Fig. 2 Initial data space (*left*), nonlinear map into the higher dimension where the objects are linearly separable/the linear separation (*right*), and corresponding nonlinear surface (*bottom*)

$$K(x,y) = \sum_{i=1}^{\infty} \lambda_i \Phi(x)_i \Phi(y)_i$$

with positive coefficients, $\lambda_i > 0$ for all i. For this expansion to be valid and for it to converge absolutely and uniformly, it is necessary that the condition

$$\int_a^b \int_a^b K(x,y)\psi(x)\psi(y)dxdy \geq 0$$

holds for all $\psi(\cdot)$ for which

$$\int_a^b \psi^2(x)dx < \infty$$

The problem with this restriction is twofold [5]. Firstly, Mercer's condition is very difficult to check for a newly constructed kernel. Secondly, kernels that fail the theorem could prove to achieve a better separation of the training samples. Applied SVMs consequently use a couple of classical kernels that had been demonstrated to meet Mercer's condition:

- the polynomial classifier of degree p: $K(x,y) = \langle x,y \rangle^p$
- the radial basis function classifier: $K(x,y) = e^{\frac{\|x-y\|^2}{\sigma}}$, where p and σ are also hyperparameters of SVMs.

However, as a substitute for the original problem solving, a direct search algorithm does not depend on the condition whether the kernel is positive (semi-)definite or not.

After the optimization problem is solved, the class of every test sample is calculated: The side of the decision boundary on which every new data example lies is determined (6).

$$class(x_i^{'}) = sgn(\langle w, \Phi(x_i^{'})\rangle - b), i = 1, 2, ..., p \, . \tag{6}$$

As it is not always possible to determine the map Φ and, as a consequence of the standard training methodology, either to explicitly obtain the coefficients, the class follows from further artifices.

The classification accuracy is then defined as the number of correctly labelled cases over the total number of test samples.

2.3   Addressing Multi-class Problems through SVMs

k-class SVMs build several two-class classifiers that separately solve the matching tasks. The translation from multi-class to two-class is performed through different systems, among which one-against-all, one-against-one or decision directed acyclic graph are the most commonly employed.

One-against-all Approach

The one-against-all (1aa) technique [10] builds k classifiers. Every j^{th} SVM considers all training samples labelled with j as positive and all the remaining as negative.

Consequently, by placing the problem in the initial space, the aim of every j^{th} SVM is to determine the optimal coefficients w^j and b^j of the decision hyperplane which best separates the samples with outcome j from all the other samples in the training set, such that (7) holds.

$$\begin{cases} \min_{w^j, b^j} \frac{\|w^j\|^2}{2} + C\sum_{j=1}^{m} \xi_i^j, \\ \text{subject to } y_i(\langle w^j, x_i\rangle - b^j) \geq 1 - \xi_i^j, \\ \xi_i^j \geq 0, i = 1, 2, ..., m, j = 1, 2, ..., k \, . \end{cases} \tag{7}$$

Once all hyperplanes are determined following the classical SVM training, the label for a test sample $x_i^{'}$ is given by the class that has the maximum value for the learning function, as in (8).

$$class(x_i^{'}) = argmax_{j=1,2,...,k}(\langle w^j, \Phi(x_i^{'})\rangle) - b^j), i = 1, 2, ..., p \, . \tag{8}$$

One-against-one Approach

The one-against-one (1a1) technique [10] builds $\frac{k(k-1)}{2}$ SVMs. Every machine is trained on data from every two classes, l and j, where samples labelled with l are considered positive while those in class j are taken as negative.

Accordingly, the aim of every SVM is to determine the optimal coefficients w^{lj} and b^{lj} of the decision hyperplane which best separates the samples with outcome l from the samples with outcome j, such that (9).

$$\begin{cases} \min_{w^{lj},b^{lj}} \frac{\|w^{lj}\|^2}{2} + C\sum_{i=1}^{m} \xi_i^{lj}, \\ \text{subject to } y_i(\langle w^{lj}, x_i \rangle - b) \geq 1 - \xi_i^{lj}, \\ \xi_i^{lj} \geq 0, i = 1, 2, ..., m, l, j = 1, 2, ..., k, l \neq j \,. \end{cases} \tag{9}$$

Once the hyperplanes of the $\frac{k(k-1)}{2}$ SVMs are found, a *voting* method is used to determine the class for a test sample $x_i', i = 1, 2, ..., p$. For every SVM, the label is computed following the sign of the corresponding decision function applied to x_i'. Subsequently, if the sign says x_i' is in class l, the vote for the l-th class is incremented by one; conversely, the vote for class j is increased by unity. Finally, x_i' is taken to belong to the class with the largest vote. In case two classes have an identical number of votes, the one with the smaller index is selected.

Decision Directed Acyclic Graph

Learning within the decision directed acyclic graph (DDAG) technique [11] follows the same procedure as in 1a1. After the hyperplanes of the $\frac{k(k-1)}{2}$ SVMs are discovered, a graph system is used to determine the class for a test sample $x_i', i = 1, 2, ..., p$. Each node of the graph has a list of classes attached and considers the first and last elements of the list. The list that corresponds to the root node contains all k classes. When a test instance x_i' is evaluated, it is descended from node to node, in other words, one class is eliminated from each corresponding list, until the leaves are reached. The mechanism starts at the root node which considers the first and last classes. At each node, l vs j, we refer to the SVM that was trained on data from classes l and j. The class of x is computed by following the sign of the corresponding decision function applied to x_i'. Subsequently, if the sign says x is in class l, the node is exited via the right edge while, conversely, through the left edge. The wrong class is thus eliminated from the list and it is proceeded via the corresponding edge to test the first and last classes of the new list and node. The class is given by the leaf that x_i' eventually reaches.

2.4 SVM Regression Hyperplanes

SVMs for regression must find a function $f(x)$ that has at most ϵ deviation from the actual targets of training samples and, simultaneously, is as flat as possible [12]. In other words, the aim is to estimate the regression coefficients of $f(x)$ with these requirements.

While the former condition is straightforward, errors are allowed as long as they are less than ϵ, the latter needs some further explanation [13]. Resulting values of the regression coefficients may affect the model in the sense that it fits current training data but has low generalization ability, which would contradict the principle of Structural Risk Minimization for SVMs [2]. In order to overcome this limitation, it is required to choose the flattest function in the definition space. Another way to interpret SVMs for regression is that training data are constrained to lie on a hyperplane that allows for some error and, at the same time, has high generalization capacity.

Suppose a linear regression model can fit the training data. Consequently, function f has the form (10).

$$f(x) = \langle w, x \rangle - b .\tag{10}$$

The conditions for the flattest function (smallest slope) that approximates training data with ϵ precision can be translated into the optimization problem (11).

$$\begin{cases} \min_{w,b} \|w\|^2 \\ \text{subject to } \begin{cases} y_i - \langle w, x_i \rangle + b \leq \epsilon \\ \langle w, x_i \rangle - b - y_i \leq \epsilon \end{cases} , \\ i = 1, 2, ..., m . \end{cases}\tag{11}$$

It may very well happen that the linear function f is not able to fit all training data and consequently SVMs will again allow for some relaxation, analogously to the corresponding situation for classification.

Therefore, the positive slack variables ξ_i and ξ_i^*, both attached to each sample, are introduced into the condition for approximation of training data and, also as before, the sum of these indicators for errors is minimized. The primal optimization problem in case of regression then translates to (12).

$$\begin{cases} \min_{w,b} \|w\|^2 + C \sum_{i=1}^m (\xi_i + \xi_i^*) \\ \text{subject to } \begin{cases} y_i - \langle w, x_i \rangle + b \leq \epsilon + \xi_i \\ \langle w, x_i \rangle - b - y_i \leq \epsilon + \xi_i^* \\ \xi_i, \xi_i^* \geq 0 \end{cases} , \\ i = 1, 2, ..., m . \end{cases}\tag{12}$$

If a linear function is not at all able to fit training data, a nonlinear function has to be chosen, as well. The procedure follows the same steps as before in SVMs for classification.

When a solution for the optimization problem is reached, the predicted target for a test sample is computed as (13).

$$f(x_i') = \langle w, \Phi(x_i') \rangle - b, i = 1, 2, ..., p .\tag{13}$$

Also as in the classification problem, the regression coefficients are rarely transparent and the predicted target actually is derived from other computations.

In order to verify the accuracy of the technique, the value of the root mean square error (RMSE) is computed as in (14).

$$RMSE = \sqrt{\frac{1}{p}\sum_{i=1}^{p}(f(x_i') - y_i')^2}\ . \tag{14}$$

2.5 Solving the Optimization Problem within SVMs

The standard algorithm of finding the optimal solution relies on an extension of the Lagrange multipliers technique. Corresponding dual problem may be expressed as (15) for classification.

$$\begin{cases} \min_{\{\alpha_i\}_{i=1,2,\ldots,m}} Q(\alpha) = \sum_{i=1}^{m}\alpha_i - \frac{1}{2}\sum_{i=1}^{m}\sum_{j=1}^{m}\alpha_i\alpha_j y_i y_j K(x_i,x_j) \\ \text{subject to} \begin{cases} \sum_{i=1}^{m}\alpha_i y_i = 0 \\ \alpha_i \geq 0 \end{cases} \\ i = 1,2,\ldots,m\ . \end{cases}$$

$$\tag{15}$$

Conversely, in the most general case of nonlinear regression, the dual problem is restated as (16). For reasons of a shorter reference, $\alpha_i(^*)$ denotes both α_i and α_i^*, in turn.

$$\begin{cases} \min_{\{\alpha_i(^*)\}_{i=1,2,\ldots,m}} Q(\alpha(^*)) = \sum_{i,j=1}^{m}(\alpha_i - \alpha_i^*)(\alpha_j - \alpha_j^*)K(x_i,x_j) \\ -\epsilon\sum_{i=1}^{m}(\alpha_i + \alpha_i^*) + \sum_{i=1}^{m} y_i(\alpha_i - \alpha_i^*) \\ \text{subject to} \begin{cases} \sum_{i=1}^{m}(\alpha_i - \alpha_i^*) = 0 \\ 0 \leq \alpha_i(^*) \leq C \end{cases} \\ i = 1,2,\ldots,m\ . \end{cases} \tag{16}$$

The optimum Lagrange multipliers $\alpha_i(^*)$s are determined as the solutions of the system by setting the gradient of the objective function to zero. For more mathematical explanation, see [3], [14].

3 Evolutionary Adaptation of the Hyperplane Coefficients to the Training Data

Apart from its emergence as a simple complementary method for solving the SVM derived optimization problem, the EA-powered determination of hyperplane coefficients had to be explored and improved with respect to runtime, prediction accuracy and adaptability.

3.1   Motivation and Aim

Despite the originality and performance of the learning vision of SVMs, the inner training engine is intricate, constrained, rarely transparent and able to converge only for certain particular decision functions. This has brought the motivation to investigate and put forward an alternative training approach that benefits from the flexibility and robustness of EAs.

The technique adopts the learning strategy of the SVMs but aims to simplify and generalize its solving methodology, by offering a transparent substitute to the initial 'black-box'. Contrary to the canonical technique, the evolutionary approach can at all times explicitly acquire the coefficients of the decision function, without any further constraints. Moreover, in order to converge, the evolutionary method does not require the positive (semi-)definition properties for kernels within nonlinear learning. Eventually, the evolutionary approach demonstrates to be an efficient tool for real-world application in vital domains like disease diagnosis and prevention or spam filtering.

There have been numerous attempts to combine SVMs and EAs, however, this method differs from the reported ones: The learning path remains the same, except that the coefficients of the decision function are now evolved with respect to the optimization objectives regarding accuracy and generalization. Several potential structures, enhancements and additions had been proposed, tested and confirmed using available benchmarking test problems.

3.2   Literature Review: Previous EA-SVM Interactions

The chapter focuses on the evolution of the coefficients for decision functions of a learning strategy similar to that of SVMs. However, there are other works known to combine SVMs and EAs. One evolutionary direction tackles model selection, which concerns the adjustment of hyperparameters within SVMs, i.e. the penalty for errors C and parameters of the kernel, and is generally performed through grid search or gradient descent methods. Alternatively, determination of hyperparameters can be achieved through evolution strategies [15]. Another aspect envisages the evolution of the form for the kernel, which can be performed by means of genetic programming [16]. The Lagrange multipliers involved in the expression of the dual problem can be evolved by means of evolution strategies and particle swarm optimization [5]. Inspired by the geometrical SVM learning, [17] also reports the evolution of w and C, while using erroneous learning ratio and lift values as the objective function. Current paper therefore extends the work in the hybridization between EAs and SVMs by filling the gap of a direct evolutionary solving of the primal optimization problem of determining the decision hyperplane, which has never been performed before, to the best of our knowledge.

3.3 Evolving the Coefficients of the Hyperplane

The evolutionary approach for support vector training (ESVM) considers the adaptation of a flexible hyperplane to the given training data through the evolution of the optimal coefficients for its equation. After the necessary revisions in the learning objectives due to a different way of solving the optimization task, the corresponding EA is adopted in a canonical formulation for real-valued problems [18]. For the purpose of a comparative analysis between ESVMs and SVMs, there are solely the classical polynomial and radial kernels that are used in this chapter to shape the decision functions.

Representation

An individual c encodes the coefficients of the hyperplane, w and b (17). Individuals are randomly generated such that $w_{i'} \in [-1,1], i' = 1, 2, ..., n$, $b \in [-1, 1]$.

$$c = (w_1, ..., w_n, b) . \tag{17}$$

Fitness Assignment

Prior to deciding on a strategy to evaluate individuals, the objective function must be established in terms of the new approach to address the optimization goals.

Since ESVMs depart from the standard mathematical treatment of SVMs, a different general (nonlinear) optimization problem is derived [19] . Accordingly, w is also mapped through Φ into H. As a result, the squared norm that is involved in the generalization condition is now $\|\Phi(w)\|^2$. At the same time, the equation of the hyperplane consequently changes to (18).

$$\langle \Phi(w), \Phi(x_i) \rangle - b = 0. \tag{18}$$

The scalar product is used in the form (19) and, besides, the kernel is additionally employed to address the norm in its simplistic equivalence to a scalar product.

$$\langle u, w \rangle = u^T w . \tag{19}$$

In conclusion, the optimization problem for classification is reformulated as (20).

$$\begin{cases} \min_{w,b} K(w, w) + C \sum_{i=1}^{m} \xi_i, C > 0 \\ \text{subject to } y_i(K(w, x_i) - b) \geq 1 - \xi_i, \\ \xi_i \geq 0, i = 1, 2, ..., m . \end{cases} \tag{20}$$

At the same time, the objectives for regression are transposed to (21).

$$\begin{cases} \min_{w,b} K(w,w) + C \sum_{i=1}^{m}(\xi_i + \xi_i^*) \\ \text{subject to} \begin{cases} y_i - K(w,x_i) + b \leq \epsilon + \xi_i \\ K(w,x_i) - b - y_i \leq \epsilon + \xi_i^* \\ \xi_i, \xi_i^* \geq 0 \end{cases} , \\ i = 1,2,...,m \,. \end{cases} \qquad (21)$$

The ESVMs targeting multi-class situations must undergo similar transformation with respect to the expression of the optimization problem [20], [21]. As 1aa is concerned, the aim of every j^{th} ESVM is expressed as to determine the optimal coefficients, w^j and b^j, of the decision hyperplane which best separates the samples with outcome j from all the other samples in the training set, such that (22) takes place.

$$\begin{cases} \min_{w^j,b^j} K(w^j,w^j) + C \sum_{i=1}^{m} \xi_i^j, \\ \text{subject to } y_i(K(w^j,x_i) - b^j) \geq 1 - \xi_i^j, \\ \xi_i^j \geq 0, i = 1,2,...,m, j = 1,2,...,k \,. \end{cases} \qquad (22)$$

Within the 1a1 and DDAG approaches, the aim of every ESVM becomes to find the optimal coefficients w^{lj} and b^{lj} of the decision hyperplane which best separates the samples with outcome l from the samples with outcome j, such that (23) holds.

$$\begin{cases} \min_{w^{lj},b^{lj}} K(w^{lj},w^{lj}) + C \sum_{i=1}^{m} \xi_i^{lj}, \\ \text{subject to } y_i(K(w^{lj},x_i) - b^{lj}) \geq 1 - \xi_i^{lj}, \\ \xi_i^{lj} \geq 0, i = 1,2,...,m, l,j = 1,2,...,k, i \neq j \,. \end{cases} \qquad (23)$$

The fitness assignment now derives from the objective function of the optimization problem and is minimized. Constraints are handled by penalizing the infeasible individuals through the introduction of a function $t : R \rightarrow R$ which returns the value of the argument, if negative, while zero otherwise.

Classification and regression variants simply differ in terms of objectives and constraints. Thus the expression of the fitness function for the former is (24), with the corresponding indices in the multi-class situations [22], [23], [24].

$$f(w,b,\xi) = K(w,w) + C \sum_{i=1}^{m} \xi_i + \sum_{i=1}^{m} [t(y_i(K(w,x_i) - b) - 1 + \xi_i)]^2. \quad (24)$$

As for the latter, the fitness assignment is defined in the form (25) as found in [25], [26], [27].

$$f(w, b, \xi) = K(w, w) + C \sum_{i=1}^{m} (\xi_i + \xi_i^*) + \sum_{i=1}^{m} [t(\epsilon + \xi_i - y_i +$$

$$K(w, x_i) - b)]^2 + \sum_{i=1}^{m} [t(\epsilon + \xi_i^* + y_i - K(w, x_i) + b)]^2. \qquad (25)$$

Selection and Variation Operators

The efficient tournament selection and the common genetic operators for real encoding, i.e. intermediate crossover and mutation with normal perturbation, are applied.

Stop Condition

The EA stops after a predefined number of generations and outputs the optimal coefficients for the equation of the hyperplane. Moreover, ESVM is transparent at all times during the evolutionary cycle, thus w and b may be observed as they adapt throughout the process.

Test Step

Once the coefficients of the hyperplane are found, the class for an unknown data sample can be determined directly following (26) .

$$class(x_i') = sgn(K(w, x_i') - b), i = 1, 2, ..., p . \qquad (26)$$

Conversely for regression, the target of test samples can be obtained through (27) .

$$f(x_i') = K(w, x_i') - b, i = 1, 2, ..., p . \qquad (27)$$

For the multi-class tasks, the label is found by employing the same specific mechanisms, only this time the resulting decision function applied to the current test sample takes the form (28) .

$$f(x_i') = K(w^j, x_i') - b^j, i = 1, 2, ..., p, j = 1, 2, ..., k . \qquad (28)$$

3.4 Preexperimental Planning: The Test Cases

Experimentation had been conducted on five real-world problems, coming from the UCI Repository of Machine Learning Databases[1], i.e. diabetes mellitus diagnosis [28] , spam detection [29] , [30], iris recognition [20] , soybean disease diagnosis [31] and Boston housing [25], [26] (see Table 1). The motivation for the choice of test cases was manifold. Diabetes and spam are

---

[1] Available at http://www.ics.uci.edu/~mlearn/MLRepository.html

Table 1 Data set properties

Data	Diabetes	Iris	Soybean	Spam	Boston
No. of samples	768	150	47	4601	506
No. of features	8	4	35	57	13
No. of classes	2	3	4	2	-

two-class problems, while soybean and iris are multi-class. Differentiating, on the one hand, diabetes diagnosis is a better-known benchmark, but spam filtering is an issue of current major concern; moreover, the latter has a lot more features and samples, which makes a huge difference for classification as well as for optimization. On the other hand, while soybean has a high number of attributes, iris has only four, but a larger number of samples. Finally, Boston housing is a representative regression task. For all reasons mentioned above, the selection of test problems certainly contained all the variety of situations that had been necessary for the objective validation of the ESVM approach. The experimental design was set to employ holdout cross-validation: For each data set, 30 runs of the ESVM were conducted – in every run, approximately 70% random cases were appointed to the training set and the remaining 30% went into test. The necessity for data normalization in diabetes, spam and iris was also observed.

4   Discovering ESVMs

While the target of the EA was straightforward, addressing several inside interactions had been not. Moreover, application of the ESVM to the practical test cases had yielded more research questions yet to be resolved. Finally, further implications of being able to instantly operate on the evolved coefficients had been realized.

4.1   A Naïve Design

As already stated, the coefficients of the hyperplane, w and b, are encoded into the structure of an individual. But, since the conditions for hyperplane optimality additionally refer the indicators for errors, ξ_i, $i = 1, 2, ..., m$, the problem becomes how to comprise them in the evolutionary solving. One simple solution could be to depart from the SVM geometrical strict meaning of a deviation and simply evolve the factors of indicators for errors. Thus, the structure of an individual changes to (29), where $\xi_j \in [0, 1], j = 1, 2, ..., m$.

$$c = (w_1, ..., w_n, b, \xi_1, ..., \xi_m) .$$
(29)

Table 2 Manually tuned SVM hyperparameter values for the evolutionary and canonical approach

	Diabetes	Iris	1a1/1aa	Soybean	Spam	Boston
ESVMs						
$p\,(\sigma)$	$p = 2$	$\sigma = 1$		$p = 1$	$p = 1$	$p = 1$
SVMs						
$p\,(\sigma)$	$p = 1$	$\sigma = 1/m$		$p = 1$	$p = 1$	$\sigma = 1/m$

Table 3 Manually tuned EA parameter values for the naïve construction

	Diabetes	Iris	1a1	Soybean	Spam	Boston
ps	100	100		100	100	200
ng	250	100		100	250	2000
cp	0.40	0.30		0.30	0.30	0.50
mp	0.40	0.50		0.50	0.50	0.50
emp	0.50	0.50		0.50	0.50	0.50
ms	0.10	0.10		0.10	0.10	0.10
ems	0.10	0.10		0.10	0.10	0.10

The evolved values of the indicators for errors can now be addressed in the proposed expression for fitness evaluation. Also, mutation of errors is now constrained, preventing the ξ_is from taking negative values.

Once the primary theoretical aspects had been completed, the experimental design had to be inspected and the practical evaluation of the viability of ESVMs had to be conducted. The hyperparameters both approaches share were manually chosen (Table 2). The error penalty C was invariably set to 1. For certain (e.g. radial, polynomial) kernels, the optimization problem is relatively simple, due to Mercer's theorem, and is also implicitly solved by classical SVMs [32]. Note that ESVMs are not restricted to using the traditional kernels, but these had been employed within to enable comparison with classical SVMs. Therefore, as a result of multiple testing, a radial kernel was used for the iris data set, a polynomial one was employed for diabetes, while for spam, soybean and Boston, a linear surface was applied. In the regression case, ϵ was set to 0.

An issue of crucial importance for demonstrating the feasibility of any EA alternative relies on the simplicity to determine appropriate parameters. The EA parameter values were initially manually determined (Table 3).

In order to validate the manually found EA parameter values and to probe the ease in their choice, the tuning method of Sequential Parameter Optimization (SPO) [33] was applied. The SPO builds on a quadratic regression model, supported by a Latin hypercube sampling (LHS) methodology and noise reduction, by incrementally increased repetition of runs. Parameter bounds were set as follows:

- Population size (ps) - 5/2000
- Number of generations (ng) - 50/300
- Crossover probability (pc) - 0.01/1
- Mutation probability (pm) - 0.01/1
- Error mutation probability (epm) - 0.01/1
- Mutation strength (ms) - 0.001/5
- Error mutation strength (ems) - 0.001/5

Since the three multi-class techniques behave similarly in the manually tuned multi-class experiments (Table 5), automatic adjustment was run only for the most widely used case of 1a1. The best parameter configurations for all problems as determined by SPO are depicted in Table 4.

Table 4 SPO tuned EA parameter values for the naïve representation

	Diabetes	Iris	Soybean	Spam	Spam +Chunks	Boston
ps	198	46	162	154	90	89
ng	296	220	293	287	286	1755
pc	0.87	0.77	0.04	0.84	0.11	0.36
pm	0.21	0.57	0.39	0.20	0.08	0.5
epm	0.20	0.02	0.09	0.07	0.80	0.47
ms	4.11	4.04	0.16	3.32	0.98	0.51
ems	0.02	3.11	3.80	0.01	0.01	0.12

Test accuracies/errors obtained by manual tuning are presented in Table 5. Differentiated (spam/non spam for spam filtering and ill/healthy for diabetes) results are also depicted.

Table 5 Accuracy/RMSE of the manually tuned naïve ESVM version on the considered test sets, in percent

	Average	Worst	Best	StD
Diabetes (overall)	**76.30**	71.35	80.73	2.24
Diabetes (ill)	50.81	39.19	60.27	4.53
Diabetes (healthy)	90.54	84.80	96.00	2.71
Iris 1aa (overall)	**95.85**	84.44	100.0	3.72
Iris 1a1 (overall)	**95.18**	91.11	100.0	2.48
Iris DDAG (overall)	**94.96**	88.89	100.0	2.79
Soybean 1aa (overall)	**99.22**	88.24	100	2.55
Soybean 1a1 (overall)	**99.02**	94.11	100.0	2.23
Soybean DDAG (overall)	**98.83**	70.58	100	5.44
Spam (overall)	**87.74**	85.74	89.83	1.06
Spam (spam)	77.48	70.31	82.50	2.77
Spam (non spam)	94.41	92.62	96.30	0.80
Boston	**4.78**	5.95	3.96	0.59

Table 6 Accuracies of the SPO tuned naïve ESVM version on the considered test sets, in percent

	LHS$_{best}$	StD	SPO	StD
Diabetes (overall)	75.82	3.27	**77.31**	2.45
Diabetes (ill)	49.35	7.47	52.64	5.32
Diabetes (healthy)	89.60	2.36	90.21	2.64
Iris (overall)	**95.11**	2.95	**95.11**	2.95
Soybean (overall)	99.61	1.47	**99.80**	1.06
Spam (overall)	89.27	1.37	**91.04**	0.80
Spam (spam)	80.63	3.51	84.72	1.59
Spam (non spam)	94.82	0.94	95.10	0.81
Boston	5.41	0.65	**5.04**	0.52

Table 6 summarizes the performances and standard deviations of the best configuration of an initial LHS sample and of the SPO .

SPO indicates that for all cases, except for the soybean data, crossover probabilities were dramatically increased, while often reducing mutation probabilities, especially for errors. However, the relative quality of SPO's final best configurations against the ones found during the initial LHS phase increases with the problem size. It must be stated that in most cases, results achieved with manually determined parameter values are only improved by SPO, if at all, by more effort, i.e. increasing population size or number of generations.

The computational experiments show that the proposed technique produces equally good results as compared to the canonical SVMs and it is further explained in subsection 4.6. Furthermore, the smaller standard deviations prove the higher stability of ESVMs.

As concerns the difficulty in setting the EA, SPO confirms: Distinguishing the performance of different configurations is difficult even after computing a large number of repeats. Consequently, the "parameter optimization potential" justifies employing a tuning method only for the larger problems, diabetes, spam and Boston. Especially for the small problems, well performing parameter configurations are seemingly easy to find. This brings evidence in support of the (necessary) simplicity in tuning the parameters inside the evolutionary alternative solving.

It must be stated that for the standard kernels, one cannot expect the ESVM to be better than the standard SVM, since the kernel transformation that induces learning is the same. However, the flexibility of the EAs as optimization tools makes ESVMs an attractive choice from the performance perspective, due to their prospective ability to additionally evolve problem-tailored kernels, regardless of whether they are positive (semi-)definite or not, which is impossible under SVMs.

4.2   Chunking within ESVMs

A first problem appears for large data sets, i.e. spam filtering, where the amount of runtime needed for training is very large. This stems from the large genomes employed, as indicators for errors of every sample in the training set are included in the representation. Consequently, this problem was tackled by an adaptation of a chunking procedure [34] inside ESVM.

A chunk of N training samples is repeatedly considered. Within each chunking cycle, the EA, with a population of half random individuals and half previously best evolved individuals, runs and determines the coefficients of the hyperplane. All training samples are tested against the obtained decision function and a new chunk is constructed based on $N/2$ randomly and equally distributed incorrectly placed samples and half randomly samples from the current chunk. The chunking cycle stops when a predefined number of iterations with no improvement in training accuracy passes (Algorithm 1).

Algorithm 1. ESVM with Chunking

Require: The training samples
Ensure: Best obtained coefficients and corresponding accuracy
  begin
  Randomly choose N training samples, equally distributed, to make a chunk;
  while a predefined number of iterations passes with no improvement **do**
    if first chunk **then**
      Randomly initialize population of a new EA;
    else
      Use best evolved hyperplane coefficients and random indicators for errors
      to fill half of the population of a new EA and randomly initialize the other
      half;
    end if
    Apply EA and find coefficients of the hyperplane;
    Compute side of all samples in the training set with evolved hyperplane coefficients;
    From incorrectly placed, randomly choose (if exist) N/2 samples, equally distributed;
    Randomly choose the rest up to N from the current chunk and add all to a
    new one;
    if obtained training accuracy if higher than the best one obtained so far **then**
      Update best accuracy and hyperplane coefficients; set improvement to true;
    end if
  end while
  Apply best obtained coefficients on the test set and compute accuracy
  return accuracy
  end

ESVM with chunking was applied to the spam data set. Manually tuned parameters had the same values as before, except the number of generations for each run of the EA which is now set to 100. The chunk size, N, was chosen

Table 7 Accuracy/RMSE of the manually tuned ESVM with chunking version on the considered test sets, in percent

	Average	Worst	Best	StD
Spam (overall)	**87.30**	83.13	90.00	1.77
Spam (spam)	83.47	75.54	86.81	2.78
Spam (non spam)	89.78	84.22	92.52	2.11

Table 8 Accuracies of the SPO tuned ESVM with chunking version on the considered test sets, in percent

	LHS_{best}	StD	SPO	StD
Spam (overall)	87.52	1.31	**88.37**	1.15
Spam (spam)	86.26	2.66	86.35	2.70
Spam (non spam)	88.33	2.48	89.68	2.06

as 200 and the number of iterations with no improvement, i.e. repeats of the chunking cycle, was designated to be 5. Values derived from the SPO tuning are presented in the chunking column from Table 4.

Results of manual and SPO tuning are shown in Tables 7 and 8. The novel approach of ESVM with chunking produced good results in a much smaller runtime; it runs 8 times faster than the previous one, at a cost of a small loss in accuracy. Besides solving the EA genome length problem, proposed mechanism additionally reduces the large number of computations that derives from the reference to the many training samples in the expression of the fitness function.

4.3 A Pruned Variant

Although already a good alternative approach, the ESVM may still be improved concerning simplicity. The current optimization problem requires to treat the error values, which in the present EA variant are included in the representation. These severely complicate the problem by increasing the genome length (variable count) by the number of training samples. Moreover, such a methodology strongly departs from the canonical SVM concept. Therefore, it had been investigated whether the indicators for errors could be computed instead of evolved.

Since ESVMs directly and interactively provide hyperplane coefficients at all times, the generic EA representation (17) can be kept and the indicators can result from geometrical calculations. In case of classification, the procedure follows [1]. The current individual, which is the current separating hyperplane, is considered and supporting hyperplanes are determined through the mechanism below. One first computes (30).

$$\begin{cases} m_1 = min\{K(w, x_i)|y_i = +1\} \\ m_2 = max\{K(w, x_i)|y_i = -1\} \,. \end{cases} \tag{30}$$

Then (31) proceeds.

$$\begin{cases} p = |m_1 - m_2| \\ w' = \frac{2}{p}w \\ b' = \frac{1}{p}(m_1 + m_2) \,. \end{cases} \tag{31}$$

For every training sample x_i, the deviation to its corresponding supporting hyperplane (32) is obtained.

$$\delta(x_i) = \begin{cases} K(w', x_i) - b' - 1, \text{ if } y_i = +1 \\ K(w', x_i) - b' + 1, \text{ if } y_i = -1 \\ i = 1, 2, ..., m \,. \end{cases} \tag{32}$$

If sign of deviation equals class, corresponding $\xi_i = 0$; else, the (normalized) absolute deviation is returned as the indicator for error. Experiments showed the need for normalization of the computed deviations in the cases of diabetes, spam and iris, while, on the contrary, soybean requires no normalization. The different behavior can be explained by the fact that the first three data sets have a larger number of training samples. The sum of deviations is subsequently added to the expression of the fitness function. As a consequence, in the early generations, when the generated coefficients lead to high deviations, their sum, considered from 1 to the number of training samples, takes over the whole fitness value and the evolutionary process is driven off the course to the optimum. The form of the fitness function remains as before (24), obviously without taking the ξ_is as arguments.

The proposed method for acquiring the errors for the regression situation is as follows. For every training sample, one firstly calculates the difference between the actual target and the predicted value that is obtained with the coefficients of the current individual (regression hyperplane), as in (33).

$$\delta_i = |K(w, x_i) - b - y_i|, i = 1, 2, ..., m \,. \tag{33}$$

Secondly, one tests the difference against the ϵ threshold, following (34).

Table 9 Manually tuned parameter values for the pruned approach

	Diabetes	Iris	Soybean	Spam	Boston
ps	100	100	100	150	200
ng	250	100	100	300	2000
pc	0.4	0.30	0.30	0.80	0.50
pm	0.4	0.50	0.50	0.50	0.50
ms	0.1	4	0.1	3.5	0.1

Table 10 SPO tuned parameter values for the pruned representation

	Diabetes	Iris	Soybean	Spam	Boston
ps	190	17	86	11	100
ng	238	190	118	254	1454
pc	0.13	0.99	0.26	0.06	0.88
pm	0.58	0.89	0.97	0.03	0.39
ms	0.15	3.97	0.08	2.58	1.36

Table 11 Accuracy/RMSE of the manually tuned pruned ESVM version on the considered test sets, in percent

	Average	Worst	Best	StD
Diabetes (overall)	74.60	70.31	82.81	2.98
Diabetes(ill)	45.38	26.87	58.57	6.75
Diabetes (healthy)	89.99	86.89	96.75	2.66
Iris 1aa (overall)	93.33	86.67	100	3.83
Iris 1a1 (overall)	95.11	73.33	100	4.83
Iris DDAG (overall)	95.11	88.89	100	3.22
Soybean 1aa (overall)	99.22	88.24	100	2.98
Soybean 1a1 (overall)	99.60	94.12	100	1.49
Soybean DDAG (overall)	99.60	94.12	100	1.49
Spam (overall)	85.68	82	88.26	1.72
Spam (spam)	70.54	62.50	77.80	4.55
Spam (non spam)	95.39	92.66	97.44	1.09
Boston	5.07	6.28	3.95	0.59

$$\begin{cases} \text{if } \delta_i < \epsilon & \text{then } \xi_i = 0 \\ \text{else} & \xi_i = \delta_i - \epsilon \\ i = 1, 2, ..., m . \end{cases} \quad (34)$$

The newly obtained indicators for errors can now be employed in the fitness evaluation of the corresponding individual, which changes from (25) to (35):

$$f(w, b) = K(w, w) + C \sum_{i=1}^{m} \xi_i . \quad (35)$$

The function to be fitted to the data is thus still required to be as flat as possible and to minimize the errors of regression that are higher than the permitted ϵ. Experiments on the Boston housing problem demonstrated that the specific method for computing the deviations does not require any additional normalization.

The problem related settings and SVM hyperparameters were kept the same as for naïve approach, except ϵ which was set to 5 for the regression problem, which reveals that the pruned representation apparently needs a more generous deviation allowance within training.

Table 12 Accuracies of the SPO tuned pruned ESVM version on the considered test sets, in percent

	LHS_{best}	StD	SPO	StD
Diabetes (overall)	72.50	2.64	**73.39**	2.82
Diabetes(ill)	35.50	10.14	43.20	6.53
Diabetes (healthy)	92.11	4.15	89.94	3.79
Iris (overall)	**95.41**	2.36	**95.41**	2.43
Soybean (overall)	**99.61**	1.47	99.02	4.32
Spam (overall)	89.20	1.16	**89.51**	1.17
Spam (spam)	79.19	3.13	82.02	3.85
Spam (non spam)	95.64	0.90	94.44	1.42
Boston	4.99	0.66	**4.83**	0.45

The EA first proceeded with the manual values for parameters from Table 9. Subsequent parameter values derived from SPO on the pruned variant are shown in Table 10.

Results obtained after manual and SPO tuning are depicted in Tables 11 and 12. The automated performance values were generated by 30 validation runs for the best found configurations after the initial design and SPO, respectively.

Results of automated tuning are similar to those of the manual regulation which once again demonstrates the easy adjustability of the ESVM. Additionally, the performance spectra of LHS was plotted in order to compare the hardness of finding good parameters for our two representations on the spam and soybean problems (Figs. 3 and 4). The Y axis represents the fractions of all tried configurations; therefore the Y value corresponding to each bar denotes the percentage of configurations that reached the accuracy of the X axis where the bar is positioned.

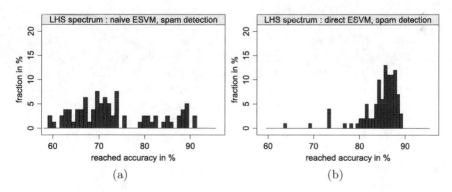

(a)   (b)

Fig. 3 Comparison of EA parameter spectra, LHS with size 100, 4 repeats, (a) for the naïve (7 parameters) and (b) the pruned (5 parameters) representation on the spam problem

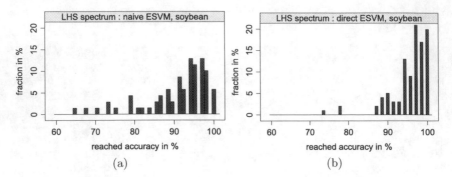

Fig. 4 Comparison of EA parameter spectra, LHS with size 100, 4 repeats, (a) for the naïve (7 parameters) and (b) the pruned (5 parameters) representation on the soybean problem

The diagrams illustrate the fact that naïve ESVM is harder to parameterize than the pruned approach: When SPO finds a configuration for the latter, it is already a promising one, as it can be concluded from the higher corresponding bars.

It is interesting to remark that the pruned representation is not that much faster. Although the genome length is drastically reduced, the runtime consequently gained is however partly lost again when computing the values for the slack variables. This draws from the extra number of scalar products that must be calculated due to (30), (32) and (33). As run length itself is a parameter in present studies, an upper bound of the necessary effort is rather obtained. Closer investigation may lead to a better understanding of suitable run lengths, e.g. in terms of fitness evaluations. However, the pruned representation has its advantages. Besides featuring smaller genomes, less parameters are needed, because the slack variables are not evolved and thus two parameters are eliminated. As a consequence, it can be observed that this representation is easier to tune.

The best configurations for the pruned representation perform slightly worse as compared to the results recorded for the naïve representation. The independent evolution of the slack variables seems to result in a better adjustment of the hyperplane as opposed to their strict computation. Parameter tuning beyond a large initial design appears to be infeasible, as performance is not significantly improved in most cases. If at all, it is successful for the larger problems of diabetes, spam and Boston. This indicates once more that parameter setting for the ESVM is rather easy, because there is a large set of good performing configurations. Nevertheless, there seems to be a slight tendency towards fewer good configurations (harder tuning) for the large problems.

Table 13 SPO tuned parameter values for the pruned representation with crowding

	Diabetes	Iris	Spam
ps	92	189	17
ng	258	52	252
pc	0.64	0.09	0.42
pm	0.71	0.71	0.02
ms	0.20	0.20	4.05

Table 14 Accuracies of the SPO tuned pruned version with crowding on the considered test sets, in percent

	LHD_{best}	StD	SPO	StD
Diabetes (overall)	74.34	2.30	**74.44**	2.98
Diabetes(ill)	43.68	6.64	45.32	7.04
Diabetes (healthy)	90.13	3.56	90.17	3.06
Iris (overall)	**95.63**	2.36	94.37	2.80
Spam (overall)	88.72	1.49	**89.45**	0.97
Spam (spam)	80.14	5.48	80.79	3.51
Spam (non spam)	94.25	1.66	95.07	1.20

4.4 A Crowding Variant

In addition to the direct pruned representation, a crowding [35] variant of the EA had also been tested. Within crowding, test for replacement is done against the most similar parent of the current population. Crowding based EAs are known to provide good global search capabilities. This is of limited value for the kernel types employed in this study, but it is important for nonstandard kernels. It is desirable, however, to investigate whether the employment of a crowding-based EA on the pruned representation would maintain the performance of the technique or not. All the other elements of the EA remained the same and the values for parameters as determined by SPO are shown in Table 13. The crowding experiment was chosen to be run only on the representative tasks for many samples (diabetes and spam), features (spam) and classes (iris).

Note that only automated tuning was performed for the pruned crowding ESVM and results can be found in Table 14.

SPO revealed that for the crowding variant, some parameter interactions dominate the best performing configurations: For larger population sizes, smaller mutation steps and larger crossover probabilities are better suited, and with greater run lengths, performance increases with larger mutation step sizes. For the original pruned variant, no such clear interactions could be attained. However, in both cases, many good configurations were detected.

Table 15 Manually tuned parameter values for the all-in-one pruned representation

	Diabetes	Iris	Boston	Spam
ps	100	50	100	5
ng	250	280	2000	480
pc	0.4	0.9	0.5	0.1
pm	0.4	0.9	0.5	0.1
hpm	0.4	0.9	0.9	0.1
ms	0.1	1	0.1	3.5
hms	0.5	0.1	0.1	0.1

4.5   Integration of SVM Hyperparameters

For practical considerations, a procedure for a dynamic choice of model hyperparameters was further included within the pruned construction. Having judged from performed experiments, the parameter expressing the penalty for errors C seemed of no significance within the ESVM technique; it was consequently dropped from the parameters pool. Further on, by simply inserting one more variable to the genome, the kernel parameter (p or $sigma$) could also be evolved. In this way, benefiting from the evolutionary solving of the primal problem, model selection was actually performed at the very same time.

The idea was tested through an immediate manual tuning of the EA parameters; the values are depicted in Table 15. For reasons of generality with respect to the new genomic variable, an extra mutation probability (hpm) and mutation strength (hms) respectively, were additionally set. The corresponding gene also had a continuous encoding, the hyperparameter being rounded at kernel application. The soybean task was not considered for this experiment anymore, as very good results had already been achieved.

The resulting values for the SVM hyperparameters were identical to our previous manual choice (Table 2), with one exception in the diabetes task, where sometimes a linear kernel is obtained.

Results of the all-inclusive technique (Table 16), similar in accuracy or regression error to the prior ones and obtained at no additional cost, sustain the inclusion of model selection and point to the next extension, the coevolution of nonstandard kernels.

4.6   ESVMs Versus SVMs

In order to validate the aim of this work, that is to offer a simpler, yet equally performing alternative to SVM training, this section compares the ESVM results with those of canonical SVMs run in R on the same data sets. The reasons for this choice of a contrast were twofold: The R software already

Table 16 Accuracy/RMSE of the manually tuned all-in-one pruned ESVM version on the considered test sets, in percent

	Average	Worst	Best	StD
Diabetes (overall)	**74.20**	66.66	80.21	3.28
Diabetes(ill)	46.47	36.99	63.08	6.92
Diabetes (healthy)	89.23	81.40	94.62	3.46
Iris (overall)	**96.45**	93.33	100	1.71
Spam (overall)	**88.92**	85.39	91.48	1.5
Spam (spam)	79.98	68.72	94.67	5.47
Spam (non spam)	94.79	84.73	96.91	2.22
Boston	**5.06**	6.19	3.97	0.5

Table 17 Accuracy/RMSE of canonical SVMs on the considered test sets, in percent, as compared to those obtained by ESVM and p-values from a Wilcoxon rank-sum test

	SVM	StD	ESVM	StD	p-value
Diabetes	**76.82**	1.84	**77.31**	2.45	0.36
Iris	**95.33**	3.16	**96.45**	2.36	0.84
Spam	**92.67**	0.64	**91.04**	0.80	0.09
Soybean	**92.22**	9.60	**99.80**	1.06	3.98×10^{-5}
Boston	**3.82**	0.7	**4.78**	0.59	1.86×10^{-6}

contains a standard package for a SVM implementation and objectivity is achieved only in similar experimental setup and test cases. However, search for the outcome of the application of other related approaches (as described in subsection 3.2) on the same data sets revealed only results on the diabetes task: A classification accuracy of 74.48% and a standard deviation of 4.30% came out of a 20-fold cross-validation within evolution of Lagrange multipliers in [32] and an accuracy of 76.88% and a standard deviation of 0.25% averaged over 20 trials was obtained through the evolution of the SVM hyperparameters in [15].

The results, obtained after 30 runs of holdout cross-validation, are illustrated in Table 17. After having performed manual tuning for the SVM hyperparameters, the best results were obtained as in the corresponding row of Table 2. It is worthy to note a couple of differences between our ESVM and the SVM implementation: In the Boston housing case, despite the employment of a linear kernel in the former, the latter produces better results for a radial function, while, in the diabetes task, the ESVMs employ a degree two polynomial and SVMs target it linearly.

The results for each problem were compared via a Wilcoxon rank-sum test. The p-values (see Table 17) suggest to detect significant differences only in the cases of Soybean and Boston data sets. However, the absolute difference is not large for Boston housing, rendering SVM a slight advantage. It may be more relevant for the Soybean task, where ESVM is better.

Although, in terms of accuracy, the ESVM approach had not achieved better results for some of the test problems, it has many advantages: The decision surface is always transparent even when working with kernels whose underlying transformation to the feature space cannot be determined. The simplicity of the EA makes the solving process easily explained, understood, implemented and tuned for practical usage. Most importantly, any function can be used as a kernel and no additional constraints or verifications are necessary.

From the opposite perspective, the training is relatively slower than that of SVM, as the evaluation always relates to the training data. However, in practice (often, but not always), it is the test reaction that is more important. Nevertheless, by observing the relationship between each result and the corresponding size of the training data, it is clear that SVM performs better than ESVM for larger problems; this is probably due to the fact that, in these cases, much more evolutionary effort would be necessary. The problem of handling large data sets is thus worth investigating deeper in future work.

5   Conclusions and Outlook

The evolutionary learning technique proposed in this chapter resembles the vision upon learning of SVMs but solves the inherent optimization problem by means of an EA. An easier and more flexible alternative to SVMs is put forward and undergoes several enhancements in order to provide a viable alternative to the classical paradigm. These developments are summarized below:

- Two possible representations for the EA (one simpler, and a little faster, and one more complicated, but also more accurate) that determines the coefficients are imagined.
- In order to boost the suitability of the new technique for any issue, a novel chunking mechanism for reducing size in large problems is also proposed; obtained results support its employment.
- The use of a crowding-based EA is inspected in relation to the preservation of performance. Crowding would be highly necessary in the immediate coevolution of nonstandard kernels.
- Finally, an all-inclusive ESVM construction for the practical perspective is developed and validated.
- On a different level, an additional aim was to address and solve real-work tasks of high importance.

Several conclusions can be eventually drawn and the potential of the technique can be further strengthened through the application of two enhancements:

- As opposed to SVMs, ESVMs are much easier to understand and use.
- ESVMs do not impose any kind of constraints or requirements.
- Moreover, the evolutionary solving of the optimization problem enables the acquirement of function coefficients directly and at all times within a run.
- SVMs, on the other hand, are somewhat faster, as the kernel matrix is computed only once.
- Performances are comparable, for different test cases ESVMs and SVMs take the lead, in turn.

Although already a suitable alternative, the novel ESVM can still be enhanced in several ways:

- The requirement for an optimal decision function actually involves two criteria: the surface must fit to training data but simultaneously generalize well. So far, these two objectives are combined in a single fitness expression. As a better choice for handling these conditions, a multicriterial approach could be tried instead.
- Additionally, the simultaneous evolution of the hyperplane and of nonstandard kernels will be achieved. This approach is highly difficult by means of SVM standard methods for hyperplane determination, whereas it is straightforward for ESVMs. A possible combination can be achieved through a cooperative coevolution between the population of hyperplanes and that of GP-evolved kernels.

Appendix - Definition of Employed Concepts

There are a series of notions that appear throughout the chapter. Their meaning, use and reference are explained in what follows:

Evolutionary algorithm (EA) [18] - a metaheuristic optimization algorithm in the field of artificial intelligence that finds its inspiration in what governs nature: A population of initial individuals (genomes) goes through a process of adaptation through selection, recombination and mutation; these phenomena encourage the appearance of fitter solutions. The best performing individuals are selected in a probabilistic manner as parents of a new generation and gradually the system evolves to the optimum. The fittest individual(s) obtained after a certain number of iterations is (are) the solution to the problem.

Support vector machine (SVM) [2] - a supervised learning method for classification and regression: Given a set of samples, the method aims for the optimal decision hyperplane to model the data and establish an equilibrium between a good training accuracy and a high generalization ability; according to [36], a possible definition of an SVM could be "a system for efficiently training linear learning machines in kernel-induced feature spaces, while respecting the insights of generalization theory and exploiting optimization theory".

Classification / regression hyperplane - the decision hyperplane whose defining coefficients must be determined; within classification, it must differentiate between samples of different classes, while, with respect to regression, it represents the surface on which the data are restrained to be positioned.

multi (k)-class SVM [10] - SVMs are implicitly built for binary classification tasks; for problems with k outcomes, $k > 2$, the technique considers the labels and corresponding samples two by two and uses common approaches like one-against-one and one-against-all to combine the obtained classifiers.

Primal problem - the direct form of the optimization task within SVMs of the determination of the decision hyperplane, while balancing between accuracy and generalization capacity. It is **dualized** in the standard SVM solving by Lagrange multipliers.

Kernel - a function of two variables that defines the scalar product between them; within SVMs, it is employed for the "kernel trick" - a technique to write a nonlinear operator as a linear one in a space of higher dimension as a result of Mercer's theorem.

Crowding-based EA [35] - a technique that was introduced as a method of maintaining diversity: new obtained individuals replace only similar individuals in the population: A percentage G (*generation gap*) of the individuals is chosen via fitness proportional selection in order to create an equal number of offspring; for each of these offspring, CF (a parameter called *crowding factor*) individuals from the current population are randomly selected – the offspring then replaces the most similar individual from these.

References

1. Bosch, R.A., Smith, J.A.: Separating hyperplanes and the authorship of the disputed federalist papers. American Mathematical Monthly 105, 601–608 (1998)
2. Vapnik, V.: The Nature of Statistical Learning Theory. Springer, Heidelberg (1995)
3. Haykin, S.: Neural Networks: A Comprehensive Foundation. Prentice-Hall, New Jersey (1999)
4. Cortes, C., Vapnik, V.: Support vector networks. Machine Learning 1, 273–297 (1995)
5. Mierswa, I.: Evolutionary learning with kernels: A generic solution for large margin problems. In: Proc. of the Genetic and Evolutionary Computation Conference, vol. 1, pp. 1553–1560 (2006)

6. Burges, C.J.C.: A tutorial on support vector machines for pattern recognition. Data Mining and Knowledge Discovery 2, 121–167 (1998)
7. Boser, D.E., Guyon, I.M., Vapnik, V.: A training algorithm for optimal margin classifiers. In: Proceedings of the 5th Annual ACM Workshop on Computational Learning Theory, vol. 1, pp. 11–152 (1992)
8. Courant, R., Hilbert, D.: Methods of Mathematical Physics. Wiley Interscience, Hoboken (1970)
9. Mercer, J.: Functions of positive and negative type and their connection with the theory of integral equations. Transactions of the London Philosophical Society (A) 209, 415–446 (1908)
10. Hsu, C.-W., Lin, C.-J.: A comparison of methods for multi-class support vector machines. IEEE Transactions on Neural Networks 13, 415–425 (2004)
11. Platt, J.C., Cristianini, N., Shawe-Taylor, J.: Large margin dags for multi-class classification. Proc. of Neural Information Processing Systems 1, 547–553 (2000)
12. Smola, A.J., Scholkopf, B.: A tutorial on support vector regression. Technical Report NC2-TR-1998-030. NeuroCOLT2 Technical Report Series (1998)
13. Rosipal, R.: Kernel-based Regression and Objective Nonlinear Measures to Access Brain Functioning. PhD thesis Applied Computational Intelligence Research Unit School of Information and Communications Technology University of Paisley, Scotland (2001)
14. Stoean, R.: Support vector machines. An evolutionary resembling approach. Universitaria Publishing House Craiova (2008)
15. Friedrichs, F., Igel, C.: Evolutionary tuning of multiple svm parameters. In: Proc. 12th European Symposium on Artificial Neural Networks, vol. 1, pp. 519–524 (2004)
16. Howley, T., Madden, M.G.: The genetic evolution of kernels for support vector machine classifiers. In: Proc. of 15th Irish Conference on Artificial Intelligence and Cognitive Science 1 (2004),
    http://www.it.nuigalway.ie/m_madden/profile/pubs.html
17. Jun, S.H., Oh, K.W.: An evolutionary statistical learning theory. International Journal of Computational Intelligence 3, 249–256 (2006)
18. Eiben, A.E., Smith, J.E.: Introduction to Evolutionary Computing. Springer, Heidelberg (2003)
19. Stoean, R., Preuss, M., Stoean, C., Dumitrescu, D.: Concerning the potential of evolutionary support vector machines. In: Proc. of the IEEE Congress on Evolutionary Computation, vol. 1, pp. 1436–1443 (2007)
20. Stoean, R., Dumitrescu, D., Preuss, M., Stoean, C.: Different techniques of multi-class evolutionary support vector machines. Proc. of Bio-Inspired Computing: Theory and Applications 1, 299–306 (2006)
21. Stoean, R., Stoean, C., Preuss, M., Dumitrescu, D.: Evolutionary multi-class support vector machines for classification. In: Proceedings of International Conference on Computers and Communications - ICCC 2006, Baile Felix Spa - Oradea, Romania, vol. 1, pp. 423–428 (2006)
22. Stoean, R., Dumitrescu, D., Stoean, C.: Nonlinear evolutionary support vector machines. application to classification. Studia Babes-Bolyai, Seria Informatica LI, pp. 3–12 (2006)
23. Stoean, R., Dumitrescu, D.. Evolutionary linear separating hyperplanes within support vector machines. Scientific Bulletin, University of Pitesti, Mathematics and Computer Science Series 11, 75–84 (2005)

24. Stoean, R., Dumitrescu, D.: Linear evolutionary support vector machines for separable training data. Annals of the University of Craiova, Mathematics and Computer Science Series 33, 141–146 (2006)
25. Stoean, R., Preuss, M., Dumitrescu, D., Stoean, C.: ϵ - evolutionary support vector regression. In: Symbolic and Numeric Algorithms for Scientific Computing, SYNASC 2006, vol. 1, pp. 21–27 (2006)
26. Stoean, R., Preuss, M., Dumitrescu, D., Stoean, C.: Evolutionary support vector regression machines. In: IEEE Postproc. of the 8th International Symposium on Symbolic and Numeric Algorithms for Scientific Computing, vol. 1, pp. 330–335 (2006)
27. Stoean, R.: An evolutionary support vector machines approach to regression. In: Proc. of 6th International Conference on Artificial Intelligence and Digital Communications, vol. 1, pp. 54–61 (2006)
28. Stoean, R., Stoean, C., Preuss, M., El-Darzi, E., Dumitrescu, D.: Evolutionary support vector machines for diabetes mellitus diagnosis. In: Proceedings of IEEE Intelligent Systems 2006, London, UK, vol. 1, pp. 182–187 (2006)
29. Stoean, R., Stoean, C., Preuss, M., Dumitrescu, D.: Evolutionary support vector machines for spam filtering. In: Proc. of RoEduNet IEEE International Conference, vol. 1, pp. 261–266 (2006)
30. Stoean, R., Stoean, C., Preuss, M., Dumitrescu, D.: Evolutionary detection of separating hyperplanes in e-mail classification. Acta Cibiniensis LV, 41–46 (2007)
31. Stoean, R., Stoean, C., Preuss, M., Dumitrescu, D.: Forecasting soybean diseases from symptoms by means of evolutionary support vector machines. Phytologia Balcanica 12 (2006)
32. Mierswa, I.: Making indefinite kernel learning practical, technical report. Technical report. Artificial Intelligence Unit, Department of Computer Science, University of Dortmund (2006)
33. Bartz-Beielstein, T.: Experimental research in evolutionary computation - the new experimentalism. Natural Computing Series. Springer, Heidelberg (2006)
34. Perez-Cruz, F., Figueiras-Vidal, A.R., Artes-Rodriguez, A.: Double chunking for solving svms for very large datasets. In: Proceedings of Learning 2004, Elche, Spain 1 (2004),
    eprints.pascal-network.org/archive/00001184/01/learn04.pdf
35. DeJong, K.A.: An Analysis of the Behavior of a Class of Genetic Adaptive Systems. PhD thesis University of Michigan, Ann Arbor (1975)
36. Cristianini, N., Shawe-Taylor, J.: An Introduction to Support Vector Machines. Cambridge University Press, Cambridge (2000)

Evolutionary Computing in Statistical Data Analysis

Roberto Baragona and Francesco Battaglia

Abstract. Evolutionary computing methods are being used in a wide field domain with increasing confidence and encouraging outcomes. We want to illustrate how these new techniques have influenced the statistical theory and practice concerned with multivariate data analysis, time series model building and optimization methods for statistical estimates computation and inference in complex systems. The distinctive features all these subject topics have in common are the large number of alternatives for model choice, parametrization over high dimensional discrete spaces and lack of convenient properties that may be assumed to hold at least approximately about the data generating process. Evolutionary computing proved to be able to offer a valuable framework to deal with complicated problems in statistical data analysis and time series analysis and we shall draw a wide though by no means exhaustive list of topics of interest in statistics that have been successfully handled by evolutionary computing procedures. Specific issues will be concerned with variable selection in linear regression models, non linear regression, time series model identification and estimation, detection of outlying observations in time series as regards both location and type identification, cluster analysis and grouping problems, including clusters of directional data and clusters of time series. Simulated examples and applications to real data will be used for illustration purpose through the chapter.

1 Introduction

Evolutionary computing is a general approach for simulating evolution on computers. In a statistical framework, evolutionary computing provides a

Roberto Baragona
Sapienza University of Rome Department of Sociology and Communication Via Salaria 113 Rome Italy
e-mail: roberto.baragona@uniroma1.it

Francesco Battaglia
Sapienza University of Rome Department of Statistics Probability and Applied Statistics Piazzale Aldo Moro 5 00100 Rome Italy
e-mail: francesco.battaglia@uniroma1.it

A. Abraham et al. (Eds.): Foundations of Comput. Intel. Vol. 3, SCI 203, pp. 347–386.
springerlink.com   © Springer-Verlag Berlin Heidelberg 2009

class of methods useful for identification, estimation, validation and prediction of models that describe relationships of interest among variables linked to real world data sets. Many evolutionary computing methods are referenced in [41] but the usual nowadays classification includes evolutionary programming, evolution strategies, genetic algorithms, estimation of distribution algorithms, differential evolution. The difference between these methods is often subtle, the unifying framework consists in assuming that several potential solutions exist that may solve a problem but the optimal one has to be discovered through an evolution process. Such a process develops along the same guidelines that drive the natural adaptation to the environment typical of the biological populations. So the optimal solution may not even be present in the set of solutions that are considered at the beginning. It is built gradually by selection, recombination and mutation of the solutions that enter the population in several iterative steps usually called generations. In this chapter we will focus on genetic algorithms (GAs) and to a minor extent to the estimation of distribution algorithms (EDAs). For an introduction to GAs see, for instance, [65] and [73], and refer to [61] for EDAs. GAs-based methods have been proposed often to solve problems in the field of statistics. Obviously we may find countless applications in the wide framework of the artificial intelligence (AI) and methods developed for AI problems have found their application in statistics as well. The meta-heuristic methods are general purpose heuristics that include as special case the GAs and the other evolutionary computing methods but extend to cover methods such as threshold accepting (TA), simulated annealing (SA), tabu search (TS) and many others ([85, 86]). Often hybridization has been proposed among meta-heuristic methods to exploit useful features of interest for the problem at hand (see [47] for an example of hybrid algorithm which combines TS and GAs, and [88] who considered SA and GAs).

As far as GAs-based heuristics are concerned, [29], [28] and [68] discussed several applications in statistics, for instance the variable selection and parameter estimation in linear regression model. The standard errors of the estimates are approximately evaluated by processing with the GA several bootstrap samples from the data. Applications of genetic algorithms were proposed as well for component and discriminant analysis ([76]), graphical model identification ([75]), model selection ([2]), subset regression ([55]), crisp and fuzzy cluster analysis ([69], [63], [54]), outlier identification for independent data ([36]), non linear optimization to determine the wavelet filter for M-band wavelet decomposition scheme ([34]), exponential power distribution parameter estimation ([83]) and mixture models investigation ([10]). Promising applications of GAs have been proposed in statistical sampling ([56]) and design of experiments ([27], [42], [51]). In machine vision and pattern recognition framework cluster algorithms were proposed based on SA by [6] and on GAs by [3], [4] and [5].

A comprehensive account on meta-heuristic methods for time series analysis may be found in [9]. Special topics for applications are autoregressive moving average (ARMA) model identification ([72]), subset ARMA (SARMA) model identification ([43], [64]) and subset vector autoregressive (VAR) model ([17]), cluster of time series ([7, 8]), outlier detection in time series ([14]), threshold autoregressive (TAR) model identification ([87]), modeling structural breaks ([37, 38]), identification of transfer function models ([33]).

The paper is organized as follows. In the next section 2 the GAs and EDAs will be outlined in some details. The remaining sections will illustrate examples of different implementations of GAs depending on the problem. Section 3 is devoted to present two examples of variable selection in multivariate linear regression and in time series SARMA models. Parameter estimation with GAs and EDAs will be considered in section 4 using as an illustration a logistic regression model on the coronary heart disease data taken from [52]. In section 5 comparison is made between meta-heuristic methods and algorithms based on gradient optimization and indirect inference methods for the exponential autoregressive (EXPAR) time series model. Then the identification and estimation of time series threshold models is considered concerned with threshold autoregressive (TAR) models and double threshold autoregressive heteroscedastic (DTARCH) and double threshold generalized ARCH (DT-GARCH) models. The outlier detection problem is accounted for in section 6. In section 7 an example of GAs-based cluster analysis is discussed concerned with time series data and directional data in comparison with TA, SA and TS methods. Section 8 outlines directions for further research and conclusions are drawn.

2 GAs and EDAs Implementations

In the next sections we shall illustrate methods essentially based on GAs. A special attention will be reserved to EDAs which is in the same domain of the GAs and similar in many aspects. For GAs in particular there is a general schema that may be implemented in several ways. The more common will be outlined in this section.

2.1 The GAs Procedure

The basic GAs procedure is usually illustrated by assuming that the potential solutions are encoded as binary strings. A real number $x \in (a, b)$ may be encoded as a binary string of length ℓ as follows

$$x = a + c(b - a)/(2^\ell - 1), \tag{1}$$

where c is the non negative integer that may be decoded from the binary string. Given a positive integer s, a set of s binary strings $(i_1, i_2, \ldots, i_\ell)$ of pre-specified length ℓ are generated at random. The s strings form the "initial population." This latter is a subset of the set of all admissible solutions. Let the initial population be the current population, and perform N iterations in each of which the s strings are allowed to change according to the "genetic operators" selection, crossover and mutation.

1. *Selection*. The objective function to be optimized is called in this context the "fitness function" and it is taken to be maximized. If the problem requires that the objective function f has to be minimized, then we may define the fitness function as $f^* = \exp(-f)$, for instance. Then, for s times, a string in the current population is chosen (with replacement) with probability proportional to its fitness function. We obtain this way s strings that are taken to replace the current population.

2. *Crossover*. For the sake of simplicity, let us assume that s is an even integer. Then, the strings in the current population are paired at random to form $s/2$ pairs. Each pair is examined in turn, and crossover takes place if $U < p_c$, where p_c is a pre-specified crossover probability and U is a uniform random number in the interval $(0, 1)$. The crossover acts as follows. (i) An integer j is chosen uniformly randomly in the interval $(1, \ell - 1)$. The number j is called the "cutting point." (ii) The bits from $j + 1$ to ℓ are exchanged between the strings that are paired.

3. *Mutation*. All chromosomes in the population are allowed to change their values with probability p_m. The choice of p_m was proven to influence the performance of the GA considerably. A high p_m may serve to maintain the diversity between the individuals into the population, but this is likely to produce premature convergence as well. The probability p_m is usually assumed rather small in the interval $(0.001, 0.1)$. A popular rule consists in choosing p_m equal to $1/\ell$, where ℓ is the chromosome length.

The "elitist strategy" (see [73] for motivation) applies, that is the best string that may be found in each of the N iterations is always maintained in the current population unless an even better string appears due to the genetic operators. If this latter is not the case, then such string replaces the string with the worst fitness function. This replacement is done so that the population retains the same number of strings s in each iteration. At the end of each iteration the strings obtained by using the genetic operators replace all the existing strings, and the new population is assumed as the current one. The best string in the population after N iterations is taken as the final solution.

2.2 The EDAs Procedure

These algorithms are best explained, and were originally derived, in the case that the chromosomes are real vectors $x = (x_1, x_2, \ldots, x_\ell)'$, though they have been extended to more general settings. In the real vector case, the problem may be formulated as that of maximizing a fitness function $f(x)$ where x is a real vector $x \in R^\ell$.

The proposal originates from the attempt of explicitly taking into account the correlation between genes of different loci (components of the vector x), that may be seen in good solutions, assuming that such correlation structure could be different from that of the less fitted individuals. The key idea is to deliver an explicit probability model and associate to each population (or a subset of it) a multivariate probability distribution.

An initial version of the estimation of distribution algorithm was originally proposed by [66], and then many further contributions developed, generalized and improved the implementation. A thorough account may be found in [57] and in a second more recent book ([61]).

The estimation of distribution algorithm is a regular stochastic population based evolutionary method, and therefore evolves populations through generations. The typical evolution process from one generation to the next may be described as follows:

1. Generate an initial population $P^{(0)} = \{x_i^{(0)}, i = 1, \ldots, N\}$; $c = 0$.
2. If $P^{(c)}$ denotes the current population, select a subset of $P^{(c)}$: $\{x_j^{(c)}, j \in S^{(c)}\}$ with $|S^{(c)}| = n < N$ individuals, according to a selection operator.
3. Consider the subset $\{x_j^{(c)}, j \in S^{(c)}\}$ as a random sample from a multivariate random variable with absolutely continuous distribution and probability density $p^{(c)}(x)$, and estimate $p^{(c)}(x)$ from the sample.
4. Generate a random sample of N individuals from $p^{(c)}(x)$: this is the population at generation $c + 1$, $P^{(c+1)}$.
5. If a stopping rule is met, stop; otherwise $c + 1 \to c$ and return to 2.

The originally proposed selection operator was the truncation selection, in other words only the n individuals with the largest fitness out of the N members of the population are selected. Later, it was proposed that other common selection mechanism such as the roulette wheel (proportional selection) or the tournament selection (choice of the best fitted inside a group of k individuals chosen at random) may be adopted.

3   Variable Selection in Linear Regression and ARMA Models

A typical issue in linear model identification is variable selection. A linear relationship is postulated between a dependent variable y and a set of independent variables $\{x_1, x_2, \ldots, x_p\}$. Let n observations be available so that we may write the usual linear regression model

$$y_i = \beta_0 + \beta_1 x_{1i} + \beta_2 x_{2i} + \ldots + \beta_p x_{pi} + u_i, \qquad i = 1, \ldots, n,$$

where $\beta = (\beta_0, \beta_1, \ldots, \beta_p)'$ is the parameter vector and $u = (u_1, \ldots, u_n)'$ is a sequence of independent and identically distributed random variables with zero mean and unknown variance σ_u^2. Let $y = (y_1, \ldots, y_n)'$ and

$X = [1, x_1, x_2, \ldots, x_p]$ where 1 denotes a column vector of ones. The linear regression model may be written in matrix form

$$y = X\beta + u.$$

If p is large it is desirable to reduce the number of independent variables to the set that includes only the variables really important to explain the variability of y. Common available methods are the iterative forward, backward and stepwise methods. However, we would be more confident about the final result if we could compare several subset alternative simultaneously. The GAs are an easy-to-use tool for performing this comparison exactly.

3.1 Subset Regression

We want to select the variables that really matter in a linear regression. The encoding for using GAs is straightforward as it suffices to define a mapping from a binary string of length p and the parameter sequence $\beta_1, \ldots, \beta_p$. The constant term β_0 is always included in the regression.

Let us illustrate a GAs-based procedure for subset regression on an example of $n = 100$ artificial data and a set of $p = 15$ variables. Both independent and dependent variables are sampled from a standard unit normal distribution. The y are generated by a model with parameters

$$\beta = (.01, 0.7, -0.8, 0.5, .01, .01, .01, -0.7, 0.6, 0.8, .01, .01, .01, .01, .01, .01)'$$

and $\sigma_u^2 = 2.25$. It is apparent that only the variables $1 - 3$ and $7 - 9$ impact y significantly.

A GA has been employed to search for the best subset model. The fitness function has been the F statistic. The chromosome is a binary string of length 15, for instance

$$000110000011100$$

is decoded to a regression model that includes only the variables $4, 5, 11, 12, 13$ as independent variables to explain y. The population size has been chosen $s = 30$, the maximum number of generations $N = 100$, the generational gap (how many new chromosomes are created) has been set equal to 90% of the past population, $p_c = 0.7$ and $p_m = 1/15$. Roulette wheel rule for selection, single cutting point crossover, binary mutation and the elitist strategy are employed. The fitness function evolution is displayed in figure 1. This is the typical fitness function behavior in the presence of elitist strategy. While fitness function improves quickly in the first iterations, then no better solutions are found and the elitist strategy prevents the fitness function from decreasing. The GA finds the best solution corresponding to $F = 29.8941$ and variables $1 - 3$ and $7 - 9$.

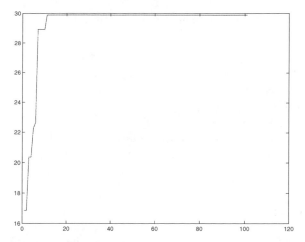

Fig. 1 Fitness function in the subset regression problem versus the number of iterations

A GAs approach to subset regression based on information criteria has been suggested by [55] while GAs have been suggested in [78] to evaluate the bias in parameter estimates produced by omitting variables from the regression model.

3.2   Autoregressive Moving Average Models

Models of the ARMA and ARIMA class are most popular in time series analysis essentially for two reasons: first, they are a natural generalization of regression models, and may be easily interpreted using similar concepts as in regression analysis; and, second, they may be seen as universal approximation for a wide class of well behaved stationary stochastic processes.

An ARMA model may be written

$$x_t - \phi_1 x_{t-1} - \phi_2 x_{t-2} - \ldots - \phi_p x_{t-p} = c + u_t - \theta_1 u_{t-1} - \theta_2 u_{t-2} - \ldots - \theta_q u_{t-q} \quad (2)$$

where $c = \mu(1 - \phi_1 - \ldots - \phi_p)$. Equation (2) is called an ARMA(p, q) model and p and q are known as the autoregressive and the moving average *orders* of the model. If the observed time series $\{x_t\}$ is not stationary, while, for some positive integer d, the differenced series $\{y_t\}$ of order d is stationary, then we have an ARIMA model by replacing x_t with y_t and setting $c = 0$ in (2).

Parsimony is universally accepted as precept among time series analysts, so that the ARMA model building problem may be seen as choosing the model with the smallest number of parameters given an approximation level. An additional way of reducing the number of parameters is considering incomplete models, where some of the parameters $\phi_1, \ldots, \phi_p, \theta_1, \ldots, \theta_q$ are constrained to zero. Such models are usually referred to as *subset ARMA* models ([30]).

The canonical model building procedure runs iteratively through the following three steps:

1. *Identification.* Selection of the orders p and q, and, if a subset model is considered, choice of which parameters are allowed to be non-zero.
2. *Parameter estimation.* Conditional on identification, the estimation of parameters may be performed through classical statistical inference.
3. *Diagnostic checking.* Once the model is completely specified and estimated, it is customary to check whether it fits the data sufficiently well. This is generally accomplished by computing the residuals. A model is generally accepted provided that the residual are approximately uncorrelated and have zero mean and constant variance.

The most difficult step of the model building procedure is identification. Two different codings have been proposed, in either case a maximum search order has to be selected, both for the autoregressive part (P say) and for the moving average part (Q). The simplest coding amounts to reserving one gene to each possible lag, filling it with 1 if the parameter is free, and with 0 if the parameter is constrained to zero. For example, if we take $P = Q = 6$, the following subset model:

$$x_t = \phi_1 x_{t-1} + \phi_4 x_{t-4} + \phi_5 x_{t-5} + u_t - \theta_2 u_{t-2} - \theta_4 u_{t-4} - \theta_6 u_{t-6}$$

is coded by means of the following chromosome

$$\underbrace{100110}_{\text{AR lags}} \qquad \underbrace{010101}_{\text{MA lags}}$$

This coding system was adopted by most authors, and has the advantage of simplicity and fixed-length chromosomes, but it is not particularly efficient.

An alternative coding based on variable-length chromosomes, was proposed by [64]. The chromosomes consist of two different gene subsets: the first one is devoted to encode the autoregressive and moving average orders, by specifying the number of relevant predictors, i. e., the number of non-zero parameters, respectively for the autoregressive part, p^*, and for the moving average part, q^*. This part consists of eight binary digits and encodes two integer numbers between 0 and 15 (therefore this coding allows for models that have up to 15 non-zero autoregressive parameters and 15 non-zero moving average parameters). The other genes subset is devoted to specifying the lags which the non zero parameters correspond to: it comprises $5(p^* + q^*)$ binary digits, and encodes, consecutively, $p^* + q^*$ integer numbers between 1 and 32. Therefore according to this implementation $P = Q = 32$ and all models containing a maximum of 15 non zero parameters, both for the AR and the MA structures, may be coded. For example. the chromosome corresponding to the previous model is:

$$0011 \ 0011 \quad 00001 \ 00100 \ 00101 \ 00010 \ 00100 \ 00110 \ .$$

The first part of the chromosome contains 8 digits, while the second part has a variable length from zero (white noise) to 150 binary digits. Encoding with the same order limitations would require, in the fixed-length scheme introduced before, 64-digit chromosomes. It appears that the advantage of using the variable length scheme might be sensible if we are searching for relatively sparse models, because the average length depends directly on the number of non zero parameters (for example, a coding admitting no more that 4 non zero parameters in each AR or MA structure, with a maximum lag of 64, would require a fixed chromosome 128 long, and a variable length chromosome with 6 to 54 binary digits).

In any case, the structure of the genetic algorithm depends on the choice of a maximum possible order: it may be based on a-priori considerations, or even on the number of available observations. A more data-dependent choice of the maximum admissible lag could also be based on an order selection criterion computed only on complete models (with all parameters free) as proposed in [22]: in this case a relatively "generous" criterion like the asymptotic information criterion (AIC) (see, for instance, [84], p. 153) seems advisable to avoid reducing the solution space excessively.

Since all proposed codings are binary, each proposal corresponds to usually slight modifications of the canonical genetic algorithm. The selection procedure is obtained by means of the roulette wheel methods, except for [17] who adopt rank selection. The mutation operator is employed in its standard form, while for the cross-over operator some authors propose a random one cut point ([43]), others a random two point cross-over ([67]) or a uniform cross-over ([17]). All authors use unit generational gap and slightly different elitist strategies: saving the best chromosome, or the two best chromosomes, or even the 10% best chromosomes in the population. When dealing with the alternative coding based on subsets of genes that encode the AR and MA order, [64] suggest that the cross-over operator should be modified in order to apply on entire fields (representing the integer numbers denoting order or lag) rather to the single binary digits. Mutation and cross-over probabilities are generally in accordance with the literature on canonical genetic algorithms, proposed values are between 0.001 and 0.1 for mutation, and from 0.6 to 0.9 for cross-over. Not many suggestions are offered concerning the population size, and the chromosomes composing the initial generation (they are generally selected at random in the solution space). It may be reasonably assumed that a good starting point, with satisfying mixing properties, would require initial individuals which may exploit the fitting ability of each single possible parameter. Therefore, advantageous chromosomes in the initial population are those encoding models with just one non zero parameter (i. e., in the first coding scheme, chromosomes with only one gene equal to one). The minimum population size that allows to fully develop this idea is obviously equal to the total number of possible parameters, or $P + Q$.

Much more relevant differences are found in the literature on ARMA model building by means of genetic algorithms, as far as the fitness function is concerned. Essentially, the fitness function is linked to some version of identification criterion or penalized likelihood. A favorite choice is adopting one of the most popular identification criteria in time series, such as the AIC or the BIC (also called Schwartz) criteria:

$$AIC(M) = N \log\{\hat{\sigma}^2_{(M)}\} + 2p(M)$$

$$BIC(M) = N \log\{\hat{\sigma}^2_{(M)}\} + p(M) \log N$$

where N is the series length, $\hat{\sigma}^2_{(M)}$ is the residual variance estimate for model M, and $p(M)$ is the number of free parameters of model M. Alternatively, [23] suggest their identification criterion called $ICOMP$. Rather than simply considering the number of non zero parameters, $ICOMP$ measures the model complexity through a generalization of the entropic covariance complexity index of [82]. The criterion $ICOMP$ is computed by estimating the Fisher's information matrix for the parameters and by adding to the likelihood- proportional term, $N \log \hat{\sigma}^2$, the quantity $C(\hat{F}^{-1})$:

$$ICOMP(M) = N \log \hat{\sigma}^2_{(M)} + C(\hat{F}^{-1})$$

where
$$C(X) = \dim(X) \log\{\text{trace}(X)/\dim(X)\} - \log|X|.$$

Finally, [64] use an AIC-like criterion where the residual variance is replaced by a prediction error variance:

$$s^2(\ell) \propto \sum_t [x_{t+\ell} - \hat{x}_t(\ell)]^2$$

where $\hat{x}_t(\ell)$ is the predictor based on the model. Obviously, for the forecast horizon $\ell = 1$ there is no difference with AIC, [64] try their criterion also for $\ell = 2$ and 3.

A common problem to all these implementations is that the proposed criteria are to be minimized, thus they cannot be employed directly as fitness function (which has, on the contrary, to be maximized). Solution of two kinds have been proposed: [23] avoid the problem of defining a fitness proportionate selection by adopting an ordered fitness selection rule: chromosomes are ordered according to the decreasing values of the $ICOMP$ criterion, and each chromosome is selected with a probability proportional to its rank. An alternative is defining the fitness function by a monotonically decreasing transformation of the identification criterion. Most natural candidates are a simple linear transformation:

$$fitness(M) = W - criterion(M)$$

which is possible if an estimate of the worst possible value of the criterion, W, is available ([43] suggests to compute W on the current population), or a negative exponential transformation:

$$fitness(M) = \exp\{-criterion(M)/d\}$$

where d is a scaling constant. A Boltzman selection procedure is proposed by [43] using the last expression for the fitness, but with a progressively decreasing "temperature" $d_k = (0.95)^k$, where k is the generation number.

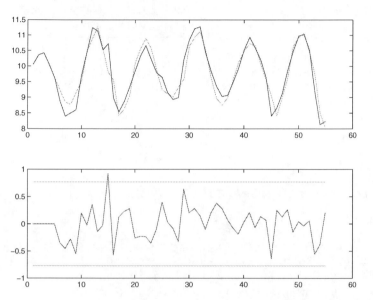

Fig. 2 Yearly number of lynx pelts sold by Hudson's Bay Company in Canada from 1857 to 1911 (top panel) and residuals from the best subset ARMA model (bottom panel)

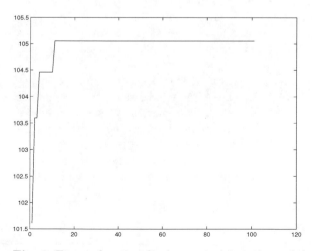

Fig. 3 Fitness function display against iterations of the genetic algorithm

Table 1 Comparison between ARMA and subset ARMA models

Model	c_0	ϕ_1	ϕ_2	ϕ_3	θ_1	θ_2	θ_3	σ_ϵ^2
ARMA	7.08	0.87	0.13	−0.73	0.15	0.35	−0.57	0.0958
	(1.30)	(0.29)	(0.46)	(0.29)	(0.32)	(0.29)	(0.17)	
SARMA	6.77	0.96		−0.64	0.23	0.27	−0.55	0.0962
	(0.70)	(0.06)		(0.06)	(0.15)	(0.15)	(0.16)	

The simulation studies presented in literature suggest satisfying results
of each implementation, with slight substantial differences, indicating that
for univariate time series with most encountered series lengths, the genetic
algorithm yields good models after just a few generations, and converges
nearly always, though after many more generations, to the true simulated
model.

3.3 An Example of Subset ARMA Fitted to a Real Data Set

Let us consider the yearly number of lynx pelts sold by Hudson's Bay Company in Canada from 1857 to 1911. The data set is composed of 55 observations and is reported in [84], p. 449, as Series W7. Following [84] (p. 150)
we use the natural logarithm of the observations for ARMA model identification. In figure 2 the logarithms of the data are displayed (solid line) in
the top panel with the predicted values (dotted line) computed from the best
subset ARMA model. In the bottom panel of figure 2 the residuals (dotted
line) are displayed with the 95% normal bounds (straight lines). The GAs are
used for identifying the best subset model. The encoding used to obtain the
best subset model is the first one. The fitness function was set equal to the
appropriate transform of the AIC criterion. In figure 3 the fitness function
evolution is displayed.

In table 1 a comparison is made between the ARMA model and the subset
ARMA model fitted to Series W7. The ARMA(3,3) model has a smaller
residual variance but its AIC (−103) is slightly greater than the subset model
(−105). The difference between the two model is concerned with a single
parameter and it seems that little is gained as for diagnostic checking values.
Nonetheless there is a considerable advantage as regards the standard errors
of the estimates which display a sharp decrease if the subset model is used.

4 The Logistic Regression Model

We give an example of application of the logistic regression model as it has
been fitted to the data by GAs by [29] and [68] and by EDAs by [74]. So it

seems of interest to use a logistic regression model to show an application of GAs and EDAs for nonlinear model parameter estimation.

Let y denote a binary dependent variable and $\{x_1, x_2, \ldots, x_p\}$ a set of independent variables. The logistic regression ([52]) assumes a non linear relationship between y, called in this context the response variable, and the covariates $x_1, x_2, \ldots, x_p$. Let Y denote a binary random variable and assume that $y = (y_1, \ldots, y_n)'$ is a sample from Y. Let π be the probability $P(Y = 1|x_1, \ldots, x_p)$ and define the logit transform

$$\text{logit}(\pi) = \log \frac{\pi}{1 - \pi}$$

As an example, we fitted a logistic regression model to a real data set, namely the coronary heart disease data from [52]. We used several algorithms to estimate the two parameters β_0 and β_1 of the logistic regression model

$$\text{logit}(\pi_i) = \beta_0 + \beta_1 x_i, \qquad i = 1, 2, \ldots, 100.$$

The observed response variable is $y_i = 1$ if insurgence of coronary heart disease in the ith patient has been recorded and $y_i = 0$ otherwise. The covariate x_i is the ith patient's age (years). The iterative re-weighted least squares (IRLS), the GAs and EDAs algorithms implemented for maximizing the logarithm of the likelihood

$$L = \sum_{i=1}^{n} y_i(\beta_0 + \beta_1 x_i) - \sum_{i=1}^{n} \log\{1 + \exp(\beta_0 + \beta_1 x_i)\}$$

are outlined as follows. Upper and lower bounds for the two parameters have been chosen $(-10, 10)$ for β_0 and $(-2, 2)$ for β_1. The input matrix is defined $X = [1, x]$ where 1 denotes a column vector of ones and the parameter vector is $\beta = [\beta_0, \beta_1]$.

- *IRLS.* This iterative algorithm implements the Newton method applied to the problem of maximizing the likelihood of a response variable y given X. Let a preliminary guess of the model parameter $\hat{\beta}^{(0)}$ be available. Then the following steps describe the move from $\hat{\beta}^{(k)}$ to $\hat{\beta}^{(k+1)}$ at iteration k.

1. Set, for $i = 1, \ldots, n$,

$$\pi_i^{(k)} = \exp\left(\hat{\beta}_0^{(k)} + \hat{\beta}_1^{(k)} x_i\right) / \left\{1 + \exp\left(\hat{\beta}_0^{(k)} + \hat{\beta}_1^{(k)} x_i\right)\right\}.$$

2. Define the weights matrix $W^{(k)} = \text{diag}(w_1^{(k)}, \ldots, w_n^{(k)})$ where $w_i^{(k)} = \pi_i^{(k)}(1 - \pi_i^{(k)})$.
3. Compute $z^{(k)} = X\hat{\beta}^{(k)} + \left(W^{(k)}\right)^{-1}(y - \pi^{(k)})$.
4. Solve with respect to $\hat{\beta}^{(k+1)}$ the weighted linear regression problem

$$\left(X'W^{(k)}X\right)\hat{\beta}^{(k+1)} = X'W^{(k)}z^{(k)}$$

5. Replace $\hat{\beta}^{(k)}$ with $\hat{\beta}^{(k+1)}$ and repeat from step 1 until some termination condition is met.

- *GA-1* (binary encoding). The potential solutions to the maximization problem have been encoded as two binary strings of length 20 each. Equation (1) has been used to obtain the value of each of the two parameters given the integer c encoded as a binary string. The Gray code has been used. The population size has been taken equal to 30, the number of iterations has been 300 and 30 bootstrap samples have been generated to compute the estimates as the average estimates and the standard errors. The stochastic universal sampling method has been used for selection, then the single point crossover with $p_c = 0.7$ and the binary mutation with $p_m = 1/20$.

- *GA-2* (real encoding). The potential solutions to the maximization problem have been encoded as floating-point numbers. The population size has been taken equal to 30, the number of iterations has been 300 and 30 bootstrap samples have been generated to compute the estimates as the average estimates and the standard errors. The stochastic universal sampling method (see, for instance, [65], p. 166) has been used for selection, then the line recombination crossover with $p_c = 0.7$ and the floating-point mutation with $p_m = 1/20$.

- *GA-3* ([29]). Equation (1) has been used to obtain the value of each of the two parameters given the integers c_0, c_1 encoded as binary strings. The chromosome is the binary string $c = [c_1, c_2]$. The population size has been taken rather large, 1000 chromosomes, and the number of generations has been chosen equal to 30. Tournament selection with $p_s = 0.7$, single point crossover with $p_c = 0.65$, and binary mutation with $p_m = 0.1$ have been used. An additional operation, the inversion ([65]), has been applied with probability $p_i = 0.75$ to each chromosome in each generation. Inversion consists in choosing two points ℓ_1 and ℓ_2 at random in the chromosome and taking the bits from ℓ_1 to ℓ_2 in reverse order. The standard errors of the parameter estimates are computed by applying the GA on 250 bootstrap samples of the data.

- *GA-4* ([68]). The potential solutions to the maximization problem have been encoded as two pairs of numbers, a real number $r \in (0, 1)$ the first one and an integer in a given interval (N_a, N_b) the second one. The parameter estimates are obtained as $\hat{\beta}_0 = r_0 N_0$ and $\hat{\beta}_1 = r_1 N_1$. The population size has been taken equal to 100, the number of iterations has been 100 and 30 bootstrap samples have been generated to compute standard errors of the parameter estimates. The binary tournament has been used as selection process. Then special crossover (modified uniform crossover) and mutation are suggested. For crossover, the chromosomes are paired at random and the integer parts exchange. If the integer parts are equal, then the exchange takes place as regards the real part of the chromosome. Only the offspring with better fit is placed in the new generation. Mutation applies, with probability $p_m = 0.1$, only to the real parts of each chromosome. This part, r say, is multiplied by a random number between 0.8 and 1.2, namely r is multiplied by $0.8 + 0.4U$, where U is an uniform random number in $(0, 1)$. If mutation yields a number greater than 1 the first component is set to 1.

- *EDA* ([74]). The potential solutions to the maximization problem are represented by s vectors of 2 real numbers, the parameters β_0 and β_1. The initial population is generated at random. Then the s chromosomes are evaluated according to the likelihood function and the better s^* are retained. Other than using the likelihood, also the AUC ([24]) criterion is suggested as an alternative fitness function. This is recommended when the logistic regression model is used as a classifier and the AUC is the area under the receiver operating characteristic (ROC) curve, a graphical device to describe the predictive behavior of a classifier. The s^* best vectors found are used to estimate a bivariate probability density function. From this latter distribution s new chromosomes are generated and a new iteration starts. We adopted a bivariate normal distribution so that only the mean vector and the variance-covariance matrix are needed to estimate the distribution in each iteration. We assumed $s = 50$ and $s^* = 30$, 300 iterations and 30 bootstrap samples.

The results are reported in table 2 for the 6 algorithms. Estimates $\hat{\beta}_0$ and $\hat{\beta}_1$ are displayed with standard errors enclosed in parentheses. As a measure of adaptation the logarithm of the likelihood L computed on the average estimates is reported.

According to the figures displayed in table 2 the best algorithm is GA-1 as it shows the largest log-likelihood. The second best would be the algorithm GA-4. In this case however the estimates are markedly biased and exhibit the largest standard errors. This result may be explained by the encoding method. Splitting each parameter in two parts has the immediate consequence of increasing the variability of the estimates. Moreover, the fitness function should be linked to the parameters as directly as possible while for algorithm

Table 2 Parameter estimates for logistic regression fitted to the CHD data by using IRLS, 4 GAs and an EDA-based algorithms

	IRLS	GA-1	GA-2	GA-3	GA-4	EDA
$\hat{\beta}_0$	-5.3195	-5.1771	-6.0010	-5.4360	-3.1658	-5.4985
	(1.1337)	(1.3775)	(1.6355)	(1.2580)	(1.9609)	(1.5062)
$\hat{\beta}_1$	0.1109	0.1070	0.1248	0.1130	1.3995	0.1147
	(0.0241)	(0.0290)	(0.0331)	(0.0270)	(0.6422)	(0.0301)
L	-53.6774	-53.1873	-53.8577	-53.6886	-53.4000	-53.6907

GA-4 there is an intermediate step that separates the chromosomes decoding from the fitness function evaluation. The other algorithms lead to slightly smaller values of the log-likelihood and yield results similar to the algorithm GA-1. The algorithms IRLS and GA-3 have the advantage to attain the smallest standard errors of both estimates $\hat{\beta}_0$ and $\hat{\beta}_1$. As an overall result it seems that binary encoding performs better than the real one for evolutionary computing algorithms.

5   Multi-regimes Model Parameter Estimation

Stationary time series do not change their characteristic features through time. This behavior is not always observed in real time series data. A model that fits well the time series in a time interval may prove inadequate in other time intervals. Often the mean of the time series changes with time but this may be found true as well for the variance, the autocorrelation function, the spectral density in some frequency intervals, for instance. On the other hand linear models are specially useful to model time series data. The ARMA specification takes advantage of a well established theoretical background and well known effective procedures for identification, estimation, diagnostic checking, validation and forecasting. The multi-regime models may take non-stationarity into account and use at the same time the ARMA models as different ARMA models apply to different subsets of data. The smooth transition autoregressive models ([81]) are an example of useful multi-regime models where switching from an AR model to another takes place gradually. The threshold models (see [79]) have been developed along the same guidelines except that a step transition replaces the smooth passage between regimes. [37, 38] employed threshold autoregressive processes for modeling structural breaks, and used a GA for identifying the model. As in each regime the ARMA modeling may be used conveniently, the main problem with multi-regime models is the transition parameter or threshold. In this section we examine some algorithms, including meta-heuristics and GAs, that aim at locating the thresholds as accurately as possible. We assume the EXPAR model as a special example in the class of the smooth transition autoregressive models. The threshold models will be examined in the next sections.

5.1   The Exponential Autoregressive Model

The EXPAR(p) model may be written

$$y_t = \{\phi_1 + \pi_1\exp(-\gamma y_{t-d}^2)\}y_{t-1} + \ldots + \{\phi_p + \pi_p\exp(-\gamma y_{t-d}^2)\}y_{t-p} + e_t, \quad (3)$$

where d is the delay parameter. Unlike linear models, a change of the error variance σ_e^2 by multiplying the $\{e_t\}$ by a constant k, say, does not imply that the $\{y_t\}$ turn into $\{ky_t\}$. The order of magnitude of $\{y_t\}$ in (3) depends on γ too, in the sense that we may obtain the time series $\{ky_t\}$ by both multiplying σ_e^2 by k^2 and dividing γ by k^2.

The ability of the EXPAR to account for limit cycles depends whether some conditions on the parameters in equation (3) be fulfilled. If the time series is believed to exhibit limit cycles behavior, then the estimation procedure needs to be constrained in some way.

A brief description of the basic parameter estimation procedure proposed by [50] for the estimation of (3) follows. It may be considered as a natural benchmark for competitive alternatives because it is quite straightforward and unlike to fail to yield a solution. It does not ensure, however, that the limit cycles conditions be fulfilled.

The algorithm requires that an interval (a, b), $a \geq 0$, be pre-specified for the γ values in (3). This interval is split in N sub-interval, so that a grid of candidate γ values is built. Let $s = (b - a)/N$ and $\gamma = a$. Then, for N times, the following steps are performed.

1. Set $\gamma = \gamma + s$
2. Estimate ϕ_j e π_j by ordinary least squares regression of y_t on y_{t-1}, $y_{t-1}\exp(-\gamma y_{t-d}^2)$, y_{t-2}, $y_{t-2}\exp(-\gamma y_{t-d}^2)$, ...
3. Compute the AIC criterion and repeat steps 1 and 2 for $N - 1$ times.

Final estimated parameters are taken that minimize the AIC.

For the existence of limit cycles, the following conditions (see, for instance, [71]) are required to hold:
(1) all the roots of

$$z^p - \phi_1 z^{p-1}... - \phi_p = 0$$

lie inside the unit circle,
(2) some of the roots of

$$z^p - (\phi_1 + \pi_1)z^{p-1}... - (\phi_p + \pi_p) = 0$$

lie outside the unit circle,
(3)

$$\frac{1 - \sum\limits_{j=1}^{p} \phi_j}{\sum\limits_{j=1}^{p} \pi_j} > 1 \text{ or } < 0.$$

Several algorithms are available for nonlinear models parameter estimation in the presence of either likelihood function or residual sum of squares difficult to maximize or minimize respectively in the presence of many local optima. We select some algorithms to perform a simulation experiment in comparison with the GAs, namely the grid search, indirect inference ([49]), TS, SA and TA.

We carried out a simulation experiment to compare the performance of these methods and of GAs for parameter estimation of the EXPAR model

Table 3 Average estimates for 100 replications from an EXPAR(2)

parameter	ϕ_1	ϕ_2	π_1	π_2	γ			
true value	1.95	−0.96	0.23	−0.24	1.0			
500 obs	$\hat\phi_1$	$\hat\phi_2$	$\hat\pi_1$	$\hat\pi_2$	$\hat\gamma$	d^2	$\hat\sigma^2$	MSE
Grid search	1.76	−.77	−.35	−.32	.71	3.99	1.64	1.64
	(.17)	(.17)	(1.65)	(.28)	(.80)		(.96)	(1.63)
Indirect inf	1.92	−.94	−.40	−.66	1.53	7.14	1.45	1.40
	(.05)	(.05)	(2.0)	(1.38)	(.59)		(.81)	(1.23)
Tabu search	1.91	−.92	.05	−.08	.79	3.73	.99	.97
	(.16)	(.15)	(1.69)	(.25)	(.82)		(.06)	(.20)
Simul anneal	1.62	−.63	.64	−.63	1.05	4.63	1.41	1.52
	(.77)	(.77)	(1.28)	(.75)	(.84)		(.54)	(1.44)
Thre accept	1.84	−.85	−.03	−.45	.88	1.44	1.25	1.25
	(.11)	(.11)	(.86)	(.19)	(.70)		(.36)	(.39)
Genetic alg	1.89	−.91	.07	−.10	.46	3.47	.99	.97
	(.20)	(.18)	(1.58)	(.26)	(.72)		(.06)	(.21)
1000 obs	$\hat\phi_1$	$\hat\phi_2$	$\hat\pi_1$	$\hat\pi_2$	$\hat\gamma$	d^2	$\hat\sigma^2$	MSE
Grid search	1.85	−.86	−.02	−.24	.79	1.03	1.25	1.35
	(.10)	(.10)	(.43)	(.15)	(.83)		(.41)	(1.02)
Indirect inf	1.94	−.95	.37	−.42	1.40	7.58	1.36	1.19
	(.04)	(.04)	(2.19)	(1.45)	(.66)		(.50)	(.41)
Tabu search	1.95	−.96	.20	−.07	.66	1.31	1.0	.99
	(.02)	(.01)	(.76)	(.12)	(.76)		(.05)	(.13)
Simul anneal	1.85	−.86	.36	−.37	1.16	1.19	1.18	1.26
	(.15)	(.15)	(.73)	(.20)	(.70)		(.32)	(.99)
Thre accept	1.85	−.86	.25	−.38	1.20	1.09	1.0	1.0
	(.20)	(.20)	(.70)	(.23)	(.62)		(.05)	(.05)
Genetic alg	1.95	−.96	.19	−.07	.45	1.32	1.0	1.0
	(.02)	(.02)	(.69)	(.12)	(.70)		(.05)	(.13)

with $\phi_1 = 1.95$, $\phi_2 = -0.96$, $\pi_1 = 0.23$, $\pi_2 = -0.24$, $\gamma = 1$ and $d = 1$. This model has been proposed as an example by [50]. We simulated 100 series of 1550 observations by using standard unit Gaussian deviates. For each series, the first 1000 observations have been discarded, and the last 50 set apart for out-of-sample one-step-ahead forecasts. So, for estimation we used 500 observations. Further, 100 series of 2100 observations were generated as well. For each series the first 1000 observations were discarded, but 100 observations were set apart for out-of-sample forecasts. The observations left for estimation purpose were 1000. The results are displayed in table 3. The parameter estimates, averaged over 100 replications, are reported, and their standard errors are enclosed in parentheses. The index d^2 is computed as the average squared Euclidean distance between the two sets of estimated and true parameters. Then, the residual variance $\hat\sigma^2$ and the mean square error forecast (MSE) have been computed. The structure of the model has been assumed known, so that the number of parameters has been held fixed.

As far as the residual variance and MSE are concerned, TS and GAs for 500 and TS, TA and GAs for 1000 observations give the best performances as their values are close to one. Parameters ϕ_1 and ϕ_2 are estimated fairly well by all methods and standard errors are small, with the only exception of SA for 500 observations. On the other hand, estimates of π_1, π_2 and γ are often severely biased, and standard errors are large. The smallest squared difference d^2 between true and estimated parameters averaged over 100 replications is obtained by using TA for 500 and grid search and TA for 1000 observations.

5.2   The Generalized EXPAR Model

We may consider the more general EXPAR model (see, for instance, [32])

$$y_t = \{\phi_1 + \pi_1 \exp(-\gamma_1 y_{t-d}^2)\} y_{t-1} + \ldots + \{\phi_p + \pi_p \exp(-\gamma_p y_{t-d}^2)\} y_{t-p} + e_t,$$

where $\gamma_1, \ldots, \gamma_p$ are positive constants and d is the delay parameter. Advantages that may come from this model may consist in greater flexibility, better fit and improved forecasts. On the other hand a grid search for estimating $\gamma_1, \ldots, \gamma_p$ is less efficient and may become infeasible if the model order is large. Then the GAs may constitute a convenient device for estimating the parameters ϕ_j, π_j and γ_j. By using a GAs-based algorithm [15] fitted several "generalized" EXPAR models to the well known Canadian lynx data and sunspot numbers (see [79], chapter 7, for a detailed analysis) and obtained satisfactory results for both time series. We shall report some results concerned with the sunspot numbers.

Computations have been performed on the mean-deleted transformed data $2\{(1 + y_t)^{1/2} - 1\}$ as suggested in [79], p. 420. We considered the AR(9) model reported by [79], p. 423, and the self-excited threshold autoregressive SETAR$(2; 11, 3)$ model proposed by [45], p. 247. Then we estimated using GAs the EXPAR(2), the EXPAR(6) and the EXPAR(9) models with one γ and with 2, 6 and 9 γ's respectively. For estimating the parameters of each model we used the observations from 1700 to 1979, while the observations from 1980 to 1995 were reserved for the multi-step forecasts. The data have been downloaded from the URL: http://www.sidc.be/sunspot-data/ (SIDC-team, Royal Observatory of Belgium, Ringlaan 3, 1180 Brussel, Belgium, The International Sunspot Number, Monthly Report on the International Sunspot Number, online catalogue, yearly data 1700-2007). The results are displayed in table 4. Models are compared by means of the residual variance and the forecasts MSE. Time origins are 1979, 1984, 1987 and lead times $1, 2, \ldots, 8$.

The best forecasts not always are obtained by using models that have the smallest residual variance. The EXPAR(9) model with 9 γ's, for instance, yields the smallest residual variance, but the SETAR$(2; 11, 3)$ model provides the best multi-step forecasts for the years 1980-1987. The results change, however, if different time intervals are considered. Thus, the least mean square forecasts error is observed for the EXPAR(9) with 9 γ's in 1985-1992, for the

Table 4 Sunspot numbers: comparison among AR, SETAR, EXPAR and generalized EXPAR

model	residual variance	*mse* 1980-87	*mse* 1985-92	*mse* 1988-95	*mse* 1980-92
AR(9)	4.05	3.60	16.5	9.01	16.19
SETAR(2; 11, 3)	3.73	1.82	33.51	17.34	22.27
EXPAR(2) one γ	4.90	7.08	65.28	31.39	32.97
EXPAR(2) 2 γ's	4.83	3.77	85.33	29.32	38.46
EXPAR(6) one γ	4.47	7.64	54.74	19.46	21.11
EXPAR(6) 6 γ's	4.34	11.85	42.01	20.62	21.89
EXPAR(9) one γ	3.66	4.99	20.43	8.21	13.02
EXPAR(9) 9 γ's	3.57	2.62	16.34	10.65	10.27

EXPAR(9) with a single γ in 1988-1995. In the wider time span 1980-1992, the EXPAR(9) with 9 γ's is able to produce the best multi-step forecasts. The cyclical behavior of this time series is changing over time, and our models may describe it better in certain years than others. It seems that the EXPAR(9) model with 9 γ's almost always yields the most accurate forecasting performance.

5.3 Threshold Autoregressive Models

A self-exciting TAR (SETAR) model may be written

$$y_t = c_j^{(i)} + \sum_{j=1}^{p} \phi_j^{(i)} y_{t-j} + \varepsilon_t \qquad \text{if} \quad y_{t-d} \in (r_{i-1}, r_i],$$

where $\{\varepsilon_t\}$ is white noise, d is a given positive integer and the k disjoint intervals $(r_{i-1}, r_i]$, $i = 1, \ldots, k$, partition the real axis .

The GAs-based TAR model identification procedure in [87] needs the preliminary specification of the maximum number of "regimes" K ($K \geq 2$), the largest autoregressive model order P, the number of candidate threshold parameters H ($H \geq K-1$) and the number of delay parameters D. If all models had to be enumerated exhaustively their number should be computed

$$(P + 1)^K \binom{H}{K - 1} D.$$

Such number of candidate solution may obviously become very large. The GA solution ([87]) is based on encoding each of the tentative models as a string which is composed of several "fragments." The first one encodes the delay parameter, the second one the candidate threshold parameters (H "percentiles" were chosen from the ordered observations), then the orders of each of the autoregressive models are encoded. For instance, if $D = 4$, $H = 3$, the number of regimes is taken equal to 2 and the maximum autoregressive order is $P = 3$, then a string of 9 bits would suffice to represent each and every potential solution. For example, the string

$$01|011|10|11$$

means that $d = 1$, the third percentile value is taken as the threshold parameter, the autoregressive order for the first regime is 2 and that for the second regime is 3. The fitness function is chosen as a modified version of the AIC. The effectiveness of the method is shown by means of both a simulation experiment and some empirical studies for investigating the changing exchange rate of Thailand.

The TAR model easily generalizes to a threshold ARMA model if at least in some regimes an MA part is specified. Obviously in some regimes the model may be a pure MA without an AR part. A GAs-based algorithm may be developed along the same guidelines by defining a chromosome augmented to take the MA part into account.

In the multivariate framework a hybrid algorithm which combines GAs and SA has been proposed by [88] for estimating a threshold vector error correction model. The GA is implemented by using the real encoding and suitable genetic operators for crossover and mutation. In addition each chromosome in the current population is updated only if the offspring is accepted according to the Metropolis rule.

5.4 Double Threshold ARCH and GARCH Models

The autoregressive conditional heteroscedastic (ARCH) models and generalized ARCH (GARCH) have been introduced for modeling volatility clustering. The self-exciting threshold autoregressive ARCH (SETAR-ARCH) is a generalization that accounts for asymmetries in levels. Asymmetries both in levels and volatility may be modeled by the double threshold ARCH (DTARCH) and double threshold GARCH (DTGARCH) models. References for ARCH and GARCH models are [39] and [20], see [58], [59] and [26] for threshold ARCH and GARCH models. The GAs have been considered by [1] for identifying the optimal structure of a GARCH model. For the identification and estimation of DTARCH and DTGARCH models GAs-based methods have been developed by [11] and [13].

A GARCH model takes the form

$$x_t = m(I^t, \theta) + \varepsilon_t \sqrt{v(I^t, \theta)}$$

$$I^t = \{x_{t-1}, x_{t-2}, \ldots, \varepsilon_{t-1}, \varepsilon_{t-2}, \ldots\} \text{ information at time t,}$$

$$m(I^t, \theta) : \text{conditional mean,}$$

$$v(I^t, \theta) : \text{conditional variance.}$$

If the conditional mean $m(I^t, \theta)$ follows a multi-regime SETAR(p) model and the conditional variance $v(I^t, \theta)$ follows a multi-regime ARMA model SETARMA(s, q) then we may specify the DTGARCH model

$$x_t = \phi_0^{(u)} + \sum_j \phi_j^{(u)} x_{t-j} + a_t \quad \text{if } x_{t-d} \in R_u$$

$$v_t = \alpha_0^{(u)} + \sum_i \beta_i^{(u)} v_{t-i} + \sum_j \alpha_j^{(u)} a_{t-j}^2 \quad \text{if } x_{t-d} \in R_u$$

where R is a partition : $(-\infty, \infty) = R_1 \cup R_2 \cup \ldots \cup R_k$, d is the *delay*, a_t are independent Gaussian zero mean random variables and $E(a_t^2) = v(I^t, \theta)$. The partition sets R_j are always intervals: $R_1 = (-\infty, r_1), R_2 = (r_1, r_2)$, $R_3 = (r_2, r_3)$, . . . , $R_k = (r_{k-1}, \infty)$, where $r_1 < r_2 < \ldots < r_{k-1}$ are the *thresholds*. In practice the DTGARCH model parameters may be distinguished in *structural parameters*, i.e.

1. delay parameter d
2. regime number k
3. thresholds $(r_1, r_2, \ldots, r_{k-1})$
4. autoregressive orders $(p_1, p_2, \ldots, p_k)$
5. GARCH orders $(q_1, q_2, \ldots, q_k, s_1, s_2, \ldots, s_k)$

and the *equation coefficients*

$$\phi_j^{(u)} , \; \alpha_j^{(u)} , \; \beta_j^{(u)}$$

subject to stationarity and variance non-negativity constraints.

If the *structural parameters* are given, the equation coefficients may be estimated by maximum likelihood:

$$\log L(x_1, x_2, \ldots, x_n | \phi, \alpha, \beta) = \text{const} - \frac{1}{2} \sum_{u=1}^{k} \{\log(v_t) + a_t^2 / v_t\} i_t^{(u)},$$

where $i_t^{(u)}$ denotes the indicator function of $x_{t-d} \in R_u$. But for the *structural parameters* there is no analytic method available. The only possible procedure which may yield an exact solution consists in:

1. enumerating all possible models
2. estimating the coefficients of each and every model
3. performing diagnostic checking and evaluating all models
4. selecting the best model.

A more viable alternative is determining sets of structural parameters by means of heuristic methods and estimating and comparing the full models.

We shall describe here a hybrid GA for estimating a DTGARCH model where the GA is used for searching for optimal structural parameters d, k,

r_j, p_j, s_j, q_j while the coefficients of the model equations $\phi_j^{(u)}$, $\alpha_j^{(u)}$, $\beta_j^{(u)}$ are estimated by maximizing the log-likelihood.

Chromosome encoding

Let K denote the maximum number of regimes and M the minimum number of observations required in each regime. The chromosome c consists in two separate parts, since orders and thresholds are separately encoded.

- *orders part*

  - The first part c_1 encodes $d, p_1, \ldots, p_k, q_1, \ldots, q_k, s_1, \ldots, s_k$,
  - the binary coding is used to represent integers and
  - mutation and crossover are assumed as in simple GA.

- *thresholds part*

  - The second part c_2 of the chromosome c encodes $(g_1, g_2, \ldots, g_{K-1})$ where g_i represents the number of observations in the ith regime and $g_i \in (M, n)$.
  - This encoding is motivated by the convenience in avoiding *legalization* problems, namely any c_2 is a valid chromosome fragment.

Chromosome decoding

Decoding c_1 is straightforward. For c_2 we have to specify a rule to compute the number of regimes k and the threshold $r_1, \ldots, r_{k-1}$ from $(g_1, g_2, \ldots, g_{K-1})$. The requirement $g_i \in (M, n)$ is needed to ensure that estimation may be performed in each regime easily. We compute the thresholds as follows:

$$r_1 = x_{(g_1)}, r_2 = x_{(g_1+g_2)}, r_3 = x_{(g_1+g_2+g_3)}, \ldots$$

where $(x_{(1)}, x_{(2)}, \ldots, x_{(n)})$ are the ordered data. The number of regimes k is computed as

$$k = \arg\max_{\kappa}\{g_1 + \ldots + g_\kappa < n - M\} + 1 \leq K.$$

In practice only the first $k - 1$ out of the $K - 1$ integers g_i's have to be computed and we may stop decoding as soon as $g_k \geq n - M$. Then we assume k regimes and thresholds $r_1, \ldots, r_{k-1}$. Note that both the chromosome fragment c_1 and c_2 have fixed length, $\ell_1 = (3k + 1)\nu$ the first one and $\ell_2 = (K - 1)\nu$ the second one, where ν is the number of bits we adopt to encode the integers as binary strings. The complete chromosome c has fixed length $\ell = \ell_1 + \ell_2$. The genes that do not contribute to the decoding procedure still belong to c and may turn useful for recombination in the later crossover steps.

Genetic operators

The crossover may be performed as usual in c assuming that the genes are integer numbers. Mutation in c_1 may be performed as binary mutation, while mutation in c_2 needs a special procedure. We adopted the following device: if gene g_i mutates then its new value is a uniform random number in the interval $(\max\{M, g_i - M/2\}, g_i + M/2)$.

Fitness function

We have adopted the AIC criterion for fitness function evaluation. Given the chromosome c which encodes the structural parameters, let us maximize the likelihood with respect to ϕ, α, β, i.e.

$$L^*(c) = \sup_{\phi,\alpha,\beta} L(x|\phi, \alpha, \beta)$$

Then the AIC is computed

$$AIC(c) = -2 \log L^*(c) + 2 \,(\text{number of parameters}).$$

In order to obtain a positive non decreasing fitness function f the following transform

$$f(c) = \exp\{-AIC(c)/n\}$$

may be used.

5.5 An Application to the Daily Hong Kong Stock Exchange (Hang Seng) Index

As a first example we considered the daily Hang Seng index data from January 1987 to December 1991. If x_t denotes the original data, the *return* series has been computed as

$$y_t = \log(x_t/x_{t-1}).$$

The data behavior suggests to fit a different model to the data recorded and transformed in each of the five years. For instance in 1987 there are 260 observations available. The time series for the year 1987 is displayed in figure 4.

The GAs-based algorithm applied to the Hang Seng index data recorded in year 1987 yields a DTGARCH model with delay parameter $d = 2$ and $k = 2$ regimes with threshold parameter $r_1 = -0.0044$. Model is

$$y_t = -0.0044 + \sum_{j=1}^{4} \phi_j y_{t-j} + a_t$$

$$v_t = 0.0153$$

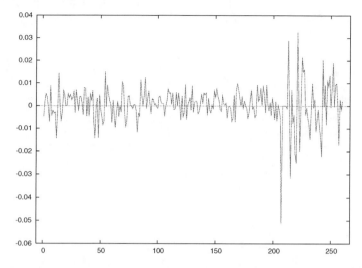

Fig. 4 Differenced logarithmic transform of the daily Hang Seng index data (1987)

if $y_{t-2} < -0.0044$, and

$$y_t = 0.01 + \sum_{j=1}^{7} \phi_j y_{t-j} + a_t$$

$$v_t = 0.002 + 0.0028 v_{t-1} + \sum_{j=1}^{6} \alpha_j a_{t-j}^2$$

if $y_{t-2} \geq -0.0044$. Note that the process is stationary AR homoscedastic when returns are low while the process is heteroscedastic when returns are high.

5.6 An Application to the Daily Exchange Rate Yen/Dollar

As a second example we consider the daily exchange rates yen/dollar from January 1, 1983 to January 28, 1985. Data have not been transformed. There are 541 observations available. The time series data are plotted in figure 5.

We obtained from the GAs-based procedure the following DTGARCH model.

- delay = 2
- two regimes : $x_{t-2} \leq 237$, $x_{t-2} > 237$
- first regime: $p = 3, q = 3, \vartheta = 3$

  - $x_t = 6.85 + \sum_{j=1}^{3} \phi_j x_{t-j} + a_t$

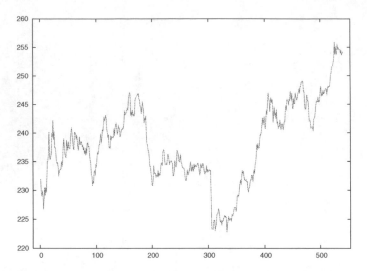

Fig. 5 Daily exchange rate Yen/Dollar from January 1, 1983 to January 28, 1985

$$- \quad v_t = 0.864 + \sum_{i=1}^{3} \beta_i v_{t-i} + \sum_{j=1}^{3} \alpha_j a_{t-j}^2$$

- second regime : $p = 1, q = 0, s = 0$

$$- \quad x_t = 2.74 + 0.989 x_{t-1} + a_t$$
$$- \quad v_t = 1.151$$

Note that when the exchange rate is low the process is AR(3) heteroscedastic while the process is a random walk when the exchange rate is high.

6 Multiple Outliers in Data Sets

Data sets are often affected by unexpected observations or gross errors that may negatively impact data analysis, model estimation and forecasting. Usually a preliminary investigation is performed to identify outlying observations. These latter are generally closely related to missing data treatment and validation procedures. The approach of robust statistics aims essentially at ensuring that reliable estimates may be obtained from the data even in the presence of outliers. We consider here the alternative approach that consists in discovering the outliers and performing some appropriate action. A comprehensive review of statistical methods for the treatment of outliers in data sets is [16]. In the next sections we shall address the identification of outliers in time series as some additional difficulties are involved due to the correlation structure. Though GAs-based methods for outlier detection have been originally introduced for independent data a great deal of work has been devoted to develop evolutionary computing methods for detecting outliers in time series.

6.1 The Outlier Problem in Time Series

Outliers in time series are observations that do not conform to the behavior of the majority of the neighboring time series data. An account may be found in [21], chapter 12. It is customary to distinguish four outlier type, i.e. the additive outlier (AO), the innovation outlier (IO), the transient change (TC) and the level shift (LS). The AO impacts the time series only at the time point where it occurs (a recording error, for instance), but it does not influence any other observation. Special methods are concerned with sequences of consecutive AO's (see, for instance, [53]). The IO affects the random shocks process underlying the time series, and its influence is present in time series points even after its occurrence. TC is a temporary deviation from the expected pattern, while LS is a permanent change in the mean. Iterative procedures such as in [31] are common practice for outlier treatment in time series. The 'skipping' approach (using the Kalman filter to eliminate from computations either AO or missing data) has been compared to the AO approach (replace missing or outlying observations with interpolation estimates) by [48] and the two approach are shown to be equivalent. This method has been implemented in the software TRAMO-SEATS (URL: http://www.bde.es/servicio/software/econome.htm)

In the multivariate framework, detecting the four types of outliers, AO, IO, TC and LS, has been considered by [80]. Vector ARMA modeling forms the basis of their procedure. A different approach consists in projecting the multivariate time series data along low-dimensional or univariate directions and using the computed low-dimensional time series to identify the potential outliers dates. Projection pursuit has been suggested by [44] and independent component analysis by [12].

Using GAs has been proposed by [14] for univariate time series and by [22] for detecting influential observations in dynamic multivariate linear models.

The outlier detection procedures are quite effective if outliers, either isolated or occurring as a "patch," are not too close each others in the time series. If it is not the case, then masking (an outlier may hide another one which is close to it) and smearing (an outlier may impact some subsequent observations so as these latter are recognized in turn as outliers, though they are not) effects, that arise because the time series observations are correlated, make particularly difficult both outlier detection and estimation. Evolutionary computing methods allow several complete outlier patterns to be considered and compared. We assume that there exists a maximum number of outliers K, say, as outliers are to be considered "rare events" and only a limited number, 5% of observations, for instance, may occur. If we want to distinguish m outlier types, the number of outlier patterns is

$$mn + m^2 \binom{n}{2} + m^3 \binom{n}{3} + \quad + m^K \binom{n}{K}.$$

Searching such a large "solution space," however, is the natural task for a evolutionary computing or meta-heuristic method. We have to adopt appropriate encoding and fitness function to obtain from an evolutionary computing algorithm the optimal (or at least near optimal) solution.

6.2   *Genetic Algorithms for Outlier Detection in Time Series*

Binary encoding is usually adopted in GAs-based algorithms for outlier detection in time series. A binary string of length $\ell = n$ has to be defined and time points are associated to the bits in the binary sequence. A bit is equal to 1 for an outlying observation and 0 otherwise. A different encoding consists in preparing a list of the time points labels where an outlying observation is supposed to occur, and make it to precede the list of the remaining time point labels, in any order. Such order-based encoding has been suggested by [36] for independent observations, but extension to correlated time series data is straightforward. If, in addition, we want to distinguish the outlier types, we may use m binary strings, each string for each type, under the constraint that in any time point there is no more than a single 1. As an alternative, we may resort to the integer encoding which has been proposed for grouping problems ([40]), by associating an integer code to any outlier type, for example 1 for AO and 2 for IO. Let, for instance, the time series $y=(y_1, y_2, \ldots, y_n)'$ have $n = 30$ observations, and let outliers (any type) be located at $t = 9, 20, 21, 22$. Then, the two encodings may look as follows

$$
\begin{array}{ll}
\text{binary} & 000000001000000000011100000000 \\
\text{order} - \text{based} & 9 \quad 20 \quad 21 \quad 22 \quad | \ldots \text{other time point labels} \ldots
\end{array}
$$

Consider instead the case that we want to distinguish between the AO and IO types. If there is a IO at $t = 9$ and there are AO's at $t = 20, 21, 22$, then we may use either the binary or integer encoding

$$
\begin{array}{ll}
\text{binary} & \left\{ \begin{array}{l} 000000000000000000011100000000 \\ 000000001000000000000000000000 \end{array} \right. \\
\text{integer} & 000000002000000000011100000000
\end{array}
$$

For evolutionary computing methods to work in a reasonable time, the fitness function has to be chosen so that it may be properly and quickly computed. We may attempt to minimize the sum of squares computed from some ARMA model fitted to the data. An identification stage, in necessarily automatic way, has to be performed, however, which requires in most cases a considerable computational effort. Then, a valid course of action consists in exploiting the relationship between the linear interpolator and the AO (see [70], p. 237; see also [35]). Only the inverse covariance function ([84], p. 123) is needed which may be easily estimated from the data. Let us consider, for the sake of simplicity, only the AO type. Let k outliers be located at $t = t_1, \ldots, t_k$. Let us define the $n \times k$ "design matrix" X, where $X_{j,h} = 1$ if $j = t_h$ (that is, the h-th outlying observation is located at time $t = j$) and 0 otherwise. Let $z = (z_1, \ldots, z_n)'$ be the observed time series and y the unobserved outlier free realization. Then, the relationship

$$z = X\omega + y$$

holds, where $\omega = (\omega_1, \ldots, \omega_k)'$ is the outlier size array. The likelihood to be maximized is approximately, under Gaussianity assumption,

$$L(X, \omega | y) = (2\pi)^{-\frac{n}{2}} \sqrt{\det(\Gamma i)} \exp\left\{ -\frac{1}{2}(z - X\omega)' \Gamma i (z - X\omega) \right\}. \qquad (4)$$

The matrix Γi of the inverse autocovariances may be estimated from the data by using robust techniques (see [46]).

For an illustration of the procedure a simulation experiment is reported. A set of 200 observations have been generated from the ARMA(0,2) model with parameters $\theta_1 = 0.7$ and $\theta_2 = -0.5$ and Gaussian white noise with mean zero and unit variance. By discarding the first 40 artificial observations, we obtain 160 outlier free observations, whose standard error is about 1.5. This time series has been modified by adding 4 to its values at $t = 60$, $t = 62$ and $t = 64$, by subtracting 5 from its value at $t = 100$ and adding 5 to its value at $t = 101$. Time series data are displayed in figure 6. This is a very difficult pattern to detect, and common procedures fail to perform the identification task correctly. In order to discover the outliers in the data, the GA has been employed with crossover probability $p_c = 0.75$ while several mutation probabilities p_m have

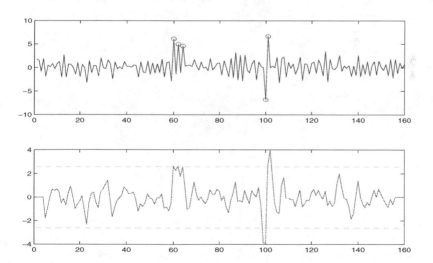

Fig. 6 Top panel, 160 observations simulated from a MA(2) model with additive outliers (circles). Bottom panel, residuals and confidence limits (straight lines)

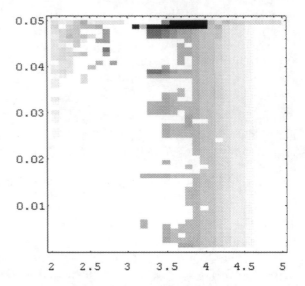

Fig. 7 Density plot of
the distance, after 500
iterations, of the fitness
from the global maxi-
mum, as a function of the
constant c and probability of mutation p_m. The
probability of crossover is
$p_c = 0.75$. The distance
(that is, the error) ranges
from 0 (white) through
13 (black). The gray lev-
els vary linearly between

been tried, from 0.001 to 0.05 with step 0.001. The inverse variance covariance matrix has been assumed known, so that, if the "right" solution is assumed, then the global maximum of the fitness function equals the logarithm
of the maximized likelihood (4), that is $\log(L) = 66.7$, minus $2ck$, where c is
a proportionality constant. For $2.52 < c < 4.67$ the maximum of the fitness
function coincides with the correct outlier identification while if $c < 2.52$, the
global maximum of the fitness function is attained by including, in addition,
the "spurious" outlier at $t = 23$. If $c > 4.67$, then the global maximum of the
fitness function is attained by considering only the observations at $t = 100$
and $t = 101$ as outlying ones. This circumstance supports the choice $c = 3$, for

Fig. 8 Fitness as a function of the number of
iterations. $p_c = 0.75$,
$p_m = 0.015$ and $c = 3.8$.
Outliers were found at
$t = 102$ (iterations $0-8$);
$t = 100, 101, 102$ and
148 (iteration 9); $t = 45$,
100, 101 and 102 (iter-
ation 10); $t = 100, 101$
and 102 (iteration 11);
$t = 100$ and 101 (itera-
tions $12-97$); $t = 60$,
62, 64, 100 and 101 (from
iteration 98 on: the elitist
strategy prevents losing
the best chromosome)

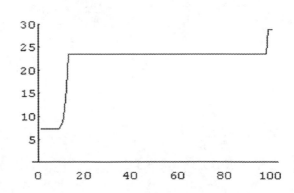

instance. In figure 7 the difference between the global optimum of the fitness function and its best value obtained in 500 generations is plotted as a function of both p_m and c, this latter varying from 2 to 5 with step 0.1. We may see that the choice of the mutation probability does not impact the quality of the solution, while the choice of c is very important for the fitness function to characterize the "right" solution. Note that in figure 7 we do not consider the "true" outlier set as the objective of the search, but only the maximization of the fitness function is taken into account. In such perspective, we may observe that the GA performance deteriorates only if c takes values in a rather narrow band. The choice of c may be done by following the usual guidelines, that is c has to be taken low to allow for "high sensitivity," and large for "low sensitivity." As far as the relationship between the fitness function and the number of iterations of the GA is concerned, we found the typical behavior displayed in figure 8. Searching for outliers has been performed by assuming $p_c = 0.75$, $p_m = 0.015$ and $c = 3.8$. The solution has been reached after 98 iterations out of the 500 iterations allowed as a maximum. The persistence of the algorithm in the local maximum corresponding to the outliers at $t = 100$ and $t = 101$ is apparent. The usefulness of the mutation operator is clearly shown, because, at this stage, the searching procedure moves towards the global maximum by mutation. The size of the population seem to be the other parameter to take under control if we want to obtain a near optimal solution in the shortest time.

7 Genetic Algorithms for Cluster Analysis

Methods for cluster analysis are the object of a large literature and are an active research field specially in connection with data mining techniques. For a comprehensive review see [18] and references therein. Let a set of n objects be given and let p measurements concerned with real variables be available for each and every object in the set. Objects and measurements define the usual $n \times p$ data matrix. A line of the matrix is an observation of p variables that characterize the corresponding object and a column is a variable. We assume that a genuine cluster structure exists in the data set. Further, we assume that a similarity (or dissimilarity) measure between each and every pair of objects may be computed from the $n \times p$ measurements. An optimality criterion is assumed to evaluate the internal cluster cohesion and the external cluster dissimilarity (see, for instance, [19]). In multiobjective cluster analysis two or more indexes are used to decide the cluster membership of an object.

A cluster analysis algorithm aims at grouping the objects so that the resulting cluster structure satisfies the optimality criterion. Every object belongs to a cluster and if none is similar to any other it forms a cluster on its own. This is a hard or crisp partition. We may consider fuzzy clustering by allowing an object to belong to more than a cluster according to some 'membership degree'.

The evolutionary computing methods and in general the methods in the class of meta-heuristics are appropriate for dealing with cluster analysis as the number of groups that may be formed with n objects is large even if n is rather small. If the number of groups may vary from 1 to a pre-specified maximum G, then

$$N(G,n) = \sum_{g=1}^{G} \frac{1}{g!} \sum_{j=1}^{g} (-1)^{g-j} \binom{g}{j} j^n$$

is the number of possible clusters.

7.1 Genetic Clustering Algorithms

Many GAs-based procedures have been developed to solve cluster analysis problems. Such procedures take several different features into account, for instance large data sets, cluster constraints and special data structures. GAs-based algorithms combined with the well known k-means and k-medoids algorithms have been considered in [69]. Variable length chromosomes have been suggested by [63] in the context of fuzzy cluster analysis. [54] deals with cluster analysis of panel data. Multi-objective GAs-based cluster analysis has been introduced in [4].

As a typical example of implementation of GAs in a cluster analysis problem we shall give a brief description of the algorithm Genetic Clustering for Unknown K (GCUK) developed by [3]. GCUK essentially combines the k-means algorithm with a GA procedure. There are two main improvements with respect to the basic k-means algorithm, namely the number of groups is unknown and does not have to be pre-specified, and the possibility that the procedure yields a local optimum as a result is greatly reduced. The GCUK algorithm requires that a suitable interval $[g_{\min}, g_{\max}]$, where $g_{\min} > 1$ and $g_{\max} \le n$ has to be pre-specified. A fixed length chromosome is assumed and $\ell = p g_{\max}$, where p is the number of measurements. The characteristic features of GCUK may be summarized as follows.

- *Encoding.* Any solution is coded as a string of $g_{\max}$ sets of centroid coordinates. These latter are vectors of p floating-point numbers each of which represents a cluster. Some of the centroids may correspond to empty clusters. In this case the symbol # ('don't care') is used to make such circumstance clear. As $g > 1$ the number of symbols # cannot be greater than $g_{\max} - 2$. In general, the chromosome will contain, arranged in any order, g centroids and $g_{\max} - g$ symbols #.
- *Fitness function.* Each chromosome is associated, as a measure of adaptation to the environment, an index of cluster validity. As evolutionary algorithms usually maximize the fitness function the reciprocal of the index is assumed if the optimum corresponds to the smallest index value.

- *Initial population.* Let s denote the population size. For $i = 1, 2, \ldots, s$ an integer g_i is generated uniformly randomly in $[g_{\min}, g_{\max}]$ and g_i objects are chosen at random. Each one of these g_i objects are assigned to a set of p consecutive genes selected at random within the chromosome. The genes that are left unassigned are marked with the symbol $\#$.

The genetic operators implemented by GCUK are the roulette wheel rule for selection, the single point crossover and mutation. However, it has to be noticed that the crossover is implemented by assuming a centroid as undivided, i.e. recombination is performed by exchanging centroids. Moreover, mutation is performed as usual in the presence of floating-point encoding. Each and every measurement may change with probability p_m of a small amount δ around its present value. δ is a number generated from the uniform distribution in $(0, 1)$ and the $+$ or $-$ sign occurs with equal probability.

7.2   Cluster of Time Series

Grouping a set of time series in smaller subsets may provide us with interesting information about the time series structure. For example, time series that follow similar models, or are strongly correlated in some sense may be assumed to belong to the same subset (cluster). Several different measures of similarity (or dissimilarity) have been proposed. A comprehensive review of cluster of time series may be found in [60]. Genetic algorithms were applied by [7, 8] for clustering time series according to either time series cross correlations or phase spectrum dissimilarities. In this latter case, the statistics for directional data introduced by [62] has been used.

The problem that we want to examine here in some detail is concerned with finding a partition of a set of time series according to their cross correlations computed after pre-whitening. Each set of the estimated partition is a cluster which groups together time series that may, for instance, be joint modeled, or are sharing properties of interest, such as correlation with some composite indicator. In this context, we shall define a cluster as a set (group) of time series that satisfy the following condition ([89]). Given a set of k stationary time series $\{x_1, \ldots, x_k\}$, where $x_i = (x_{i,1}, \ldots, x_{i,n})'$, $i = 1, \ldots, k$, a subset C which includes k' series $(k' < k)$ is said to form a group if, for each of the $k'(k' - 1)/2$ cross-correlations $\rho_{i,j}(\tau)$, we have

$$|\rho_{i,j}(\tau)| > c(\alpha) \tag{5}$$

for at least a lag τ between $-m$ and m, and $i, j \in C, i \neq j$. A positive integer m has to be pre-specified which denotes the maximum lag. The cross-correlations $\rho_{i,j}(\tau)$ have to be computed from the pre-whitened time series (see, for instance, [25], p. 232). If all time series have n as a common number of observations, then choosing the significance level $\alpha = 0.05$, say, gives the figure $c(\alpha) = 1.96/\sqrt{n}$ in (5). The previously stated definition does not exclude that a time series may belong to more than a single group. Then there are possibly several allowable partitions to consider, and their number may be very large. In [7] a GA is developed to find the optimal partition that fulfills equation (5) in each cluster.

The fitness function is based on the k-min cluster criterion ([77]) and may be defined

$$f^+(C_1, C_2, \ldots, C_g; g) = \sum_{\omega=1}^{g} \sum_{i,j \in C_\omega, i \neq j} d_{i,j}^+, \qquad (6)$$

where

$$d_{i,j}^+ = \max_{\tau \in (-m, m)} (1 - |\rho_{i,j}(\tau)|)$$

and g is assumed unknown. When using (6) each and every cluster needs to be a group, according to (5), for, otherwise, any algorithm, unless prematurely ended, will put together all time series into a single cluster.

It looks convenient to code any admissible time series partition in permutation form. Each time series is labeled with a positive integer number between 1 and k. Then, let $(i_1, i_2, \ldots, i_k)$ be a permutation of $(1, 2, \ldots, k)$. Given the significance level α, the permutation will be given its proper meaning as follows.

1. The first time series, labeled i_1, is taken as the first element of the first cluster.
2. Let i_2 be considered. If the maximum absolute value cross-correlation between the time series i_1 and i_2, computed after pre-whitening, is greater than $c(\alpha)$, then i_2 joins i_1 into the first cluster. Otherwise, the time series i_2 is to become the first element of the second cluster.
3. The i_j-th time series joins an existing cluster if (5) turns true for all pairs belonging to it. If such a circumstance applies for more than one cluster, then the cluster ω is chosen for which $\sum_{i \in C_\omega} d_{i,i_j}^+$ is greatest.
4. The decoding procedure ends as soon as each time series belongs to a cluster.

The choice criterion included into step 3 may look somewhat arbitrary, but it proved necessary, for if, for instance, the time series were assigned so as to maximize the overall criterion, then some undesirable penalization of small clusters would be introduced.

In [7] three procedures were proposed for solution each of which was designed by implementing TS, SA and GAs respectively and several simulation experiments were carried out. Results showed that the three algorithms may be considered effective in recovering correctly the cluster of time series. Further computations on the same artificial data sets by using an implementation of TA produce similar results (not shown here).

8 Concluding Remarks

Evolutionary computing methods provide useful tools for handling many difficult problems that arise in statistical data analysis. However, as for the other fields of application, their usefulness is best exploited if the particular problem involves the search in a finite, but very large, set of discrete parameters. In order that a problem may really require that an evolutionary computing method be implemented for its solution, essentially three circumstances have to be verified.

1. The space of the solutions is quite large.
2. The problem may be coded directly in a natural meaningful way.
3. The objective function to be optimized has to be readily and quickly computed.

In general, it is convenient to resort to evolutionary computing when the objective function to optimize does not meet the usual mathematical requirements, such as continuity, differentiability and convexity.

In statistical data analysis, we could see that there are problems that are suited for use with evolutionary computing techniques, whilst others had better solved by gradient-based techniques. For instance, it is not advisable to employ an evolutionary computing method to estimate the parameters of an ARMA model, but it is convenient to use evolutionary computing or other meta-heuristics if subset ARMA models have to be identified. We reviewed some important problems that are commonplace in statistical data analysis and may need evolutionary computing techniques for reliable solution. These are the estimation of some special non linear models, the identification of threshold parameters in AR and ARCH models, the identification and estimation of subset ARMA and VAR models, detection of location and type of outlying observations, cluster of time series.

Other topics may be envisaged where evolutionary computing methods may turn useful, though not always specific and detailed approach have been fully developed. These are, for instance, the identification and estimation of more general time series state dependent models, the filter design and wavelet filtering, the detection of outliers in vector time series and in non linear time series, the development of new methodological tools for statistical design of experiments. For some of these problems guidelines were provided, however. The algorithms that have been designed for threshold autoregressive model identification and estimation may be extended to include multivariate models. The filter design by genetic algorithm may be extended to wavelet filtering. In the GA framework, development of symbolic regression systems has been considered. In symbolic regression the algorithm is designed to find both the functional specification from a given set of suited functions and the parameter estimates. This same principle applies for selecting wavelets and parameters to optimize the fitness function. Another interesting application for wavelet filtering design is using evolutionary computing-based techniques to select the coefficients to be set to zero in the wavelet signal expansion. Moreover, a promising field for applications may be the outlier detection in vector time series linear models, and in non linear time series, either univariate or multivariate. Some procedures exist that may be investigated, generalized, and checked by simulation studies. Needless to say, even the fields where evolutionary computing proved to be particularly useful in dealing with statistical data were not yet fully studied. Better understanding is needed on how the evolutionary computing-based techniques work in some specific problems, such as clustering time series, and better encoding and design are likely to be able to greatly improve their performance.

Acknowledgements. Support from EU Commission under contract MRTN-CT-2006-034270 Marie Curie Research and Training Network "COMISEF" Computational Optimization Methods in Statistics, Econometrics and Finance, and from Sapienza University of Rome is gratefully acknowledged.

References

1. Adanu, K.: Optimizing the garch model - An application of two global and two local search methods. Computational Economics 28, 277–290 (2006)
2. Balcombe, K.G.: Model selection using information criteria and genetic algorithms. Computational Economics 25, 207–228 (2005)
3. Bandyopadhyay, S., Maulik, U.: Genetic clustering for automatic evolution of clusters and application to image classification. Pattern Recognition 35, 1197–1208 (2002)
4. Bandyopadhyay, S., Maulik, U., Mukhopadhyay, A.: Multiobjective Genetic Clustering for Pixel Classification in Remote Sensing Imagery. IEEE Transactions on Geoscience and Remote Sensing 45, 1506–1511 (2007)
5. Bandyopadhyay, S., Mukhopadhyay, A., Maulik, U.: An improved algorithm for clustering gene expression data. Bioinformatics 23, 2859–2865 (2007)
6. Bandyopadhyay, S., Saha, S., Maulik, U., Deb, K.: A Simulated Annealing Based Multi-objective Optimization Algorithm: AMOSA. IEEE Transaction on Evolutionary Computation 12, 269–283 (2008)
7. Baragona, R.: A simulation study on clustering time series with metaheuristic methods. Quaderni di Statistica 3, 1–26 (2001)
8. Baragona, R.: Further results on Lund's statistic for identifying cluster in a circular data set with application to time series. Communications in Statistics – Simulation and Computation 32(3) (2003)
9. Baragona, R.: General local search methods in time series. Contributed paper at the International Workshop on Computational Management Science, Economics, Finance and Engineering, Limassol, Cyprus, March 28-30, 2003, vol. 2003(10), pp. 28–59 (October 2003),
   http://www.sciencedirect.com/preprintarchive
10. Baragona, R., Battaglia, F.: Multivariate mixture models estimation: a genetic algorithm approach. In: Schader, M., Gaul, W., Vichi, M. (eds.) Between Data Science and Applied Data Analysis, Series: Studies in Classification, Data Analysis and Knowledge Organization, pp. 133–142. Springer, Berlin (2003)

11. Baragona, R., Battaglia, F.: Genetic algorithms for building double threshold generalized autoregressive conditional heteroscedastic models of time series. In: Rizzi, A., Vichi, M. (eds.) Compstat 2006 Proceedings in Computational Statistics, 17th Symposium Held in Rome, Italy, pp. 441–452. Springer, Berlin (2006)

12. Baragona, R., Battaglia, F.: Outliers detection in multivariate time series by independent component analysis. Neural Computation 19, 1962–1984 (2007)

13. Baragona, R., Cucina, D.: Double threshold autoregressive conditionally heteroscedastic model building by genetic algorithms. Journal of Statistical Computation and Simulation 78, 541–559 (2008)

14. Baragona, R., Battaglia, F., Calzini, C.: Genetic algorithms for the identification of additive and innovation outliers in time series. Computational Statistics & Data Analysis 37, 1–12 (2001)

15. Baragona, R., Battaglia, F., Cucina, D.: A note on estimating autoregressive exponential models. Quaderni di Statistica 4, 71–88 (2002)

16. Barnett, V., Lewis, T.: Outliers in Statistical Data, 3rd edn. John Wiley & Sons, Chichester (1994)

17. Bearse, P., Bozdogan, H.: Subset selection in vector autoregressive models using the genetic algorithm with informational complexity as the fitness function. Systems Analysis Modelling Simulation 31, 61–91 (1998)

18. Berkhin, P.: Survey of clustering data mining techniques. Technical Report, Accrue Software, San Jose, California (2002),
http://citeseer.nj.nec.com/berkhin02survey.html

19. Bezdek, J.C., Pal, N.R.: Some new indexes of cluster validity. IEEE Transactions on Systems, Man and Cybernetics – Part B: Cybernetics 28, 301–315 (1998)

20. Bollerslev, T.: A generalized autoregressive conditional heteroskedasticity. Journal of Econometrics 31, 307–327 (1986)

21. Box, G.E.P., Jenkins, G.M., Reinsel, G.C.: Time Series Analysis: Forecasting and Control, 3rd edn. Prentice Hall, Englewood Cliffs (1994)

22. Bozdogan, H.: Information complexity criteria for detecting influential observations in dynamic multivariate linear models using the genetic algorithm. Journal of Statistical Planning and Inference 114, 31–44 (1988)

23. Bozdogan, H., Bearse, P.: ICOMP: A new model-selection criterion. In: Bock, H.H. (ed.) Classification and Related Methods of Data Analysis, pp. 599–608. Elsevier Science Publishers, Amsterdam (2003)

24. Bradley, A.P.: The use of the area under the ROC curve in the evaluation of machine learning algorithms. Pattern Recognition 30, 1145–1159 (1997)

25. Brockwell, P.J., Davis, R.A.: Introduction to Time Series and Forecasting. Springer, New York (1996)

26. Brooks, C.: A double-threshold GARCH model for the French Franc Deutschmark exchange rate. Journal of Forecasting 20, 135–143 (2001)

27. Broudiscou, A., Leardi, R., Phan-Tan-Luu, R.: Genetic algorithms as a tool for selection of D-optimal design. Chemometrics and Intelligent Laboratory Systems 35, 105–116 (1996)

28. Chatterjee, S., Laudato, M.: Genetic algorithms in statistics: procedures and applications. Communications in Statistics – Theory and Methods 26(4), 1617–1630 (1997)

29. Chatterjee, S., Laudato, M., Lynch, L.A.: Genetic algorithms and their statistical applications: an introduction. Computational Statistics & Data Analysis 22, 633–651 (1996)

30. Chen, C.W.S.: Subset selection of autoregressive time series models. Journal of Forecasting 18, 505–516 (1999)
31. Chen, C., Liu, L.-M.: Joint estimation of model parameters and outlier effects in time series. Journal of the American Statistical Association 88, 284–297 (1993)
32. Chen, R., Tsay, R.S.: Functional-coefficient autoregressive models. Journal of the American Statistical Association 88, 298–308 (1993)
33. Chiogna, M., Gaetan, C., Masarotto, G.: Automatic identification of seasonal transfer function models by means of iterative stepwise and genetic algorithms. Journal of Time Series Analysis 29, 37–50 (2008)
34. Chitre, Y., Dhawan, A.P.: M-band wavelet discrimination of natural textures. Pattern Recognition 32, 773–789 (1999)
35. Choy, K.: Outlier detection for stationary time series. Journal of Statistical Planning and Inference 99, 111–127 (2001)
36. Crawford, K.D., Wainwright, R.L.: Applying genetic algorithms to outlier detection. In: Eshelman, L.J. (ed.) Proceedings of the Sixth International Conference on Genetic Algorithms, pp. 546–550. Morgan Kaufmann, San Mateo (1995)
37. Davis, R.A., Lee, T.C.M., Rodriguez-Yam, G.A.: Structural break estimation for nonstationary time series models. Journal of the American Statistical Association 101, 223–239 (2006)
38. Davis, R.A., Lee, T.C.M., Rodriguez-Yam, G.A.: Break detection for a class of nonlinear time series models. Journal of Time Series Analysis 29, 834–867 (2008)
39. Engle, R.F.: Autoregressive conditional heteroscedasticity with estimates of the variance of United Kingdom inflation. Econometrica 50, 987–1007 (1982)
40. Falkenauer, E.: Genetic Algorithms and Grouping Problems. Wiley, New York (1998)
41. Fogel, D.B.: Evolutionary computation: toward a new philosophy of machine intelligence. IEEE Press, New York (1998)
42. Forlin, M., Poli, I., De March, D., Packard, N., Gazzola, G., Serra, R.: Evolutionary experiments for self-assembling amphiphilic systems. Chemometrics and Intelligent Laboratory Systems 90, 153–160 (2008)
43. Gaetan, C.: Subset ARMA model identification using genetic algorithms. Journal of Time Series Analysis 21, 559–570 (2000)
44. Galeano, P., Peña, D., Tsay, R.S.: Outlier detection in multivariate time series by projection pursuit. Journal of the American Statistical Association - Theory and Methods 101, 654–669 (2006)
45. Ghaddar, D.K., Tong, H.: Data transformation and self-exciting threshold autoregression. Applied Statistics 30, 238–248 (1981)
46. Glendinning, R.H.: Estimating the inverse autocorrelation function from outlier contaminated data. Computational Statistics 15, 541–565 (2000)
47. Glover, F., Kelly, J.P., Laguna, M.: Genetic algorithms and tabu search: hybrids for optimization. Computers and Operations Research 22, 111–134 (1995)
48. Gomez, V., Maravall, A., Peña, D.: Missing observations in ARIMA models: Skipping approach versus additive outlier approach. Journal of Econometrics 88, 341–363 (1999)
49. Gourieroux, C., Monfort, A., Renault, E.: Indirect inference. Journal of Applied Econometrics 118, S85–S118 (1993)
50. Haggan, V., Ozaki, T.: Modelling nonlinear random vibrations using an amplitude-dependent autoregressive time series model. Biometrika 68, 189–196 (1981)

51. Heredia-Langner, A., Carlyle, W.M., Montgomery, D.C., Borror, C.M., Runger, G.C.: Genetic algorithms for the construction of D-optimal designs. Journal of Quality Technology 35, 28–40 (2003)
52. Hosmer, D.W., Lemeshow, S.: Applied Logistic Regression, 2nd edn. John Wiley & Sons, Hoboken (2000)
53. Justel, A., Peña, D., Tsay, R.S.: Detection of outlier patches in autoregressive time series. Statistica Sinica 11, 651–673 (2001)
54. Kapetanios, G.: Cluster analysis of panel data sets using non-standard optimisation of information criteria. Journal of Economic Dynamics and Control 30, 1389–1408 (2006)
55. Kapetanios, G.: Variable selection in regression models using nonstandard optimisation of information criteria. Computational Statistics & Data Analysis 52, 4–15 (2007)
56. Keskinturk, T., Er, S.: A genetic algorithm approach to determine stratum boundaries and sample sizes of each stratum in stratified sampling. Computational Statistics & Data Analysis 52, 53–67 (2007)
57. Larrañaga, P., Lozano, J.A.: Estimation of distribution algorithms: a new tool for evolutionary optimization. Kluwer, Boston (2002)
58. Li, W.K., Lam, K.: Modelling asymmetry in stock returns by threshold autoregressive conditional heteroscedastic model. The Statistician 44, 333–341 (1995)
59. Li, C.W., Li, W.K.: On a double-threshold autoregressive heteroscedastic time series model. Journal of Applied Econometrics 11, 253–274 (1996)
60. Liao, T.W.: Clustering of time series data - a survey. Pattern Recognition 38, 1857–1874 (2005)
61. Lozano, J.A., Larrañaga, P., Inza, I., Bengoetxea, G.: Towards a new evolutionary computation. Advances in estimation of distribution algorithms. Springer, Berlin (2006)
62. Lund, U.: Cluster analysis for directional data. Communications in Statistics – Simulation and Computation 28(4), 1001–1009 (1999)
63. Maulik, U., Bandyopadhyay, S.: Fuzzy Partitioning Using Real Coded Variable Length Genetic Algorithm for Pixel Classification. IEEE Transactions on Geosciences and Remote Sensing 41, 1075–1081 (2003)
64. Minerva, T., Poli, I.: Building ARMA models with genetic algorithms. In: Boers, E.J.W., Gottlieb, J., Lanzi, P.L., Smith, R.E., Cagnoni, S., Hart, E., Raidl, G.R., Tijink, H. (eds.) EvoIASP 2001, EvoWorkshops 2001, EvoFlight 2001, EvoSTIM 2001, EvoCOP 2001, and EvoLearn 2001. LNCS, vol. 2037, pp. 335–342. Springer, Heidelberg (2001)
65. Mitchell, M.: An introduction to genetic algorithms. MIT Press, Cambridge (1996)
66. Mühlenbein, H., Paas, G.: From Recombination of Genes to the Estimation of Distributions I. Binary Parameters, Proceedings of the 4th International Conference on Parallel Problem Solving from Nature, September 22-26, 1996, pp. 178–187 (1996)
67. Ong, C.S., Huang, J.J., Tzeng, G.H.: Model identification of ARIMA family using genetic algorithms. Applied Mathematics and Computation 164, 885–912 (2005)
68. Pasia, J.M., Hermosilla, A.Y., Ombao, H.: A useful tool for statistical estimation: genetic algorithms. Journal of Statistical Computation and Simulation 75, 237–251 (2005)
69. Paterlini, S., Minerva, T.: Evolutionary approaches for cluster analysis. In: Bonarini, A., Masulli, F., Pasi, G. (eds.) Soft Computing Applications, pp. 167–178. Springer, Berlin (2003)

70. Peña, D.: Influential observations in time series. Journal of Business & Economic Statistics 8, 235–241 (1990)
71. Priestley, M.B.: Non-linear and Non-stationary Time Series Analysis. Academic Press, London (1988)
72. Qian, G., Zhao, X.: On time series model selection involving many candidate ARMA models. Computational Statistics & Data Analysis 51, 6180–6196 (2007)
73. Reeves, C.R., Rowe, J.E.: Genetic algorithms - Principles and Perspective: A Guide to GA Theory. Kluwer Academic Publishers, London (2003)
74. Robles, V., Bielza, C., Larrañaga, P., González, S., Ohno-Machado, L.: Optimizing logistic regression coefficients for discrimination and calibration using estimation of distribution algorithms. TOP (2008) (published on line) doi:10.1007/s11750-008-0054-3
75. Roverato, A., Poli, I.: A genetic algorithm for graphical model selection. Journal of the Italian Statistical Society 7, 197–208 (1998)
76. Sabatier, R., Reyne's, C.: Extensions of simple component analysis and simple linear discriminant analysis using genetic algorithms. Computational Statistics & Data Analysis 52, 4779–4789 (2008)
77. Sahni, S., Gonzalez, T.: P-Complete approximation problems. Journal of the Association for Computing Machinery 23, 555–565 (1976)
78. Sessions, D.N., Stevans, L.K.: Investigating omitted variable bias in regression parameter estimation: A genetic algorithm approach. Computational Statistics & Data Analysis 50, 2835–2854 (2006)
79. Tong, H.: Non Linear Time Series: A Dynamical System Approach. Oxford University Press, Oxford (1990)
80. Tsay, R.S., Peña, D., Pankratz, A.E.: Outliers in multivariate time series. Biometrika 87, 789–804 (2000)
81. van Dijk, D., Terasvirta, T., Franses, P.H.: Smooth transition autoregressive models - A survey of recent developments. Econometric Reviews 21, 1–47 (2002)
82. Van Emden, M.H.: An analysis of complexity, vol. 35, Mathematical Centre Tracts, Amsterdam (1971)
83. Vitrano, S., Baragona, R.: The genetic algorithm estimates for the parameters of order p normal distributions. In: Bock, H.-H., Chiodi, M., Mineo, A. (eds.) Advances in Multivariate Data Analysis, Series: Studies in Classification, Data Analysis and Knowledge Organization, pp. 133–143. Springer, Berlin (2004)
84. Wei, W.W.S.: Time Series Analysis. Addison-Wesley, Redwood (1990)
85. Winker, P.: Optimization Heuristics in Econometrics: Application of Threshold Accepting. John Wiley & Sons, Chichester (2001)
86. Winker, P., Gilli, M.: Applications of optimization heuristics to estimation and modelling problems. Computational Statistics & Data Analysis 47, 211–223 (2004)
87. Wu, B., Chang, C.-L.: Using genetic algorithms to parameters (d, r) estimation for threshold autoregressive models. Computational Statistics & Data Analysis 38, 315–330 (2002)
88. Yang, Z., Tian, Z., Yuan, Z.: GSA-based maximum likelihood estimation for threshold vector error correction model. Computational Statistics & Data Analysis 52, 109–120 (2007)
89. Zani, S.: Osservazioni sulle serie storiche multiple e l'analisi dei gruppi. In: Piccolo, D. (ed.) Analisi Moderna delle Serie Storiche, Franco Angeli, Milano, pp. 263–274 (1983)

Meta-heuristics for System Design Engineering

Rachid Chelouah, Claude Baron, Marc Zholghadri, and Citlalih Gutierrez

Abstract. Industries have to design and produce performing and reliable systems. Nevertheless, designers suffer from the diversity of methods, which are not really adequate to their needs. Authors highlight the need of close interactions between product and project design, often treated either independently or sequentially, necessary to improve system design, and logistics in this context. Strengthening the links between product design and project management processes is an ongoing challenge, and this situation relies on perfect control of methods, tools and know-how, both on the technical side as well as on the organizational side. The aim of our work is to facilitate the project manager's decision making, thus allowing him to define, follow and adapt a working plan, while still considering various organizational options. From these options, the project manager chooses the scheme that best encompasses the project's objectives with respect to costs, delay and risks, without neglecting performance and safety. To encourage the project manager to explore various possibilities, we developed and tested a heuristic based on ant colony optimization and evolutionary algorithm adapted for multi-objective problems. Its hybridization with a tabu search and a greedy algorithm were performed in order to accelerate convergence of the research study and to reduce the cost engendered by the evaluation process. The experiments carried out reveals that it was possible to offer the decision maker a reduced number of solutions that he can evaluate more accurately in order to choose one according to technical, economic and financial criteria.

Rachid Chelouah
Laboratoire des Sciences des Systèmes d'Information
EISTI, avenue du Parc, 95011 Cergy-Pontoise Cedex
Tel.: 01.34.25.10.10; Fax : 01.34.25.10.00
e-mail: rachid.chelouah@eisti.fr

Claude Baron and Citlalih Gutierrez
LAboratoire Toulousain des Technologies et Ingénierie des Systèmes (LATTIS),
INSA, 135 avenue de Rangueil,
31077 Toulouse cedex 4, France
Tel.: +33 (0)5 61 55 98 26
e-mail: claude.baron@insa-toulouse.fr

Marc Zholghadri
Laboratoire de l'Intégration du Matériau au Système, Université de Bordeaux,
351, Cours de la Libération, 33 405 TALENCE Cedex
e-mail: marc.zolghadri@ims-bordeaux.fr

A. Abraham et al. (Eds.): Foundations of Comput. Intel. Vol. 3, SCI 203, pp. 387–423.
springerlink.com © Springer-Verlag Berlin Heidelberg 2009

Keywords: co-design, System Engineering, Model Driven Engineering, product design, project organization, multi-objective optimization, ant colony optimization, hybrid algorithms, tabu search, greedy search, decision support.

1 Introduction

The growing complexity and diversity of technologies lead to design always more performing and reliable systems in shorten times without decreasing for that their quality. This phenomenon involves a development and a renewal of methodologies and processes of System Design, in order to obtain a better organized contribution of all the available information and an optimal satisfaction of the requirements defined in terms of delays, economy, safety and quality constraints.

The steps of evaluation and decision in these processes are unanimously recognized as being essential to ensure performances and quality of the process results. Nevertheless, researchers, as well as decision makers in companies, are often extremely deprived in front of the quantity and the variety of the methods and practices. The instrumentation of the activity of design was, in a first phase, focused on the product by integrating collaborative processes for example in the paradigm of concurrent engineering.

A certain maturity in this field having been reached, we think that it is the interaction between the object of the design (the product) and the process of design (the project) which must be now studied and optimized. Indeed, product and project are often treated either in an independent way (what causes inconsistencies) or in a sequential way (what causes additional delays). An analysis of the current practices in these fields highlights the need to bring closer the two processes of design product and project organization, on a methodological level but also by means of dedicated tools, in order to obtain more coherence in the decisions and thus more efficiency.

Thus, the general scientific objective is to associate the various processes that participate to the global design process, to identify the exchanges between these processes, their nature and their level, and to make them closely collaborate in order to guide and to optimize the process of design. In front of the complexity of the problems to be solved, we will rely on the one hand on the standards of System Engineering and on the other hand on the recommendations of the Model Driven Engineering, MDE in short [5], as methodological bases.

In the end, we propose an original approach that integrates product design and project management processes. In this context, our first objective is to help the project manager define, follow and adapt a working plan.

Our second objective is to make the project management process more robust and adaptive in the face of technical, social or economic disturbances. Obviously as regards the project, our goal is to respect the defined target as closely as possible. We have several possibilities for achieving our goal. These possibilities are called scenarios of a project. To find these good scenarios of a project, we have a multitude of possibilities. As we can not test all these possibilities, otherwise we will have a combinatorial explosion, so we rely on heuristics such as ACO (Ant Colony Optimization) and EA (Evolutionary Algorithm).

The goal of this study is to experiment with an hybrid evolutionary algorithm with tabu and greedy searches and to compare these results with those obtained with ant colony optimization and its hybridizations with tabu and greedy searches presented in [8].

This work is original, because there is no work that deals with the coupling between the design and implementation of industrial projects, and the using heuristics to find a good scenario for a project, hence the title of the paper.

The paper is structured as follow: Section 2 presents an analysis of the possible connections between the decomposition into several processes proposed by the EAI 632 (Enterprise Application Integration) standard to model the global design process and the model transformation approach recommended by System Engineering organisms. Section 3 focuses on the System Design and Technical Management processes and illustrates the connections we see between these two processes (level and content). Section 4 shows how to generate and select the best scenarios. The obtained results are presented in Section 5, and Section 6 concludes with the contribution of this methodology to project management and indicates future work envisaged.

2   The Multi-process Point of View of the Ieee 15288 and Eia 632 Standards

Initial planning for the ISO/IEC 15288:2002(E) standard started in 1994 when the need for a common Systems Engineering process framework was recognised. In 2004 this standard was adopted as IEEE 15288. The ISO/IEC 15288 is a Systems and software Engineering standard covering processes and life cycle stages(processes).

According to AFIS (Association Française d'Ingénierie Système), system engineering is "an interdisciplinary collaborative approach to derive, evolve, and verify a life-cycle balanced solution which satisfies customer expectations and meets public acceptability".

2.1   The IEEE Standard

The **ISO/IEC 15288** is a Systems and software Engineering standard covering processes and life cycle stages(processes). The standard defines processes divided into four categories: technical processes, project processes, agreement processes, and enterprise processes. Each description contains a purpose, outcomes, and activities. Example life cycle stages described in the document are: concept, development, production, utilisation, support, and retirement.

Agreement processes

The first step in the systems engineering process is to establish an agreement with the customer, an order to build a new software product.

Project processes

After an agreement is made, the planning of the project will begin and this results in a project plan, which can be modified during the technical processes. In fact, the project processes are not sequential processes and go in parallel with the whole project, because at each moment a project needs planning, assessment and control

Technical processes

The technical processes cover the design, development and implementation phase of a system life cycle. In the earlier agreement phase top level (or customer) requirements have been established

Evaluation processes

Another loop in the model is the evaluation loop. During and after the creation of a software product the following questions have to be answered: Does the product do what it is intended to do? Are the requirements met? and as mentioned in the previous paragraph: Are the requirements valid and consistent? How is the requirements prioritization?

2.2   The EIA 632 Standard

This section presents the main principles of System Engineering about model transformations on which our research relies, coupled to the recommendations of the EIA 632 standard on the decomposition of the global design process into several partner processes.

The EIA 632 standard [1] covers all the processes involved in the development modeling and thus goes beyond the pure technical processes to define a system; it covers the different stages from product realization till their use by the customer. Moreover, it takes into account the contractual supply and acquisition processes (see Figure 1). In this standard, the technical process is also connected to two other partner processes technical management and technical evaluation processes.

The EIA-632 standard associates five partner processes into the general system development process: technical management, acquisition and supply, system design, product realization and technical evaluation. Each process is then decomposed into several tasks that the standard recommends to plan and realize in order to lead a good development process. Each task of the project presents several *options* of realization, each option requiring specific resources, as the Section 4 clarifies. The various possibilities to organize a process are called *scenarios*. The generation of these scenarios must be performed on the basis of technical and non technical considerations related to project management (costs,

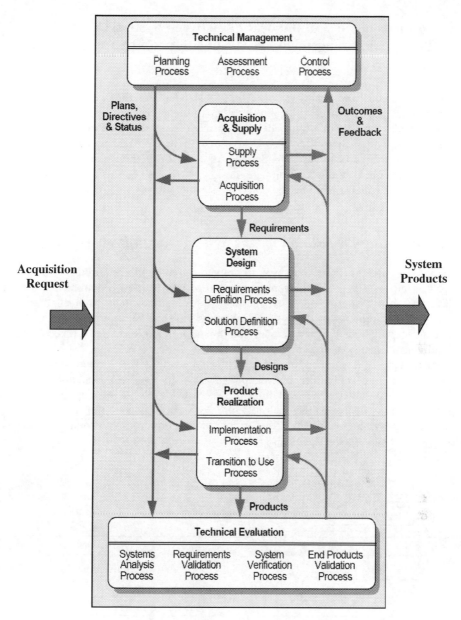

Fig. 1 Processes and their relationship for engineering a system (the EIA 632 standard)

quality, certification, times and supply constraints, risks, etc.). We propose to obtain the different scenarios from a generic representation of the process, using a graph connecting the project tasks and by collecting the several options to achieve them.

3   Towards a Close Collaboration between System Design, Technical Management and Acquisition, and Supply Processes

We simultaneously refer now to the EIA 632 standard for the decomposition of the general design process into multi collaborating processes and to the MDE recommendations to represent model transformations; to illustrate our researches, this paper only consider the System Design and Technical Management processes (as defined in Figure1). Design is defined as "a set of methods and processes to build a specified system". Prototype design work makes easier to develop, integrate, test and maintain a projected system: it is also part of project management work.

3.1   Modeling Proposition for the System Design Process

The System Design Process is used to convert agreed-upon requirements of the acquire into a set of realizable products that satisfy acquire requirements. The developer has to define logical solutions representations that conform to the technical requirements of the system as well as physical solution representations that agree with the assigned logical solution representations, derived technical requirements, and system technical requirements.

As previously mentioned, one of our objectives is to define a reference design process that connects both representations. Even if the EIA-632 doest not exactly define a Process (steps and guidelines by which to develop a system), a kind of generic design process, as the one on Figure 2, can be extracted, thus leading to a hierarchical 'system' architecture that is consistent with an implementation taking into account suppliers and reuses.

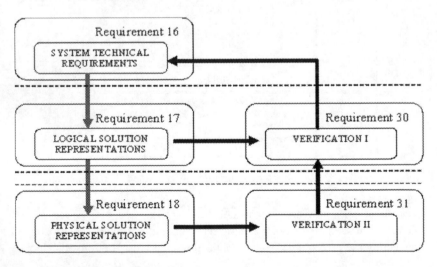

Fig. 2 Reference process

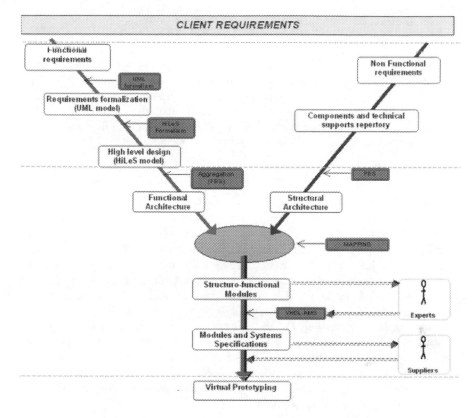

Fig. 3 Implemented process

The survey methodologies for designing systems used in industry leads us to do the following: there are very few methodologies based UML and taking into account the "formal verification". The idea first defended in this paper is based on the idea of linking the appearance design by building a UML model with formal verification.

The second idea advocated in this paper is to add an additional treatment requirements before the design cycle to make explicit the distinction between functional requirements and non-functional.

The diagrams SysML requirements allow the expression of these two types of demands by opposition UML which use cases deal only functional requirements.

The implementation of this generic process (see Figure 3) supposes to join some methods and tools to it, and to detail and complete it if necessary. Our proposition is to build a step-by-step formalization by successive transformations of UML/SysML diagrams that would lead to a logico-temporal architecture of the system. Thus, it is necessary to envisage developing a logical verification interface from this very step. A prototype of tool, HiLeS, has been developed in that purpose; it allows the description of this type of architecture and its verification with Petri nets (tool TINA) [7, 13, 15].

With these complementary data, decomposition into blocks, modules, or building-blocks (to adopt the EIA-632 vocabulary) can be elaborated. That's what we call the partitioning operation.

The general process of Figure 3 illustrates the mapping of the functional design with the structural design to lead to modules. This representation is very interesting because it illustrates the progressive modelling process by a succession of models transformations on the one hand, and the mapping of two complementary models (functional and structural) to obtain a detailed and effective modelling solution on the other hand.

3.2   Mapping of the System Design Process with the Technical Management Process

The term 'Management' defines all the tasks and the organisation for conducting an operation from the starting phase to the 'successfully' final phase. Technical Management is one of the five basic development processes defined in EIA-632 standard. This standard distinguishes three Technical Management sub-processes: these are to be used to plan, assess, and control the technical work efforts required to satisfy the established agreement. The relationship between them is shown in Figure 4.

For example, the Planning Process is used to support enterprise and project decision making and to prepare necessary technical plans that support and complement project plans to:
–   arrive at a decision to supply services according to an external solicitation;
–   determine whether to proceed with an internal enterprise project for a new product or a product improvement;
–   guide the work efforts that will meet the requirements of an established agreement, etc.

This means that if any decisions should be made in terms of potential collaboration with partners, this sub-process have to provide sufficient and necessary data for the management purposes. The connection between this phase and the design of network of partners will be clearly defined in the next section.

Considering the connections between System Design and Technical Management processes, we identified the following difficulties (roughly represented on Figure 5):
–   mapping of the functional specifications with the structural architecture of the product into components,
–   mapping between the product structure and the project organization.

Our proposal is to establish the coupling of the two processes at the stage of high level design where the system is represented with functional blocks, in a hierarchical architecture. The designer has already checked the conformity of the description and the logical architecture, on the HiLeS platform, by the TINA procedure, based on temporal Petri nets verification. Then, (s)he has to envisage the technological choices to take to materialize the design, the arbitrations between hardware and software, the geographical and physical distribution of the embedded systems in the complete system, etc.

Acquisition documents, Agreement,
Outcomes and Feedback

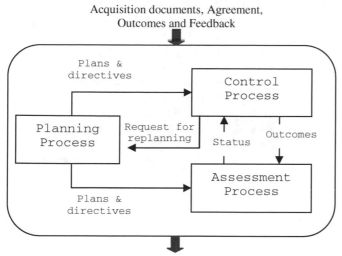

Plans, Directives and Status

Fig. 4 Technical Management sub-processes

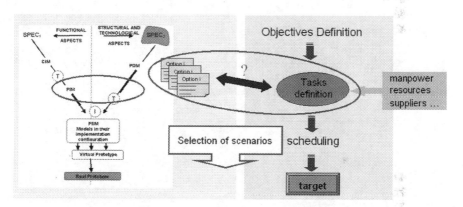

Fig. 5 Connections between System Design and Technical Management processes

The precondition to all these choices is to know the exact composition of each multifunctional module and its physical positioning in the system: it's the partitioning operator, introduced here above. Once the modules thus defined, it becomes possible to specify the design needs to the suppliers in the form of "supplies" and to compare several technologies or several options inside a given technology. These supplies can be assimilated to tasks and used as basic models to schedule the project.

We start the project management process at the logical functional architecture which we organize until the definition of the tasks to plan. To define these tasks, we propose to use the supplies definition, obtained by the fusion of the functional architecture derived by the design product process with the organic decomposition of the system. This fusion must also illustrate the definition of the reusable modules, defining a functional operation, a test, etc. It supposes an expert technical knowledge, formalized and supplemented by the options defined in the specifications. It must be described in the structural model: constitutive elements of the system, active elements supporting the information (data processing architecture), requirements of assembly and conditioning.

The non-functional considerations are brought in confrontation with the functional requirements to lead to a total representation of the System. But the first consequence of this fusion of models is the definition of supplies (partitioning) and of tasks to get to a true coordination between design and project management (Figure 6).

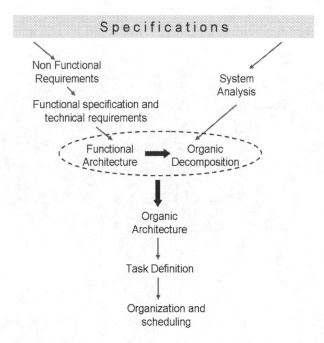

Fig. 6 Project Management process

We can summarize and visualize the whole methodology by the association of two processes in "Y" shown in Figure 6, with the integration of the technical side (conceptual and support information) and the management side (tasks or activities and the available means). These two "Y" process are represented in parallel in Figure 7.

Once this association done, the project manager needs to choose between the different technical and non-technical options to organize his project. He thus needs the help of decision support methodologies and tools. We offer him to

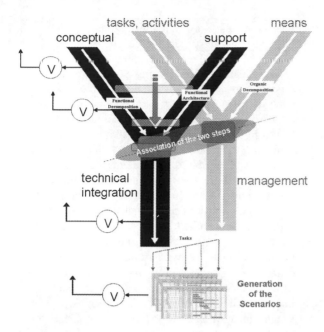

Fig. 7 Association of the processes

automatically generate a reduced set of scenarios, between which he could choose some that allow him to reach his objectives best, thanks to the use of advanced evolutionary algorithms and simulations, with the GESOS tool [3]. In this paper, to present the best scenarios to the project manager, we implemented an aided decision platform using ant colony optimization hybridized with tabu and gready search. This algorithm is presented in Section 4.

3.3 Mapping with the Design of Network of Partners

In the past section, two main difficulties were addressed: functional-structural architecture mapping and product structure-project mapping. However, one the main possibilities that companies have is to co-work with others in order to accomplish their project. Till now, this part of problem is often postponed to business strategy or marketing strategy. But, the collaboration with partners would bring complicated problems to solve in the daily tasks of every worker and manager of firms. This means that a crucial attention should be paid to this aspect, even in the engineering part of the problem.

One may see intuitively, that the product structure is linked to the partners' selection. More precisely, there are mutual constraints that all the actors should take account in their specific decisions: project managers, product designers and network of partners' managers too.

Let us look deeply on the way that the product structural architecture is related to the network of partners.

This question is closely connected to the way that the modules are defined. During any concept development phase of the product lifecycle, engineers look first for technological solutions for any functional requirements of an acquire. Somehow they answer to the very classical question of "make or buy" and this for any module. We can roughly distinguish two classes of items for every single block of a product: in-house made item (make) or out-house item (buy). If in a given structure, every building block of the product has items that belong to these two classes, it means roughly that long-term, mid-term and short-term problems have to be solved. In the strategic level, any out-house item would mean that at least one new partner should participate to the development first and to the manufacturing then. At the mid-term and short-term, it means that manufacturing management system has to take account of explicit constraints (so anticipative elements of the collaboration such as the quality of made items) and also implicit constraints (hardly defined and therefore controllable aspects of collaboration) of every single supplier of these out-house items. To resume, one can talk about the synchronization situations. The Figure 8 illustrates the way that the architecture of a product and therefore the decisions made concerning the in-house and out-house items can modify the complexity of synchronization situations. The product at the left has four building blocks and each of them has both in/out kinds of items. Therefore, there are at least 4 synchronization situations (between in-house items and coming from outside and partners out-house items). If the product can be modified according to the structure proposes in the right, there will remain at least 3 synchronization situations. As a consequence, when any structure of the product will be determined and used for the future development, the managers and engineers should be aware of these problems.

One of the great issues of the co-design paradigm is to study and model these connections and provide recommendations for both engineers (for their design tasks) and managers (for their negotiation and collaboration tasks with partners).

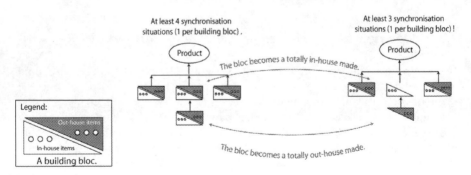

Fig. 8 Illustration of synchronisations

What is discussed in this section show that the subject is quite complicated and further works must be done by determining how the product structural architecture should takes account of the chosen partners' constraints and possibilities.

After having presented the diagram in Y to study the couplings possible between the integration of the technical side and the management side, and as we announced at the end of paragraph 3.2, this section presents two adaptations of an ant colony optimization and evolutionary algorithm as applied to selection a good scenarios. Two possible hybridizations of this methodology are considered: combining first a global search by ant colony (ACO) and evolutionary algorithm (EA) and a local search using the tabu search (TS) or the greedy search (GS) algorithms. The obtained hybrid algorithms are called "HACOTS" and "HACOGS", respectively for ACO, and "HEATS" and "HEGS" respectively for EA..

4 Generation and Selection of the Scenarios

With any a project, the project manager has a limited set of resources and must respect imperative delivery constraints. He needs to establish an initial planning, on the basis of an a priori scheduling of tasks. In addition, should a risk occur during the project, he must be able to readjust this planning in order to maintain the project objectives as closely as possible. Considering other options to carry out some of the tasks is often an option. This section presents how we propose to include these elements in a representation of the project starting from of graph.

4.1 Generic Representation of Scenarios

The representation that we propose is very close to the formalism suggested by Beck [6] for scheduling with alternate activities. This section introduces the graph elements step by step.

The first types of handled constraints are the temporal relations between tasks, specifically the relationships of precedence. Figure 9 presents an example of a graph project including 4 tasks; tasks are presented by numbered squares and relations of precedence by arcs. A possible Gantt diagram can be associated to this graph: Task 1 and Task 2 must be achieved before Task 3, Task 4 can be executed independently of the others.

The expression of the relationships of precedence allows us to consider classical problems of scheduling, based on the assumption that the tasks and their relations are entirely defined. However, in our problem, the final organization of the project is unknown: uncertainties subsist regarding the tasks to be achieved.

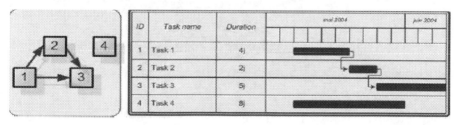

Fig. 9 Example of a project graph with its associated GANTT diagram

We have to deal with several possible options to achieve the tasks, but we also have to make some choices regarding which tasks (or set of tasks) to achieve. That is why the notion of choice exists in the project and why *decision nodes*, represented by a pair of nodes, appear in the project graph. The first circle, with a double outline, represents an opening node and the second, with a simple outline, a closing node. A single sequence of tasks (*a path*) is then allowed between these nodes; they are the equivalent to the "exclusive" or operator for tasks sequences. We define *a path* inside a choice as a set of elements (tasks or nodes) belonging to the same sequence for a choice.

To complete this representation we introduce the "and", useful to represent two tasks belonging to the same choice that appear just after an opening node, the equivalent of the logic "and". To symbolize this relation we use double vertical lines (see Figure 10).

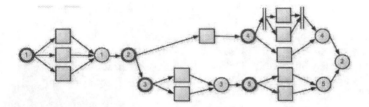

Fig. 10 Example of a project graph with "or" and "and"

This formalism represents all the decisions what can be taken during a project. Note that each task can be implicitly developed by a choice between several options of achievement, as seen on Figure 11, in which the option Y of task X is noted TXOY.

This allows a synthetic representation of the different project progress possibilities. From this model, it is easy to represent the structures of complex projects, from which different scenarios will be generated.

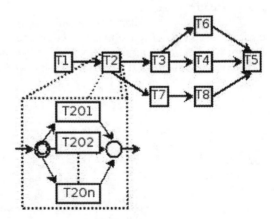

Fig. 11 Explicit development of a task in the graph project: each task represents an elementary choice to be made among different options

4.2   Searching for "Good" Scenarios

Project design generally engenders several different options, leading to numerous scenarios for managing the project. These scenarios can be optimized according to several criteria (cost, time, performance, etc). This section defines what an optimum scenario is and how to obtain it. The main difficulty comes from the multi-objective characteristic of the problem: the optimum is not a simple scalar, but a vector of scalars with one component by criterion. These optimized vectors represent the best compromise solution, called *the Pareto optimal set*, and constitute the *Pareto front* [20] in the objective space.

Before defining what we call a *good* scenario, let us present the optimal criterion which defines this Pareto front. Let us consider a problem of minimization, with *v* and *u*, two vectors of decision. If all components of *v* are strictly lower or equal to those of *u*, with at least one strictly lower component, then the vector *v* dominates the vector *u* in the sense of Pareto (see Equation 1):

$$v \overset{p}{\prec} u \Leftrightarrow \forall i \in [1; n], v_i \leq u_i \land \exists j \in [1; n]: v_j < u_j \qquad (1)$$

The Pareto front is constituted by the set of vectors that are not dominated in the sense of Pareto. In the rest of this article, every time we speak about dominance, we shall use the dominance in the sense of Pareto. We will use this notion of dominance to filter the scenarios and to preserve only those belonging to the population of Pareto optimal or (Pareto optimal scenarios) called P_{fp}.

Having defined the procedure which allows us to select the optimum scenarios, we now explain how we answer the project leader's needs, by offering a method to obtain the set of optimal solutions constituting this Pareto front .

4.3   Hybrid Methods to Select the Best Scenarios

4.3.1   Ant Colony Optimization

In 1999, Di Caro and Gambardella [9] defined the Ant Colony Optimization metaheuristic. Behavior of real ant colonies inspired Ant Colony Optimization of artificial systems, which was subsequently used to solve discrete optimization problems. More generally, research on the collective behaviors of social insects seems to provide methods for the design of combinatory optimization algorithms. In our case, we adapt this collective behavior to the problems of multi-objective optimization.

In nature, real ants are able to find the shortest path from a food source to their nest without using visual signals: the ants lay down "pheromone" (chemical signal) trails on their path outward towards the food source and again up on return towards the nest. At the beginning, the path is randomly chosen, but among all the paths from the nest to the food source, the shortest path quickly becomes the most marked, because the ants taking it arrive more quickly at the nest and statistically have a greater chance of taking it when they return towards the food source.

Hence, the shorter path receives a higher amount of pheromone and this incites a greater number of ants to choose the shorter path. Due to this positive feedback process, very soon all the ants choose the shorter path.

In classical Ant Colony Optimization, after the initialization phase, one can find three sequential operators (appearing with a white background in Figure 12a): (1) the ant colony generating paths from nest to food source, (2) evaporation of the pheromone trail, thus reducing its attractive strength, and (3) compensation of the best ants, laying down pheromone trails on the paths.

To adapt this ant colony algorithm to a multi-objective optimization, we added a local memory that keeps track of the best paths that are not dominated (we will refer to this memory in Section 4.3.1), and that will be rewarded. Finally a procedure of classification of the solutions constituting this Pareto front is used; these modifications appear in a grey background on Figure 12a.

For every ant colony generation, an intensification procedure is applied to all new paths belonging to the Pareto front to improve their quality; it appears as a hatched background in Figure 12a. This procedure is implemented by tabu search or by greedy algorithm. Finally, the mechanism decides if the new front of the Pareto obtained after intensification will be kept or deleted. Before providing the list of the obtained best ants, the algorithm proceeds to a classification of these ants. The pseudocode of the resulting algorithm is described in Figure 12b.

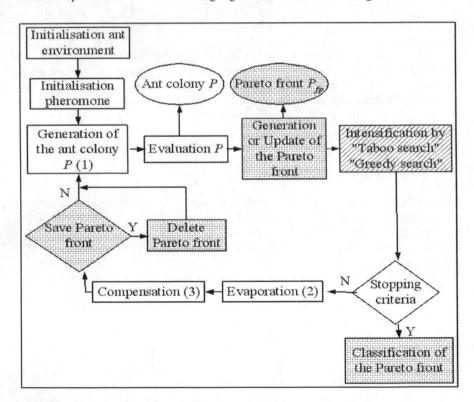

Fig. 12a Classical and modified ant colony algorithms

```
P     : Population of individuals I with size μ
P_fp  : Population of the Pareto front with size η
Generate (Q_P)
Until Stopping criteria are reached Do
  Generate colony (P)
  Evaluate (P)
  Generate (P_fp)
  Repeat η times
    I' ← Intensification (I') (including the evaluation)
  End of Repeat
  Repeat μ times
    Evaporation (Q_E)
  End of Repeat
  Repeat η times
    Compensation (Q_R)
  End of Repeat
  P ← P_fp
End of until
  Classification (n, P_fp)
```

Fig. 12b Pseudocode of the hybrid algorithm with ant colony optimization

Section 4.4 specifies upon which bases the hybrid algorithms are set up. The detailed hybrid algorithm will be described in Section 4.5. The intensification procedures by tabu and greedy searches will be described separately in Section 4.6. The stopping criteria and the classification of the found scenarios are given in Section 4.7 and Section 4.8 respectively.

4.3.2 Evolutionary Algorithm

A first algorithm, inspired by genetic algorithms, has already been proposed [16, 17]. The fundamental principles of genetic algorithms were exposed by Holland in [12]: such algorithms work on a set of solutions, represented by a population of individuals that follow a process of evolution during which they try to adapt to their environment. In the classical genetic procedure, after the generation and the evaluation of an initial population, can be found the sequencing of three operators (appearing with a white background on the figure 13a): (1) selection of individuals for the reproduction and the constitution of the next generation, (2) crossover between parents individuals to build a new offspring of individuals, and (3) mutation, which generally performs minor modifications on a few individuals, in order to introduce some diversity into the population.

It is during this stage of reproduction (crossover and mutation) that new individuals are created from the previously selected individuals. Finally, a last process reinserts some individuals from the offspring into the population in replacement of the least adapted parents. To adapt this genetic algorithm to a

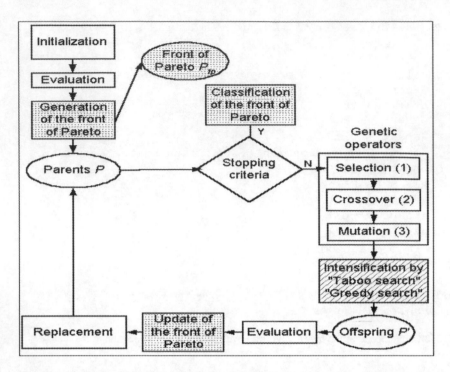

Fig. 13a Classical genetic algorithm and modified one

P : *Population of individuals I with size* μ
P_{fp} : *Population of the front of Pareto*
s : *a scenario or a reduced graph of an individual I*

```
Initialize(P)
Evaluate(P)
Generate(P_fp)
Until Stopping criteria are reached Do
   P' ← Selection (P) (best individuals I of P)
   Repeat μ times
       I' ← Crossover (P_x, I_i; I_j)
       I' ← Intensification(I')(including the evaluation)
       Replacement(I', P')
   End of Repeat
   P ←P'
   Generate(P_fp)
End of until
Classification(n, P_fp)
```

Fig. 13b Pseudocode of the hybrid algorithm

multi-objective optimization, we showed in [3] how we added a memory that keeps a trace of the individuals that are not dominated in the sense of Pareto (to which we will refer to this memory in section 4.3.1), and a procedure of classification of the solutions constituting this front of Pareto; these modifications appear in a grey background on the figure 13a.

In our hybrid algorithm, at each generation, a given number of crossovers and mutations are performed. A procedure of intensification is then applied to each new individual thus generated to improve its quality; it appears in a hatched background on the figure 13a. This procedure can be implemented by an algorithm of taboo search or by greedy algorithm. Finally, the mechanism of replacement decides if the new individual obtained after intensification will be introduced into the parent population or not. Before providing the list of the obtained scenario to the project leader, the algorithm proceeds to a classification of these scenarios. The pseudocode of the resulting algorithm is proposed by the figure 13b.

4.4 Setting Up the Algorithm

Before detailing the implementation of the hybrid algorithm, this section describes how the individuals are coded, how the project scenarios are rebuilt from the coded individuals, and finally how their fitness is calculated. We will illustrate the different steps using a simple example of a project graph (see Figure 14) of 11 tasks and 2 decision nodes; each node has 2 or 3 different choices and each task has up to 10 options of realization.

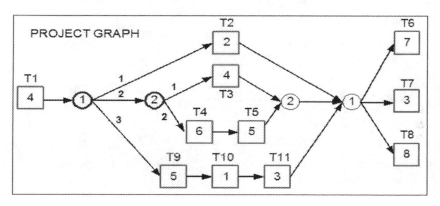

Fig. 14 Example of a project graph

4.4.1 Coding and Reconstructing the Scenarios

The ants are represented by two vectors of integers. The first and the second vectors are the number of tasks and the number of decision nodes in the project, respectively. The values of the first vector represent the chosen options for the project at each task, and the second specifies the selected arc at each decision node. The set of chosen options and selected arcs represent one path in the project graph,

as illustrated in Figure 15. In Figure 16, for example, option T4O6 is selected for
Task 4, option T6O7 for Task 6, the first arc is selected for the choice n°1, etc.

From this representation, it is possible to deduce a planning, by a simple and
automatic procedure, whose characteristics are clearly identified thanks to task
options and chosen paths. We now detail this procedure.

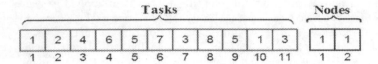

Fig. 15 Representation of an ant by 2 vectors of integer

The first step (Figure 16) consists of translating the considered individual into a
scenario that represents one of the possible organizations to manage the project. The
second step consists of finding a possible planning for this organization (Figure 17).
From the reduced graph of the project, we choose and apply a classical algorithm of
scheduling, which allows us to plan a solution with respect to temporal constraints,
resources constraints, etc... The algorithm currently used is inspired by the PERT
method; basically it considers the precedence constraints between tasks and those
due to the use of cumulative resources. It is possible to complete the algorithm by
integrating of more constraints, thus enriching the model.

This procedure is implicitly used in our algorithm at two levels: each time it is
necessary to evaluate a scenario in order to obtain its performances in the progress
of the algorithm and also to offer some good scenarios to the project manager
under a form that he is accustomed to.

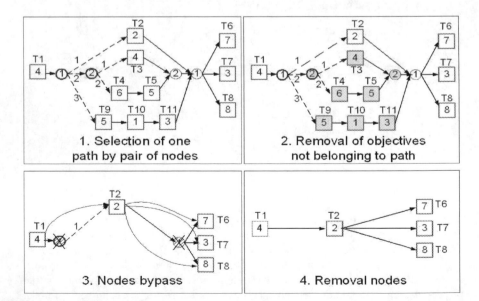

Fig. 16 Reduction of the project graph to obtain a scenario

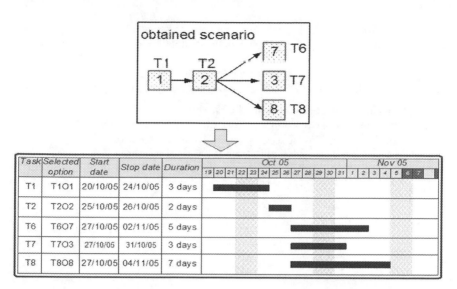

Fig. 17 Tasks scheduling

To evaluate an ant, it is necessary to transform it completely into a representation where the tasks not belonging to the scenario are removed. To perform this operation, we associate, with each ant, a reduced graph of the project in which the tasks are determined from the first vector. Then, we successively reduce this graph by carrying out for each pair of nodes, the following stages: (1) decoding of the ant to determine the selected path, (2) suppression of the set of tasks and nodes between the opening node and the closing node that do not belong to the selected path, (3) creation of precedence links between the tasks before the closing nodes and after the opening nodes, and finally (4) removal of the current pair of nodes. Once this transformation is completed, we obtain a scenario to realize the project.

4.4.2 Objective Function

To measure a scenario's performance, we propose to measure the distance which separates it from the reached project objectives; these must be minimized to reach the objectives as closely as possible. The evaluation of the objective function must allow finding the scenarios of which variations from the objectives are minimal. In our project, we worked with two essential criteria in project management: the difference between the budget and the cost envisaged by the scenario and the difference between the desired duration and the planned one.

4.4.3 Fitness Function

The evaluation must permit the determination of scenarios whose variations compared to the objectives are minimum, while favoring the diversity of scenarios. For calculating of the adaptation function, we chose a method inspired by *Strength Pareto Evolutionary Algorithm* presented in [22]. We proceed with two sets of ants: P the colony and P_{fp} the set of ants composing the Pareto front, $P_{fp} \subseteq P$. If a new ant of the colony P appears to be Pareto optimal, then we create its copy into P_{fp}. The ants of *Pfp* which could not be Pareto optimal any more are removed. The calculation of the fitness is decomposed into two steps:

Step 1: To each ant I of P_{fp} is associated a value s_i representing its force, which is equal to α the number of solutions dominated by I in the colony P divided by the size μ of the colony P increased by one, see Equation (2).

Step 2: The fitness *fi* of each ant I belonging to P is equal to the reverse of the sum of forces of ants of P_{fp} which dominate it, increased by one, see Equation (3).

$$ s_i = \frac{\alpha}{\mu + 1} \tag{2} $$

$$ f_i = \frac{1}{1 + \sum\limits_{\substack{P \\ i, i < j}} s_i} \tag{3} $$

We thus value the individuals of the Pareto front; this favors the convergence towards the good scenarios. The separation of ants belonging to the Pareto front from the rest of the colony permits filtering available solutions and introduces a memory. A method of clustering resulting from *the average linkage method* [19] is introduced to reduce this set: if the number of ants in the Pareto front exceeds a fixed threshold, then we replace a group of close ants by only ones one, the center of the group (we can choose the barycentre for example), and this as many times as necessary to obtain the desired number of ants in P_{fp}. We thus obtain a stabilization of the subpopulation size into niches.

4.5 Detailed Description of the Hybrid Algorithm with Ant Colony Optimization

The algorithm starts with a colony of ants representing a set of scenarios. A scenario is directly rebuilt by using the information stored in its corresponding ant (see Section 4.4.1). The pheromone trail progressively dies down after each iteration, two methods of evaporation are implemented: a linear method and a geometrical method. We save the best ants in a Pareto front, and these "elite" ants representing the best scenarios, are recompensed for depositing pheromone on the chosen options and arcs, representing tasks and decision nodes taking part in the project's scenario. As regards evaporation, pheromone deposition is achieved in two ways: linear or geometrical.

This paragraph describes how the generic hybrid procedure with ant colony algorithm is implemented (Figures 12 and 13).

4.5.1 Generation of Ant Colony

At the beginning, there is the same quantity of pheromone at each decision node arc and at each task option. These quantities are updated during the ant colony

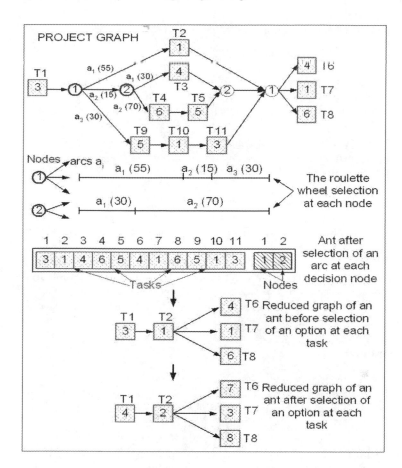

Fig. 18 Ant selection

generation by the evaporation and the reward operations. At each generation, to select a set of scenarios (ants) for a project realization amongst all possible scenarios, we opted for a stochastic procedure using the *Roulette Wheel Selection*, also called stochastic sampling with replacement, such as represented in Figure 18. This method is directly derived from the "proportional selection". Its principle was used to model the behavior of the ant in front of various pheromone trails. The stronger the trail (i.e. the numbers between brackets), the more likely the ant will choose this trail. At each decision node, the available arcs are mapped to contiguous segments of a line, such that each arc's segment is equal in size to the value of the quantity of pheromone deposited on it. A number is randomly generated at each decision node, and the arc whose segment contains the random number is selected. After selecting the arc at each node, we constitute the scenarios, and we repeat the same procedure at each task constituent to this scenario. This process is repeated until the desired number of ants is obtained.

4.5.2 Evaluation of the Ant Colony

The procedure of evaluation follows four steps:

- decode the ant, i.e. find the technical features of the product and rebuild the adopted scenario for the project progress,
- evaluate the objective function of each scenario resulting from the ants' decoding of the colony,
- manage the list of Pareto optimum ants,
- assign a performance mark to scenarios favouring the appearance of good diagrams in the colony (see Section 4.4.3).

4.5.3 Update of the Pareto Front

In the developed application, we decide whether to preserve the elites (Pareto front) during the iterations. We can thus continue to cumulate the recompenses of the best ants as long as they are not replaced by better ones. Starting the algorithm with bad ants, however, can precipitate the research towards a local optimum. To avoid this premature convergence towards a local optimum, it is better to empty this Pareto front at each iteration, in order to give the other paths (ants) a chance to belong to this Pareto front and to be rewarded.

4.5.4 Evaporation

Evaporation is an operator that allows reducing uniformly the quantity of pheromone available on each arc at each decision node, and on each realization option of each task. In general, the quantity of pheromone available is normalized; it lies between 0 and 1. In this algorithm, we implemented two various operations to reduce the quantity of pheromone: *absolute* and *proportional* evaporation.

In the case of absolute evaporation, after each generation the same pheromone quantity is evaporated at each project vertex (decision node or task). The quantity of pheromone which will be evaporated at each arc of a given decision node or at each option of a given task, is equal to the evaporated pheromone quantity at this given vertex (decision node or task), divided by the number of arcs at this decision node or by the number of options of this task.

4.5.5 Compensation

After each ant colony generation, and after evaluation, we reward only the ants belonging to the Pareto front. We add the pheromone on the arcs and on the option of promising ants constituting the Pareto front. In the recompense operator, the deposited quantity of the pheromone is normalized between 0 and 1, and we use the same kind of operators as in the evaporation operator.

In the evaporation phase, if we use the absolute evaporation, and after some ant generation, if a given ant was never generated up to now, then it will never be recompensed, we risk to annul its pheromone trail at its arcs or at its options. To avoid eliminating a set of ants during the search process of the algorithm, we replace the absolute evaporation by the proportional one in the evaporation phase. In this case, we shall always have a pheromone trail at arcs and at options, even if they have never been chosen up to now. Consequently, ants composed by these arcs and options shall always retain a probability of being chosen in the next generation, even if their pheromone trail is weak. It is more interesting to use an absolute recompense to favor the ants belonging to Pareto front, to use a proportional evaporation to encourage the search towards new directions and to give a chance to the news ants to be chosen, even if their pheromone trail is weak.

From an elite ant, we build a partially reduced graph by keeping the nodes belonging to the project graph (see Figure 19). From this partially reduced graph, we reward the options chosen at each task, and the arcs chosen at each decision node taking part in this project graph. For the ant of Figure 11, only the arc a_1 of the node 1 is rewarded, and options 4, 2, 7, 3, 8 of the respective tasks 1, 2, 6, 7, 8 are recompensed.

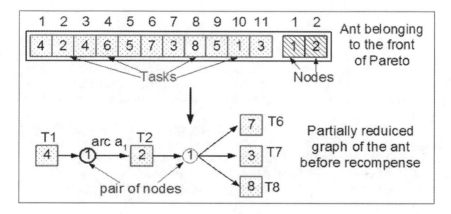

Fig. 19 Ant recompense

4.6 Detailed Description of the Hybrid Algorithm

This paragraph describes the implemented hybrid procedure; it develops the algorithm presented on the figure 5 step by step. The following sub sections describe these steps

4.6.1 Generation of the Initial Population

The general principle of the initial population generation is to sample the search domain uniformly in order to effectively cover it. As we do not have any indication allowing orientating the search, we randomly generate a first individual. For the next individuals, we proceed by a permutation of the choices at each node; for each given choice, we permute the options of some tasks that are randomly chosen. This allows covering the search space at best.

4.6.2 Evaluation of the Population

The procedure of evaluation follows four steps:

- decode the individual chromosome, i.e. find the technical features of the product and rebuild the adopted scenario for the project progress,
- evaluate the objective function of each scenario resulting from the individuals decoding of the population,
- manage the list of Pareto optimum individuals,
- assign a performance mark to scenarios favouring the appearance of good diagrams in the population (see section 3.2.4).

4.6.3 Selection

There are two categories of procedures to choose the parent population for the reproduction starting from the initial population: *deterministic* procedures and *stochastic* procedures.

In deterministic procedures, the best individuals are selected (according to the defined performance function). This method supposes a sorting of the population, which requires a great computing time, particularly if the population is of a large size. Moreover, the least powerful individuals are completely eliminated from the population, and the best individuals are always selected. This selection leads to an impoverishment of the population, and the algorithm risks to be trapped in a local minimum.

In stochastic procedures, the selection always consists in favouring the best individuals, but in a stochastic way, which leaves a chance to the less powerful individuals. Besides, it can happen that the best individual is not selected, and that the children obtained after the evolutionary operations do not reach a performance as good as the best parent contained in the population of the preceding generation.

We opted for a stochastic procedure using the *Roulette Wheel Selection*, called stochastic sampling with replacement, such as defined in [2]. This method is directly derived from the "proportional selection" [18]: the individuals are mapped to contiguous segments of a line, such that each individual's segment is equal in size to the value of its fitness. A number is randomly generated and the individual whose segment contains the random number is selected. The process is repeated until the desired number of individuals is obtained (called mating population i.e. a population ready for the reproduction). To palliate the problem previously mentioned, we integrated a procedure of replacement into the algorithm in order to always select the best individuals between parents and children.

4.6.4 Crossover

The crossover operator recombines pairs of individuals with a given probability to produce offspring. To favour many crossover operations and well cover the solution space, we select the parents from the population with a strong probability of crossover Px. To generate an offspring, we used a crossover binary generator. This generator is run at random and the parity of the result indicates which parent will supply the offspring. The offspring is produced by taking the gene from Parent 1 if the corresponding generator bit is 1 or the gene from Parent 2 if the corresponding generator bit is 0. We repeat this procedure for each gene of the offspring. The figure 10 show the individual offspring obtained from its parents; it also exhibits the scenarios representing each parent and the one representing the offspring individual obtained after the crossover operation.

We specify that if a given arc at one choice or if a given option at one task is not present on any chromosome in the initial population then we never can obtain it with the crossover operator. The mutation operators and the intensification by local search are there to palliate these limitations respectively.

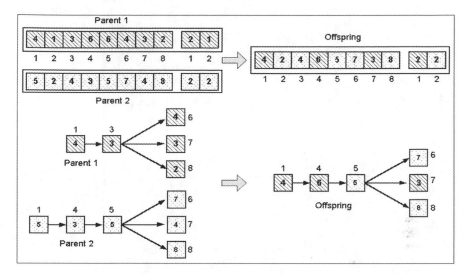

Fig. 20 Crossover

4.6.5 Mutation

This operator allows to provide the paths which were not produced during the generation operation of the initial population. Indeed, this operator only assigns genes coding the performed choices at each node. The mutation operator generates other paths at each node with a probability equal to *Pmut*. The figure 21 considers the offspring individual of the figure 20, obtained after crossover, and shows how the mutation takes place: at the node n° 1 of the reduced graph of the figure 9 appears the arc n° 1 instead of the arc n° 2 after mutation. The graph that corresponds to the individuals before and after mutation also appears on the same figure.

4.7 Intensification

In the algorithm proposed in Figure 5, after the operations of evaluation and generation of the Pareto front, the algorithm builds the reduced graph, and then applies the tabu or the greedy procedure, thus performing a local search on the basis of this reduced graph.

This section now introduces these two different intensification procedures.

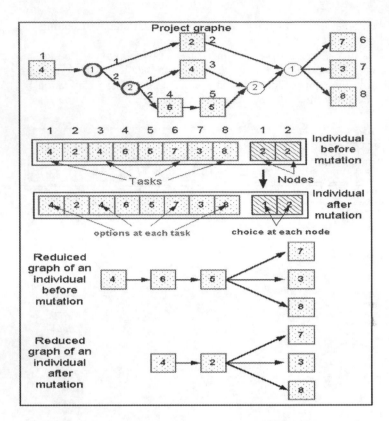

Fig. 21 Mutation

4.7.1 Intensification by the Tabu Search Procedure

The tabu search was developed by Glover [11] and applied with efficiency to the multi-objective optimization [10]. This method uses a tabu list of length L to memorize the N last visited solutions. Starting from any solution x belonging to the solutions X, this solution is momentarily the best solution x^*. We perform a movement towards a solution $v(x)$ located in the neighborhood $[V(x)]$ of x and which does not belong to the list L. The algorithm evaluates the objective function f in each point $v(x)$ of $V(x)$, and retains the best neighbor even if it degrades the objective function f. This allowed degradation of the objective function avoids the algorithm being trapped in a local minimum. This best neighbor becomes the new current solution, it is added to the list, and replaces x^* if it dominates it in sense of Pareto.

The management of this tabu list is voluminous, because it is necessary to remember all the elements defining a solution. To compensate this, we replace the list of prohibited solutions by a list of prohibited movements. The replacement of the list of visited solutions by the elementary transformations $\{x, v(x)\}$ list not only leads to a prohibition to return towards preceding solutions, but also to go towards a set of solutions of which several of them are not visited up to now. To implement this Tabu list, we used a two dimensional list of length L. Each time a movement $\{s, v(s)\}$ is applied to move from s to $v(s)$, the indexes i and j of the changed variable s are saved in the Tabu list. Thus, the opposite movement is prohibited for the next L iterations of the algorithm. At any time, it is easy to check if a given move is marked as Tabu or not.

To permit acceptation of a movement even if it belongs to the Tabu list, a criterion of aspiration is introduced. We accept a prohibited movement after a given number of iterations, and if the solution corresponding to this movement dominates the best solution x^* found up to now by the tabu search procedure in sense of Pareto.

Let S be the set of possible scenarios to realize the project, and $V(s)$ the neighborhood of s. Two scenarios s and $v(s)$ belonging to S are neighbors (i.e. $v(s) \in V(s)$) if they differ by the value of only one alternative. More formally, $V(s)$ = $\{v(s) \in S \mid$ distance $(s; v(s)) = 1\}$. It results that from a scenario s, it is possible to obtain a neighbor $v(s)$ by modifying for one given task i the alternative number j such as the scenario $v(s)$ still is realizable. The movement of s to $v(s)$ is then characterized by whole numbers i and j representing indexes of the variable s_{ij} which was changed. Those indexes which will be considered as attributes of the movement (see Figure 22). If the scenario s is composed by N tasks, and if only one task option is modified, then this scenario s owns N neighboring scenarios. The neighbor s' to replace s must be the scenario dominating s. If several neighbors are dominant, then we randomly choose a neighbor among all the dominant scenarios. If no neighbor is Pareto optimum, then we randomly choose a scenario among all the neighbors.

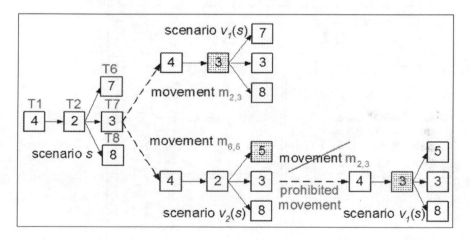

Fig. 22 Neighborhood

4.7.2 Intensification by Greedy Search Procedure

The principle of the greedy search is to locally choose the optimal option at each iteration without ever reconsidering this choice at next time. This strategy does not always lead to a global optimal solution, because it may be that the option choice for a given task offers a good solution to the scenario evaluation at this time, but then makes possible to obtain a better scenario evaluation when we change the options of the next tasks. This method is acceptable in our case because it is hybridized with an evolutionary algorithm which makes it possible to explore the search space; the greedy search is useful only to intensify the search locally. We conclude that these two combined algorithms are complementary; the imperfections of the first one are compensated by the performances of the other.

Figure 23 shows the different steps of the greedy algorithm on the scenario containing 5 tasks obtained after update Pareto list in ant colony optimization. As each task has one or several options, during the visit of each task, the algorithm modifies its options, evaluates the objective function and keep, for this task, the option which supplies the best scenario until now, taking into account the already selected options for the tasks previously visited, then envisages the next task until it covers all the tasks.

4.8 Stopping Criterion

In the hybrid procedures with genetic and ant colony algorithm, we use a strict limitation of the number of generation. In the tabu search procedure, we stop the intensification after a given number of iterations without improvement of the current solution, i.e. without founding a neighbor that dominates the scenario built from an offspring individual obtained after crossover and mutation operators, in sense of Pareto.

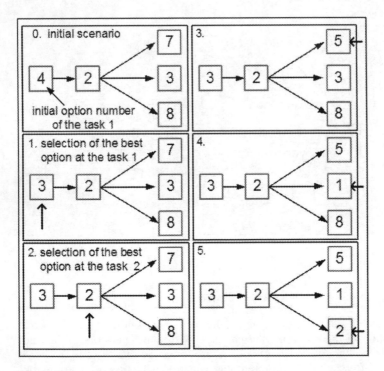

Fig. 23 The greedy search

4.9 Classification of the Found Scenarios

Before presenting the results to the project manager and to help him choose, we present the best found scenarios, classified according to a preference scale. Once the cost of the scenarios known, it is indeed easy to classify the scenarios belonging to the Pareto front before proposing them to the project leader. The method used to classify these scenarios is the approach of Taguchi presented in [4, 8, and 21]. Taguchi's Loss Function is a weighted sum of criteria, based on a formula of evaluation of socio-economic criteria (see Equations 4 and 5). This formula measures with precision when the target is reached with a scenario and analyzes at which point an exact conformity with the technical requirements (execution, quality, reliability, testability...) is obtained. This function allows determining the cost of each scenario selected by aggregating the costs of parameters. Let C be the vector of the costs of a scenario, in which O is the vector of the optimal costs and Δ the vector of the additional costs (Equation 4). The total cost G of a scenario is then given by the Equation 5.

$$C = O + \Delta = \begin{pmatrix} c_1 \\ c_2 \end{pmatrix} = \begin{pmatrix} o_1 \\ o_2 \end{pmatrix} + \begin{pmatrix} \delta_1 \\ \delta_2 \end{pmatrix} = \begin{pmatrix} o_1 \\ o_2 \end{pmatrix} + \begin{pmatrix} k_1(e_1 - t_1)^2 \\ k_2(e_2 - t_2)^2 \end{pmatrix} \qquad (4)$$

$$G = \sum_{i=1}^{2} c_i = \sum_{i=1}^{2} o_i + \sum_{i=1}^{2} \delta_i = \sum_{i=1}^{2} o_i + \sum_{i=1}^{2} k_i(e_i - t_i)^2 \qquad (5)$$

As in our project, we worked with two essential criteria, the difference between the budget and the cost envisaged by the scenario and the difference between the desired duration and the planned one. We set the budget e_1 and the desired duration e_2 equal to 5000, the taguchi constants k_1 and k_2 are equal to the unit and all the components of the vector O are null. In this configuration, the best scenario is the one where the aggregated cost G is null.

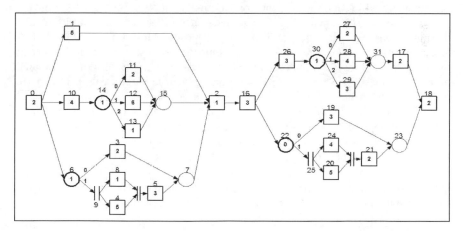

Fig. 24 Example of a benchmark project control

5   Experimental Results

An example of a performed a benchmark graph project is given in Figure 24. This project contains twenty-two tasks (each task has up to ten options of realization) and four nodes (each node has up to three choices).

We cascaded twice then 3 times this initial benchmark, to obtain two new benchmarks. The obtained results are similar with the three benchmarks. In what follows we shall present you only the results associated to the initial benchmark.

The efficiency of the two hybridized ant colony optimization HACOTS and HACOGS are compared. To avoid any misinterpretation of the optimizing results, related to the choice of the neighborhood in the tabu search or any particular initial ant colony, evaporation and recompense operators, we performed each test 100 times, starting from various randomly selected ants, inside the search domain for the following parameters: a colony with 15 ants, a Pareto front with 5 ants, the maximum number of ants colony generations is 30, and finally the maximum number of successive ants colony generations, without improvement of the best solution is 10. We noticed that the control parameters of ACO are as follows: ant population size and ants Pareto size are equal to 15 and 5 individuals, respectively; the maximal number of generation is equal to 30. In the tabu search procedure, we set the Tabu list size to 10, and the maximum number of successive iterations without any improvement of the best solution is of 5.

5.1   Measure of Quality

To study this dispersal of the solutions and compare the quality of the algorithms, several authors ask the question "why quality assessment of multiobjective optimizers is difficult", and how should one evaluate its quality [22]. The solution quality in multi-objective optimization can be assessed in different manner. Some authors compare their obtained front with the optimal Pareto front, and the others propose using different criteria as the contribution, S-metric, Entropy [17 and 23].

Many metrics for measuring the convergence of a set of non-dominated solutions towards the Pareto front have been proposed. Almost all of these metrics were constructed in order to directly compare two sets of non-dominated solutions. There are also approaches which compare a set of non-dominated solutions with a set of Pareto optimal solutions if the true Pareto front is known. In what follows we review some existing metrics for convergence.

The metrics were introduced by Zitzler in [24, 25]. The S metric measures how much of the objective space is dominated by a given non-dominated set A. Using C-metric, two sets of non-dominated solutions can be compared to each other. The D-metric can be used to solve the inconvenience met in C-metric.

Consider the following notations. Let $A, B \subseteq X$ be two sets of decision vectors. The size of the space dominated by A and not dominated by B (regarding the objective space) is denoted D-metric (A, B) and is given by the Equation 6.

$$D\text{-}metric(A, B) = S\text{-}metric(A + B) - S\text{-}metric(B) \qquad (6)$$

5.2 Interpretation of Our Results

In our study, as we don't know the true Pareto front, in order to determine the best control parameters and to compare the efficiency of hybrid ACO, EA, and theirs hybridizations, we are interested to the percentage of efficiency solutions and to D-metric.

In order to assess the results, we execute each algorithm ten times and each run during forty iterations. In Figure 25 the Pareto fronts obtained by the two algorithms ACO and EA are very close.

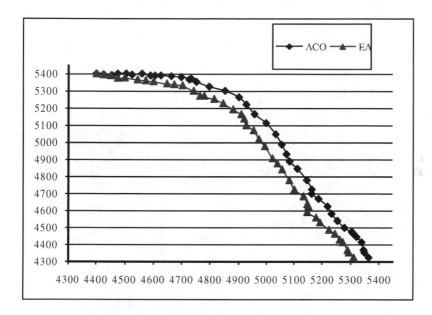

Fig. 25 Comparison of Pareto fronts

Figure 26 shows the evolution of the percentage of efficiency solution with the iteration numbers. The shown values on this figure represent the mean values obtained for one hundred executions. In means, after forty iterations, the percentage of efficiency solutions for HACOTS and HACOGS are close to 100 percent, contrary to those of the ACO which does not exceed 60%. The same comment can be done on the percentage of efficiency solutions for EA and its hybridizations HEATS and HEAGS.

Evolution of the D-metric with the number of iterations is represented in Figure 27. To measure D-metrics, we compare two by two the results of different algorithms (HACOTS-HACOGS and ACO-HACOTS, ACO-HACOGS). The curves provided by D-metric between the algorithms ACO-HACOTS and ACO-HACOGS are very close, contrary to that supplied by the D-metric

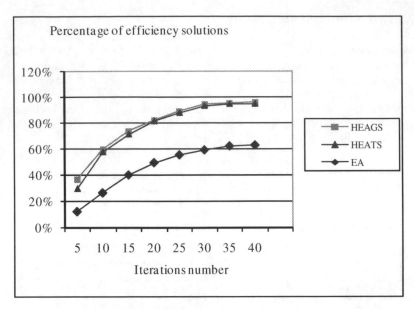

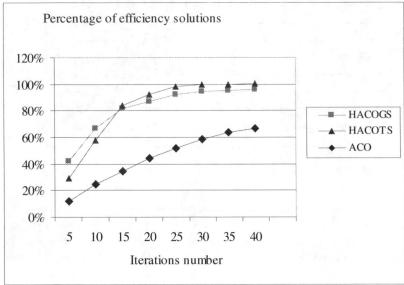

Fig. 26 Evolution of the percentage of efficiency solutions with the number of iterations

HACOTS-HACOGS. The D-metric between HACOTS and HACOGS do not come down below 5. At new iteration, we obtain a reduction of D-metric. After forty iterations, the D-metric (ACO-HACOGS) is close to zero.

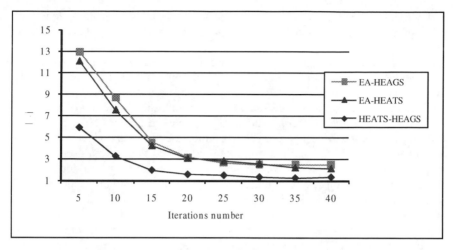

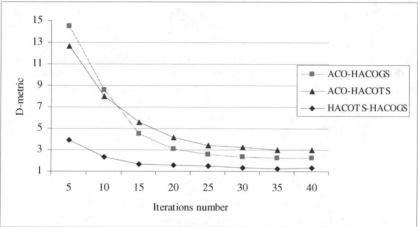

Fig. 27 Evolution of the D-metric with the number of iterations

6 Conclusion

Due to the growing requirements in manufacturing, solutions must be found to design always more performing and reliable systems in shorten times without decreasing their quality for that. This phenomenon progressively involves a development and a renewal of methodologies and processes of System Design, in order to obtain a better organized contribution of all the available information and an optimal satisfaction of the requirements defined in terms of delays, economy, safety and quality constraints.

In order to obtain more coherence in the decisions during project management and thus more efficiency, this paper promoted the idea that there is a real need to bring closer the processes of design product and of project organization, on a methodological level but also by means of dedicated tools. It analyse how both processes could be associated, trying to identify the exchanges between these processes, their nature and their level, to make them closely collaborate in order to guide and to optimize the process of design. For that, we relied on the one hand on the standards of System Engineering and on the other hand on the recommendations of the Model Driven Engineering as methodological bases. The methodology proposed and experimented on an industrial case study [4, 14] is thus based on multiple and consecutive model transformations. The first section dealt with how we model a project to obtain a representation of integrating functional and non functional data and then introduced the criteria used to select the project scenarios. To optimize these scenarios, to find the "best" ones and to fully satisfy the project constraints, we implemented a hybrid algorithm to improve the first results obtained by a previous evolutionary algorithm. This ACO, combined with tabu algorithm and greedy search, has been effectively applied to a multi-objective optimization problem, in a specific project management context. The hybrid method with ACO accelerated the convergence of the algorithm and, with a refinement process performed by the method of tabu algorithm or greedy search, allowed us to obtain a solution of better quality. This faster processing gives the hybrid algorithm more time to explore the solutions' space during the diversification phase. Indeed, this algorithm allows the rapid selection of scenarios belonging to Pareto front and thus considerably reduces the possible number of solutions before their classification. Thus, it becomes possible to offer the decision maker a reduced number of solutions that he can evaluate more accurately in order to choose one according to technical, economic and financial criteria.

References

[1] ANSI/GEIA EIA-632, Standard Processes for Engineering a System, Government Electronics and Information Technology Association (1998)
[2] Baker, J.E.: Reducing Bias and Inefficiency in the Selection Algorithm. In: Proceedings of the Second International Conference on Genetic Algorithms and their Applications, New Jersey, USA, pp. 14–21 (1987)
[3] Baron, C., Rochet, S., Esteve, D.: Gesos: a multi-objective genetic tool for project management considering technical and non-technical constraints. In: Artificial Intelligence Applications and Innovations (AIAI), Toulouse, IFIP World Computer Congress France (2004)
[4] Baron, C., Rochet, S., Gutierrez, C.: Proposition of a methodology for the management of innovative design projects. In: 5th annual International Symposium of the International Council on Systems Engineering (2005)
[5] Blanc, X.: MDA en action, Ingénierie logicielle guidée par les modèles, Eyrolles (2005)
[6] Beck, J.: Texture Measurement as a Basis for Heuristic Commitment Techniques in Constraint-Directed Scheduling, PhD thesis, University of Toronto Department of Computer Science (1999)
[7] Berthomieu, B., Ribet, P., Vernadat, F.: The tool TINA - Construction of Abstract State Spaces for Petri Nets and Time Petri Nets. International Journal of Production Research 42(4) (2005)

[8] Chelouah, R., Baron, C.: Ant colony algorithm hybridized with tabu and greedy searches as applied to multi-objective optimization in project management. Journal of Heuristic (September 21, 2007) ISSN 1381-1231 (Print) 1572-9397 (Online)

[9] Dorigo, M., Socha, K.: Ant Colony Optimization. In: Gonzalez, T.F. (ed.) Handbook of Approximation Algorithms and Metaheuristics, 26.1–26.14. Chapman & Hall/CRC, Boca Raton, FL (2007)

[10] Gandibleux, X., Mezdaoui, N., Freville, A.: A multi-objective tabu search procedure to solve combinatorial optimization problems. Lecture Notes in Economics and Mathematical Systems, vol. 455, pp. 291–300. Springer, Heidelberg (1997)

[11] Glover, F., Hanafi, S.: Tabu Search and Finite Convergence. Discrete Applied Mathematics 119(1-2), 3–36 (2002)

[12] Holland, J.H.: Building Blocks, Cohort Genetic Algorithms, and Hyperplane-Defined Functions. Evolutionary Computation 8(4), 373–391 (2000)

[13] Hamon, J.C., Esteve, D., Pampagnin, P.: HiLeS Designer: A tool for systems design. In: Int. Symposium Convergence 2003: Aeronautics, Automotive & Space, Paris (2003)

[14] Hamon, J.C.: Méthodes et outils de la design amont pour les systèmes et microsystèmes, Thèse de doctorat, LAAS-CNRS, Toulouse, France (2005)

[15] Deb, K.: Multi-Objective Optimization using Evolutionary Algorithms. John Wiley & Sons, Chichester (2001)

[16] HileS Designer, Version 0.9 (November 2005), http://www2.laas.fr/toolsys/hiles.htm

[17] Knowles, J.D., Come, D.W., Oates, M.J.: On the Assessment of Multiobjective Approaches to the Adaptive Distributed. In: Proceedings of the Sixth International Conference on Parallel Problem Solving from Nature, pp. 869–878 (September 2000)

[18] Michalewicz, Z., Schmidt, M.: Parameter Control in Practice. Parameter Setting in Evolutionary Algorithms, 277–294 (2007)

[19] Morse, J.: Reducing the size of the non dominated set: Pruning by clustering. Computers and Operations Research 7(1-2), 55–66 (1980)

[20] Zinflou, A., Gagne, C., Gravel, M., Price, W.L.: Pareto memetic algorithm for multiple objective optimization with an industrial application. Journal of Heuristics, 1381–1231 (August 2008) (Print) 1572-9397 (Online)

[21] Steele, S., et al.: Proceedings of ANTEC 1988 Conference, An Analysis of Injection Molding by Taguchi Methods (1988)

[22] Zitzler, E., Thiele, L.: Multi-objective Evolutionary Algorithms: A comparative Case Study and the Strength Pareto Approach. IEEE Trans. On Evolutionary Computation 3(4), 257–271 (1999)

[23] Zitzler, E., Laumanns, M., Thiele, L., Fonseca, C.M., Fonseca, V.G.: Why Quality Assessment of Multiobjective Optimizers Is Difficult. In: Proceedings of the Genetic and Evolutionary Computation Conference, GECCO 2002, New-York, July 9-13, 2002, pp. 666–674 (2002)

[24] Zitzler, E., Laumanns, M., Thiele, L., Fonseca, C.M., Fonseca, V.G.: Performance assessment of multiobjective optimizers: an analysis and review. IEEE Trans. Evolutionary Computation 7(2), 117–132 (2003)

[25] Zitzler, E., Thiele, L., Bader, J.: On Set-Based Multiobjective Optimization. Technical Report 300, Computer Engineering and Networks Laboratory, ETH Zurich (February 2008)

Transgenetic Algorithm: A New Endosymbiotic Approach for Evolutionary Algorithms

Elizabeth F. Gouvêa Goldbarg and Marco C. Goldbarg

Summary. This chapter introduces a class of evolutionary algorithms whose inspiration comes from living processes where cooperation is the main evolutionary strategy. The proposed technique is called Transgenetic Algorithms and is based on two recognized driving forces of evolution: the horizontal gene transfer and the endosymbiosis. These algorithms perform a stochastic search simulating endosymbiotic interactions between a host and a population of endosymbionts. The information exchanging between the host and ensosymbionts is intermediated by agents, called transgenetic vectors, who are inspired on natural mechanisms of horizontal gene transfer. The proposed approach is described and a didactic example with the well-known Traveling Salesman Problem illustrates its basic components. Applications of the proposed technique are reported for two NP-hard combinatorial problems: the Traveling Purchaser Problem and the Bi-objective Minimum Spanning Tree Problem.

1 Introduction

The inspiration in natural processes in the construction of computational methods has led to the emergence of various classes of algorithms. One of these classes is constituted by the evolutionary algorithms, which search the space of solutions of optimization problems through operators inspired on

Elizabeth F. Gouvêa Goldbarg
Universidade Federal do Rio Grande do Norte, Campus Universitário Lagoa Nova, Natal, Brazil
e-mail: beth@dimap.ufrn.br

Marco C. Goldbarg
Universidade Federal do Rio Grande do Norte, Campus Universitário Lagoa Nova, Natal, Brazil
e-mail: gold@dimap.ufrn.br

A. Abraham et al. (Eds.): Foundations of Comput. Intel. Vol. 3, SCI 203, pp. 425–460.
springerlink.com   © Springer-Verlag Berlin Heidelberg 2009

biological mechanisms. Evolutionary algorithms are defined to be stochastic search methods that operate on a population of candidate solutions that evolves iteratively by means of variation and selection until a stopping criterion is satisfied [44]. Due to their characteristic of performing the search with populations of solutions, these algorithms have an intrinsic parallelism where many different possibilities are explored simultaneously. The search operators used by a major part of these algorithms are inspired on neo-Darwinian evolutionary mechanisms whose typical rules are: selection, crossover and mutation [45]. The information is transmitted from generation to generation based on the notion of vertical inheritance of genetic characters, the one that is passed from parents to offspring, such as in Genetic Algorithms.

Recent research in evolution has shown that the information sharing among individuals regarding evolutionary mechanisms is much more flexible and dynamic than previously thought. In addition to the vertical inheritance of genes, new mechanisms have been discovered and taken as strong evolutionary forces. This chapter introduces a new class of evolutionary algorithms, named Transgenetic Algorithms, whose inspiration comes from two major recognized evolutionary forces: the horizontal gene transfer and the endosymbiosis.

Horizontal or lateral gene transfer refers to the acquisition of foreign genes by organisms. It occurs extensively among prokaryotes (living forms whose cells do not have nucleus) and provides organisms with access to genes in addition to those that can be inherited [30]. Researches in genomics have increasingly shown that many genes have been acquired by horizontal transfer [29]. Sequencing of multiple, complete genomes of diverse organisms have changed the picture of evolution, bringing to light the horizontal gene transfer as a significant evolutionary force [47]. Those researches show how the horizontal gene transfer compliments the modular nature of genetic information, making it feasible to swap whole sets of genetic code - like the genes that allow bacteria to defeat antibiotics.

The term "endosymbiosis" specifies the relationship between organisms which live one within another (symbiont within host). Present-day intracellular associations include a range of parasites, mutualists and commensal symbionts that play important roles in the ecology and physiology of their hosts. Endosymbiosis, a biotic interaction wherein a symbiont inhabits inside its host cell or digestive organs, is universally recognized in nature [9]. The first works on endosymbiosis date back to the period among 1880 and 1926 and are due to Sachs, Altmann and Mereschkowsky [38]. Ivan Wallin in his book "Symbionticism and the Origin of Species" published in 1927 [38] proposed what became later known as the Theory of Symbiogenesis. Margulis [41] explains the difference between the ecological concept of symbiosis and the term "symbiogenesis". Symbiosis refers to the living together of organisms of different species. Endosymbiosis is a kind of symbiosis where one partner, the endosymbiont, lives inside of another, the host. Symbiosis is not an evolutionary process "per se". It refers to physiological, temporal or topological associations with environmentally determined fates. Symbiogenesis, however, implies the appearance of new tissues, new organs, physiologies or other new features that result from protracted symbiotic association. In the endosymbiotic evolution, a new mutualistic system emerges from phylogenetically distinct organisms. The discovery of those systems is considered one of the major transitions in evolution [43].

The scenario described above provides an instigating alternative to the concept of evolution based on the logic of the "selfish DNA" [15] and opens the possibility of using agents or operations inspired on intra, inter and extracellular natural vectors in order to guide and accelerate the search performed by evolutionary algorithms. Inspired on those natural mechanisms, the Transgenetic Algorithms perform a stochastic search proposing an evolutionary process which the basis is the information exchanging among individuals of different natures: host and endosymbionts. In these algorithms the evolution is thought to occur among a population of endosymbionts and its host (a cell). The Transgenetic Algorithms imitate the evolutionary moment of assimilation of the endosymbionts by their host. The population of endosymbionts is the base for the search process representing problem solutions. The host contains information about the problem ("a priori" information) and the search process ("a posteriori" information). The information exchanging between the host and the endosymbionts is intermediated by agents, called transgenetic vectors, who are inspired on natural mechanisms of horizontal gene transfer. These vectors are the main tools for search intensification and diversification. Unlike other evolutionary approaches, in Transgenetic Algorithms endosymbionts (chromosomes) do not share genetic material directly by means of crossover or recombination.

Mechanisms of horizontal gene transfer and endosymbiotic interactions, separately, were sources of inspiration for the development of other evolutionary algorithms that are briefly presented in Sec. 3. The approaches of these algorithms, however, are fundamentally different from the one proposed in this chapter, which incorporates these two biological concepts simulating what is known by the biologists as endosymbiotic gene transfer [63]. The previous approaches dealing with the concept of horizontal transfer have a single context of information: the population of chromosomes. The basic difference between these algorithms and standard genetic algorithms is that the formers use operations for the exchange of genes between the chromosomes of the population based on mechanisms of horizontal transfer in order to avoid crossover operations. The algorithms proposed previously in the literature that use the concept of endosymbiotic interactions are based on the ecological concept of symbiosis, being a variation of cooperative symbiotic algorithms. These approaches are quite different from the Transgenetic Algorithms, which use other contexts of information than the population of chromosomes and that incorporate such information in the chromosomes through operators based on mechanisms of horizontal gene transfer.

The remainder of this chapter is organized as follows. The biological fundamentals concerning the horizontal gene transfer and endosymbiosis are

presented in Sec. 2. A brief review of evolutionary algorithms based on correlated biological concepts is presented in Sec. 3. The Transgenetic Algorithms
are introduced in Sec. 4. A didactic example with the Traveling Salesman Problem illustrates the main features of the proposed approach. In Secs. 5 and 6 the
proposed approach is applied to a single objective and to a bi-objective combinatorial problem, respectively. Peculiarities of the Transgenetic Algorithms
regarding other evolutionary approaches are presented in Sec. 7. Finally, some
conclusions and future directions for research are pointed out in Sec. 8.

2   Biological Fundamentals

The horizontal gene transfer is defined to be the movement of genetic material between organisms other than by descent in which information travels
through the generations as the cell divides. During bacterial evolution, the
ability of bacteria to adapt to new environments most often results from the
acquisition of new genes through horizontal transfer rather than by the alteration of gene functions through numerous point mutations. Horizontal gene
transfer between unrelated species is thought to be one of the leading creative
forces driving bacterial evolution [16].

Bacterial species have acquired several mechanisms by which to exchange
genetic materials. The classification of such mechanisms is a hard task due
to the sheer diversity of mobile elements that are known to exist [67]. Some
researchers think about those mobile genomic elements as the sum of a set
of relatively independent and exchangeable functional units that can broadly
be categorized as promoting intercellular or intracellular mobility, or replication, selection, stability and maintenance [67]. Two types of those mobile elements are called plasmids and transposons. Plasmids are mobile genetic particles that can replicate independently of the chromosome. They are composed
of DNA and can be thought of as "mini-chromosomes". By means of genetic
engineering, a plasmid can be built with an artificially created DNA. That artificial DNA is originated from two or more sources and is called "recombinant
DNA". The plasmids formed with recombinant DNA are often referred as "recombinant plasmids". Transposons or "jumping genes" are genetic elements
that can spontaneously move from one position to another in a DNA molecule.
Transposons are DNA sequences that are part of other genetic elements such
as chromosomes or plasmids. Transposons are classified into two classes based
on their mechanism of transposition: Retrotransposons and DNA transposon.
Retrotransposons work by copying themselves and pasting copies back into the
genome in multiple places. DNA transposons usually move by a mechanism
analogous to cut and paste, rather than copy and paste. The mechanisms of
transposons can result on an effect that is similar to insertions, deletions, inversion, fusion and translocations of genetic material [56].

The means bacteria use to acquire new genetic information are known
as: transformation, conjugation and transduction [12]. Transformation is the

uptake of foreign genetic material that can mediate the exchange of any part of a chromosome. The information is acquired and incorporated spontaneously. Conjugation is the method where the transference of the mobile elements occurs between bacterial cells that are in physical contact. The transduction is a method in which the transport of the mobile element between organisms involves the mediation of viruses. In the case of conjugation and transduction, for the occurrence of the horizontal transfer, the genetic information should be selected, copied or detached from the donor cell's DNA, and then directly transported or spontaneously acquired by the receptor cell where the genomic element is incorporated. A detailed summary of these mechanisms can be found in the paper of Zaneveld et al. [67].

Another important point to note is that in the intracellular context of the recipient cell the imported genetic information may be recirculated. This mechanism allows the genetic material acquired by the recipient cell be mixed with plasmids of the cell resulting in the possibility of complex patterns of recombination.

Besides the horizontal gene transfer mechanisms, the metaphor of the Transgenetic Algorithms also relies on the endosymbiosis as an evolutionary force. The endosymbiotic theory of evolution was popularized by Lynn Margulis with the disclosure of her essay "Symbiosis in Cell Evolution" which was further incorporated in a book [40]. Margulis states that cooperation, interaction, and mutual dependence among life forms were significant driving forces in evolution. According to Margulis and Sagan [42], "Life did not take over the globe by combat, but by networking".

The emergence of integrated organisms occurs with each of the subunits providing some adaptive benefit to its partner [17]. The endosymbiosis is directed to the endosymbiont absorption on the long run. As a result of that absorption the endosymbiont genome is reduced while the shared genetic material is augmented. The endosymbiont becomes more specialized while the host takes some of the functions of the endosymbiont [65]. The phenomenon involves different genetic rearrangements in both organisms [64].

The living world is divided into two cell categories: prokaryotes (cells without nucleus) and eukaryotes (cells with nucleus). Two great classes of eukaryotic cell organelles, plastids and mitochondria, evolved symbiogenetically in two independent events [41]. Plastids, such as mitochondria, once were prokaryotic cells that entered a host and became endosymbionts. Today, they are organelles, compartments of eukaryotic cells. During evolution, organelles relinquished many genes to the chromosomes of their host, but they also learned to reimport the nuclear-encoded products of transferred genes [62]. Horizontal gene transfer between the endosymbionts and the host plays an important role in endosymbiosis. Within the endosymbiosis, the horizontal gene transfer has its own characteristics. This form of horizontal gene transfer is called endosymbiotic gene transfer and as noted by Timmis et al. [63], comparative genome analyses show that gene transfers have occurred at different times in the past and indicate that the process is continuing.

The endosymbiotic gene transfer processes are complex and vary with the nature of the cell. Basically they are composed of systems that allow the identification and acquisition of genes in the donor cells, processes that allow the genetic information to be transported in the cellular environment without being destroyed or corrupted, and systems that incorporate this DNA to the receptor cell.

3 Evolutionary Algorithms Based on Correlated Biological Concepts

Several concepts discussed previously inspired the design of different types of non standard evolutionary algorithms. The algorithms that exploit the mechanisms of intracellular cooperative co-evolution were called Endosymbiotic Algorithms. The algorithms that employ techniques based on the mechanisms of horizontal gene transfer did not constitute a specific algorithmic class. They are accepted as belonging to the class of Genetic Algorithms equipped with special operators. Some of these methods are reviewed in the following.

Usually the endosymbiotic algorithms are classified as cooperative co-evolutionary algorithms [31, 32]. Co-evolutionary algorithms are search algorithms that imitate the biological symbiosis. Although there are many variants of co-evolutionary algorithms, they are typically classified into two main forms: cooperative co-evolutionary algorithms and competitive co-evolutionary algorithms. Cooperative algorithms are based on positive fitness interactions between individuals of different populations. In these algorithms, a success on one individual improves the chances of survival of the other. The populations may reciprocally enhance the adaptability to complex environments by the cooperative relationships and co-evolution [10]. On the other hand, competitive algorithms are based on inverse (negative) fitness interactions between individuals of the different populations [28, 54]. A success on one side implies a failure of the other side to which one must respond in order to maintain one's chances of survival. Daida *et al.* [14] presents a study of the features of symbiotic interactions in evolutionary computing.

As proposed by Kim *et al.* [31, 32], the endosymbiosis is imitated in the computational context through the formation of three different populations: P_a, P_b and P_c. P_a and P_b consist of partial solutions of the investigated problem. P_c consists of complete solutions. Initially, P_a and P_b join together to form P_c. Then, during the algorithm execution, the three populations interact to form the individuals of P_c. Each population evolves, separately, by means of a standard genetic algorithm. The cooperative co-evolutionary aspect of the endosymbiotic algorithms allows the approach, with the necessary adaptations, to incorporate other techniques practiced in the cooperative evolutionary computing.

Several papers addressed the issue of horizontal gene transfer in the development of genetic algorithms with special strategies. The transduction is addressed in the paper of Kubota *et al.* [37]. The author proposes an evolutionary algorithm called VEGA (Virus-Evolutionary Genetic Algorithm) where a population of chromosomes (hosts) is attacked by a population of viruses. The viruses are formed by sub-chains of DNA (pieces of the solutions represented in the chromosomes). The viruses are vectors capable of copying DNA fragments from the population of hosts, transport those DNA fragments and transcribe the genetic information in other chromosomes. Basically, the VEGA algorithm is a Genetic Algorithm with a stage of viral transduction [18, 36].

Harvey [27] proposes a microbial genetic algorithm with a type of conjugation based on tournament selection. Parents are selected on a random basis. The tournament is used to define, among the two chromosomes selected for crossover, the donor (winner) of genetic material and the recipient (loser) cell. After defining two points, just like in crossover, the genes enclosed in the donor are injected in the recipient individual. The winner is not changed. Although a mimetism of a conjugation process is not reported explicitly, the recombination method proposed by Harvey [27] is generalized in the paper of Mühlenbein and Voigt [46]. The paper opens the possibility of obtaining the genetic material necessary for the conjugation from several different chromosomes. Smith [61] uses conjugation as a method of genetic recombination to solve hard satisfiability problems. Perales-Graván and Lahoz-Beltra [48] present a genetic algorithm that uses an operator based on conjugation. In their algorithm a pair of chromosomes is chosen at random in the current population. The fittest chromosome is considered the donor cell while the other chromosome is the receptor. Part of the donor chromosome is copied to the receptor and the necessary adjustments to maintain feasibility are done. The efficiency of the bacterial conjugation operator is illustrated on designing an AM radio receiver where the main features of the electronic components of the AM radio circuit, as well as those of the radio enclosure, are optimized.

Simões and Costa [59, 60] present genetic algorithms with operations based on transformation for the 0/1 Knapsack Problem. Instead of crossover, they use transformation mechanisms as the main genetic operators to generate variation in the population. Initially, fragments of DNA are obtained in the population, which are aggregated to chromosomes along the evolutionary process. These DNA fragments consist of binary strings of different lengths and form the gene segment pool. Transformation is applied every generation instead of a standard crossover operator. Every generation, the gene segment pool is updated with new genetic information obtained in the current population. The idea is very similar to the method presented by Fukuda *et al.* [18], without the presence of a virus as an intermediary agent.

Transposons also motivate operators for a number of genetic algorithms. The first genetic algorithms that used transposition operations were presented by Simões and Costa [57, 58] who utilized transposition instead of crossover operations. Chan *et al.* [11] and Yeung *et al.* [66] include transposition in a multiobjective genetic algorithm. The transposition process considers that the transposons are sequences of consecutive genes in the chromosomes. The algorithm uses cut-and-paste and copy-and-paste transpositions. In the former, a sequence of genes is cut from its original position and pasted into a new position in the chromosome. In the later operation, the DNA fragment that forms the transposon replicates itself and the copy is inserted into a new location of the chromosome, whereas the original sequence of genes remains unchanged. Transpositions occur between the selection and the crossover. In the transposition operation, the choice between the cut-and-paste and copy-and-paste transposition is done randomly. The transpositions can occur within the same chromosome or between different chromosomes. Whether the transposition is done within the same chromosome or between different chromosomes is also a random choice.

4 Transgenetic Algorithms

Transgenetic algorithms are evolutionary computing techniques based on living processes where cooperation is the main evolutionary strategy. Those processes contain the movement of genetic material between living beings and endosymbiotic interactions. The basic components of Transgenetic Algorithms are presented in Sec. 4.1. A didactic example that illustrates the concepts introduced in Sec. 4.1 is given in Sec. 4.2 with the well-known Traveling Salesman Problem.

4.1 Basic Components of the Transgenetic Algorithms

The Transgenetic Algorithms are based on three premises. The first premise consists in contextualizing the computational evolution in a host cell populated by endosymbionts. In nature, the chromosomes of the host and the endosymbionts are different. In addition to its chromosomes, other genetic information exists in the host's cytoplasm. In Transgenetic Algorithms the endosymbiont chromosomes represent solutions. The genetic material of the host consists of information related to the problem, clones of the endosymbiont chromosomes or fragments of the endosymbiont chromosomes that are transferred to the host along the evolutionary process. The information related to the problem is obtained "a priori". This information simulates the evolutionary past of the host. The clones and fragments of the endosymbiont chromosomes are information obtained "a posteriori". As noted in Sec. 2, the extent to which the endosymbionts transmit genetic information to the nucleus of the host cell, they also learn to re-import the nuclear-encoded products of transferred genes [62]. This postulate firms the computational endosymbiotic interactions as a compact system with genetic material cycling.

The second premise consists in guiding the artificial evolution through the imitation of the process of optimization of the genetic code of endosymbionts to the host. In nature the process of endosymbiosis promotes the formation of a hybrid individual through genetic changes of the host and the endosymbionts.

In this process the genetic code of the endosymbionts is optimized with the aim of eliminating metabolic redundancies and to allow using the functions of the host. In the Transgenetic Algorithms, the optimization of the code of the endosymbiont represents the optimization of the problem that is being tackled. The process ends, similarly to the natural, when the exchange of genetic information between host and endosymbionts do not result in any useful changes for such optimization process.

The third premise consists in mimicking the mechanisms of information exchange between the host and the population of endosymbionts. The host has units of genetic information (that are thought to be in its cytoplasm). The genetic information contained in the host is stored in a repository. The information of the host's repository is transferred to the endosymbiont chromosomes by means of vectors that mimic natural mechanisms of gene transfer. The entities that manipulate the genetic code of the endosymbiont chromosomes are called *transgenetic vectors*. In analogy with the terminology employed by microbiologists, there are four types of transgenetic vectors: plasmid, recombinant plasmid, virus and transposon. A detailed explanation of the transgenetic vectors is given further.

Information about the problem to be tackled by the algorithm, called "a priori" information, can be obtained from a number of sources, such as: upper or lower bounds, heuristic solutions, information obtained with statistical analysis of the problem (instance) structure, among others. "A posteriori" information is obtained during the algorithm execution, such as solutions or parts of solutions. This information is carried by the transgenetic vectors from the host to the endosymbiont chromosomes.

The variation in the endosymbiont chromosomes that is obtained with the manipulation of the transgenetic vectors generates information that re-feeds the host's repository.

The flow of information in Transgenetic Algorithms is illustrated in Fig. 1 where the host cell and one endosymbiont cell are represented.

Plasmids, viruses and one type of recombinant plasmids receive genetic information from the host to manipulate the endosymbiont chromosomes. Another type of recombinant plasmid and the transposons do not receive genetic information from the host, once they have their own methods to generate their information. After manipulation, the chromosomes are modified, and, as a result, new information appears in the context of the population of endosymbionts. The feedback of the process is done with the new information being sent to the host context. Three distinct contexts are considered in Transgenetic Algorithms:

- a population of endosymbiont chromosomes;
- a population of transgenetic vectors that represent vehicles for transferring genes and editing the genetic information of the endosymbionts;
- the genetic information contained in the host, that here, is referred as the genetic information repository.

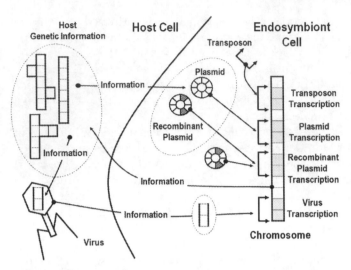

Fig. 1 Flow of information in a Transgenetic Algorithm

A transgenetic vector, λ, is a pair $\lambda = (I, \phi)$, where I is an information string and ϕ is a manipulation method, $\phi = (p_1, ..., p_s)$, p_j, $j = 1, ..., s$, are procedures that define the vector's action. Two types of information can be carried by the transgenetic vectors: DNA fragments obtained in the genetic information repository or abstract rules for genetic rearrangement. The manipulation method determines how that information is transcribed into chromosomes. The action of a transgenetic vector over an endosymbiont chromosome is referred as manipulation. This action causes change in the code of the manipulated chromosome. As a consequence, the fitness of the manipulated chromosome is also altered. The transgenetic vectors differ from each other according to the type of information they carry and according to the distinct procedures their manipulation methods are constituted. The procedures that compose the manipulation methods of the transgenetic vectors are summarized in Table 1.

A vector is called plasmid when its information is encoded in the same way as in the chromosomes (a DNA fragment) and its method is constituted by procedures p_1 and p_2. The recombinant plasmids are differentiated from the plasmids regarding the method their information strings are built. In the biological context, the recombinant plasmids mix information obtained from various sources of genetic information. The computational recombinant plasmids undertake these mixtures through own selection and recombination methods of genetic information. The information of recombinant plasmids can be built in three ways. First, the information is built by grouping distinct units of genetic information of the host's repository (typical recombinant plasmid). Second, the information is built by a constructive method (autonomous recombinant plasmid). Finally, the information can be built with a mixture of units of

Table 1 Procedures of the transgenetic vectors

Procedure Name		Description
p_1	Attack	Verifies, according to a given criterion, whether chromosome C is susceptible to be manipulated by the transgenetic vector λ.
p_2	Transcription	Defines how the information string carried by λ is transcribed in C. The transcription is executes only if procedure p_1 returns "true".
p_3	Blocking	Establishes a period of time (e.g. number of iterations) in which the transcribed information cannot be altered in C.
p_4	Identification	Identifies positions in C that will be utilized to limit λ's operation.

genetic information of the host's repository and a constructive method (mixed-information recombinant plasmid).

A vector is called virus if its information string is encoded like in the plasmids and its method is constituted by procedures p_1, p_2 and p_3.

The information string of transposons are rules for rearranging the genes of the endosymbiont chromosomes. They utilize procedures p_1, p_2 and p_4. The vector transposon requires the demarcation of a certain area of the chromosome it will manipulate (procedure p_4). In the proposed approach the action of transposons occur exclusively on the information within the endosymbiont chromosomes. Three possibilities of manipulation are considered for transposons:

- Permutation between the genes within the range of action of the vector in the chromosome, called *jump and swap* trasnsposon;
- Exchange between the genes of the delimited interval and the remaining genes of the manipulated chromosome, called *erase and jump* transposon;
- Random alteration of the genes within the defined range. This type of manipulation is motivated by the fact that, in the natural context, transposons are also mutagens. Mutagens are physical or chemical agents that change the genome of their host cell in different ways. In the computational context, this type of transposon is called *mutagen* transposon.

It is important to note that the transposons in the Transgenetic Algorithms are substantially different from the ones presented in correlated works. In the works described in Sec. 3, transposons are considered as sequences of genes that can be moved inside the chromosome. In the proposed approach, transposons implement rules to rearrange the genes of the endosymbiont chromosomes. Therefore, the transposons proposed here generalize the methods presented in the correlated works. In the Transgenetic Algorithms, transposons that operate with more than one chromosome are not considered.

A general framework of Transgenetic Algorithms is presented in *Algorithm AlgoTrans*. An initial population of endosymbiont chromosomes, *Pop*, is created. Like in other evolutionary algorithms a fitness value is assigned to each chromosome. The repository of genetic information, GIR, is created in step 2. This repository is initialized with "a priori" information. The steps 3-9 are repeated while a stop condition is not satisfied. The set of transgenetic vectors, TV, is generated in step 4. The information in GIR is used in the transgenetic vectors generation. A subset of chromosomes of Pop, $SubPop$, is selected to be manipulated by the transgenetic vectors. The selection of chromosomes that compose $SubPop$ can be done with any classical selection method utilized in Evolutionary Algorithms or simply, the whole population is selected to be manipulated. The repository of genetic information is updated with new information ("a posteriori") generated during the evolutionary process (step 8), such as the best endosymbiont chromosomes or pieces of solutions represented in these chromosomes.

Algorithm AlgoTrans
1. $Pop \leftarrow$ `initial_population( )`
2. $GIR \leftarrow$ `genetic_information( )`
3. `repeat`
4. $TV \leftarrow$ `transgenetic_vectors(`GIR`)`
5. $Subpop \leftarrow$ `select_chromosomes(`Pop`)`
6. $NewSubpop \leftarrow$ `manipulate_chromosomes(`$Subpop, TV$`)`
7.   `Update(`$Pop, SubPop$`)`
8.   `Update(`GIR`)`
9. `until a stopping criterion is satisfied`

A fitness value may also be assigned to the transgenetic vectors. This fitness can be used to give the fittest transgenetic vectors a higher probability to be chosen to manipulate chromosomes in a given iteration. The fitness can be computed as a success rate, for instance, regarding the total number of chromosomes that were susceptible to the vector's manipulation up to a given moment. For the vectors that carry genetic information, the fitness can be a function of the "quality" of the information they carry. The vector's fitness can also be a function of the difference between the fitness value of a chromosome before and after the manipulation by that transgenetic vector. Finally, the fitness can be a function of the evolutionary stage. For example, at early stages of the algorithm execution some transgenetic vectors may be expected to give better contributions to the search than others. Thus, the fitness assigned to the former transgenetic vectors may be higher than the value assigned to later ones.

The aim of including "a priori" information is to accelerate the evolutionary process. The mixture of information from the context of the host with the information in the population of endosymbionts produces a good potential for diversification. Moreover, depending on the algorithm designer, the manipulation of transposons and autonomous recombinant plasmids may be directed towards diversification or towards intensification tasks. These transgenetic vectors do not depend on the genes that are currently present in the computational evolutionary process. The proposed form for evolutionary algorithms dispenses the use of reproduction, since it is based on natural mechanisms that immediately incorporate any improvement that is reached in the endosymbiotic co-evolution.

When the information in the host's context has high quality, the plasmid, the typical recombinant plasmid and the viruses are the transgenetic vectors more appropriate to be used. So, if there is good information available prior to the start of the evolutionary search, these agents can be used to speed up the search process. In more advanced stages of the evolutionary search, when the "a priori" information has already been incorporated to the endosymbionts or even surpassed by the search context, the transposons and autonomous recombinant plasmids are fundamental agents in order to promote innovation. It is important to note, however, that at every stage of the search process the composition of various types of vectors tends to balance efforts of exploitation and exploration.

Transgenetic algorithms were first proposed by Gouvêa [24] who applied them to the Quadratic Assignment Problem. Researches addressing the use of Transgenetic algorithms to real world problems concern the scheduling of workover rigs [22], video distribution under demand conditions [39], pipe sizing problems [20], distribution of products of petroleum [4], configuration of co-generation systems [23] and protein folding [2], among others. Ramos *et al.* [49] applied logistic regression for parameter tuning in Transgenetic Algorithms. A hybrid Transgenetic Algorithm is applied to the Prize Collecting Steiner Tree Problem [55].

4.2 The Traveling Salesman Problem: A Didactic Example

The well-known Traveling Salesman Problem, TSP [25] is used to illustrate the algorithmic ideas of the Transgenetic Algorithms. Given a graph $G = (N, E)$, where $N = \{1,...,n\}$ is the set of nodes, $E = \{1,...,m\}$ is the set of edges, and costs, c_{ij}, associated with each edge linking vertices i and j, the problem consists in finding the minimal total length Hamiltonian cycle in G. The length of a cycle is calculated by adding the costs of the edges in the cycle. The nodes are also called cities due to the fact that they represent cities a traveling salesman has to visit, the edges represent streets or roads between the cities and the edge costs are thought as taxes that must be paid to cross the roads.

Chromosomes

In the Transgenetic Algorithms, the endosymbiont chromosomes represent solutions of the investigated problem, as the chromosomes of other evolutionary algorithms. The sequence of cities the traveling salesman visits is a usual

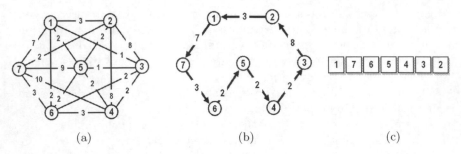

Fig. 2 (a) TSP instance, (b) one solution for the TSP instance and (c) the correspondent chromosome

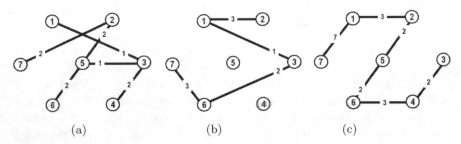

Fig. 3 Sources of genetic information:(a) Minimum spanning tree; (b) shortest paths between cities 3-2 and cities 3-7; (c) Hamiltonian path

representation of TSP solutions in evolutionary algorithms. These sequences are represented as permutations of the n cities. Thus, the endosymbiont chromosomes represent tours and the fitness assigned to them is the correspondent tour cost. An instance of the TSP is shown in Fig. 2(a). Figure 2(b) shows a solution for the instance of Figs. 2(a) and 2(c) shows the chromosome that represents the solution of Fig. 2(b).

Units of genetic information

In the case of the TSP, the units of genetic information may be obtained from structures of the graph that represents the TSP instance such as minimum spanning trees, shortest paths, Hamiltonian paths, matchings, etc. Figures 3(a), (b) and (c) exemplify, for the graph of Fig. 2(a), some sources for obtaining units of genetic information.

Plasmid, recombinant plasmid and virus

Plasmids, recombinant plasmids and viruses for the TSP can obtain their information from the sources of genetic information shown in Fig. 3. Figure 4 illustrates two units of genetic information obtained from the minimum spanning tree and the Hamiltonian path shown in Figs. 3 (a) and (c),

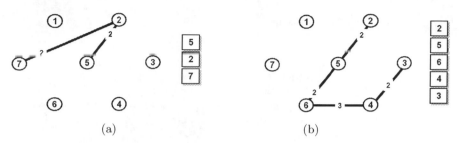

Fig. 4 Units of genetic information from: (a) the Minimum spanning tree; (b) the Hamiltonian path

respectively. Figures 4(a) and (b) show paths with lengths equal two and four, respectively. The length of the information string is a parameter that can be fixed by the designer or can be chosen at random in a pre-defined range.

Table 2 shows the example of a plasmid for the TSP where the type of unit of genetic information the plasmid carries, I, and procedures p_1 and p_2 are defined. Suppose that the host's repository contains more than one source of information, which usually is the case. From which source does the transgenetic vector obtain its information? Given the quality of the information in the repository, the designer defines the best strategy for choosing the information source. The designer can simply define that one of those sources is chosen at random with equal probability or establish another strategy that can, inclusively, vary during the execution of the algorithm. The plasmid exemplified in Table 2, transcribes information I in a chromosome C with the insertion operation defined in procedure p_2. Let C represent the solution shown in Fig. 5(a), $k = 2$ and I be the unit of genetic information of Fig. 4(a). The last allele of I is 7. Then, as defined in procedure p_2, this allele is paired with the allele 7 of C, as shown in Fig. 5(b). The information is inserted and the repeated alleles are removed from C. The result is chromosome C' that represents the solution shown in Fig. 5(c). Procedure p_1 establishes the condition which chromosome C is susceptible to the manipulation of the plasmid. This condition establishes that the manipulation is "accepted" if the value of the solution represented in C', $fit(C')$ is lower than the value of the solution represented in C, $fit(C)$. In this case, the new chromosome replaces the original one in the population. The cost of the solutions of Figs. 5(a) and (c) are 34 and 15, respectively. Thus, in this case, chromosome C' replaces chromosome C in the population of endosymbionts.

An alternative for procedure p_2 of Table 2 is to test all possibilities of pairing the alleles of the plasmid with the correspondent ones in the chromosome and assume the best manipulation result as the resultant chromosome C'.

The difference between the plasmid and the computational virus is that the latter blocks the transcribed information in the manipulated chromosome for a given number of iterations. It means that the sequence of genes transcribed

Table 2 Example of a plasmid for the TSP

Procedure	Description
I	A sequence of vertices defining a path with length at most k
p_1	$fit(C) - fit(C') > 0$, then return "true", otherwise return "false"
p_2	Starting at the position of the last allele of I in chromosome C, insert I and remove repeated vertices that are not in the positions altered with the manipulation.

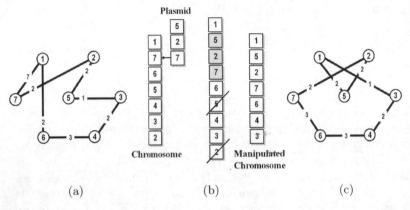

Fig. 5 (a) Solution represented in the original chromosome; (b) plasmid's transcription; (c) solution represented in the manipulated chromosome

by a virus has to remain unchanged in the manipulated chromosome during a given number of iterations. Those genes are not modified by latter transcriptions unless the pre-specified period of time has elapsed. This "protection" for the transcribed genes can be very useful when these genes represent part of a high quality solution or if they are likely to belong to an optimal solution.

The typical recombinant plasmid obtains it information in more than one source. Figure 6(a) shows the information carried by a typical recombinant plasmid that is a combination of two units of genetic information obtained from distinct sources: path 3-5-2 from the minimum spanning tree (Fig. 3(a)) and path 3-6-7 from the shortest path (Fig. 3(b)). The resultant path has length $k = 4$. Figure 6(b) shows the example of a unit of genetic information of an autonomous recombinant plasmid. It was obtained with the nearest neighbor heuristics [6]. The greedy iterative process starts at vertex 3 and the path has length $k = 3$.

Transposon

Table 3 summarizes the features of a *jump and swap* transposon for the TSP. Figure 7(a) shows a solution and the correspondent chromosome. Let vertices 7 and 3 be selected in procedure p_4. The transposon examines all

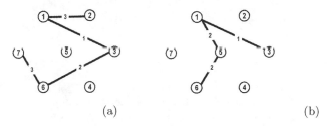

Fig. 6 Units of genetic information of: (a) a typical recombinant plasmid; (b) an autonomous recombinant plasmid

Table 3 Features of a jump and swap transposon for the TSP

Procedure	Description
I	To swap vertices
p_1	$fit(C) - fit(C') > 0$, then return "true", otherwise return "false"
p_2	To swap two adjacent vertices. To examine all two swaps and output the one that yields the best result.
p_4	To choose 2 vertices at random with distance, at least, k in the represented tour

possible swaps between two neighboring vertices within the range defined by the application of procedure p_4 as illustrated in Fig. 7(b). The transcription operator (procedure p_2) returns the configuration that represents the best solution found with the swaps. In this case, the best chromosome is formed by the sequence (1 6 7 2 4 3 5) representing a solution with cost 20. Once the manipulated chromosome represents a better solution than the one represented in the original chromosome, the latter is replaced by the former in the current population.

Table 4 summarizes the features of an erase and jump transposon for the TSP. The information is to remove vertices of the delimited interval in the

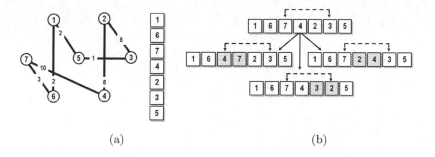

Fig. 7 (a) One TSP solution and its representation in a chromosome ; (b)action of the jump and swap transposon in the delimited sequence of genes

Table 4 Features of an erase and jump transposon for the TSP

Procedure	Description
I	To remove vertices
p_1	$fit(C) - fit(C') > 0$, then return "true", otherwise return "false"
p_2	To replace the removed vertex by another vertex chosen at random out of the range delimited by procedure p_4. To examine all two swaps and output the one that yields the best result.
p_4	To choose 2 vertices at random with distance, at least, k in the represented tour

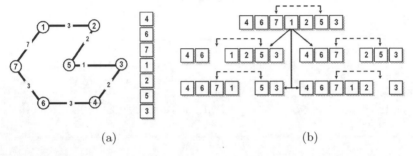

(a)   (b)

Fig. 8 (a) One TSP solution and its representation in a chromosome; (b) action of the erase and jump transposon in the delimited sequence of genes

target chromosome. In order to do this, procedure p_2 defines a transcription method that swaps the removed vertex with another vertex out of the delimited range. Figure 8(b) illustrates the action of an erase and jump transposon on the chromosome of Fig. 8(a). Figure 8(b) shows that the vertices in the delimited interval are considered one at a time. The removed vertex is replaced by another one chosen at random among vertices 4, 6 and 3 that are out of the defined interval. Let the pair (x, y) represents the swap between vertices x and y, in and out the delimited range, respectively. If swaps (7,3), (1,4), (2,6) and (5,3) are performed, then the best chromosome is obtained with the swap of vertices 2 and 6. The resultant chromosome represents a solution with cost 24. Once the original chromosome represents a solution with cost 21, procedure p_1 returns "false" and the manipulated chromosome does not replace the original chromosome in the population.

The random alterations done by mutagen transposons may be concerned with diversification tasks. In those cases, the algorithm may accept that the manipulated chromosomes represent solutions that are worse than the original ones. Table 5 summarizes the features of a mutagen transposon for the TSP. The information is to swap vertices and procedure p_1 always return "true". It means that whatever the chromosome generated by the manipulation, it replaces the original one in the current population.

Table 5 Features of a mutagen transposon for the TSP

Procedure	Description
I	To swap k pairs of vertices
p_1	return "true"
p_2	To swap a vertex chosen at random in the range delimited by procedure p_4 with another vertex also chosen at random.
p_4	To choose 2 vertices at random with distance, at least, k in the represented tour

In examining various possibilities to make their transpositions, the transposons (mainly the jump-and-swap transposon and the erase-and-jump transposon) perform a form of restricted local search with the elements in the delimited range.

5 Application to the Traveling Purchaser Problem

In this section an application to the NP-hard problem named Traveling Purchaser Problem (TPP), a generalization of the Traveling Salesman Problem is presented. In this variant there is a set of m markets, vertices of a graph G, and a set of n products that must be purchased. Each product is available, with different quantities, on a subset of markets and the unit cost of a product depends on the market where it is available. The objective of the purchaser is to buy all the products, departing and returning to a domicile (location v_0), with the least possible cost. The cost is defined as the summation of the weights of the edges in the tour plus the price paid to acquire the products. Thus, there is no need of including all the markets in the tour. The problem can be stated as follows. Given a domicile, v_0, a set of markets $M = \{v_1, v_2, ..., v_m\}$ and a set of products $K = \{f_1, f_2, ..., f_n\}$, the problem is represented in a graph $G = (V, E)$ where $V = \{v_0\} \cup M$ and $E = \{[i, j] : v_i, v_j \in V, i < j\}$. A demand d_k is assigned to each product f_k. The number of units of product f_k at market v_i is denoted by q_{ki} and M_k denotes the set of markets where the product f_k is available, $M_k \subseteq M$. The cost of product f_k at market v_i is denoted by b_{ki} and the cost of traveling from market v_i to market v_j is given by c_{ij}. The objective is to determine a minimum cost tour in G starting and ending at v_0 through a subset of markets so that the demand for products f_k is satisfied. The uncapacitated version of the TPP (UTPP) is considered here. In the UTPP version, it is assumed that if a product is available at a given market, its quantity is sufficient to satisfy the demand [7], that is $d_k = 1$ and $q_{ki} \in \{0, 1\}, 1 \leq k \leq n, 1 \leq i \leq m$.

Algorithm TransTPP

```
1.  Pop ← initial_population(#sizeP)
2.  GIR ← genetic_information( )
3.  j ← β
4.  repeat
5.     i ← 1
6.     repeat
7.        u ← random(η)
8.        if (u ≥ j) then
9.           λ ← choose_best_plasmid(r)
10.          for each C ∈ Pop do
11.             C' ← manipulate_plasmid(C, λ)
12.          else
13.             for each C ∈ Pop do
14.                C' ← manipulate_transposon(C)
15.          if procedure_p₁(C, C') thenC ← C'
16.          if (C than the current best solution) then
17.             update(GIR, C)
18.          i ← i + 1
19.       until(i = β)
20.       j ← β
21.   until(j > η)
```

A general framework of the algorithm for the UTPP is shown in *Algorithm TransTPP* [21]. The population of endosymbiont chromosomes is generated with $\#sizeP$ individuals. Each chromosome contains the subset of markets visited by the traveling purchaser. The sequence begins and ends at the domicile, v_0, but this information is omitted in the chromosome. Each product is purchased in the market that offers the lowest price. The fitness is given by the cost of the tour represented in the chromosome plus the cost of acquisition of all products on the markets of the represented tour. In order to generate the sequence of markets of a given chromosome, an iterative method is performed. At each iteration step, a market is chosen at random, with equal probability, among the markets that are not yet in the sequence. The market is included in the sequence and the process continues until a feasible sequence has been generated. In order to be feasible, each product must be available in, at least, one market. After the sequence is built the Lin and Kernighan algorithmic version of Applegate *et al.* [3] is applied to the sequence in order to improve the cost of the tour.

In this algorithm two types of transgenetic vectors are utilized: plasmids and transposons. The repository of sources of genetic information stores "a priori" and "a posteriori" information. The "a priori" information is a Hamiltonian cycle of G obtained with the Lin and Kernighan algorithm. "A posteriori" information is obtained from the four best solutions generated up to the current iteration. In this transgenetic algorithm the evolutionary process

Table 6 Example of a plasmid for the UTPP

Procedure	Description
I	A sequence of markets with length k. The value for k is chosen at random in the interval $[3, \lfloor m/8 \rfloor]$.
p_1	$fit(C) - fit(C') > 0$, then return "true", otherwise return "false"
p_2	Remove from C the markets that are both in I and C. Test the best position for inserting I in C, verifying the insertion of I in every position of C. Remove from C the markets where no products are purchased. Improve the tour with the Lin and Kernighan heuristics.

is developed in levels. It means that different probabilities are assigned to the type of the transgenetic vectors chosen during the execution depending on the current "evolutionary level". Those levels are simulated in the nested loops of the algorithm. In the inner loop the probability of choosing a plasmid or a transposon remains fixed. Four input parameters are passed to the algorithm: $\#sizeP$, the population size; r, the number of plasmids generated at each iteration step where plasmids are chosen to manipulate the endosymbiont chromosomes; β, the number of iterations of the inner loop; and η, the total number of iterations of the Transgenetic Algorithm.

Plasmids are more likely to be chosen in the initial iterations. The extent to which the algorithm is executed, the probability changes and transposons become more likely. At each iteration step, a type of transgenetic vector, plasmid or transposon, is chosen with a probability that depends on the evolutionary process stage. The counter j controls that tendency, being initialized at step 3 and updated at step 20. Its effect is determined on the comparison of step 8.

If a plasmid is selected, r vectors are generated. They are evaluated and the best one among the r plasmids attacks all endosymbiont chromosomes, as shown in steps 10-11. The evaluation criterion is described further. The unit of genetic information carried by the plasmid is a sequence of markets obtained in one of the sources stored in GIR. One element of GIR is chosen at random, with uniform probability, to be the source of information of each plasmid. The features of the computational plasmid implemented for the UTPP are summarized in Table 6. An initial point of the selected source is randomly chosen, then starting on that point, k successive markets form the sequence that is carried by the plasmid.

At each iteration where the plasmid is chosen as the manipulation vector, $r = 30$ plasmids are generated. Those plasmids are evaluated in order to choose the best one to manipulate the endosymbiont chromosomes. This evaluation is done with the units of genetic information they carry. Three parcels make up the basis of evaluation:

1. The weights of the edges of the path correspondent to the sequence of markets;

2. The lowest prices of the products available on some market in the sequence;
3. The highest prices of the products which are not available in any market of the sequence.

The plasmid with the lowest associated value is chosen as the transgenetic vector of the current iteration. The insertion of the unit of genetic information is done with the method described in p_2 as illustrated in Fig. 9. First, the markets that are both in C and in the unit of genetic information of the plasmid are removed. In the example shown in Fig. 9, market 1 is in this case. The positions where the unit of genetic information can be inserted are illustrated in the chromosome with dashed arrows. In the example, the unit of genetic information is inserted between markets 4 and 2. After that, the markets where products are no longer purchased are removed. Suppose, that the products previously purchased in markets 4 and 6 are available at lower prices in markets 8 and 3. Then, those products are purchased in the new inserted markets and markets 4 and 6 are in the tour, but no products are purchased there. Thus, markets 4 and 6 are removed from the chromosome. After these operations the manipulated chromosome is (8 3 1 2 5). Finally, the Lin and Kernighan algorithm is applied to this tour and the resultant chromosome is (2 1 3 8 5). This is implemented in procedure manipulate_plasmid() in step 11 of the algorithm.

Fig. 9 Example of the transcription of the plasmid with information string (8 3 1) in the UTPP chromosome (4 2 1 6 5)

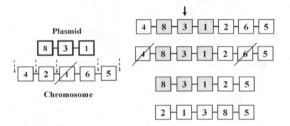

The manipulation with the transposon is implemented in procedure manipulate_transposon() in line 14. This is a mutagen transposon. The features of the transposon are summarized in Table 7. Two positions of the chromosome are selected at random in procedure p_4. The markets between those positions constitute the set of markets that will be manipulated. The information of the transposon is to remove markets. Thus, each market of that sequence is considered to be removed, one at a time. When a market is removed, the solution represented in the chromosome may become infeasible, that is, it can occur that one or more products are not available in the remaining markets. Thus the transcription operator has to restore the solution feasibility. This is done with the addition of new markets to the manipulated chromosome.

Let S be the set of markets that are not in the manipulated chromosome C. A value is associated with each market v of S. The value associated to v is calculated by the sum of two parcels: the least increase its addition brings

Table 7 Features of a transposon for the UTPP

Procedure	Description
I	To remove markets
p_1	$fit(C) - fit(C') > 0$, then return "true", otherwise return "false"
p_2	Remove one market and add new markets until reaching feasibility. If the chromosome is not improved, then repeat the operation removing two markets
p_4	To choose 2 positions in the chromosome at random

to the cost of the tour represented in C and the total of the highest prices of the products that, even with the inclusion of v, are still not purchased. The least increase of the tour cost is calculated with the nearest insertion heuristics [53]. The market with the least associated value, is inserted in the chromosome in the position determined by the nearest insertion heuristics. If the solution is still infeasible, the process is repeated until a feasible solution is built. Finally, the markets where no products are purchased are removed and the tour of the markets in the chromosome is improved with the Lin and Kernighan heuristics. The markets are removed one at a time, and the chromosome that represents the best solution among all generated by the transposon's manipulation is taken as the resultant chromosome. If the resultant chromosome does not represent a better solution than the original chromosome, then the procedure is repeated with the removal of two markets.

The proposed approach for the UTPP was tested in a computational experiment with 141 benchmark instances: 89 with known optimal solutions and 52 for which optimal solutions are not known. The tests were run on a Pentium IV 2.8 GHz, 512 MB of RAM, Ubuntu Linux operational system and gcc compiler. The parameters $r = 30$, $\eta = 40$ and $\beta = 4$ were fixed after preliminary experiments. Among the instances with known optimal solutions, m varies between 50 and 250, and n varies between 50 and 200. The results of the Transgenetic Algorithm, TA, were compared with the results of the Local Search presented by Riera-Ledesma and Salazar-González [50] and the results of the Ant Colony algorithm presented by Bontoux and Feillet [8].

The results concerning the instances with known optimal solutions are summarized in Table 8. The instances are divided into classes established by the number of markets and the number of products. The results correspond to the average of the best solutions found by each algorithm for the instances of the correspondent group. The results are shown in terms of the percent deviation of the average from the optimal solution. Two hundred independent runs of the Transgenetic Algorithm were performed for each instance.

Concerning the instances grouped by number of markets, except for the group with 150 markets where a percent deviation of 0.01 is found, the TA finds percent deviation 0 for all groups. A similar result is observed for the results grouped by the number of products. Regarding quality of solution

Table 8 Results for UTPP instances grouped by markets, m, and by products, n

Methods				m					n		
		50	100	150	200	250		50	100	150	200
RL-SG	Gap	0.07	0.14	0.03	0.32	0.06		0.07	0.24	0.10	0.08
	T(s)	3	10	14	19	25		5	13	20	21
BF	Gap	0	0	0.08	0.02	0.01		0	0.05	0	0.03
	T(s)	2	20	172	232	154		37	154	96	165
TA	Gap	0	0	0.01	0	0		0	0	0	0.01
	T(s)	4	25	44	43	64		12	37	39	50

Table 9 Results for UTPP instances with optimal solution not known

			BF		TA					BF		TA	
m	n	Id	Sol	T(s)	Sol	T(s)	m	n	Id	Sol	T(s)	Sol	T(s)
200	150	4	2419	1216.92	2419	23.98	300	150	5	1816	309.25	1816	41.16
200	200	4	2344	527.03	2344	99.19	300	200	1	1815	488.15	1803	575.39
200	200	4	2344	527.03	2344	99.19	300	200	2	1791	1918.52	1790	627.73
250	100	1	1301	33.84	1301	143.19	300	200	3	2442	2852.05	2437	184
250	100	4	1673	10.23	1673	3.55	300	200	4	1815	2946.79	1815	113.82
250	100	5	1641	550.24	1641	1.84	300	200	5	2022	1577.83	2014	605.39
250	150	4	1836	45.24	1836	2.27	350	50	1	723	46.04	723	1.7
250	150	5	1531	21.1	1531	151.43	350	50	2	736	25.71	736	13.02
250	200	2	2785	1137.65	2786	246.31	350	50	3	942	6	942	1.82
250	200	3	1924	281.88	1924	16.45	350	50	4	805	379.39	805	5.01
250	200	4	2116	83.83	2116	3.06	350	50	5	1125	26.35	1225	1.67
250	200	5	1797	930.03	1797	38.97	350	100	1	1317	1698.48	1317	229.99
300	50	1	1477	160	1477	1.5	350	100	2	962	155.48	962	2.37
300	50	2	813	116.01	813	1.41	350	100	3	796	839.65	796	2.43
300	50	3	1117	20	1117	1.46	350	100	4	1059	13.94	1059	9.14
300	50	4	1176	2.11	1176	1.44	350	100	5	1566	464.86	1566	41.76
300	50	5	1257	276	1256	1.57	350	150	1	1457	1986.42	1459	319.67
300	100	1	1035	55.54	1035	2.29	350	150	2	1315	159.12	1315	16.31
300	100	2	1179	617.22	1180	3.98	350	150	3	2553	257.69	2558	597.74
300	100	3	1498	103.42	1498	2.25	350	150	4	1239	595.85	1239	3.06
300	100	4	1749	312.16	1749	37.49	350	150	5	2288	8.93	2288	229.27
300	100	5	1774	2.74	1774	2.27	350	200	1	1503	1033.39	1498	25.34
300	150	1	1457	756.71	1457	98.66	350	200	2	1374	3085.09	1369	56.07
300	150	2	1656	483.32	1656	3.02	350	200	3	1873	368.66	1873	59.05
300	150	3	2485	663.24	2484	6.34	350	200	4	1385	122.24	1356	32.88
300	150	4	1801	95.93	1801	8.17	350	200	5	2336	2385.65	2336	204.53

the RL-SG is outperformed by the TA in all instance classes. The BF is outperformed by the TA in three classes grouped by number of markets and 2 classes grouped by number of products. The TA is not outperformed by any algorithm regarding quality of solution.

Table 9 shows a comparison between the solutions found by the TA and the BF for the remaining 52 instances. Optimal solutions are not known for those instances. The results of the RL SG are not reported because it does not present better solutions than the other two algorithms for any instance. The results shown for TA correspond to the best solution found in 5 independent executions for each instance. The stopping criterion was a maximum of 200 iterations. The results shown by BF are reported by Bountoux and Feillet [8]. The TA finds new best solutions for 9 of the 51 instances. BF reports the best solutions of 4 benchmark instances. Both algorithms find the best known solutions for the remaining instances.

Although the platforms of BF and TA are different, some conclusions about the processing times can be drawn. In order to this, we can consider the processing times of BF divided by a factor of 2. It gives the BF some advantage over the TA, once the difference in speed between the two platforms is smaller than this factor. Even dividing the processing times of BF shown in Table 9 by 2, the TA is, for the 51 instances, 3 times faster, in average, than the other algorithm.

6   Application to the Bi-objective Minimum Spanning Tree

A spanning tree of a connected undirected graph $G = (N, E)$ is an acyclic spanning subgraph of G with $n - 1$ edges, where $n = |N|$. If G is a weighted graph, a minimum spanning tree, MST, of G is spanning tree for which the summation of the weights of its edges is minimum over all spanning trees of G. A survey of the MST is presented by Bazlamaçci and Hindi [5]. Although, the MST problem is polynomial, constraints often render it NP-hard [19]. Examples include the degree-constrained minimum spanning tree, the maximum-leaf spanning tree, and the shortest-total-path-length spanning tree problems. Another difficult variant of this problem is the multi-criteria Minimum Spanning Tree, mc-MST [1]. In the q-objective MST Problem, a vector of non negative weights $w_{ij} = (w_{ij}^1, ..., w_{ij}^q)$, $q > 1$, is assigned to each edge $(i, j) \in E$. Let S be the set of all possible spanning trees, $T = (N_T, E_T)$, of G and $W = (W_1, ..., W_q)$, where

$$W^t = \sum_{ij \in E_T} w_{ij}^t, t = 1, ..., q. \tag{1}$$

The problem seeks $S^* \subseteq S$, such that $T^* \subset S^*$ if and only if $\nexists T \in S$, such that T dominates T^*. In order to understand the domination concept, consider the general multi-objective minimization problem (with no restrictions) that can be stated as:

$$\text{``minimize''} f(x) = (f_1(x), ..., f_q(x)), \tag{2}$$

subjected to $x \in X$, where x is a discrete value solution vector and X is a finite set of feasible solutions. Function $f(x)$ maps the set of feasible solutions X in Z, the q-dimensional objective space, $q > 1$. Then, $f : X \to Z$ is a function that assigns an objective vector $z = f(x) \in Z$ to each solution $x \in X$. Let $z^1 = (z_1^1, ..., z_q^1)$ and $z^2 = (z_1^2, ..., z_q^2)$, $z^1, z^2 \in Z$ be two objective vectors, then $z^1 \succ z^2$ (z^1 dominates z^2) if z^1 is not worse than z^2 in any objective and is better in at least one. The goal is to discover solutions that are not dominated by any other in the objective space. The non-dominated solutions are said also to be efficient solutions. To solve a multi-criteria problem, one is required to find the set of efficient solutions. Solutions of this set can be divided in two classes: the supported and non-supported efficient solutions. The supported efficient solutions can be obtained by solving the minimization problem with a weighted sum of those objectives. More formally,

$$minimize \sum_{j=1,...,q} \alpha_j f(x_j), \qquad (3)$$

where

$$\sum_{i=1,...,q} \alpha_i = 1, \alpha_i > 1. \qquad (4)$$

The non-supported efficient solutions are those which are not optimal for any weighted sum of objectives. Once the single objective MST is solved in polynomial time, then the set of non-supported solutions is a major challenge for researchers that deal with the multi-criteria MST.

In this work the bi-objective problem is considered, although the proposed algorithm can be adapted to consider $q > 2$ objectives. Evolutionary algorithms for the bi-objective MST are presented by Zhou and Gen [68], Knowles and Corne [34], and Rocha *et al.* [51, 52].

The pseudo-code of the transgenetic algorithm is shown in *Algorithm TransMcMST* [52]. The algorithm that utilizes plasmids, recombinant plasmids and transposons runs a fixed number of iterations, *#maxgen*. At first the endosymbiont chromosomes are manipulated by a transposon. There are two types of transposons and one of them is chosen at random at each iteration step with uniform probability. Both are mutagen transposons. The first type transposon is called cycle-transposon. The rule of this transposon is to insert an edge in the part of the tree (represented in the target endosymbiont chromosome) specified in procedure p_4. A cycle is formed when a new edge is inserted in a tree. Therefore, another edge of that cycle must be removed in order to maintain feasibility. At first, a scalarizing vector and a node i in the delimited range are selected at random. The edge with the lowest scalarized cost, with one extremity in i, that is not in the tree is chosen to be inserted. The edge of the cycle with the greatest scalarized cost is removed. The operation is repeated for all nodes in the specified interval. An edge that was withdrawn is not allowed to be re-inserted in the tree for 3 iterations of the transposons action. In procedure p_1 the dominance condition of the new chromosome is verified. If the new chromosome, C', dominates the original chromosome, C, or if the former is not dominated by any solution in G_A, then the manipulation is accepted. The same procedure p_1 is used for the other transgenetic vectors.

Algorithm TransMcMST
```
1.   Pop ← initial_population(#sizeP)
2.   create_archive(G_A, Pop)
3.   for i ← 1 to #maxgen do
4.     for j ← 1 to #sizeP do
5.       type_t ← random(1, 2)
6.       C' ← manipulate_transposon(C_j, type_t)
7.       if procedure_p₁(C', C_j)
8.          C_j ← C'
9.       else cont[j] ← cont[j] + 1
10.      if (cont[j] = 2)
11.         λ ← create_plasmid( )
12.         C' ← manipulate_plasmid(C_j, λ)
13.         cont[j] ← 0
14.         if procedure_p₁(C', C_j)
15.            C_j ← C'
16.      update (G_A, C_j)
17.    end_for_j
18.  end_for_i
```

The rule of the second type transposon is to remove two edges. Two edges (i, j) and (r, s), $i \neq j \neq r \neq s$, are removed from the tree. There is only one way to re-link those terminal vertices in order to maintain feasibility. It is checked whether the inclusion of edge (i, r) creates a cycle or not. If a cycle is not created then edges (i, r) and (j, s) are added, otherwise the edges included in the solution are (i, s) and (j, r). As in the cycle-transposon, the removed edges are not allowed to be re-inserted in the tree for the next 3 iterations of the transposon's action.

Each chromosome of the population of a given iteration step is manipulated by one of the two transposons chosen at random in accordance with a uniform probability (steps 5-8). The algorithm maintains a counter for each chromosome that is initially set to zero. If a chromosome is not improved by the manipulation of the chosen transposon, its counter is incremented, otherwise its counter is set to zero. If the counter of a given chromosome reaches value 2, then the chromosome is manipulated by a plasmid or by a recombinant plasmid and its counter is set again to zero. During the creation of the transgenetic vector in step 11, the type of transgenetic vector (plasmid or recombinant plasmid) is chosen at random with uniform probability. As noted previously, the difference between a plasmid and a recombinant plasmid is in the construction of the unit of genetic information carried by the vector. Both vectors utilize the same procedures p_1 and p_2 in this application.

The genetic information utilized in the plasmids is obtained from the solutions in G_A. At each iteration step, one solution of G_A is chosen to be the source of information of the plasmid. One of the cells of the multidimensional grid with the lowest number of solutions is chosen at random with a uniform distribution. Then, a solution of that cell is randomly selected. A fragment of that tree with k edges is randomly chosen to be the unit of genetic information of the plasmid. The transcription operator (procedure p_2) begins with an empty tree. It inserts all the edges of the unit of genetic information of the plasmid. After, all edges of the original tree that do not form cycle with the edges of I, are inserted in the tree. Finally, if the structure is not yet a tree, random edges are inserted until a tree is formed.

This implementation uses an autonomous recombinant plasmid. The unit of genetic information carried by the recombinant plasmid is generated with the random version of the Kruskal's algorithm described previously. As in the plasmid, the length of the unit of genetic information is chosen at random in the interval $[0.30n, 0.60n]$.

The Transgenetic Algorithm was applied to 40 instances generated in accordance with the method described in the work of Knowles [33]. The results obtained with the Transgenetic Algorithm were compared with the ones obtained by the Memetic Algorithm proposed by Rocha *et al.* [51]. Two groups of twenty instances belonging to the classes concave and correlated were generated as complete graphs with two objectives. Each class has two instances with $n = 100, 200, 300, 400, 500, 600, 700, 800, 900$ and 1000. To be generated, the correlated instances require a correlation factor, β, and the concave instances require two parameters, ζ and η. Table 10 summarizes the parameters utilized to generate the set of instances. Both algorithms were implemented in C and ran on a Pentium IV (2.8 GHz and 512 Mb of RAM) with Linux 2.6, gcc 4. Forty independent runs of each algorithm were executed for each instance. The maximum number of iterations of the Transgenetic Algorithm, $\#maxgen$, is 30, and the size of the population, $\#sizeP$, is 150. The MA was implemented as described in the paper of Rocha *et al.* [51].

Three quality indicators were utilized in the comparison: the binary epsilon indicator (Eps) [70], the hypervolume indicator (Hyp) [69], and the R2 indicator [26]. The R2 indicator utilized the augmented Tchebycheff function [35]. The reference sets contain the non-dominated solutions obtained by both algorithms. The test of Mann-Withney (U-test) is utilized to verify the statistical significance of the results obtained with these indicators [13]. The results are shown in Table 11. The column "Id" shows the instance name, columns "MA T(s)" and "TA T(s)" show the average processing time of the Memetic Algorithm (MA) and the Transgenetic Algorithm (TA), respectively. The p-values shown in Table 11 were calculated under the assumption that the TA outperforms the MA. Therefore, considering a significance level of 0.05, values less than 0.05 indicate that there is statistical evidence that the

Table 10 Parameters of the bi-objective MST instances

	Concave		Correlated			Concave		Correlated
Id	ζ	η	β	Id	ζ	η	β
100_1	0.01	0.02	0.3	600_1	0.0016	0.1	0.125
100_2	0.02	0.1	0.7	600_2	0.002	0.02	0.95
200_1	0.05	0.2	0.3	700_1	0.0014	0.03	0.35
200_2	0.08	0.1	0.7	700_2	0.001	0.008	0.7
300_1	0.03	0.1	0.3	800_1	0.00125	0.035	0.45
300_2	0.05	0.125	0.7	800_2	0.0015	0.03	0.05
400_1	0.025	0.125	0.3	900_1	0.0011	0.009	0.15
400_2	0.04	0.2	0.7	900_2	0.002	0.01	0.85
500_1	0.02	0.1	0.3	1000_1	0.001	0.2	0.4
500_2	0.03	0.15	0.7	1000_2	0.0005	0.1	0.9

Table 11 Comparison of the performance of the Memetic and the Transgenetic Algorithms

	Correlated					Concave				
Id	Eps	Hyp	R2	MA T(s)	TA T(s)	Eps	Hyp	R2	MA T(s)	TA T(s)
100_1	0.99	0	0	67.0	11.0	0.70	0	0	61.4	8.8
100_2	1	0.38	0	72.2	9.2	0	0	0	56.3	21.6
200_1	0	0	0	133.3	50.7	0	0	0	123.8	35.7
200_2	0.61	0.09	0.17	130.2	43.5	0.98	0	0	153.2	20.2
300_1	0.03	0	0	214.8	89.3	0.01	0	0	224.6	59.9
300_2	0.01	0	0	198.6	81.9	0	0	0	226.2	56.8
400_1	0	0	0	296.5	119.2	0	0	0	335.6	98.4
400_2	0.99	0	0	289.6	122.3	0	0	0	354.2	101.7
500_1	0	0	0	412.0	164.8	0	0	0	417.3	149.1
500_2	0.68	0	0	401.3	160.5	0	0.01	0	413.0	140.4
600_1	0.10	0	0	536.0	215.6	0.01	0	0	558.1	202.5
600_2	0	0.01	0.02	484.6	211.7	0	0	0	527.3	197.2
700_1	0	0	0	658.6	284.0	0.55	0	0	671.6	264.6
700_2	0.97	0	0	657.3	272.8	0	0	0	670.9	258.8
800_1	0.07	0	0.02	777.0	353.6	0.55	0	0	793.5	331.4
800_2	0	0	0	815.2	354.9	0.51	0	0	794.3	331.6
900_1	0	0	0	932.3	425.2	0	0	0	910.0	405.6
900_2	0.36	0	0	915.0	416.4	0	0	0	911.3	386.9
1000_1	0.38	0	0	1113.9	511.9	0.87	0	0	1723.6	483.9
1000_2	0	0	0	1036.8	496.9	0.94	0	0	1151.1	493.1

TA outperforms the MA, according to the correspondent indicator. Values higher than 0.95 indicates that the MA outperform the TA.

Concerning the correlated instances, the Transgenetic Algorithm obtains better approximation sets than the Memetic Algorithm in 11, 18 and 19 instances according to the epsilon, hypervolume and R2 indicators, respectively.

The MA outperforms the TA on 4 instances according to the epsilon indicator and does not outperform the TA on any instance according to the other two indicators. A similar result is verified for the concave instances, where the TA outperforms the MA on 13, 20 and 20 instances, according to the epsilon, hypervolume and R2 indicators, respectively. The MA outperforms the TA on 1 concave instance, according to the epsilon indicator. The columns correspondent to the average processing time for the two algorithms show that the TA outperforms the MA in all tested instances, being, in average, 2 times faster.

7   Peculiarities of the Transgenetic Algorithms

In this section, some differences between the Transgenetic Algorithms and other evolutionary algorithms are presented.

The first difference concerns the existence of distinct contexts of information in the Transgenetic Algorithms. Besides the endosymbiont chromosomes that represent a short-term memory of the search, information is also stored in the host's context and in the transgenetic vectors. Although these three components of the Transgenetic Algorithms are interdependent, they are autonomous and are equivalently important for the search process. A complex behavior is created in the algorithm through the nesting of these components.

The information of the host is not necessarily encoded on chromosomes, nor necessarily represents solutions of the optimization problem being tackled. This information represents a long-term memory, not exclusively associated with the evolutionary process being performed by the algorithm. The information can be evaluated regarding the expectation of producing transgenetic vectors that are successful in manipulating the chromosomes. The information of the host is updated with the new genetic information that arises from the population of chromosomes during the algorithm execution. This feedback features up a co-evolutionary spiral of convergence of the genetic material shared between the endosymbiont chromosomes and the host.

The transgenetic vectors are dynamic and volatile elements without a perfect match to the elements of traditional evolutionary algorithms. They may be submitted to various types of selective pressure. They cooperate with the evolution of the system host/endosymbiont, being guided by the information of the host in the task of accomplishing their transcriptions in the chromosomes.

The endosymbiont chromosomes do not reproduce or share genetic material directly. They are uniquely subjected to the pressure that results from the manipulation performed by the transgenetic vectors. The genes of a given chromosome are modified only if a successful transcription is done by a transgenetic vector. To be successful the transcription has to be evaluated in a positive way concerning the susceptibility criterion (implemented in procedure p_1 of the transgenetic vector). Otherwise, the genes remain unchanged over the iterations of the algorithm. The changes in the code of chromosomes are, therefore, directed to the improvement of the solutions represented in the population. Furthermore, the variation on the code of the endosymbiont chromosomes is restricted by the information of the transgenetic vectors and is guided by pre-existing genetic information of the host.

Mutations in the endosymbiotic theory of evolution have a different role than the one assigned to them in other theories of evolution. This difference is so remarkable that mutations can be completely avoided in the proposed computational context, as in the examples presented in Secs. 5 and 6. The mixture of information from the host's context with those existing in the population of endosymbiont chromosomes has the potential to produce, in many cases, the diversification needed to escape from local minima. The process ends when the exchange of information between the host and the population of endosymbionts does not result in further changes (improvements) in the fitness of the endosymbionts.

The characteristics listed above can achieve results in the algorithmic search that can not be easily simulated by other evolutionary methods. Due to the guidance of the search and the impossibility of reproduction, the algorithm has the potential to preserve the good quality building blocks which emerge during its execution.

The transposons can be designed to simulate a peculiar form of restricted local search providing an excellent source of intensification for the algorithm. The approach allows modulating, simultaneously, the sharing of genetic information between the host and the population of endosymbionts and the intensification effort promoted with these transposons. This modulation can be implemented in such a way that the most promising vector is utilized in different stages of the search. For instance, if the "a priori" information stored in the host has proven quality, it is likely that the manipulation with plasmids using such information in the early stages of the search is more beneficial, while the transposons designed for intensification tasks will be more useful in the final stages. Otherwise the modulation follows the reverse logic.

The decentralized and cooperative architecture proposed in the Transgenetic Algorithms propitiates the use, within the evolutionary search, of information obtained with constructive methods, heuristics, or other procedures that produce structures associated with the problem even where such structures do not represent solutions to the problem examined.

8 Conclusions and Future Works

The biological processes have been a valuable source of inspiration for the development of algorithms to deal with complex problems. In this chapter, a theory of evolution that considers cooperative interactions is adopted as a source of inspiration for the development of evolutionary algorithms. This theory deals with the integration of the genomes of endosymbionts and a host cell. In nature, this integration is responsible for many macroevolutionary events and biological phenomena. The endosymbiotic interactions change the focus of the evolution by competition to the evolution by a radical form of cooperation: the integration.

The algorithms introduced in this chapter, called transgenetic algorithms, perform a stochastic search with elements that are motivated by endosymbiotic interactions and natural mechanisms of lateral gene transfer. In the proposed approach, the operations based on mechanisms of horizontal gene transfer comply with tasks that accelerate and make more effective the genetic changes of the elements involved in the endosymbiotic fusion. Since the proposed endosymbiotic evolutionary process occurs in a micro-system and is subjected to a guidance imposed by the genetic repository of the host, a wide range of options for the development of bio-inspired computational methods is opened.

This chapter presented the outlines for the development of Transgenetic Algorithms and reported the implementation of these algorithms to a single and to a bi-objective combinatorial problem. The mono objective problem is the uncapacitated version of Traveling Purchaser Problem, where the proposed algorithm managed to find nine new best solutions for benchmark instances.

The other application was given for the Bi-objective Minimum Spanning Tree Problem, where a computational experiment showed that the proposed approach finds better approximation sets and lower processing times than a Memetic Algorithm presented previously for the same problem.

In future works the full process of serial endosymbiosis will be considered. It will consist of successive events of engulfment and integration of populations of endosymbionts. Such events can progressively enrich the information of the host. The successive integrated populations may represent solutions in different regions of the search space of the problem being tackled, mono-objective solutions of a multi-objective problem, etc. In addition, the different integration events can occur with different manipulation strategies. The future research will also extend the algorithms in order to consider other types of "a posteriori" information, such as statistical analysis of the population of endosymbionts and transgenetic vectors. The algorithms will also be applied to other combinatorial problems with one or more objectives.

References

1. Aggarwal, V., Aneja, Y., Nair, K.: Minimal spanning tree subject to a side constraint. Computers & Operations Research 9, 287–296 (1982)
2. Almeida, C.P., Goldbarg, E.F.G., Gonçalves, R.A., Regattieri, M.D., Goldbarg, M.C.: TA-PFP: A transgenetic algorithm to solve the protein folding problem. In: Proceedings of ISDA 2007 Seventh International Conference on Intelligent Systems Design and Applications, vol. 1, pp. 163–168 (2007)

3. Applegate, D., Bixby, R., Chvatal, V., Cook, W.: Finding tours in the TSP. Technical Report TR99-05. Department of Computational and Applied Mathematics: Rice University (1999)
4. Barboza, A.O.: Simulação e técnicas da computação evolucionária aplicadas a problemas de programação linear inteira mista. D.Sc. Thesis, Universidade Tecnológica Federal do Paraná, Brazil (2005)
5. Bazlamaçci, C.F., Hindi, K.S.: Minimum-weight spanning tree algorithms: A survey and empirical study. Computers & Operations Research 28, 767–785 (2001)
6. Bellmore, M., Nemhauser, G.L.: The traveling salesman problem: A survey. Operations Research 16, 538–582 (1968)
7. Boctor, F.F., Laporte, G., Renaud, J.: Heuristics for the traveling purchaser problem. Computers & Operations Research 30, 491–504 (2003)
8. Bontoux, B., Feillet, D.: Ant colony optimization for the traveling purchaser problem. Computers & Operations Research 35, 628–637 (2008)
9. Buchner, P.: Endosymbiosis of animals with plant microorganisms. Wiley Interscience, New York (1965)
10. Bull, L., Fogarty, T.C.: Artificial symbiogenesis. Artificial Life 2, 269–292 (1995)
11. Chan, T.-M., Man, K.-F., Tang, K.-S., Kwong, S.A.: Jumping gene algorithm for multiobjective resource management in wideband CDMA. The Computer Journal 48(6), 749–768 (2005)
12. Chen, I., Dubnau, D.: DNA uptake during bacterial transformation. Nature Reviews Microbiology 2, 241–249 (2004)
13. Conover, W.J.: Practical nonparametric statistics, 3rd edn. John Wiley & Sons, Chichester (2001)
14. Daida, J.M., Grasso, C.S., Stanhope, S.A., Ross, S.J.: Symbionticism and complex adaptive systems I: Implications of having symbiosis occur in nature. In: Proceedings of the Fifth Annual Conference on Evolutionary Programming, pp. 177–186 (1996)
15. Doolittle, W.F.: Lateral genomics. Trends in Genetics 15(12), M5–M8 (1999)
16. Dutta, C., Pan, A.: Horizontal gene transfer and bacterial diversity. Journal of Biosciences 27, 27–33 (2002)
17. Eigen, M., Schuster, P.: The hypercycle: A principle of natural selforganization. Naturwissenschafter 64(11), 541–565 (1977)
18. Fukuda, T., Kubota, N., Shimojima, K.: Virus-evolutionary genetic algorithm and its application to traveling salesmam problem. In: Yao, X. (ed.) Evolutionary Computation, Theory and Applications. World Scientific, Singapore (1999)
19. Garey, M.R., Johnson, D.S.: Computers and intractability: A guide to the theory of NP-completeness. Freeman, New York (1979)
20. Goldbarg, E.F.G., Castro, M.P., Goldbarg, M.C.: A transgenetic algorithm for the gas network pipe sizing problem. Computational Methods 1, 893–904 (2006)
21. Goldbarg, E.F.G., Goldbarg, M.C., Bagi, L.B.: Transgenetic algorithm: A new evolutionary perspective for heuristics design. In: Proceedings of GECCO 2007 Genetic and Evolutionary Computation Conference, pp. 2701–2708 (2007)
22. Goldbarg, E.F.G., Goldbarg, M.C., Costa, W.E.: Evolutionary algorithms applied to the workover rigs schedule problem. Annals of XI Latin-Iberian American Congress of Operations Research (2002)

23. Goldbarg, M.C., Goldbarg, E.F.G., Medeiros Neto, F.D.: Algoritmos evolucionários na determinação da configuração de custo mínimo de sistemas de co-geração de energia com base no á natural. Pesquisa Operacional 25(2), 231–259 (2005)
24. Gouvêa, E.F.: Transgenética computacional: Um estudo algorítmico. Ph.D. Thesis, Universidade Federal do Rio de Janeiro, Brazil (2001)
25. Guttin, G., Punnen, A.: The traveling salesman problem and its variations. Kluwer Academic Publishers, Dordrecht (2002)
26. Hansen, M.P., Jaszkiewicz, A.: Evaluating the quality of approximations to the non-dominated set. Technical Report IMM-REP-1998-7, Technical University, Denmark (1998)
27. Harvey, I.: The microbial genetic algorithm (unpublished manuscript) (1996), http://citeseer.ist.psu.edu/13824.html
28. Hillis, D.W.: Co-evolving parasites improve simulated evolution in an optimization procedure. Physica D 42, 228–234 (1999)
29. Jain, R., Rivera, M.C., Lake, J.A.: Horizontal gene transfer among genomes: The complexity hypothesis. Proceedings of the National Academy of Sciences USA 96, 3801–3806 (1999)
30. Jain, R., Rivera, M.C., Moore, J.E., Lake, J.A.: Horizontal gene transfer accelerates genome innovation and evolution. Molecular Biology and Evolution 20(10), 1598–1602 (2003)
31. Kim, J.Y., Kim, Y., Kim, Y.K.: An endosymbiotic evolutionary algorithm for optimization. Applied Intelligence 15, 117–130 (2001)
32. Kim, Y.K., Kim, J.Y., Kim, Y.: An endosymbiotic evolutionary algorithm for the integration of balancing and sequencing in mixed-model U-lines. European Journal of Operational Research 168, 838–852 (2006)
33. Knowles, J.D.: Local-search and hybrid evolutionary algorithms for Pareto optimization, PhD Thesis, Department of Computer Science, University of Reading, Reading, UK (2002)
34. Knowles, J.D., Corne, D.W.: A comparison of encodings and algorithms for multiobjective spanning tree problems. In: Proceedings of the 2001 Congress on Evolutionary Computation (CEC 2001), pp. 544–551 (2001)
35. Knowles, J.D., Thiele, L., Zitzler, E.: A tutorial on the performance assessment of stochastic multiobjective optimizers, TIK 214, Computer Engineering and Networks Laboratory (TIK), Swiss Federal Institute of Technology (ETH), Zurich (2006)
36. Kubota, N., Arakawa, T., Fukuda, T., Shimojima, K.: Trajectory generation for redundant manipulator using virus evolutionary genetic algorithm. In: Proceedings of the IEEE International Conference on Robotics and Automation, pp. 205–210 (1997)
37. Kubota, N., Shimojima, K., Fukuda, T.: Virus-evolutionary genetic algorithm - coevolution of planar grid model. In: Proceedings of the Fifth IEEE International Conference on Fuzzy Systems (FUZZIEEE 1996), vol. 1, pp. 8–11 (1996)
38. Kutschera, U., Niklas, K.J.: Endosymbiosis, cell evolution, and speciation. Theory in Biosciences 124, 1–24 (2005)
39. Leite, L.E., Souza Fillho, G., Goldbarg, M.C., Goldbarg, E.F.G.: Comparando algoritmos genéticos e transgenéticos para otimizar a configuração de um serviço de distribuição de Vídeo baseado em replicação móvel. Anais do SBRC2004 22 Simpósio Brasileiro de Redes de Computadores 1, 129–132 (2004)

40. Margulis, L.: Symbiosis in cell evolution: Microbial communities in the archean and proterozoic eons. W.H. Freeman, New York (1992)
41. Margulis, L.: Serial endosymbiotic theory (SET) and composite individuality Microbiology Today 31, 172–174 (2004)
42. Margulis, L., Sagan, D.: Microcosmos. Summit Books, New York (1986)
43. Maynard-Smith, J., Szathmáry, E.: The major transitions in evolution. W.H. Freeman, Oxford (1995)
44. Michalewicz, Z., Fogel, D.B.: How to solve it: Modern heuristics. Springer, Heidelberg (2000)
45. Mitchell, M.: An introduction to genetic algorithms. MIT Press, Cambridge (1998)
46. Mühlenbein, H., Voigt, H.-M.: Gene pool recombination in genetic algorithms. In: Proceedings of the Sixth International Conference on Genetic Algorithms, pp. 104–113 (1995)
47. Novozhilov, A.S., Karev, G.P., Koonin, E.V.: Mathematical modeling of evolution of horizontally transferred genes. Molecular Biology and Evolution 22(8), 1721–1732 (2005)
48. Perales-Graván, C., Lahoz-Beltra, R.: An AM radio receiver designed with a genetic algorithm based on a bacterial conjugation genetic operator. IEEE Transactions on Evolutionary Computation 12(2), 1–29 (2008)
49. Ramos, I.C.O., Goldbarg, M.C., Goldbarg, E.F.G., Dória Neto, A.D.: Logistic regression for parameter tuning on an evolutionary algorithm. In: Proceedings of the IEEE CEC 2005 Congress on Evolutionary Computation, vol. 2, pp. 1061–1068 (2005)
50. Riera-Ledesma, J., Salazar-González, J.J.: A heuristic approach for the traveling purchaser problem. European Journal of Operational Research 162, 142–152 (2005)
51. Rocha, D.A.M., Goldbarg, E.F.G., Goldbarg, M.C.: A memetic algorithm for the biobjective minimum spanning tree problem. In: Gottlieb, J., Raidl, G.R. (eds.) EvoCOP 2006. LNCS, vol. 3906, pp. 222–233. Springer, Heidelberg (2006)
52. Rocha, D.A.M., Goldbarg, E.F.G., Goldbarg, M.C.: A new evolutionary algorithm for the bi-objective minimum spanning tree. In: Proceedings of ISDA 2007 Seventh International Conference on Intelligent Systems Design and Applications, pp. 735–740 (2007)
53. Rosenkrantz, D.J., Stearns, R.E., Lewis II, P.M.: An analysis of several heuristics for the traveling salesman problem. SIAM Journal on Computing 6, 563–581 (1977)
54. Rosin, C.D., Belew, R.K.: New methods for competitive coevolution. Evolutionary Computation 5(1), 1–29 (1997)
55. Schmidt, C., Goldbarg, E.F.G., Goldbarg, M.C.: A hybrid transgenetic algorithm for the prize collecting Steiner tree problem. In: Proceedings of ISDA 2007 Seventh International Conference on Intelligent Systems Design and Applications, vol. 1, pp. 271–276 (2007)
56. Shapiro, J.A.: Transposable elements as the key to a 21st century view of evolution. Genetica 107, 171–179 (1999)
57. Simões, A.B., Costa, E.: Transposition: A biologically inspired mechanism to use with genetic algorithms. In: Proceedings of the Fourth International Conference of Neural Networks and Genetic Algorithms, pp. 178–186 (1999)
58. Simões, A.B., Costa, E.: Transposition versus crossover: An empirical study. In: Proceedings of the Genetic and Evolutionary Compuation Conference (GECCO 1999), pp. 612–619 (1999)

59. Simões, A.B., Costa, E.: On biologically inspired genetic operators: Transformation in the standard genetic algorithm. In: Proceedings of the Genetic and Evolutionary Computation Conference (GECCO 2001), pp. 584–591 (2001)
60. Simões, A.B., Costa, E.: An evolutionary approach to the zero/one knapsack problem: Testing ideas from biology. In: Proceedings of the Fifth International Conference on Neural Networks and Genetic Algorithms (ICANNGA 2001), pp. 22–25 (2001)
61. Smith, P.W.H.: Finding hard satisfiability problems using bacterial conjugation. In: Proceedings of the AISB96 Workshop on Evolutionary Computing, pp. 236–244 (1996)
62. Theissen, U., Martin, W.: The difference between organelles and endosymbionts. Current Biology 16(24), R1016–R1017 (2006)
63. Timmis, J.N., Ayliffe, M.A., Huang, C.Y., Martin, W.: Endosymbiotic gene transfer: organelle genomes forge eukaryotic chromosomes. Nature Reviews Genetic 5, 123–135 (2004)
64. Vothknecht, U.C., Soll, J.: Protein import: The hitchhikers guide into chloroplasts. Biological Chemistry 381, 887–897 (2000)
65. Wernegreen, J.J.: For better or worse: genomic consequences of intracellular mutualism and parasitism. Genetics & Development 15, 572–583 (2005)
66. Yeung, S.-H., Ng, H.-K., Man, K.-F.: Multi-criteria design methodology of a dielectric resonator antenna with jumping genes evolutionary algorithm. International Journal of Electronics and Communication (AEÜ) 62, 266–276 (2008)
67. Zaneveld, J.R., Nemergut, D.R., Knight, R.: Are all horizontal gene transfers created equal? Prospects for mechanism-based studies of HGT patterns. Microbiology 154, 1–15 (2008)
68. Zhou, G., Gen, M.: Genetic algorithm approach on multi-criteria minimum spanning tree problem. European Journal of Operational Research 114, 141–152 (1999)
69. Zitzler, E., Thiele, L.: Multiobjective evolutionary algorithms: A comparative case study and the strength Pareto approach. IEEE Transactions on Evolutionary Computation 3(4), 257–271 (1999)
70. Zitzler, E., Thiele, L., Laumanns, M., Fonseca, C.M., Fonseca, V.G.: Performance assessment of multiobjective optimizers: An analysis and review. IEEE Transactions on Evolutionary Computation 7(2), 117–132 (2003)

Multi-objective Team Forming Optimization for Integrated Product Development Projects

Hisham M.E. Abdelsalam

Summary. Integrated product development (IPD) is a holistic approach that helps to overcome problems that arise in complex product development environments. This paper presents a model that aims to support the optimal formulation and assignment of multi-functional teams in IPD organizations - or any project-based organization. The model accounts for limited availability of personnel, required skills, team homogeneity, and, further, maximizes organization's payoff by formulating and assigning teams to projects with higher expected payoffs. A Pareto multi-objective particle swarm optimization approach was used to solve the model. It allows personnel to work in several concurrent projects and considers both person-job and person-team fit.

1 Introduction

By the mid-1990's, the importance of early introduction of new products to both market share and profitability became fully understood. Reducing product time-to-market, thus, became an essential requirement for continuous competition. Knowing that about 70% of the life cycle cost of a product is committed at early design phases, the motivation for developing and implementing more effective methodologies for managing the design process of new product development projects became very strong.

The difficulties in designing complex engineering products do not arise simply from their technical complexity but rather from the managerial complexity necessary to manage the interactions between the different engineering disciplines, which impose additional challenges on the design process [1]. The basic disciplines for making progress in that context belong not only to mechanical engineering but also to industrial engineering, mathematics, management science, and computer science. As a result, a systems level solution must be determined and deployed.

Hisham M.E. Abdelsalam
Decision Support Department, Cairo University, Cairo, Egypt
e-mail: h.abdelsalam@fci-cu.edu.eg

A. Abraham et al. (Eds.): Foundations of Comput. Intel. Vol. 3, SCI 203, pp. 461–478.
springerlink.com ⓒ Springer-Verlag Berlin Heidelberg 2009

Integrated product development (IPD) is a holistic approach that helps to overcome problems that arise in complex product development environments. IPD was defined as: a process whereby all functional groups that are involved in a product life cycle participate as a team in the early understanding and resolution of key product development issues including quality, manufacturability, reliability, maintainability, environment, and safety [2]. *IPD* is based on concurrent engineering, but goes beyond Concurrent Engineering (CE) with regard to the level of integration. In the scope of IPD, designers, assembly planners and production planners, as well as persons responsible for quality or testing not only consult themselves while they are working simultaneously on their tasks, but exchange interconnected intermediate results in a continuous interplay [3, 4]. In the light of the importance of new product development timeliness to organizational success, IPD teams must be composed of cross functional members who work well together.

Following the introduction section, the rest of this chapter is organized as follows: Section 2 reviews literature related work. Then Section 3 provides background to the Myers-Briggs Type Indicator - that will be the basis for evaluating team efficiency - followed by problem description in Section 4. The model formulation is presented in Section 5, and solution algorithm is detailed in Section 6. An illustrative example is given in section 7 and, finally, conclusions in section 8.

2   Related Work

Team development, generally, goes through five stages: forming, storming, norming, performing, and adjourning [5]. Numerous literatures can be found on team development. Dalziel and Sommervill [6] study team forming to building a project team applicability of self perception inventory. Schneider [7] discusses culture crossing management of task for team forming stage.

The concept of forming multi-functional teams has gained increasing attention within product development organizations. Some of the basic conditions for the use of multi-functional teams in product development were indicated by [8]. Askin and Sodhi [9] have presented an approach to organizing teams in concurrent engineering. Dietz and Rosenshine [10] developed an analytical method for determining an optimal specialization strategy for a maintenance workforce. The method developed can be specifically applied to optimize the maintenance manpower structure for a small unit of military tactical aircraft. Zakarian and Kusiak [11] presented a conceptual framework for prioritizing team members based on customer requirements and product characteristics. Sethi and Nicholson [12] explored structural and contextual correlates of charged behavior in product development teams. Tseng et al. [13] applied the fuzzy and grey approaches to form a good multi-functional team and when dealing with insufficient information at the team forming stage. For further discussions on the general aspects of team formation and team performance, the reader is referred to [14, 15, 16]. These models, however, did not consider: (1) possible limitation on personnel availability in the organization; and (2) interactions among team members as a result of different personalities' characteristics.

To ensure sufficient breadth and depth of technical skills and team synergy, Fitzpatrick and Askin [17] developed a mathematical model for formation of effective human teams. Team synergy was measured and included in the model using Kolbe Conative Index [18, 19]. Their model, however, assumed: (1) an individual possesses only one skill; (2) that there exist sufficient individuals in the various skill categories to meet the requirements for the teams; and (3) each team has approximately the same importance. Abdelsalam, et. al. [20] presented and solved a mathematical model that tries to support the optimal formulation of multi-functional teams in an organization with limited availability of personnel and skills. Their model, still, did not account for the effects of inter-personalities issues among team members.

The model presented in this chapter aims to support the optimal formulation of multi-functional teams in integrated product development organizations. The model, however, can be generalized to any project-based organization. The model builds on Abdelsalam, et. al. [20] work as it: (1) accounts for limited availability of personnel and skills; (2) maximizes organization's payoff (revenue) by formulating and assigning teams to projects with higher expected payoffs; and (3) maximizes team homogeneity and optimal formulation from personalities' point of view. One other extension was the use of a multi-objective particle swarm optimization algorithm to develop a Pareto-front for the objective functions.

3 Myers-Briggs Type Indicator (MBTI)

Personality has been identified as a potentially helpful selection variable in the determination of optimal team composition [21]. This research uses a very popular and well established test of personality; the Myers-Briggs Type Indicator (MBTI) that was developed by Isabel Briggs Myers and her mother, Katherine Briggs during World War II [22]. The MBTI is based on the theory that each person's personality fits into only one of sixteen types. These types are based on four features of personality, each consisting of two opposite preferences.

The theory claims that personality types determine how people will behave in all situations. The four dimensions are [23]: **Extroversion (E) vs. Introversion (I)**. This dimension reflects the perceptual orientation of the individual. **Sensing (S) vs. Intuition (N)**. People with a sensing preference rely on that which can be perceived and are considered to be oriented toward that which is real. **Thinking (T) vs. Feeling (F)**. A preference for thinking indicates the use of logic and rational processes to make deductions and decide upon action. **Judgment (J) vs. Perception (P)**. Indicates if rational or irrational judgments are dominant when a person is interacting with

the environment. The preferences are normally notated with the initial letters of each of their four features, for instance: ISTJ is (Introverted, Sensing,
Thinking, Judging) and so on for all sixteen possible combinations.

4   Problem Description

4.1   Context

The situation considered by the presented model is that at a given point of
time the company has a set of candidate projects to be carried out. However,
as there is a fixed capacity of resources (personnel) pool, not all projects will
start working concurrently; only projects that have their complete personnel
requirements will work.

The decision whether to start a project or not will depend on two major factors: (1) the ability to compose the team required by that specific
project, and (2) the expected project payoff. Thus, there is a set of competing projects, and a limited (finite) pool of personnel. It is required to form
and assign teams - from the pool - to different projects based on technical
skills and time requirements by each project. MBIT will be used to ensure
team efficiency. The problem is, thus, of three-fold: (1) assigning persons with
specific technical skills required by a certain project to that project (team
forming); (2) increasing team efficiency by composing a team of members
with complementary personality types; and (3) increasing company payoff
through increased number of project to start working. Consequently, formation of multi-functional teams for IDP projects becomes a challenge. It should
be mentioned here that the model allows company payoff to be given in any
form; financial, market share, etc. The model, further, assumes that projects
can work concurrently, thus, one person can be assigned to more than one
project as long as his/her available time is less than the threshold value.

4.2   Personnel Characteristics

In addition to personal characteristic (gender, age, etc.), each person in the
pool has the following set of characteristics: " Technical skills. A person is
assumed to have at most five skills with different levels of competence. "
Availability; the %time that person is available to be assigned to different
project at the teams formation stage. " Personality profile; as indicated by
MBTI. The model assumes that a person has a certain level of competence
in at most five technical skills. These levels are: excellent (best level); very
good; good; and accepted (least level). The competence level of a person in a
certain skill is, further, assumed to be deterministic and constant throughout
one project and in all projects.

4.3 Project Requirements

The problem considers a set of competing projects - each with its own expected payoff - over a limited personnel pool. For its execution, a project would, typically, require a team with different functions/skills. These skills' requirements are assumed to be clearly defined to the satisfaction of management. A project can not start unless all needed technical requirements are available and its team is formed. We assume that a successful team must at least have members with the following roles:

(1) leader: non-stop worker to make project successful;
(2) visionary: the map reader who knows where the team has been and where to go;
(3) marketer: the steering wheel - understands customer's requirements;
(4) architect: understands how the intended product will satisfy customer's needs, and
(5) backstop: answers any questions and codes the team out of almost any bind.

The model assumes that the project manager (leader) is assigned by company executives and thus was removed from the mathematical formulation. However, the model can easily accommodate for it when needed. The model is based on the following personality types for each of the rest four main members; visionary-ENXP; marketer-EXTP; architect-XNXP; and backstop-XSTJ. The X refers to that either dimension value is accepted. These types were selected based on our experience. However, further research can be conducted to validate the composition.

4.4 Objectives

At a certain point of time, given a list of candidate projects and a pool of personnel, it is required to determine the list of project that will be accepted by the company to be carried out and the team associated with each of these projects. This is a multi-objective optimization problem with two main objectives:

- Maximizing effectiveness; maximizing company payoff (revenue) by allowing as many as possible projects - with different projects' payoffs - to start working, and
- Maximizing efficiency; this is further divided into:

    Minimizing the deviation in skill levels of team members selected for different projects. This deviation results as a consequence of assigning one person, to a skill requirement in a project, with a level different than the level needed by this requirement.

    Minimizing the deviation from optimal team composition (as discussed early in project requirements subsection)

5   Model Formulation

Let:

C_{iu} Technical skill number u that person i has

L_{iu} Level of technical skill number u for person i

R_{jk} Technical requirement number k that project j needs

V_{jk} Levels of technical requirement number k for project j

T_{jk} %Time required by requirement k in project j

A_i Available time for person i

P_j expected payoff from project j

$j = 1, 2, ..., n$ where n is the number of projects

$i = 1, 2, ..., l_j$ where m is the number of persons available

$k = 1, 2, ..., m$ where l_j is the number of requirements in project j

$W \in \{1, 2, 3, 4\}$ set of levels that a skill can assume

$u \in \{1, 2, 3, 4, 5\}$ index of the number of skills a person has

$Z \in \{001, 003, , .., 050\}$ set of codes of available skills in the company (assumed to be 50)

Model decision variables are:

$x_{ijk} =$

$$\begin{cases} 1 & \text{, if person } i \text{ is assigned to project } j \text{ in requirement } t, \\ 0 & \text{, otherwise.} \end{cases}$$

$y_{jk} =$

$$\begin{cases} 1 & \text{, if } C_{iu} \text{ - } R_{jk} = 0 \text{ (assign person skill satisfies project requirement } t, \\ 0 & \text{, otherwise.} \end{cases}$$

$$\phi_j =$$

$$\begin{cases} 1 & \text{, if } \sum_{k=1}^{l_j} y_{jk} = l_j \text{ (team is formed for project } j), \\ 0 & \text{, otherwise.} \end{cases}$$

$$MA_i =$$

$$\begin{cases} 1 & \text{, if person } i \text{ can be a marketer,} \\ 0 & \text{, otherwise.} \end{cases}$$

$$VI_i =$$

$$\begin{cases} 1 & \text{, if person } i \text{ can be a visionary,} \\ 0 & \text{, otherwise.} \end{cases}$$

$$AR_i =$$

$$\begin{cases} 1 & \text{, if person } i \text{ can be an architect,} \\ 0 & \text{, otherwise.} \end{cases}$$

$$BI_i =$$

$$\begin{cases} 1 & \text{, if person } i \text{ can be backstop,} \\ 0 & \text{, otherwise.} \end{cases}$$

The model has two main objective functions:

The first is concerned with maximizing the payoff generated from projects to be carried out by the company.

$$\max f_1 = \sum_{j=1}^{n} \phi_j P_j \tag{1}$$

The second one is concerned with maximizing team efficiency. This research assumes that team efficiency can be enhanced through minimizing the deviation between: (1) the levels of the required skills in a team and the competency levels of persons assigned to this team; and between (2) optimal team composition and soft characteristics of persons assigned to this team. Thus, the second objective function can be defined as:

$$\min f_2 = \sum_{j=1}^{n} \phi_j (w_1 D_{jt} + w_2 D_{js}) \qquad (2)$$

Where w_1 and w_2 are user defined weights and $w_1 + w_2 = 1$ and D_{jt} is the deviation with respect to technical requirement.

$D_{jt} =$

$$\begin{cases} 1 & \sum_{k=1}^{l_j} \sum_{i=1}^{m} \sum_{u=1}^{5} (V_{jk} - L_{iu}) x_{(ijk} y_{jk}), \text{ if } (V_{jk} - L_{iu}) > 0, \\ 0 & \text{, otherwise.} \end{cases}$$

D_{js} is the deviation with respect to team composition (soft requirements).

$$D_{js} = \sum_{j=1}^{n} (D_{jMA} + D_{jAR} + D_{jVI} + D_{jBI}) \qquad (3)$$

Where

$D_{jMA} =$

$$\begin{cases} 1 & \text{,if } \sum_{k=1}^{l_j} \sum_{i=1}^{m} x_{ijk} MA_i \geq 1, \\ 0 & \text{, otherwise.} \end{cases}$$

$D_{jAR} =$

$$\begin{cases} 1 & \text{,if } \sum_{k=1}^{l_j} \sum_{i=1}^{m} x_{ijk} AR_i \geq 1, \\ 0 & \text{, otherwise.} \end{cases}$$

$D_{jVI} =$

$$\begin{cases} 1 & \text{,if } \sum_{k=1}^{l_j} \sum_{i=1}^{m} x_{ijk} VI_i \geq 1, \\ 0 & \text{, otherwise.} \end{cases}$$

$D_{jBI} =$

$$\begin{cases} 1 & \text{,if } \sum_{k=1}^{l_j} \sum_{i=1}^{m} x_{ijk} BI_i \geq 1, \\ 0 & \text{, otherwise.} \end{cases}$$

These objectives are subject to the following constraints:

Sum of all the requirements' times that person i is assigned to is less than or equal to his/her total available time.

$$\sum_{i=1}^{n}\sum_{k=1}^{l} X_{ijk}(T_{jk}) \leq A_i \qquad\qquad \forall(i) \qquad\qquad (4)$$

A personal skill must belong to the set of skills available in the company

$$C_{iu} \in Z \qquad\qquad \forall(i,u) \qquad\qquad (5)$$

A project technical requirement must belong to the set of skills available in the company

$$R_{jk} \in Z \qquad\qquad \forall(j,k) \qquad\qquad (6)$$

Level of any skill assumes a value from the set only w

$$L_{iu} \in W \qquad\qquad \forall(i,u) \qquad\qquad (7)$$

Level of any technical requirement assumes a value from the set only w

$$V_{jk} \in W \qquad\qquad \forall(j,k) \qquad\qquad (8)$$

6   Solution Algorithm

6.1   *Multi-Objective Particle Swarm Optimization (MOPSO)*

Particle swarm optimization (PSO) is a population-based optimization method originally designed by Kennedy and Eberhart in 1995 inspired by observing the bird flocking or fish school. Since 2002, PSO has been growing rapidly with over 100 published papers every year [24]. All the related research has totaled over 300 papers until 2004 [25]. Most of the applications have been concentrated on solving continuous optimization problems, but the studies of PSO on discrete optimization problems are relatively few [26]. In PSO, potential solutions are called particles.

To reach an optimal solution, each particle adjusts its position and velocity according to its own experience and to other particles' experience. Kennedy and Eberhart [27] developed a discrete version of PSO. Discrete PSO essentially differs from the original (or continuous) PSO in two characteristics. First, the particle is composed of the binary variable. Second, the velocity must be transformed into the change of probability, which is the chance of the binary variable taking the value one [25]. This research uses the algorithm developed in Kennedy et al. [28] with the addition of the inertia weight proposed by Shi and Eberhart [29]. Due to the success of PSO in single objective optimization, in recent years, more and more attempts have been made to

Fig. 1 Representation of
one particle (solution)

	Projects * Requirements					
	Project 1			Project 2		
Persons	1	2	3	1	2	3
1	1	0	0	0	1	0
2	0	0	0	0	0	1
3	0	0	1	0	0	0
4	0	1	0	1	0	0
5						
6						
7						
8						

extend PSO to the domain of multi-objective problems [30, 31, 32]. The main challenge in multi-objective particle swarm optimization (MOPSO) is to select the global and local attractors such that the swarm is guided towards the Pareto optimal front and maintains sufficient diversity [30]. A MOPSO starts with a set of uniformly distributed random initial particles defined in the search space. A set of particles are considered as a population at certain generation.

In addition to the population, another set (called Archive) is defined in order to store the obtained non-dominated solutions. The particles are evaluated and the non-dominated solutions are added to the archive in every generation, while dominated solutions are pruned. In the next step, the particles are moved to a new position in the space using the mechanism presented in [28]. For more details on Pareto MOPSO the reader is kindly referred to [30].

6.2 Particle Definition

The particle (solution) is represented in a matrix form with rows representing the number of the available persons in the company, and columns representing different requirements in different candidate projects. For example, Fig. 1 shows a particle definition for a problem in which the company has 4 persons and there are two candidate projects each with 3 requirements. The cells assume binary values (0 or 1) representing the variable x_{ijk}. So, the solution in Fig. 1 indicates that:

- Person 1 is assigned to project 1 in requirement 1 also assigned to project 2 in requirement 2
- Person 2 is assigned to project 2 in requirement 3
- Person 3 is assigned to project 1 in requirement 3
- Person 4 is assigned to project 1 in requirement 2 also assigned to project 2 in requirement 1

The position matrix (population) consists of a certain number of solutions (the size of the population). Fig. 2 shows a position matrix with three particles.

Fig. 2 Position Matrix
in PSO

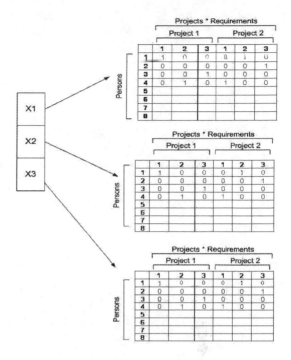

6.3 Initialization

A heuristic is used to generate an initial feasible solution (particle). This
particle is, and then used to fill all the population. Velocity matrix is initialized by generating random numbers between two specified numbers (min,
max). All parameters are either set to default values or to values specified by
the user. Finally, the best position for each particle is the same as the first
initialized solution at the first iteration.

6.4 Fitness Calculation

Both objectives were normalized as follow:

f_1 was changed to a percentage through Eq. (9)

$$\max f_1 = \frac{\text{payoff generated from project to be carried out}}{(\text{payoff when all candidate project are carried out}) - (\text{payoff when no project ar carried out})}$$

$$\max f_1 = \frac{\sum_{j=1}^{n} \phi_j P_j}{\sum_{j=1}^{n} P_j} \qquad (9)$$

Table 1 Parameters setting

Parameter		Value
Number of iterations		100
Pareto set size		10
Maximum inertia		1.2
Maximum velocity		0.4
Minimum velocity		-4
w_1		0.5
w_2		0.5

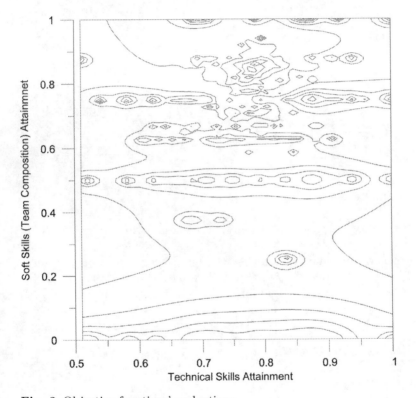

Fig. 3 Objective functions' evaluations

On the other hand, f_2 was changed to a percentage using Eq. (10)

$$\max f_2 = w_1 \left(\frac{1}{\sum_{j=1}^n \phi_j} \left(\sum_{j=1}^n \phi_j \left(1 - \frac{D_{jt}}{l_j} \right) \right) \right) + w_2 \left(\frac{1}{\sum_{j=1}^n \phi_j} \left(\sum_{j=1}^n \phi_j \left(\frac{4 - D_{js}}{4} \right) \right) \right)$$

$$(10)$$

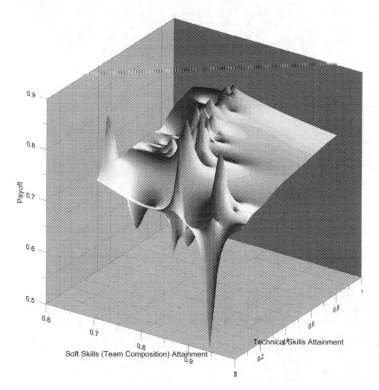

Fig. 4 Objective functions' evaluation (3D)

6.5 Generating New Solutions

For generating new solutions, we use the update rule of the PSO algorithm that updates the velocity of each particle *solution* then using the updated velocity to generate new position *that represents the new solution.* Particle velocity is adjusted (updated) according using Eq. 11

$$v_{id}^{t+1} = wv_{id}^t + \phi_1 r_1 (pbest_{id})^t - x_{id}^t + \phi_2 r_2 (pbest_{gd}^t - x_{id}^t) \qquad (11)$$

The velocity value is constrained to the interval [0,1] using the sigmoid function; Eq. (12).

$$s(v_{id}^t) = \frac{1}{1 + exp(-v_{id}^t)} \qquad (12)$$

Where $s(v_{id}^t)$ denotes that the probability of $x^t(id)$ taking 1. To avoid $s(v_{id}^t)$ approaching 0 or 1, a constant v_{max} is used to limit the range of $v^t(id)$. In practice, v_{max} is often set at 4, i.e., $v^t(id)$ belongs to (v_{max}, v_{min}). For

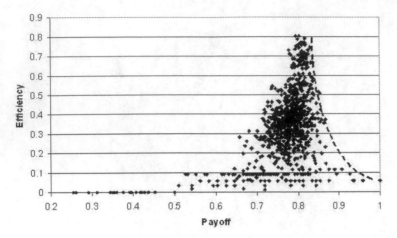

Fig. 5 Pareto front

updating the position matrix, for each particle we generate a random number and then if that number is less than $s(v_{id}^t)$ then $x^t(id) = 1$; otherwise $x^t(id) = 0$.

6.6 Convergence

Generally, the algorithm is stopped whenever it reaches either of two criteria: (1) the number of generations exceeds a threshold value, or (2) the objective function improvement is below some defined value. This research adopts the first rule.

6.7 Feasibility

Infeasible solutions are not tolerated within the algorithm. Whenever a constraint is violated, the solution is considered infeasible. There are three to handle infeasible solutions: (1) removing this infeasible solution from the population without replacing it with another solution, (2) replacing this infeasible solution with one that is guaranteed to be feasible, and this would be possible if we used the heuristic used for initializing the PSO population, and (3) repairing this infeasible solution to make it feasible. This research used the third method; infeasible solutions stay in the population after being repaired to be feasible.

6.8 Detailed Computational Flow

The following provides the detailed steps for the proposed algorithm.

Algorithm 1. Clinical parameters assessment algorithm

Input: Input parameters of the problem.

Output: a set of the Pareto-optimal solutions from the archive

1. Generate an initial feasible solution and initialize randomly the position and velocity of each particle
2: For each particle of the population, calculate its fitness functions, and evaluate each of the particles in the population.
3: Store the positions of the particles that are non-dominated in the archive.
4: Initialize the memory of each particle in which a single local best for each particle is contained (this memory serves as a guide to travel through the search space; pbest).
5: Update iteration counter
6: Determine the best global particle (gbest)
7: Compute the velocity and the new position of each particle as described in section 6.5, and maintain the particles within the search space in case they go beyond its boundaries as discussed in section 6.7.
8: Evaluate each particle in the population by its fitness functions.
9: Update the contents of the archive. This update consists of inserting all the currently non-dominated locations into the archive. Any dominated locations from the archive are eliminated in the process.
10: Update the contents of the pbest. If the current position of the particle is dominated by the position in the pbest, then the position in the pbest is kept; otherwise, the current position replaces the one in memory; if neither of them is dominated by the other, one of them is randomly selected.
11: Update the contents of the gbest accordingly
12: If the maximum number of iterations is reached then **Stop**. Otherwise, go to Step 5.

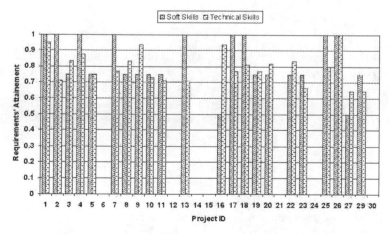

Fig. 6 Working projects and their corresponding efficiencies

7 Illustrative Example

This example presents a case in which a company has 100 staff member (personnel pool) and there are 30 candidate projects to start with different expected revenue (payoff) and different technical requirements (4 to 5 requirements). There are 7 technical skills available and the total revenue from all 30 projects is 1560. Opt-TF (an Excel Add-in that deploys the presented model and algorithm) was used to solve the problem.

Algorithm parameters are given in Table 1. 1000 feasible solutions were evaluated. Fig. 3 and Fig. 4 illustrate the three objective functions' values of these solutions in 3D and contour diagram respectively. The figures depicts the hard nature of the problem as there are many scattered sharp peaks which makes it very difficult to use traditional optimization techniques to solve this problem.

Fig. 5 illustrates the evaluation of solutions with respect to the two main objective functions. A clear Pareto front exists which allows decision maker to choose the solution that better fits his needs. Assuming that the decision maker has chosen (from the Pareto set) the solution with $f_1 = 0.788$; this solution has 24 working projects with total revenue of 1230 and $f_2 = 0.801$. The technical efficiency of the chosen solution is (80%) while team efficiency is (83%).

Fig. 6 shows different projects' efficiencies for this solution. Opt-TF provides detailed reports for the selected solution; (1) personnel assignment to different working projects with total time a person dedicates to a certain project; (2) total personnel assignment time; (3) personnel remaining available time; (4) personnel assignment per requirement per project; and (5) team composition.

8 Conclusions

Success of projects depends not only on the technical competency of its team, but also on synergy among personalities of team members. Integrated both dimensions - technical and personality - along profitability through a mathematical model would support project-based organizations in their quest for success. This paper presented and tested a multi-objective optimization model for the formulation and assignment of multi-functional teams in project-based organizations.

For an organization with limited pool of personnel and a number of candidate projects to be carried out, the model forms and assigns teams to different projects in such a way that: (1) maximizes total payoff (revenue) for the organization, (2) minimizes deviation between team member skills and project technical requirements, and (3) minimizes deviation between the formed team and the optimal team composition (based on personality profiles). A Pareto particle swarm intelligence optimization (PSO) algorithm was used to solve the model. The model was applied a hypothetical example that demonstrates the efficiency of the proposed solution algorithm The model allows personnel to work in several concurrent projects and considers both person-job and person-team fit. Extensions to the work presented would, typically, include adding scenarios (options) for overtime and/or outsourcing of personnel.

Acknowledgements. This research has been partially supported by funds from Cairo University, Egypt (Young Researcher Annual Grants).

References

1. Yassine, A., Chelst, K., Falkenburg, D.: Engineering Design Management: An Information Structure Approach. International Journal of Production Research 37(13), 2957–2975 (1999)
2. Fiksel, J.: Design for Environment: Creating Eco-Efficient Products & Processes. McGraw-Hill, New York (1991)
3. Lindemann, U., Bichlmaier, C., Stetter, R., Viertlböck, M.: Enhancing the Transfer of Integrated Product Development in Industry. In: Lindemann, U., Birkhofer, H., Meerkamm, H., Vajna, S. (eds.) Proc. of the 12th Intern. Conference on Engineering Design ICED 1999, München, TU, August 24-26, vol. 1, pp. 373–376 (1999) (Schriftenreihe WDK 26)
4. Lindemann, U., Stetter, R., Viertlböck, M.: A Pragmatic Approach for Supporting Integrated Product Development. Transactions of the Society for Deign and Process Science 5(2), 39–51 (2001)
5. Tuckman, B., Jensen, N.: Stage of small group development revisited. Group and Organizational Studies 2, 419–427 (1977)
6. Dalziel, S., Sommerville, J.: Project team building-the applicability of Belbin's team-role self-perception inventory. International Journal of Project Management 16(3), 165–171 (1998)
7. Schneider, A.: Project management in international teams: instruments for improving cooperation. International Journal of Project Management 13(4), 247–251 (1995)
8. Lawrence, P., Lorsch: Organization and Environment: Managing Differentiation and Integration. Harvard Business School, Boston (1967)
9. Askin, R.G., Sodhi, M.: Organization of teams in concurrent engineering. In: Dorf, R.D., Kusiak, A. (eds.) Handbook of Design, Manufacturing, and Automation, pp. 85–105. John Wiley & Sons, New York (1994)
10. Dietz, D.C., Rosenshine, M.: Optimal specialization of a maintenance workforce. IIE Transactions 29, 423–433 (1997)
11. Zakarian, A., Kusiak, A.: Forming teams: an analytical approach. IIE Transactions 31, 85–97 (1999)
12. Sethi, R., Nicholson, C.Y.: Structural and contextual correlates of charged behavior in product development teams. The Journal of Product Innovation Management 18, 154–168 (2001)
13. Tseng, T.L., Huang, C.-C., Chu, H.-W., Gung, R.: Novel approach to multi-functional project team formation. International Journal of Project Management 22, 147–159 (2004)
14. Barrick, M.R., Stewart, G.L., Neubert, M.J., Mount, M.K.: Relating member ability and personallty to work-team processes and team effectiveness. Journal of Applied Psychology 83, 377–391 (1998)

15. Yeatts, D.A., Hyten, C.: High-performing Self-managed Work Teams: A Comparison of Theory and Practice. Sage Publications, Thousand Oaks (1998)
16. Molleman, E., Slomp, J.: Functional flexibility and team performance. International Journal of Production Research 37, 1837–1858 (1999)
17. Fitzpatrick, E.L., Askin, R.G.: Forming effective worker teams with multifunctional skill requirements. Computers & Industrial Engineering 48, 593–608 (2005)
18. Kolbe, K.: The conative connection. Addison-Wesley, New York (1989)
19. Kolbe, K.: Pure instinct. Random House, New York (1993)
20. Abdelsalam, H.M., Akram, S., Magdy, A.: A Particle Swarm Optimization Approach for Multi-functional Teams Formation. In: Proceedings of The 9th Cairo University International Conference on Mechanical Design and Production (MDP-9), Cairo, Egypt, January 8-10, 2008, pp. 1665–1678 (2008)
21. Kichuk, S.L., Wiesner, W.H.: The big five personality factors and team performance: implications for selecting successful product design teams. Journal of Engineering and Technology Management 14, 195–221 (1997)
22. Myers, I.B., Myers, P.B.: Gifts Differing: Understanding Personality Type. Davies-Black Publishing, Mountain View (1995)
23. Myers, I.B., McCaulley, M.H.: Manual: A Guide to the Development and Use of the Myers-Briggs Type Indicator. Consulting Psychologists Press, Palo Alto (1985)
24. Kennedy, J., Eberhart, R.C.: Particle swarm optimization. In: Proceedings of IEEE international conference on neural networks, NJ, Piscataway, pp. 1942–1948 (1995)
25. Hu, X., Shi, Y., Eberhart, R.C.: Recent advances in particle swarm. In: Proceedings of the IEEE congress on evolutionary computation, Oregon, Portland, vol. 1, pp. 90–97 (2004)
26. Liao, C.-J., Tseng, C.-T., Luarn, P.: A discrete version of particle swarm optimization for flowshop scheduling problems. Computers & Operations Research 34, 3099–3111 (2007)
27. Kennedy, J., Eberhart, R.C.: A discrete binary version of the particle swarm algorithm. In: Proceedings of the world multiconference on systemics, cybernetics and informatics, NJ, Piscatawary, pp. 4104–4109 (1997)
28. Kennedy, J., Eberhart, R.C., Shi, Y.: Swarm intelligence. Morgan Kaufmann, San Francisco (2001)
29. Shi, Y., Eberhart, R.C.: A modified particle swarm optimizer. In: Proceedings of the IEEE congress on evolutionary computation, NJ, Piscataway, pp. 69–173 (1998)
30. Mostaghim, S., Teich, J.: Strategies for finding good local guides in multiobjective particle swarm optimization. In: IEEE Swarm Intelligence Symposium, Indianapolis, USA , pp. 26–33 (2003)
31. Alvarez-Benitez, J.E., Everson, R.M., Fieldsend, J.E.: A MOPSO algorithm based exclusively on pareto dominance concepts. In: Coello-Coello, C., et al. (eds.) Evolutionary Multi-Criterion Optimization, pp. 459–473. Springer, Heidelberg (2005)
32. Mostaghim, S., Branke, J., Schmeck, H.: Multi-Objective Particle Swarm Optimization on Computer Grids. Technical Report 502, Institute AIFB University of Karlsruhe (2006)
33. Parsopoulos, K.E., Vrahatis, M.N.: Particle swarm optimization method in multiobjective problems. In: Proceedings of the 2002 ACM Symposium on Applied Computing (SAC 2002), pp. 603–607. ACM Press, Madrid (2002)

Genetic Algorithms for Task Scheduling Problem

Fatma A. Omara and Mona M. Arafa

Abstract. The scheduling and mapping of the precedence-constrained task graph to the processors is considered one of the most crucial NP-complete problems in the parallel and distributed computing systems. Several genetic algorithms have been developed to solve this problem. The primary distinction among most of them is being the used chromosomal representation for a schedule. However, these existing algorithms are monolithic as they attempt to scan the entire solution space without consideration how to reduce the complexity of the optimization. In this chapter, two genetic algorithms have been developed and implemented. Our developed algorithms are genetic algorithms with some heuristic principles have been added to improve the performance. According to the first developed genetic algorithm, two fitness functions have been applied one after another. The first fitness function is concerned with minimizing the total execution time (schedule length) and the second one is concerned with the load balance satisfaction. The second developed genetic algorithm is based on task duplication technique to overcome the communication overhead. Our proposed algorithms have been implemented and evaluated using benchmarks. According to the evolution results, it found that our algorithms always outperform the traditional algorithms.

1 Introduction

The problem of scheduling task graphs of a parallel program onto parallel and distributed computing systems is a well-defined NP-complete problem

Fatma A. Omara
Computer Science Department, Faculty of Computer and Information
Cairo University
e-mail: f.omara@ffci-cu.edu.eg

Mona M. Arafa
Mathematics Dept., Faculty of Science Banha University
e-mail: m-h-banha@yahoo.com

A. Abraham et al. (Eds.): Foundations of Comput. Intel. Vol. 3, SCI 203, pp. 479–507.
springerlink.com   © Springer-Verlag Berlin Heidelberg 2009

that has received a large amount of attention, and it is considered one of the most challenging problems in parallel computing [1]. This problem involves mapping a Directed Acyclic Graph (DAG), of collection of computational tasks and their data precedence, onto a parallel processing system. The goal of a task scheduler is to assign tasks to available processors such that precedence requirements between tasks are satisfied and in the same time the overall execution length (i.e., make span) is minimized [2]. Generally, the scheduling problem exists in two types: static and dynamic.

According to the static scheduling, the characteristics of a parallel program such as task processing times, communication, data dependencies, and synchronization requirement are known before execution [3] . According to the dynamic scheduling, a few assumptions about the parallel program can be made before execution, then, scheduling decisions have to be made on-the-fly [4]. The work in this chapter concerns static scheduling problem. One the other hand, a general taxonomy for static scheduling algorithms has been reviewed and discussed by Kwong and Ahmad [3]. Many task scheduling techniques have been developed with moderate complexity as a constraint, which is a reasonable assumption for general purpose development platforms [5, 6, 7, 8]. Generally, the task scheduling algorithms may be divided in two main classes; greedy and non-greedy (iterative) algorithms [9]. The greedy algorithms attempt to minimize the start time of the tasks of a parallel program only. This is done by allocating the tasks into the available processors without back tracking. On the other hand, the main principle of the iterative algorithms is that they depart from an initial solution and try to improve it.

The greedy task scheduling algorithms might be classified into two categories: algorithms with duplication and algorithms without duplication. One of the common algorithms in the first category is the duplication scheduling heuristic (DSH) algorithm [1], the main principles of the DSH algorithm are: the nodes are arranged in a descending order according to their static b-level and the start-time of the node on the processor without duplication of any ancestor is determined. After that the ancestors of the node is tried to duplicate into the duplication time slot until the slot is used up or the start-time of the node does not improve. One the other hand, one of the best algorithms in the second category is the Modified Critical Path (MCP) algorithm [10]. The MCP algorithm first computes the ALAPs of all the nodes, then create ready list containing ALAP times of the nodes in an ascending order. The ALAP of a node is computed by first computing the length of the Critical Path (CP) and then subtracting the b-level of a node from it. Ties are broken by considering min ALAP time of the children of a node. If the min ALAP time of the children is equal, ties are broken randomly.

According to MCP algorithm, the highest priority node in the list is picked and assign to a processor that allows the earliest start time using insertion approach. Recently, Genetic Algorithms (GAs) have been widely reckoned as a useful vehicle for obtaining high quality solutions or even optimal solutions for a broad range of combinatorial optimization problems including task scheduling problem [2, 3]. Another merit of a genetic search is that its inherent parallelism can be exploited so as to further reduce its running time. The basic principles of GAs were firstly laid down by Holland [11], and after that they are well described in many texts. The Gas operate on a population of solutions rather than a single solution. The genetic search begins by initializing a population of individuals. Individual solutions are selected from the population then mate to form new solutions. The mating process implemented by combining or crossing over genetic material from two parents to form the genetic material for one or two new solutions, confers the data from one generation of solutions to the next. Random mutation is applied periodically to promote diversity. The individuals in the population are replaced by the new solutions. A fitness function, which measures the quality of each candidate solution according to the given optimization objective, is used to help determine which individuals are retained in the population as successive generations evolve [12]. There are two important but competing themes exist in a GA search; the need for selective pressure so that the GA is able to focus the search on promising areas of the search space, and the need for population diversity so that important information (particular bit values) is not lost [13, 14].

Recently, several GAs have been developed for solving the task scheduling problem, the primary distinction among them being the chromosomal representation of a schedule [15, 16, 17, 18, 19, 20, 2]. Two hybrid genetic algorithms called Critical Path Genetic Algorithm (CPGA) and Task Duplication Genetic Algorithm (TDGA) have been proposed in this chapter. Our developed algorithms show the effect of the amalgamation of the greedy algorithms with the genetic one. The first algorithm CPGA is based on how to use the ideal time of the processors efficiently, and reschedule the critical path nodes to reduce their start time. Finally, two fitness functions have been applied, one after another. The first fitness function is concerned with how to minimize the total execution time (schedule length), and the second one is concerned with the load balance satisfaction. The second algorithm TDGA is based on task duplication principle to minimize the communication overheads.

The reminder of this chapter is organized as follows: Section 2 gives a description for the model for task scheduling problem. An implementation of the standard GA is presented in Section 3. Our developed CPGA is introduced in section 4. Section 5 produces the details of our TDGA algorithm. A comparative study of our developed algorithms, MCP algorithm, DSH algorithm, and SGA algorithm are presented in Section 6. Conclusion is presented in Section 7.

2 Task Scheduling Problem Model

The model of the underline parallel system to be considered in this research work could be described as follows [3]: The system consists of a limited number of fully connected homogeneous processors. Let a task graph G be a Directed, Acyclic Graph (DAG) composed of N nodes $n_1, n_2,..., n_N$, each node termed a task of the graph which in turn is a set of instruction that must be executed sequentially without preemption in the same processor. A node has one or more inputs. When all inputs are available, the node is triggered to execute. A node with no parent is called an entry node and a node with no child is called an exit node. The weight is called the computation cost of a node ni and is denoted by (n_i) weight. The graph also has E directed edges representing a partial order among the tasks. The partial order introduced a precedence-constrained DAG and implies that if $n_i \rightarrow n_j$, then n_j is a child, which cannot start until its parent n_i finishes. The weight on an edge is called communication cost of the edge and is denoted by $c(n_i, n_j)$. This cost is incurred if ni and nj are scheduled on different processors and is considered to be zero if n_i and n_j are scheduled on the same processor. If a node n_i is scheduled to processor P, the start time and finish time of the node are denoted by $ST(n_i, p)$ and $FT(n_i, p)$ respectively. After all nodes have been scheduled, the schedule length is defined as $\max FT(n_i, p)$ across all processors. The objective of the task scheduling problem is that how to find an assignment and the start times of the tasks to processors such that the schedule length is minimized and, in the same time, the precedence constrains are preserved. A Critical Path (CP) of a task graph is defined as the path with the maximum sum of node and edge weights from an entry node to an exit node. A node in CP is denoted by CP Nodes (CPNs). An example of a DAG is represented in Figure1 with CP is drawn in bolt.

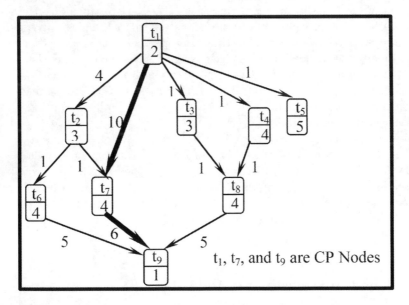

Fig. 1 Example of DAG, where t_1, t_7, and t_9 are CP Nodes

Table 1 Selected Benchmark Programs

Benchmarks programs	No tasks	Source	Note
Pg1	100	[22]	Random Graphs
Pg2	90	[22]	Robot Control program
Pg3	98	[22]	Sparse Matrix Solver

3 The Developed Genetic Algorithms

Before presenting the details of our developed algorithms, some principles which are used in the design are discussed.

Definition 1. *(Data Arrival Time) Any task cannot be start unit all parents have been finished. Let P_j be the processor on which the $k-th$ parent task t_k of task t_i is scheduled. Data Arrival Time (DAT) of t_i on a processor P_i is defined as:*

$$DAT = \max(FT(t_k, P_j) + c(t_i, t_k)), k = 1, 2, ..., No - parent \qquad (1)$$

Where, $No - parent$ is the number of parents of t_i,

If $(i = j)$ then $c(t_i, t_k) = 0$

The parent task that maximizes the above expression is called the favorite predecessors of t_i and it is denoted by $favpred(t_i, P_j)$. The benchmark programs which have been used to evaluate our algorithms are listed in Table (1).

3.1 *Standard Genetic Algorithm - SGA*

The SGA has been implemented first. This algorithm is started with an initial population of feasible solution. Then, by applying some operators, the best solution could be finding through some generations. The selection of the best solution is determined according to the value of fitness function. According to this SGA, the chromosome is divided into two sections; mapping and scheduling. The mapping section contains the processors indices where tasks to be run on it. The schedule section determines the sequence for processing of the

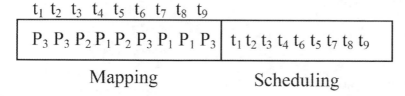

Mapping   Scheduling

Fig. 2 Representation of a Chromosome

tasks. Figure 2 shows an example of such representation of the chromosome. Where, tasks t_4, t_7, t_8 will be scheduled on processor P_1, tasks t_3, t_5 will be scheduled on processor P_2, and the tasks t_1, t_2, t_6 and t_9 will be scheduled on processor P_3. The length of the chromosome is linear proportional to the number of tasks.

Genetic Formulation of SGA

Initial Population

The initial population is constructed randomly. The first part of the chromosome (i.e. mapping) is chosen randomly from 1 to No-Processors, where the No-Processors is the number of processors in the system. The second part (i.e. schedule) is generated randomly such that the topological order of the graph is preserved. The Pseudo Code of *The Task Schedule* using SGA is as follow:

Fitness Function

The main objective of the scheduling problem is to minimize the schedule length of a schedule.

$$Fitness - Function = (\frac{a}{Slength}) \tag{2}$$

Where a is a constant and $Slength$ is the schedule length which is determined by the following equation:

$$Slength = \max(FT[t,]), i = 1, ..., K_{noTask} \tag{3}$$

Function Schedule length

1. $\forall RT[P_j] = 0$ //RT is the ready time of the processors.

2. Let LT be a list of tasks according to the topological order of DAG.

3. **For** i=1 to NoTasks **Do**

// NoTasks is number of tasks

(a) Remove the first task t_i form list LT.

(b) **For** j = 1 to NoProcessors **Do**
// NoProcessors is number of Processors.

If t_i is scheduled to processor P_j

$$ST[t_i] = \max(RT[P_j], DAT(t_i, P_i))$$

$$FT[t_i] = ST[t_i] + weight[t_i]$$

$$RT[P_j] = FT[t_i]$$

Endif, Endfor, Endfor.

$Slength = \max(FT)$.

Example: By considering the chromosome represented in Figure 2 as a solution of a DAG represented in Figure 1, the Fitness Time function defined by equation 3 has been used to calculate the schedule length (see Figure 3).

Genetic Operators

In order to apply crossover and mutation operators, the selection phase should be applied firstly. This selection phase used to allocates reproductive trials to

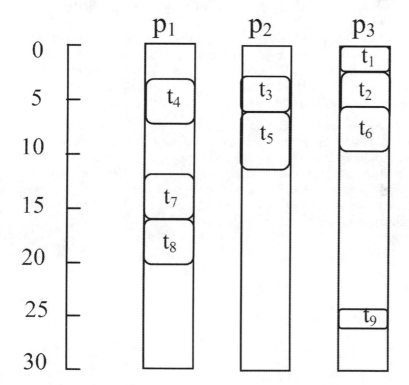

Fig. 3 The Schedule Length

Table 2 A comparison between roulette wheel and tournament selection

Benchmarks programs	Roulette Wheel Selection	Tournament Selection
Pg1	301.6	283.7
Pg2	1331.6	969
Pg3	585.8	521.8

chromosomes according to their fitness. There are different approaches could be applied in the selection phase. According to the work in this chapter, fitness-proportional roulette wheel selection [23] and tournament selection [24] are compared such that the best method is used (i.e., produce the shortest schedule length). In the roulette wheel selection, the probability of selection is proportional to an chromosome's fitness. The analogy with a roulette wheel arises because one can imagine the whole population forming a roulette wheel with the size of any chromosome's slot proportional to its fitness. The wheel is then spun and the figurative **ball** thrown in. the probability of the ball coming to the rest in any particular slot is proportional to the arc of the slot and thus to the fitness of the corresponding chromosome. In binary tournament selection, two chromosomes are picked at random from the population. Whichever has the higher fitness is chosen. This process is repeat number of population size.

Table (2) contains the comparing results between these two selection methods using 4 processors for each benchmark program listed in Table1. According to the results listed in Table 2, the tournament selection method produce schedule length is smaller than the roulette wheel selection. Therefore, the tournament selection method is used in the work of this chapter.

Crossover Operator. Each chromosome in the population is subjected to crossover with probability μ. Two chromosomes are selected from the population, and a random number $RN \in [0, 1]$ is generated for each chromosome. If $RN < \mu$, these chromosomes are applied using one of the two kinds of the crossover operators; single point crossover and order crossover operators. Otherwise, these chromosomes are not changed. The pseudo code of the crossover function is as follows.

Function Crossover

1. Select two chromosomes chrom1 and chrom2

2. Let P a random real number between 0 and 1

3. **If** $P < 0.5$ /* operators probabilty

Crossover-Map(chrom1, chrom2)

Else

Crossover-Order(chrom1, chrom2).

According to the crossover function, one of the crossover operators is used.

Crossover Map. When the single crossover is selected, it is applied to the first part of the chromosome. By given two chromosomes a random integer number called the crossover point is generated from 1 to No-Tasks. The portions of the chromosomes lying to the right of the crossover point are exchanged to produce two offsprings (see Figure 4).

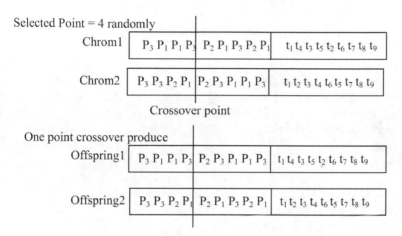

Fig. 4 One point crossover operator

Order Crossover. When the order crossover operator is applied to the second part of the chromosome, a random point is chosen. First pass the left segment from the chrom1 to the offspring, and then construct the right fragment of the offspring according to the order of the right segment of chrom2 crossover operator is given in (see Figure 5 as an example).

Mutation Operator. Each position in the first part of the chromosome is subjected to mutation with probability . Mutation involves changing the

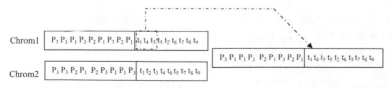

Fig. 5 Order crossover operator

Before mutation

After mutation

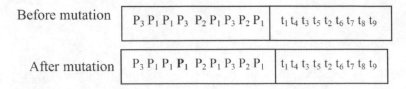

Fig. 6 Mutation Operator

assignment of a task from one processor to another. Figure 6 illustrate the mutation operation on chrom1. After the mutation operator is applied, the assignment of t_4 is changed from processor P_3 to processor P_1.

4   The Critical Path Genetic Algorithm (CPGA)

Our developed CPGA algorithm is considered a hybrid of GA principles and heuristic algorithms principles (e.g., given priority of the nodes according to ALAPlevel). On the other hand, the same principles and operators which are used in SGA algorithm have been used in the CPGA algorithm. The encoding of the chromosome is the same as in SGA, but in the initial population the second part (schedule) of the chromosome can be constructed using one of the following ways:

1. The schedule part is constructed randomly as in SGA.
2. The schedule part is constructed using ALAP.

These two ways have been applied using benchmark programs listed in Table 1 with four processors. According to the comparative results listed in Table (3), it is found that the priority of the nodes by ALAP method outperforms the random one in the most cases.

Table 3 A comparison between Random and Order ALAP Order methods

Benchmarks programs	Random Order	ALAP Order
$Pg1$	183.4	152.3
$Pg2$	848.5	826.4
$Pg3$	301.8	293.8

By using ALAP, the second part of the chromosomes is become static along the population. So, the crossover operators are restricted to the one point crossover operator. Three modifications have been applied in the SGA to improve the scheduling performance. These modifications are: (1) Reuse idle time, (2) Priority of the CPNs, and (3) Load balance.

Function Test-Slots

1. Let LT be a list of ready tasks
2. Initially the deal-time list is empty, S-ideal-time=0, and E-ideal-time=0
3. While the list LT is not empty, get a task t_i from the head of the list
(a) $Min = ST = \inf$
(b) **For** each processor P_j

If t_i is scheduled to P_j.

Let $thisST =$ the start time of t_i on P_j

If $thisST > MinST$ **Then** $MinST = thisST$

If the idealtime list of P_j is not empty

For each timeslot of the idealtime list

If $(Eidealtime - Sidealtime) <= weight[t_i]$ & $DAT(t_i, P_j) > Sidealtime$

Then schedule t_i in the idealtime and update the Sidealtime and Eideal-time

Let sttime be the start time of the task t_i equal to Sidealtime.

End If

(c) **If** $sttime > MinST$ **Then**

MinST =sttime.

Example. Suppose the schedule represented in Figure (3). The processor P_1 has an ideal time slot; the start of this ideal time (S-ideal-time) is equal to 7 while its end time (E-ideal-slot) is equal to 12. On the other hand, the weight (t_S) =4 and $DAT(t_S, P_1) = S - ideal - slot = 7$. By applying the modification, t_S can be rescheduled to start at time 7. The final schedule length according to this modification becomes 23 instead of 26 (see Figure 7).

Priority of CPNs Modification

According to the second modification, another optimization factor is applied to recalculate the schedule length after giving high priorities for the (CPNs) such that they can start as early as possible. This modification is implemented using a function called Reschedule-CPNs Function. The pseudo code of this function is as follows:

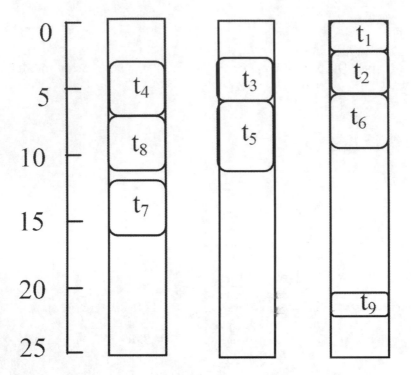

Fig. 7 The schedule after applying the test slots function is reduced from 26 to 23

Function Reschedule-CPNs

    1. Determine the CP and make a list of CPNs

    2. **While** the list of CPNs is not empty **DO**

    - Remove the task t_i from the list

    - Let $VIP = favpred(t_i, P_j)$

    If VIP is assigned to processor P_j

    Then The task t_i is assigned to processor p_j

    End If

Example. We apply the Reschedule-CPNs Function in the scheduling presented in Figure 7. According to the DAG in presented in Figure 1, it is found that the CPNs are t1, t7, and t_9. t_1 is the entry node and it has no predecessor and the favpred of the t_7 is the task t_1. The task t_7 is scheduled

Fig. 8 The schedule after applying the reschedule of the CPNs function is reduced from 23 to 17

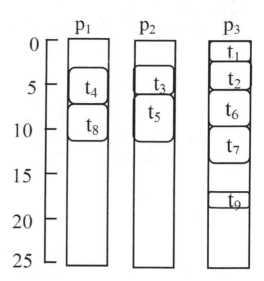

to processor P_1. Also the favpred of t_9 is t_8, but in the same time it starts early on the processor P_3, so t_9 is not moved. The final schedule length is reduced to 17 instead of 23(see Figure 8).

Load Balance Modification

Because the main objective of the task scheduling is to minimize the schedule length, it is found that several solutions can give the same schedule length, but load balance between processors might be not satisfied in some of them. The aim of load balance modification is that how to obtain the minimum schedule length and, in the same time, the load balance is satisfied. This has been satisfied by using two fitness functions one after another instead of one fitness function. The first fitness function concerns with minimizing the total execution time, and the second fitness function is used to satisfy load balance between processors. This function is proposed in [25] and it is calculated by the ratio of the maximum execution time (i.e. schedule length) to the average execution time over all processors.

If the execution time of processor P_j is denoted by $Etime[P_j]$, then the average execution time over all processors is:

$$avg = \sum_{j=1}^{NoProcessor} \frac{Etime[P_j]}{NoProcessors} \qquad (4)$$

So, the load balance is calculated as:

$$LoadBalance = \frac{Slength}{Avg} \qquad (5)$$

Supposing two task scheduling solutions are given in Figure 9 (a,b). The schedule length of both solutions is equal to 23.

Solution a: $Avg = \frac{12+17+23}{3}$,

Loadbalance $= \frac{23}{17.33} \approx 1.326$

Solution b: $Avg = \frac{9+11+23}{3} \approx 14.33$,

$Loadbalance = \frac{23}{14.33} \approx 1.604$.

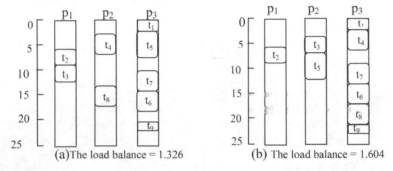

(a)The load balance = 1.326   (b) The load balance = 1.604

Fig. 9 according to balance fitness function solution (a) is better than solution (b)

According to the balance fitness function as shown in Figure (9), the solution (a) is better than the solution (b).

Adaptive μ_c and μ_m Parameters

Srinivas and patnaik [26] have proposed an adaptive method to tune crossover rate μ_c and mutation rate μ_m on the fly based on the idea of sustaining in diversity in a population without affecting its convergence properties. Therefore; the rate μ_c as:

$$\mu_c = \frac{k_c(f_{max} - f_c)}{(f_{max} - f_{avg})} \qquad (6)$$

And the rate μ_m is defined as:

$$\mu_m = \frac{k_m(f_{max} - f_m)}{(f_{max} - f_{avg})} \qquad (7)$$

Where, f_{max} is the maximum fitness value, f_{avg} is the average fitness value f_c is the fitness value of the fitter chromosome for the crossover f_m is the fitness value of the chromosome to be mutated k_c and k_m are positive real constant less than 1.

Table 4 A comparison between static and dynamic μ_c, μ_m parameters

Benchmarks programs	Dynamic parameters	Static parameters
$Pg1$	148	152.3
$Pg2$	785.6	826.4
$Pg3$	288.2	293.8

The CPGA algorithm has been implemented into two versions: the first version is done using static parameters ($\mu_c = 0.8$ and $\mu_m = 0.02$) and the second version is done using adaptive parameters. Table 4 represents the comparison results between these two versions. According to the results, it found that using adaptive parameters (μ_c and μ_m) can help preventing a GA from getting stuck at local minima. So the adaptive method is batter than using static values of μ_c and μ_m.

5   The Task Duplication Genetic Algorithm (TDGA)

Even with an efficient scheduling algorithm, some processors might be ideal during the execution of the program because the tasks assigned to them might be waiting to receive some data from the tasks assigned to other processors. If the idle time slots of the waiting processor could be used effectively by identifying some critical tasks and redundantly allocating them in these slots, the execution time of the parallel program could be further reduced [27].

According to our proposed algorithm, a good schedule based on task duplication has been proposed. This proposed algorithm called Task Duplication Genetic Algorithm (TDGA) employs a genetic algorithm for solving the scheduling problem.

Definition 2. *At a particular scheduling step; for any task t_i on a processor P_i, if $STF(favpred(t_i, p_j)) + weight(favpred(t_i, p_j)) \leq EST(t_i, p_j)$ Then $EST(t_i, p_j)$ can be reduced by scheduling $favpred(t_i, p_j)$ to p_j. Therefore, this definition could be applied recursively upward the DAG to reduce the schedule length.*

Example. To clearfy the effect of the task duplication technique, consider a schedule presented in Figure 10(a) for DAG in Figure (1), the schedule length is equal to 21. If t_1 is duplicated to processor p_1 and p_2 the schedule length is reduced to 18 (see Figure 10(b)).

Genetic Formulation of The TDGA

According to our TDGA algorithm, each chromosome in the population consists of a vector of order pairs (t, p) indicates that task t is assigned to processor p. The number of order pairs in a chromosome may vary in length.

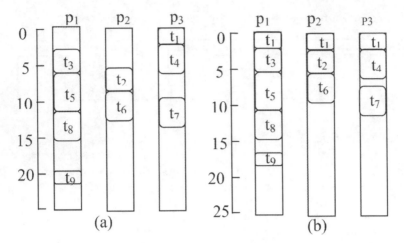

Fig. 10 (a) before duplication (schedule length=21) (b) After duplication (schedule length=18)

Fig. 11 An Example of
the Chromosome $\qquad (t_2,P_1)(t_3,P_2)(t_4,P_1)(t_4,P_2)$

An example of a chromosome is shown in Figure 11. The first order pair shows that task t_2 is assigned to processor P_1, and the second one indicates that task t_3 is assigned to processor P_2, etc.

According to the duplication principles, the same task may be assigned more than once to different processors without duplicating it in the same processor. If a task processor pair appears more than once on the chromosome, only one of the pairs is considered. According to Figure 11, the task t2 is assigned to processor P1 and P2.

Definition 3. *(Invalid chromosomes) Invalid chromosomes are the chromosomes that not contain all DAG tasks. These invalid chromosomes might be generated.*

Initial Population. According to our TDGA algorithm, two methods to generate the initial population are applied. The first one, called Random Duplication (RD) and the second one called Heuristic Duplication (HD). According to RD, the initial population is generated randomly such that each task can be assigned to more than one processor.

According to HD, the initial population is initialized with randomly generated chromosomes, while each chromosome consists of exactly one copy of each task (i.e. no task duplication). Then, each task is randomly assigned to a processor. After that a duplication technique is applied by a function called *Duplication-Process*. The pseudo code of the *Duplication-Process* function is as follows:

Table 5 A comparison between the methods (HD and RD)

Benchmarks programs	HD	RD
$Pg1$	493.9	494.1
$Pg2$	1221	1269.5
$Pg3$	641.2	616.2

Chrom1 (t_3, P_2) (t_1, P_2) (t_4, P_1) | (t_1, P_1) | (t_2, P_2)
Chrom 2 (t_1, P_1) (t_3, P_1) (t_4, P_2) | (t_4, t_2) | (t_2, P_1) (t_1, P_2)

      Crossover point1   Crossover point2

Two points 2 and 4 are generated randomly, two point crossover operator produce

Offspring1 (t_3, P_2) (t_1, P_2) | (t_4, P_2) (t_4, t_2) | (t_2, P_2)
Opsspring2 (t_1, P_1) (t_3, P_1) | (t_4, P_1) (t_1, P_1) | (t_2, P_1) (t_1, P_2)

      Crossover point1   Crossover point2

Fig. 12 Example of two point crossover operator

(t_3, P_2) $(t_1, \mathbf{P_2})$ (t_4, P_1) (t_1, P_1) (t_2, P_2) Chrom

(t_3, P_2) $(t_1, \mathbf{P_1})$ (t_4, P_1) (t_1, P_1) (t_2, P_2) Offspring

Fig. 13 Example of Mutation Operator

Function Duplicatin-Process

1.Compute SL for each task in the DAG
2.Make a list Slist of the tasks according to SL in descending order
3.Take the task t_i from Slist
4. **While** Slist is not empty.

    If is assigned to processor ρ_i
      if $favpred(t_i, \rho_i)$ is not assigned to ρ_i
   if $(timeslot \geq weight(favpred(t_i, \rho_i))$

    assigned $favpred(t_i, \rho_i)$ to ρ_i

According to the implementation results using two methods, it is found that the methods give nearly results. Therefore, the first method (HD) has been considered in our TDGA algorithm.

Fitness Function. Our fitness function is defined as 1/Slength, where Slength is defined as the maximum finishing time of all tasks of the DAG. The proposed GA assigns zero to an invalid chromosome as its fitness value.

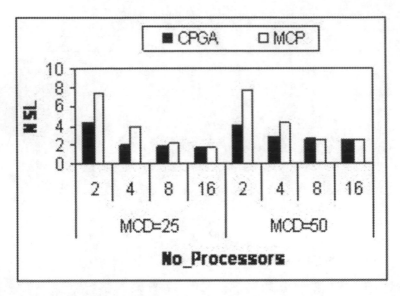

Fig. 14 NSL for $Pg1$ and MCD 25, 50

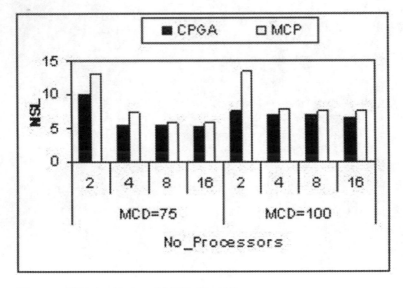

Fig. 15 NSL for $Pg1$ and MCD 75, 100

Genetic Operators: Crossover Operator. Two point crossover operator is used. Since each chromosome consists of a vector of task processor pair, crossover exchange substrings of pairs between two chromosomes. Two points are randomly chosen and the partitions between the points are exchanged between two chromosomes to form two offsprings. The crossover probability

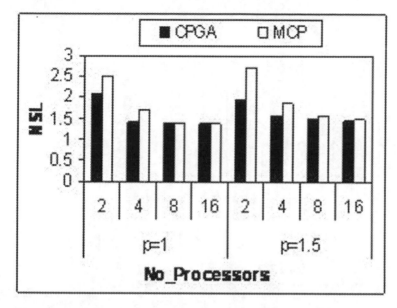

Fig. 16 NSL for $Pg2$ and two values of ρ

gives the probability that a pair of chromosome will undergo crossover. An example of two point crossover is shown in Figure 12.

Mutation Operator

The mutation probability indicates the probability that an order pair will be changed. If a pair is selected to be mutated, the processor number of that pair will be randomly changed. An example of mutation operator is shown in Figure 13.

6 Comparative Study and Performance Evaluation

To evaluate our proposed algorithms, we have implemented them using an Intel processor (2.6 GH) using c++ language and it is applied using different task graphs of specific benchmark applications programs as well as, a random one without communication delays which are listed in Table (1). All benchmark programs are taken from a Standard Task Graph (STG) archive [22]. The first two programs of This STG set consists of task graphs generated randomly $Pg1$, the second program is the robot control ($Pg2$) as an actual application programs and the last program is the sparse matrix solver ($Pg3$). Also, we consider the task graphs with random communication costs. These communication costs are distributed uniformly between 1 and a specified maximum communication delay (MCD). Also, the population size is considered 200 and the number of generations is considered 500 generation.

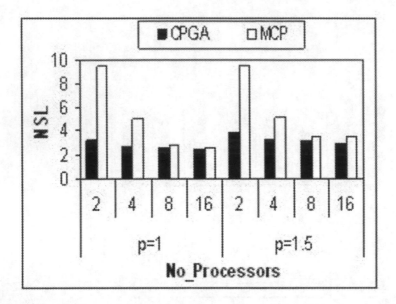

Fig. 17 NSL for $Pg3$ and two values of ρ

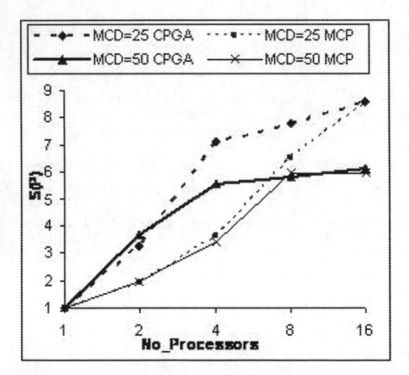

Fig. 18 Speedup for $Pg1$ and MCD 25 and 50

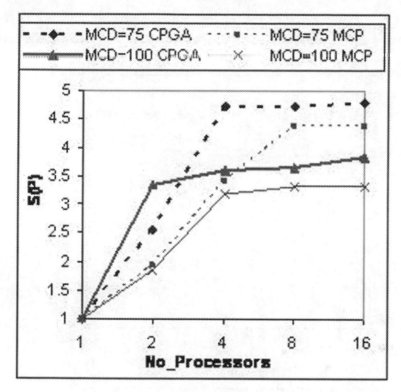

Fig. 19 NSpeedup for $Pg1$ and MCD 75and 100

6.1 The Developed CPGA Evaluation

The comparison has been done among our algorithm CPGA, SGA and one of the best greedy algorithms is called MCP algorithm. Firstly, a comparison among the CPGA, SGA and MCP algorithms with respect to the Normalized Schedule Length (NSL) with different number of processors has been done. The NSL is defined as [28]:

$$NSL = \frac{Slength}{\sum_{x \in CP}(Weight(n_i))} \tag{8}$$

Where SLength is the schedule length and weight (n_i) is the weight of the node n_i. The sum of computation costs on the CP represents a lower bound on the schedule length. Such lower bound may not always be possible achieve, and the optimal schedule length may be larger than this bound. Secondly, the performance of the CPGA, SGA and MCP are measured with respect to speedup [29]. The speedup is can be estimated as:

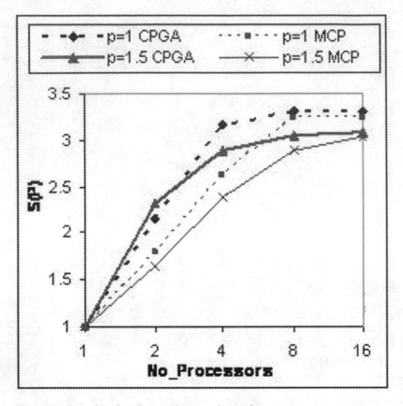

Fig. 20 Speedup for $Pg2$ and two values of ρ

$$S(p) = \frac{T1}{Tp} \tag{9}$$

where, $T(1)$ is the time required for executing a program on a uniprocessor computer and $T(P)$ is the time required for executing the same program on a parallel computer containing P processors. The NSL for CPGA and MCP algorithms using 2, 4, 8, and 16 processors for $Pg1$ and different MCD (25, 50, 75, and 100) are given in Figures (14 and 15). Also the NSL for Pg2 and Pg3 graphs with two different number of μ are given in Figures 16 and 17 respectively.

Figures (14, 15, 16, and 17) show that the performance of our proposed CPGA algorithm is always outperformed SGA and MCP algorithms. According to the obtained result, it is found that the NSL of all algorithms is increased when processor number is increased. Although, our CPGA is always the best, and it achieves lower bound when the communication delay is small.

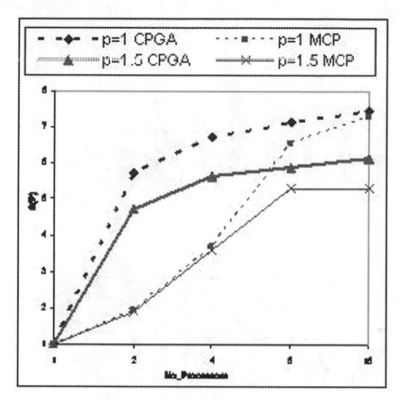

Fig. 21 Speedup for $Pg3$ and two values of ρ

6.2 The Developed TDGA Evaluation

To measure the performance of the TDGA, a comparison among the TDGA algorithm, SGA, and one of well known heuristic algorithm based on task duplication called DSH algorithm has been done with respect to NSL and speedup. To clarify the effect of task duplication in our TDGA algorithm, the same benchmark random and application programs $Pg1$, $Pg2$, and $Pg3$ listed in Table (1) have been used with high communication delay.

The NSL for TDGA, SGA, and DSH algorithms using 2, 4, 8, and 16 processors for $Pg1$ with two value of Communication Delay (CD) 100 and 200 is given in Figure 22. Also the NSL for bench mark application programs $Pg2$, and $Pg3$ is given in Figures 23, and 24.

According to the results in Figures (22, 23, and 24), it is found that our TDGA algorithm outperforms SGA and DSH algorithms especially when the number of communication, as well as, the number of processor increases.

The speedup of TDGA algorithm and DSH algorithm is given in Figures (25, 26, and 27) for $Pg1$, $Pg2$, and $Pg3$ programs respectively.

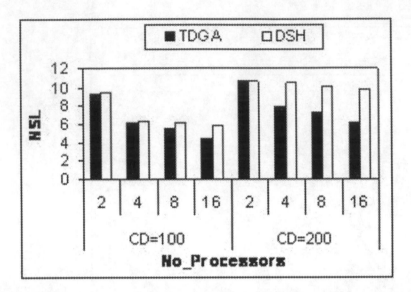

Fig. 22 NSL for Pg1 and CD =100 and 200

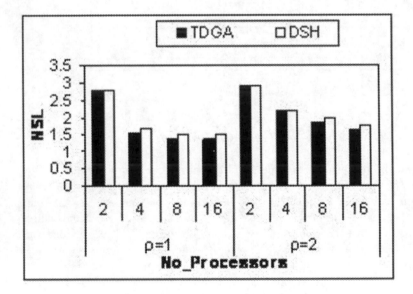

Fig. 23 NSL for Pg2 and ρ=1 and 2

The results reveal that the performance of the TDGA algorithm is always outperformed the DSH algorithm. Also, the TDGA speedup is nearly linear especially for random graphs

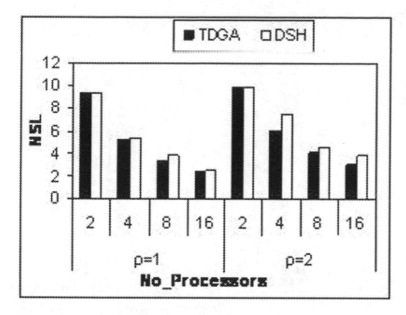

Fig. 24 NSL for Pg3 and $\rho=1$ and 2

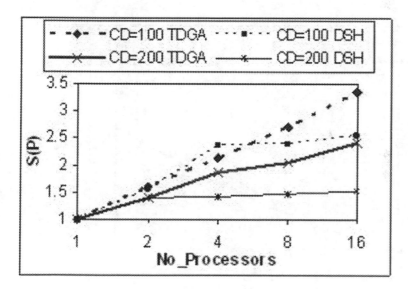

Fig. 25 Speedup for Pg1 and $\rho=1$ and 2

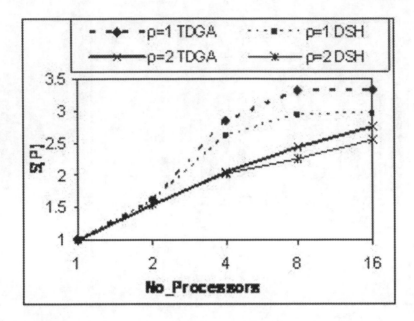

Fig. 26 Speedup for Pg2 and ρ=1 and 2

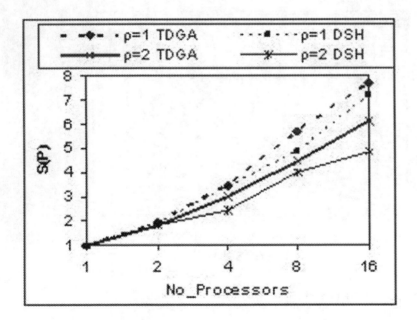

Fig. 27 Speedup for Pg3 and ρ =1 and 2

7 Conclusion

In this chapter, an implementation of a standard GA (SGA) to solve the task scheduling problem has been presented. Some modifications have been added to this SGA to improve the scheduling performance. These modifications are based on amalgamating heuristic principles with the GA principles. The new developed algorithm called Critical Path Genetic Algorithm (CPGA) is based on rescheduling the critical path nodes (CPNs) in the chromosome and then through different generations. Also, two modifications have been added. The first one concerns with how to use the idle time of the processors efficiently, and the second one concerns about to satisfy load balance among processors. The last modification is applied only when there are two or more scheduling solutions with the same schedule length are produced.

A comparative study among our CPGA, SGA algorithms and one of the standard heuristic algorithm called MCP algorithm have been presented using standard task graphs with considering random communication costs. The experimental studies show that the CPGA always outperform the SGA as well as the MCP algorithm in most cases. Generally, the performance of our CPGA is better than the SGA and MCP algorithms. According to task duplication technique, the communication delays are reduced and then minimizing the overall execution time, in the same time, the performance of the genetic algorithm is increased. The performance of the TDGA is compared with a traditional heuristic scheduling technique: DSH and SGA. The TDGA outperforms the DSH algorithm and SGA in most cases.

Acknowledgements. This research has been partially supported by funds from Cairo University, Egypt (Young Researcher Annual Grants).

References

1. El-Rewini, H., Lewis, T.G., Ali, H.H.: Task Scheduling in Parallel and Distributed Systems. Prentice-Hall International Editions (1994)
2. Wu, A.S., Yu, H., Jin, S., Lin, K.-C., Schiavone, G.: An Incremental Genetic Algorithm Approach to Multiprocessor Scheduling. IEEE Trans. Parallel and Distributed Systems 15, 824–834 (2004)
3. Kwok, Y., Ahmad, I.: Static Scheduling Algorithms for Allocating Directed Task Graphs to Multiprocessors. ACM Computing Survey 31, 406–471 (1999)
4. Palis, M.A., Liou, J.C., Rajasekaran, S., Shende, S., Wei, S.S.L.: Online Scheduling of Dynamic Trees. Parallel Processing Letter 5, 635–646 (1995)
5. Sih, G.C., Lee, E.A.: A Compile-Time Scheduling Heuristic for Interconnection-Constrained Heterogeneous Processor Architectures. IEEE Trans. Parallel and Distributed Systems. 4, 75–87 (1993)
6. Kwok, Y., Ahmad, I.: Dynamic Critical Path Scheduling: An Effective Technique for Allocating Task Graphs to Multi-processors. IEEE Trans. Parallel and Distributed Systems. 7, 506–521 (1996)

7. Omara, F.A., Allam, A.: An Efficient Tasks Scheduling Algorithm for Distributed Memory Machines With Communication Delays. Information Technology Journal (ITJ) 4, 326–334 (2005)
8. Radulescu, A., van Gemund, A.J.C.: Low Cost Task scheduling for Distributed Memory Machines. IEEE Trans. Parallel and Distributed Systems 13, 648–658 (2002)
9. Bouvry, P., Chassin, J., Trystram, D.: Efficient Solutions for Mapping Parallel Programs. CWI-Center for Mathematics and computer science, Amsterdam, The Netherlands (1995) (published in Euro-Par)
10. Wu, M., Gajski, D.D.: Hypertool: A Programming aid for message-passing systems. IEEE Trans. Parallel Distributed Systems 1, 381–422 (1990)
11. Corman, T.H., Leiserson, C.E., Rivests, R.L.: Introduction to Algorithms. MIT Press, Cambridge (1990)
12. Holland, J.H.: Adaptation in Natural and Artificial Systems. Univ. Of Michigan Press, Ann Arbor (1975)
13. Levine, D.: A Parallel Genetic Algorithm for The Set Partitioning Problem, Ph.D. thesis in computer science, Department of Mathematics and computer science, IIlinois Institute of Technology, Chicago, USA (1994)
14. Back, T., Hammel, U., Schwefel, H.-P.: Evolutionary Computation: Comments on the History and Current State. IEEE Trans. Evolutionary Computation 1, 3–17 (1997)
15. Talbi, E.G., Muntean, T.: A new Approach for The Mapping Problem: A Parallel Genetic Algorithm (1993), citessr.ist.psu.edu
16. Ali, S., Sait, S.M., Benten, M.S.T.: GSA: Scheduling And Allocation Using Genetic Algorithm. In: Proceedings of the Conference on EURO-DAC with EURO WDHL 1994, Grenoble, pp. 84–89 (1994)
17. Hou, E.H., Ansari, N., Ren, H.: A Genetic Algorithm for Multiprocessor Scheduling. IEEE Trans. Parallel Distributed Systems. 5, 113–120 (1994)
18. Ahmed, I., Dhodhi, M.K.: Task Assignment using a Problem-Space Genetic Algorithm. Concurrency. Pract. Exper. 7, 411–428 (1995)
19. Kwok, Y.: High performance Algorithms for Compile-time Scheduling of Parallel Processors, Ph.D. Thesis, Hong Kong University (1997)
20. Tsuchiya, T., Osada, T., Kikuno, T.: Genetic-Based Multiprocessor Scheduling Using Task Duplication. Microprocessors and Microsystems 22, 197–207 (1998)
21. Alaoui, S.M., Frieder, O., EL-Ghazawi, T.A.: Parallel Genetic Algorithm for Task Mapping On Parallel Machine. In: Proc. of the 13th International Parallel Processing Symposium & 10th Symp. Parallel and Distributed Processing (IPPS/SPDP) Workshops, San Juan, Puerto Rico (April 1999)
22. Haghighat, A.T., Nikravan, M.: A Hybrid Genetic Algorithm for Process Scheduling in Distributed Operating Systems Considering Load Balancing. In: The IASTED Conference on Parallel and Distributed Computing and Networks (PDCN), Innsbruck, Austria (2005)
23. Blickle, T., Thiele, L.: A Mathematical Analysis of Tournament Selection. In: Proc. of the 6th International Conf. on Genetic Algorithms (ICGA 1995). Morgan Kaufmann, San Francisco (1995)
24. Kumar, S., Maulik, U., Bandyopadhyay, S., Das, S.K.: Efficient Task Mapping on Distributed Heterogeneous Systems for Mesh Applications. In: Proceedings of the International Workshop on Distributed Computing, Kolkata, India (2001)

25. Ahmad, I., Kwok, Y.: A New Approach to Scheduling Parallel Programs Using Task Duplication. In: Proc. of the 23rd International Conf. on Parallel Processing, North Carolina State University, NC, USA (August 1994)
26. http://www.Kasahara.Elec.Waseda.ac.jp/schedule/
27. Ahmad, I., Kwok, Y.: Benchmarking and Comparison of the Task Graph Scheduling Algorithms. Journal of Parallel and Distributed Computing 95, 381–422 (1999)
28. Akl, S.G.: Parallel Computation: Models and Methods. Prentice-Hall, Inc., Englewood Cliffs (1997)
29. Wilkinson, B., Allen, M.: Parallel Programming: Techniques and applications using Networked Workstations and Parallel Computers. Pearson Prentic Hall, London (2005)

PSO_Bounds: A New Hybridization Technique of PSO and EDAs

Mohammed El-Abd and Mohamed S. Kamel

Abstract. Particle Swarm Optimization (PSO) is a nature inspired population-based approach successfully used as an optimization tool in many application. Estimation of distribution algorithms (EDAs), are evolutionary algorithms that try to estimate the probability distribution of the good individuals in the population. In this work, we present a new PSO algorithm that borrows ideas from EDAs. This algorithm is implemented and compared to previous PSO and EDAs hybridization approaches using a suite of well-known benchmark optimization functions.

1 Introduction

Particle Swarm Optimization (PSO) [1, 2] is an optimization method widely used to solve continuous nonlinear functions. Although, the original intent was to simulate the movement of a flock of birds or a school of fish looking for food, It was soon realized that the associated equations of motion could be used as a very powerful optimization tool.

Estimation of distribution algorithms (EDAs) [3] are evolutionary algorithms that solve the problem in hand by trying to build a probabilistic model that estimates the distribution of good regions in the search space. These algorithms work by continuously updating the generated model and using it to produce new solutions. One of the early works in this are is Population-Based Incremental Learning (PBIL) proposed in [4]. PBIL is an optimization method similar to Genetic algorithms but with maintaining a probabilistic model rather than a population of solutions. This model was updated in every generation and was used to produce the next population.

Mohammed El-Abd and Mohamed S. Kamel
ECE Dept., University of Waterloo, 200 University Av. W., Waterloo, Ontario, Canada, N2L3G1

A. Abraham et al. (Eds.): Foundations of Comput. Intel. Vol. 3, SCI 203, pp. 509–526.
springerlink.com   © Springer-Verlag Berlin Heidelberg 2009

In the past few years, two hybrid models that mix the PSO algorithm with EDAs have been proposed in the literature [5, 6]. Both approaches use the same probabilistic model to describe the search space but differ in the way the information is gathered and used to build the model. They also differ in the way the model is used to generate new solutions.

In this work, we propose a new model that is based on PBIL. This model continuously use the distribution of the particles during the search to update the bounds of the PSO search space. This in turn affects the particles movement as it affects both the bounds of allowable movement and the maximum allowable velocity. The proposed algorithm is compared to other hybrid techniques and is shown to outperform them on the more difficult multimodal functions.

The chapter is organized as follows: a brief background on PSO is given in Section 1.2. This is followed by an introduction to EDAs in Section 1.3. A literature review of previous PSO and EDAs hybridization techniques are covered in Section 1.4. The new algorithm is proposed in Section 1.5. Results and discussions are presented in Section 1.6. The chapter is concluded in Section 1.7.

2 Particle Swarm Optimization

PSO [1, 2] is regarded as a population-based method, where the population is referred to as a swarm. The swarm consists of a number of individuals called particles. Each particle i in the swarm holds the following information:

- The current position x_i,
- The current velocity v_i,
- The best position, the one associated with the best fitness value the particle has achieved so far $pbest_i$,
- The global best position, the one associated with the best fitness value found among all of the particles $gbest$.

In every iteration, each particle adjusts its own trajectory in the space in order to move towards its best position and the global best according to the following equations:

$$v_{ij}^{t+1} = w v_{ij}^t + c_1 r_{1j}^t (pbest_{ij}^t - x_{ij}^t)$$
$$+ c_2 r_{2j}^t (gbest_j^t - x_{ij}^t), \tag{1}$$

$$x_{ij}^{t+1} = x_{ij}^t + v_{ij}^{t+1}, \tag{2}$$

for $j \in 1..d$ where d is the number of dimensions, $i \in 1..n$ where n is the number of particles, t is the iteration number, w is the inertia weight, r_1 and r_2 are two random numbers uniformly distributed in the range [0,1], and c_1 and c_2 are the acceleration factors.

Afterwards, each particle updates its personal best using the equation (assuming a minimization problem):

$$pbest_i^{t+1} = \begin{cases} pbest_i^t & \text{if} \quad f(pbest_i^t) \leq f(x_i^{t+1}) \\ x_i^{t+1} & \text{if} \quad f(pbest_i^t) > f(x_i^{t+1}) \end{cases} \tag{3}$$

Finally, the global best of the swarm is updated using the equation (assuming a minimization problem):

$$gbest^{t+1} = \arg \min_{pbest_i^{t+1}} f(pbest_i^{t+1}), \tag{4}$$

where $f(.)$ is a function that evaluates the fitness value for a given position. This model is referred to as the *gbest* (global best) model.

Another model is the *lbest* (local best) model [7], in each particle does not hold the global best position. Instead, each particle only holds the best position achieved by its own neighborhood. Different neighborhood structures were previously examined for such a model [8] including the ring topology and the Von Neumann model.

3   Estimation of Distribution Algorithms

Estimation of distribution algorithms (EDAs) are evolutionary algorithms that try to estimate the probability distribution of the good individuals in the population. EDAs try to estimate this probability distribution by using selected individuals, from the current population, to construct a probabilistic model. This model is consequently used to generate the offspring. The new population is generated by selecting individuals from both the offspring and the current population in a proportionate manner. Finally, the new population replaces the current one. Hence, EDAs maintain a continuously updated probabilistic model from one generation to the next. Although, it has been originally introduced to tackle combinatorial optimization problems, recent numerical applications have been proposed as well [9, 10, 11, 12]. The general steps for an EDA is shown in Algorithm 1.

EDAs are categorized based on the degree of dependencies, allowed by the probabilistic model used, between the problem variables:

- No dependency: the problem variables are assumed to be independent,
- Bivariate dependency: the dependencies are only assumed between two variables at a time,
- Multivariate dependency: the dependencies could be modeled between any number of variables.

For a complete survey of the different optimization techniques adopted using building probabilistic models, the interested reader could refer to [13].

Algorithm 1. Estimation of Distribution Algorithm (EDA)

1: $P \Leftarrow$ Initialize the population
2: Evaluate the initial population
3: **while** $iter\_number \leq Max\_iterations$ **do**
4:    $P_s \Leftarrow$ Select the top s individuals
5:    $M \Leftarrow$ Estimate a new Model from P_s
6:    $P_n \Leftarrow$ Sample n individuals from M
7:    Evaluate P_n
8:    $P \Leftarrow$ Select n individuals from $P \cup P_n$
9:    iter_number = iter_number + 1
10: **end while**
11: **return** Best_Individual

---

4 PSO Based on Probabilistic Models

This section surveys the two previous attempts to introduce the concepts of EDAs into PSO in order to improve its performance.

4.1 EDPSO

An estimation of distribution particle swarm optimizer (EDPSO) was proposed by Iqbal and Montes de Oca [5]. The method borrowed some ideas from a development in ACO for solving continuous optimization problems [14, 15]. The approach relies on estimating the joint probability distribution for one dimension at a time using mixtures of weighted Gaussian functions. The Gaussian functions are defined through an archive of k solutions (*pbests* of the particles). For each dimension d, the dimension is either updated using PSO equations or by sampling a Gaussian distribution selected from the archive. The values of this dimension d across all the solutions in the archive compose the vector $\boldsymbol{\mu_d}$, which is the vector of means for the univariate Gaussian distributions:

$$\boldsymbol{\mu_d} = < pbest_{1d}, pbest_{2d}, ..., pbest_{kd} > \tag{5}$$

To select one of these distributions, the weights vector $\mathbf{w}$, which holds the weights associated with each distribution, is calculated. This is done by sorting the solutions according to their fitness, with the best solution having a rank of 1. A weight is calculated for each solution as follows:

$$\mathbf{w} = < w_1, w_2, ..., w_k > \tag{6}$$

$$w_l = \frac{1}{qk\sqrt{2\pi}} e^{-\frac{(l-1)^2}{2q^2k^2}} \tag{7}$$

which is a Gaussian function with mean l and standard deviation qk, where q is a constant that determines how much we prefer good solutions and l is the solution rank.

The Gaussian function to be used is selected probabilistically. The probability of selecting a certain Gaussian function is proportional to its weight. This probability is calculated as follows:

$$\mathbf{p} =< p_1, p_2, ..., p_k >$$
$$p_l = \frac{w_l}{\sum_{r=1}^{k} w_r} \tag{8}$$

After selecting a certain Gaussian function G_d denoted by its mean $pbest_{gd}$, where $1 \leq g \leq k$, the standard deviation for this functions is calculated as:

$$\sigma_{gd} = \xi \sum_{i=1}^{k} \frac{|pbest_{id} - pbest_{gd}|}{k-1} \tag{9}$$

which the average distance between the selected mean and the other entries of the archive. ξ is a parameter to balance the exploration-exploitation behaviors. if ξ is small, this will lead to having a smaller value for σ_{gd} and the search will tend to search in a closer range around the chosen mean.

Finally the selected Gaussian function is evaluated (not sampled) to generate a value r in order to probabilistically move the particle. This is done by generating a uniformly distributed random number $U(0,1)$. If it is less than r, the particle moves using the normal PSO equations. Otherwise, the Gaussian function is sampled to move the particle. The steps are shown in Algorithm 2.

4.2 EDA-PSO

A hybrid EDA-PSO approach was proposed in [6]. The algorithm works by sampling an independent univariate Gaussian distribution based on the best half of the swarm. The mean and standard deviation of the model is calculated in every iteration as:

$$\boldsymbol{\mu} = \frac{1}{M} \sum_{i=1}^{M} \mathbf{x}_i \tag{10}$$

$$\sigma_j = \sqrt{\frac{1}{M} \sum_{i=1}^{M} (\mathbf{x}_{ij} - \mu_j)^2}, \tag{11}$$

where $M = N/2$ for a swarm with N particles and i is the particle number.

The choice of whether to update the particle using the normal PSO equations or to sample the particle using the estimated distribution is made with a probability p, referred to as the *participation ratio*. If $p = 0$, the algorithm will behave as a pure EDA algorithm. On the other hand, if $p = 1$, it will be a pure PSO algorithm. In the hybrid approach, where $0 < p < 1$, each

---

Algorithm 2. The EDPSO algorithm.

Require: Max_Function_Evaluations
1: Initialize the swarm
2: Max_Iterations $= \frac{Max\_Function\_Evaluations}{Num\_Particles}$
3: iter_number $= 1$
4: **while** $iter\_number \leq Max\_Iterations$ **do**
5:     Update the swarm
6:     Rank the particle's using pbests information
7:     Calculate weights vector **w**
8:     Calculate probabilities vector **p**
9:     **for** every particle i **do**
10:       **for** each dimension d **do**
11:         Update v_{id} and x_{id}
12:         Select a Gaussian function according to p_i
13:         Calculate σ_{gd}
14:         Prob_move $= \sigma_{gd}\sqrt{2\pi}G_d(x_{id})$
15:         **if** $U(0,1) < Prob\_move$ **then**
16:           continue
17:         **else**
18:           $x_{id} = \text{Gauss}(s_{gd}, \sigma_{gd})$
19:         **end if**
20:       **end for**
21:     **end for**
22:     iter_number $=$ iter_number $+ 1$
23: **end while**
24: **return** *gbest*

---

particle is either totally updated by the PSO equations or totally sampled from the estimated distribution (not on a dimension-by-dimension basis as in EDPSO). Finally, the particle gets updated only if its fitness improves.

The authors also proposed different approaches in order to adaptively set the parameter p. These approaches depend on the success rate of both the PSO and EDA parts in improving the particles' fitness:

- The *Generation based*, where the success rates are calculated based on the information gathered during the last generation,

$$p^{t+1} = \frac{\frac{sum\_PSO^t}{num\_PSO^t}}{\frac{sum\_PSO^t}{num\_PSO^t} + \frac{sum\_EDA^t}{num\_EDA^t}} \tag{12}$$

- The *All historical information*, where the success rates are calculated based on the information gathered during the entire search,

$$p^{t+1} = \frac{\sum_{i=1}^{t} \frac{sum\_PSO^i}{num\_PSO^i}}{\sum_{i=1}^{t} \frac{sum\_PSO^i}{num\_PSO^i} + \sum_{i=1}^{t} \frac{sum\_EDA^i}{num\_EDA^i}} \tag{13}$$

Algorithm 3. The EDA-PSO algorithm

Require: Max_Function_Evaluations

1: Initialize the swarm
2: Max_Iterations = $\frac{Max\_Function\_Evaluations}{Num\_Particles}$
3: iter_number = 1
4: **while** $iter\_number \leq Max\_Iterations$ **do**
5:     Calculate $\boldsymbol{\mu}$ and $\boldsymbol{\sigma}$ using top $\frac{N}{2}$ particles
6:     **for** every particle i **do**
7:       **if** $U(0,1) < p$ **then**
8:         candidate_particle = PSO equations
9:       **else**
10:         candidate_particle = Gauss($\boldsymbol{\mu}$,$\boldsymbol{\sigma}$)
11:       **end if**
12:       **if** candidate_particle has a better fitness **then**
13:         particle i = candidate_particle
14:       **end if**
15:     **end for**
16:     iter_number = iter_number + 1
17: **end while**
18: **return** $gbest$

- The *Sliding window*, where the success rates are calculated considering only the information in the last m generations.

$$p^{t+1} = \frac{\sum_{i=t-m+1}^{t} \frac{sum\_PSO^i}{num\_PSO^i}}{\sum_{i=t-m+1}^{t} \frac{sum\_PSO^i}{num\_PSO^i} + \sum_{i=t-m+1}^{t} \frac{sum\_EDA^i}{num\_EDA^i}} \tag{14}$$

In all the previous equations sum_PSO^t and num_PSO^t refers to the sum of improvements and number of improvements done by the PSO component at iteration t. While sum_EDA^t and num_EDA^t refers to the sum of improvements and number of improvements done by the EDA component at iteration t. Finally, m is the window size.

The complete algorithm for EDA-PSO is shown in Algorithm 3.

5 PSO with Varying Bounds

A PBIL approach for continuous search spaces was proposed in [10]. The algorithm explored the search space by dividing the domain of each gene into two equal intervals referred to as the *low* and *high* intervals. A probability h_d, which is initially set to 0.5, is the probability of gene number d being in the *high* interval as shown:

$$x_d \in [a, b], h_d - Probability(x_d > \frac{a + b}{2}) \tag{15}$$

After each generation, this distribution is updated according to the gene values of the best individual using the following formula:

$$p = \begin{cases} 0 & \text{if} \quad x_d^{max} < \frac{a+b}{2} \\ 1 & otherwise \end{cases} \tag{16}$$

$$h_d^{t+1} = (1 - \alpha) * h_d^t + \alpha * p \tag{17}$$

where α is the *relaxation factor* and t is the iteration number. If h_d gets below h_{dmin} or above h_{dmax}, the population gets re-sampled in the corresponding interval, $[a, \frac{a+b}{2}]$ or $[\frac{a+b}{2}, b]$, respectively.

In this work, we propose a new PSO algorithm, referred to as PSO_Bounds, which borrows concepts from PBIL. At the beginning, the particles are initialized in the predefined domain. After every iteration, the probability h_d of each dimension d gets adjusted according to the probability of this dimension value being in the *high* interval of the defined domain. This probability is calculated using information from all the particles and not only *gbest* to prevent premature convergence. Hence, the original equations of PBIL are changed as follows:

$$p_{id}^t = \begin{cases} 0 & \text{if} \quad pbest_{id}^t < \frac{a+b}{2} \\ 1 & otherwise \end{cases} \tag{18}$$

$$p_d^t = \frac{\sum_i^n p_{id}^t}{n} \tag{19}$$

$$h_d^{t+1} = (1 - \alpha) * h_d^t + \alpha * p_d^t \tag{20}$$

where $i \in 1..n$ where n is the number of particles, t is the iteration number, and d is the dimension. Please note that these equations are applied for each dimension d separately.

In PBIL, the probabilities were updated using the value of the best individual, which is analogous to the current position of the particles in PSO. However, in our implementation, we use the values of *pbests* instead as it reflects the best experience of the swarm and would guide the search towards better solutions.

When h_d^{t+1} gets specific enough, the domain of dimension d is adjusted accordingly and h_d^{t+1} gets re-initialized to 0.5. In this model, different dimensions

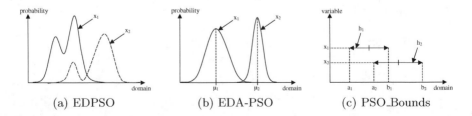

(a) EDPSO   (b) EDA-PSO   (c) PSO_Bounds

Fig. 1 Probabilistic models

Algorithm 4. The PSO_Bounds algorithm

Require: Max_Function_Evaluations, $h_{dmin}, h_{dmax}, \alpha$

1. Initialize the swarm
2. Max_Iterations $= \frac{Max\_Function\_Evaluations}{Num\_Particles}$
3. iter_number $= 1$
4. **while** $iter\_number \leq Max\_Iterations$ **do**
5.   Update the swarm
6.   **for** each dimension d **do**
7. $p_d = 0$
8.   **for** every particle i **do**
9.   Calculate p_{id}
10. $p_d = p_d + p_{id}$
11.   **end for**
12. $h_d = (1 - \alpha)h_d + \alpha p_d$
13.   **if** $h_d < h_{dmin}$ **then**
14. $x_{dmax} = b = \frac{a+b}{2}$
15.   Update v_{dmin} and v_{dmax}
16. $h_d = 0.5$
17.   **else if** $h_d > h_{dmax}$ **then**
18. $x_{dmin} = a = \frac{a+b}{2}$
19.   Update v_{dmin} and v_{dmax}
20. $h_d = 0.5$
21.   **end if**
22.   **end for**
23.   iter_number $=$ iter_number $+ 1$
24. **end while**
25. **return** $gbest$

---

might end up having different domains and different velocity bounds which does not happen in normal PSO.

Figure 1 illustrates the approaches taken by the different PSO and EDAs hybridization techniques in order to model the distribution of good solutions across the search space in every dimension.

The steps taken by PSO_Bounds is shown in Algorithm 4, where x_{dmin} and x_{dmax} refer to the minimum and maximum search bounds for dimension d while v_{dmin} and v_{dmax} refer to the minimum and maximum velocity bounds.

6 Results and Discussions

6.1 Experimental Settings

Table 1 shows the parameter settings used for applying the algorithms under study. For all experiments, all the particles have been randomly initialized in the specified domain using uniform distribution. The values for q and ξ are

Table 1 Parameter settings

Model	Parameter	Value
Normal PSO	w	0.9 to 0.1
	c1 and c2	2
EDPSO	q	0.1
	ξ	0.85
EDA-PSO	p	Adaptive - all historical information
PSO_Bounds	α	0.1
	h_{dmin}	0.2
	h_{dmax}	0.8

the same as was proposed in [5] and the value for p is set adaptively using the *allhistoricalinformation* approach, as it was found to be the best one based on our experiments. The values for $(\alpha, h_{dmin}, h_{dmax})$ are changed from (0.01, 0.1, 0.9) in [10] to (0.1, 0.2, 0.8) to allow a faster process of varying the bounds. The experiments are conducted for a problem dimensionality of 10, 30, and 50 with 40 particles in the swarm performing 100000, 100000, and 200000 function evaluations, respectively. The results reported are the averages taken over 30 runs.

The experiments are run using the benchmark test functions shown in Table 2.

The experiments are also conducted using the benchmark functions f6-f14 proposed in CEC2005, available at [16] and shown in Table 3. In order to constrain the particles movement within the specified domain for the CEC05 functions, any violating particle gets its position randomly re-initialized inside the specified domain. The error values $f(x) - f(x*)$ are reported, where $x*$ is the global optimum.

In [6], the values for μ and σ are calculated using the best half of the swarm. The authors in [17] proposed calculating σ using the whole population instead, which is found to produce better results due to the induced diversity avoiding premature convergence. The same approach is used in this work when applying the EDA-PSO algorithm.

Table 2 Benchmark functions

Function	Equation	Domain
Spherical	$f(x) = \sum_{i=1}^{n} x_i^2$	100
Rosenbrock	$f(x) = \sum_{i=1}^{n/2} \left(100(x_{2i} - x_{2i-1}^2)^2 + (1 - x_{2i-1})^2\right)$	2.048
Griewank	$f(x) = \frac{1}{4000} \sum_{i=1}^{n} x_i^2 - \prod_{i=1}^{n} \cos\left(\frac{x_i}{\sqrt{i}}\right) + 1$	600
Ackley	$f(x) = 20 + e - 20 \exp\left(-0.2\sqrt{\frac{1}{n}\sum_{i=1}^{n} x_i^2}\right)$	
	$- \exp\left(\frac{1}{n}\sum_{i=1}^{n} \cos 2\pi x_i\right)$	30
Rastrigin	$f(x) \sum_{i=1}^{n} \left(x_i^2 - 10\cos 2\pi x_i + 10\right)$	5.12

Table 3 CEC05 Benchmark Functions

Benchmark Function	Description	Lower Domain	Upper Domain
f6	shifted Rosenbrock	-100	100
f7	shifted rotated Griewank	0	600
f9	shifted Rastrigin	-5	5
f10	shifted rotated Rastrigin	-5	5
f11	shifted rotated Weierstrass	-0.5	0.5
f12	Schwefel	-100	100
f13	expanded extended Griewank plus Rosenbrock	-3	1
f14	shifted rotated expanded Scaffer	-100	100

Table 4 Results of all the algorithms for the classical functions

Function	Dim.	EDPSO		EDA-PSO		PSO_Bounds	
		Mean	Std.	Mean	Std.	Mean	Std.
Spherical		**9.881e-324**	**0**	8.400e-266	0	5.087e-03	2.786e-02
Rosenbrock		**5.519e-06**	**1.044e-05**	7.827e-02	8.422e-02	7.744e-01	5.857e-01
Griewank	10	2.084e-02	1.447e-02	**7.882e-03**	**7.325e-03**	1.229e-01	5.988e-02
Ackley		**5.887e-16**	**2.006e-31**	1.268e+00	2.258e-15	8.606e-02	4.708e-01
Rastrigin		**3.051e+00**	**1.609e+00**	4.013e+00	1.998e+00	7.131e+00	2.172e+00
Spherical		3.698e-67	2.026e-66	**4.234e-141**	**1.425e-140**	5.416e+02	3.674e+02
Rosenbrock		**9.562e-01**	**2.042e-01**	1.123e+00	4.552e-01	1.707e+01	4.633e+00
Griewank	30	1.479e-03	3.462e-03	**0**	**0**	4.871e+00	2.021e+00
Ackley		**4.378e-015**	**9.014e-016**	1.586e+00	9.034e-16	5.467e+00	1.137e+00
Rastrigin		**1.791e+01**	**4.222e+00**	3.4067e+01	2.922e+01	6.799e+01	1.339e+01
Spherical		1.104e-59	3.644e-59	**2.811e-103**	**1.539e-102**	2.979e+03	1.131e+03
Rosenbrock		**2.078e+00**	**3.954e-01**	1.565e+00	2.745e+00	3.131e+01	7.791e+00
Griewank	50	**3.286e-04**	**1.800e-03**	2.132e-03	6.314e-03	2.697e+01	8.899e+00
Ackley		**7.094e-15**	**1.319e-15**	1.641e+00	2.258e-16	9.068e+00	9.036e-01
Rastrigin		**4.016e+01**	**8.593e+00**	4.630e+01	1.410e+01	1.457e+02	1.891e+01

The best results highlighted in bold in all the tables are selected based on a two-sample t-test where the null hypothesis is rejected with a 95% confidence level.

6.2   Experimental Results

Table 4 shows the results obtained by applying EDPSO, EDA-PSO and PSO_Bounds to the classical functions for different problem sizes.

As shown in Tables 4 for the classical functions, both EDPSO and EDA-PSO outperform PSO_Bounds. The reason for this is that the global optimum is at the center of the search space and the Gaussian model adopted by these algorithms along with the uniform distribution used in initializing the particles make it very easy for these algorithms to reach better results.

Table 5 Results of all the algorithms for the CEC05 benchmark functions

Function	Dim.	EDPSO		EDA-PSO		PSO_Bounds	
		Mean.	Std.	Mean	Std.	Mean	Std.
f6		1.375e+00	4.557e+00	**1.123e-02**	**1.626e-02**	1.451e+02	2.218e+02
f7		2.687e-01	2.258e-01	**1.927e-01**	**1.905e-01**	-	-
f9		**3.217e+00**	**1.604e+00**	4.046e+00	2.277e+00	3.454e+00	1.471e+00
f10	10	1.989e+01	6.327e+00	**4.819e+00**	**3.642e+00**	7.543e+00	4.528e+00
f11		3.868e+00	3.859e+00	6.588e+00	1.340e+00	**3.529e+00**	**1.730e+00**
f12		2.919e+04	7.054e+03	1.616e+04	6.334e+03	**4.243e+03**	**5.001e+03**
f13		1.194e+00	5.372e-01	8.465e-01	3.968e-01	**6.904e-01**	**1.770e-01**
f14		2.429e+00	5.255e-01	2.667e+00	5.991e-01	**2.365e+00**	**5.792e-01**
f6		7.522e+01	1.007e+02	**1.716e+01**	**2.011e+01**	6.602e+05	1.841e+06
f7		**8.700e-03**	**5.920e-03**	1.300e-02	7.589e-03	-	-
f9		**1.175e+00**	**2.044e+00**	2.789e+00	6.498e+00	3.315e+01	7.072e+00
f10	30	1.850e+02	1.348e+01	1.187e+02	6.191e+01	**5.556e+01**	**2.068e+01**
f11		4.028e+01	1.676e+00	3.494e+01	2.674e+00	**2.849e+01**	**3.897e+00**
f12		1.129e+06	1.266e+05	9.219e+05	2.060e+05	**2.941e+05**	**2.155e+05**
f13		1.489e+01	1.497e+00	7.942e+00	4.688e+00	**4.333e+00**	**7.852e-01**
f14		1.334e+01	2.309e-01	1.325e+01	2.933e-01	**1.245e+01**	**6.541e-01**
f6		1.429e+02	2.023e+02	**3.725e+01**	**4.515e+01**	3.458e+07	4.913e+07
f7		**3.000e-03**	**5.813e-03**	9.867e-03	1.374e-02	-	-
f9		**1.282e+01**	**6.519e+00**	4.232e+01	1.080e+01	7.047e+01	1.338e+01
f10	50	3.765e+02	1.520e+01	2.931e+02	8.820e+01	**1.222e+02**	**2.553e+01**
f11		7.393e+01	1.266e+00	6.744e+01	3.031e+00	**5.778e+02**	**6.800e+00**
f12		5.760e+06	3.738e+05	3.965e+06	1.259e+06	**1.254e+05**	**1.167e+05**
f13		3.057e+01	2.701e+00	1.696e+01	1.109e+01	**9.327e+00**	**2.030e+00**
f14		2.310e+01	2.551e-01	2.282e+01	3.451e-01	**2.237e+01**	**4.455e-01**

Table 6 Comparison between all the algorithms using the *gbest* model

Algorithm	Classical Functions		CEC05 Functions		Total Number of Cases
	No. of Cases	Best in	No. of Cases	Best in	
PSO_Bounds	-	-	15	f11, f12 f13, f14	15
EDA-PSO	7	-	5	f6	12
EDPSO	11	Rosenbrock Ackley, Rastrigin	5	-	16

On the other hand, for the more difficult CEC05 benchmark functions shown in Table 5, PSO_Bounds has the best performance across the different problem sizes.

Table 6 summarizes the comparison between all the algorithms based on the results shown in Tables 4 and 5. The upper bound for the number of cases is 15 (5 functions in 3 problem sizes) in the classical functions and 21 (7 functions in 3 problem sizes) in the CEC05 functions.

Please note that PSO_Bounds is not applied for f7 as this function is not bounded by a specified domain (the bounds shown in Table 3 are only used as an initialization range).

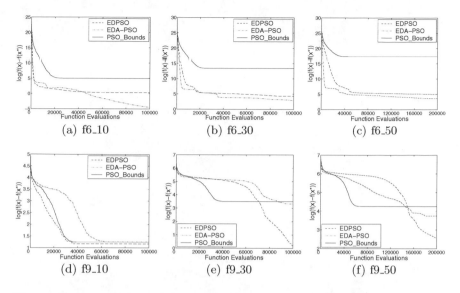

Fig. 2 Convergence behavior of all the algorithms for the CEC05 functions

The convergence behavior shown in Figure 2 and Figure 3 illustrates that PSO_Bounds usually has a slow speed of convergence compared with the other algorithms. It only has the fastest speed of convergence in both f6 and f9 where it does not produce good results.

Convergence figures also show that both EDPSO and EDA-PSO have a very similar behavior on most of the functions. This could be due to the fact that both algorithms use the same Gaussian model for sampling the search space.

6.3 Changing the Population Topology

In [18], the authors stated that "modern research performed using only swarms with a global topology is incomplete at best". For this reason, the experiments are rerun again for all the algorithms using the *lbest* model. Table 7 and Table 8 show the obtained results.

The results show that PSO_Bounds still has a deteriorated performance in the classical functions while outperforming other algorithms on the more difficult multimodal functions. This means that all the algorithms exhibit the same performance compared to each other as in the case of using the *gbest* model.

Table 9 summarizes the comparison between all the algorithms based on the results shown in Tables 7 and 8. The results emphasize that the performance of these algorithms (compared to each other) is the same regardless of the underlying population topology.

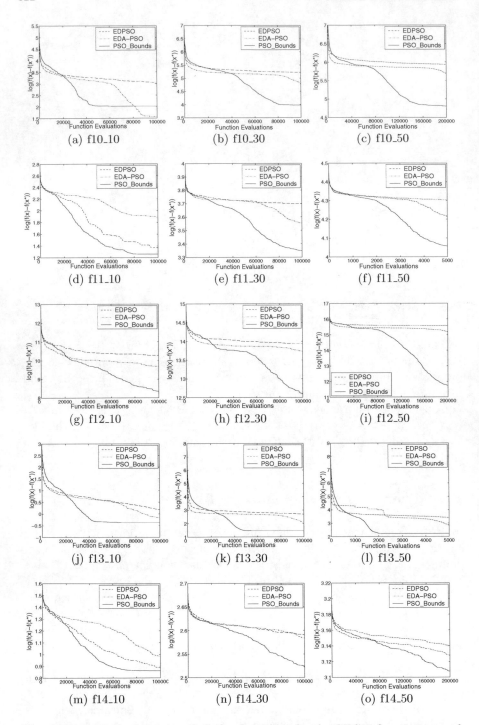

Fig. 3 Convergence behavior of all the algorithms for the CEC05 functions, contd

Table 7 Results of all the algorithms using the *lbest* model for the classical functions

Function	Dim.	EDPSO_L Mean	EDPSO_L Std.	EDA-PSO_L Mean	EDA-PSO_L Std.	PSO_Bounds_L Mean	PSO_Bounds_L Std.
Spherical		0	0	3.850e-267	0	3.194e-17	1.722e-16
Rosenbrock		4.019e-03	5.097e-03	1.029e-01	5.276e-02	2.390e-01	1.781e-01
Griewank	10	1.682e-02	1.194e-02	4.959e-03	7.767e-03	4.881e-02	1.517e-02
Ackley		5.887e-15	2.006e-31	1.268e+00	2.258e-16	5.037e-11	9.260e-11
Rastrigin		3.118e+00	1.5180e+00	3.263e+00	1.885e+00	3.798e+00	1.444e+00
Spherical		6.031e-94	3.205e-95	7.020e-141	9.280e-141	5.183e-01	1.811
Rosenbrock		1.076e+00	1.779e-01	1.742e+00	2.524e+00	1.085e+01	3.416e+00
Griewank	30	2.052e-03	4.970e-03	3.288e-02	1.801e-03	7.657e-02	8.055e-02
Ackley		4.141e-015	0	1.586	3.651e-16	6.166e-01	7.217e-01
Rastrigin		1.523e+01	3.999e+00	4.472e+01	3.057e+01	4.192e+01	7.900e+00
Spherical		3.055e-82	1.256e-81	3.700e-11	2.026-10	14.233	33.711
Rosenbrock		2.104e+00	2.490e-01	6.237e+00	1.035e+01	2.356e+01	4.209e+00
Griewank	50	1.232e-03	3.284e-03	2.919e-03	1.465e-02	6.929e-01	3.780e-01
Ackley		6.865e-15	1.528e-15	1.641e+00	2.258e-16	2.200e+00	6.118e-01
Rastrigin		3.270e+01	7.202e+00	7.481e+01	5.061e+01	9.140e+01	1.469e+01

Table 8 Results of all the algorithms using the *lbest* model for the CEC05 benchmark functions

Function	Dim.	EDPSO_L Mean	EDPSO_L Std.	EDA-PSO_L Mean	EDA-PSO_L Std.	PSO_Bounds_L Mean	PSO_Bounds_L Std.
f6		6.554e+00	1.936e+01	2.092e-01	7.899e-01	1.497e+01	25.785
f9		2.919e+00	1.566e+00	3.310e+00	1.259e+00	8.025e-01	9.603e-01
f10		1.946e+01	6.939e+00	1.105e+01	5.716e+00	6.712e+00	3.003e+00
f11	10	7.292e+00	3.473e+00	6.196e+00	8.1308e-1	4.480e+00	1.027e+00
f12		2.729e+04	7.468e+03	1.877e+04	6.712e+3	6.535e+03	2.841e+03
f13		1.435e00	4.549e-01	1.224e+00	3.724e-01	6.422e-01	1.390e-01
f14		2.204e+00	5.145e-01	2.910e+00	2.684e-01	2.777e+00	3.261e-01
f6		8.592e+01	1.305e+02	7.063e+01	4.586e+01	8.883e+03	3.632e+04
f9		1.605e+01	5.372e+00	4.208e+01	2.742e+01	2.536e+01	4.694e+00
f10		1.778e+02	9.953e+00	1.608e+02	1.719e+01	1.384e+02	1.864e+01
f11	30	4.043e+01	1.148e+00	3.641e+01	2.107e+00	3.163e+01	2.479e+00
f12		1.140e+06	1.148e+05	9.571e+05	1.696e+05	4.978e+05	1.443e+05
f13		1.447e+01	1.328e+00	1.156e+01	2.639e+00	4.755e+00	9.558e-01
f14		1.347e+01	1.873e-01	1.327e+01	2.282e-01	1.302e+01	2.674e-01
f6		6.550e+01	5.446e+01	6.789e+01	4.228e+01	1.967e+06	1.012e+07
f9		3.250e+01	6.450e+01	5.334e+01	2.326e+01	5.547e+01	9.813e+00
f10		3.629e+02	1.624e+01	3.359e+02	1.660e+01	2.879e+02	4.048e+01
f11	50	7.381e+01	1.911e+00	6.926e+01	2.552e+00	6.184e+01	4.231e+00
f12		5.631e+06	4.676e+05	4.725e+06	5.695e+05	1.896e+06	3.675e+05
f13		2.738e+01	4.029e+00	2.446e+01	4.217e+00	1.062e+01	2.183e+00
f14		2.316e+01	1.693e-01	2.292e+01	2.184e-01	2.261e+01	2.103e-01

Table 9 Comparison between all the algorithms using the *lbest* model

Algorithm	Classical Functions		CEC05 Functions		Total Number of Cases
	No. of Cases	Best in	No. of Cases	Best in	
PSO_Bounds	2	-	16	f10, f11 f12, f13	18
EDA-PSO	7	Spherical	5	f6	12
EDPSO	14	Spherical, Rosenbrock Ackley, Rastrigin	6	-	20

7 Conclusion and Discussion

This chapter gives a brief introduction to Particle Swarm Optimization and Estimation of Distribution Algorithms (EDAs). The chapter surveys the different methods previously adopted to combine PSO and EDAs.

The chapter introduces a new algorithm, PSO_Bounds, which is a PSO algorithm that borrows ideas from PBIL. The new algorithm uses the same equations of motion as PSO while using the current distribution of the particles during the search to continuously update the allowable search domain.

Along with the proposed algorithm, all the approaches covered are implemented and compared using a suite of well-known benchmark optimization functions with different properties. It is shown that PSO_Bounds outperforms other PSO and EDAs hybridization techniques on the more difficult shifted and/or rotated multimodal functions. It is also shown that the new proposed algorithm has in general a slower speed of convergence when compared to other algorithms.

Moreover, the relative performance of all the algorithms is shown to be independent of the underlying population topology used by the PSO component.

Many future directions could be followed to further improve on the performance of such algorithms. The deteriorated performance of PSO_Bounds in some functions could be due to the fact that the width of the allowable domain for the different dimensions becomes smaller and smaller as the search progresses. It would eventually get to the point of being very close to zero (or zero, even). Once this happens, the particles will stop moving as the allowable movement domain for the particles is very small as well as the allowable maximum velocity, hence, the search stagnates. One way to improve this is to re-initialize those domains again, by re-setting them to the initial search ranges, if the width drops under a pre-determined threshold.

A similar approach could be adopted for EDA-PSO by re-initializing the current positions of the particles, while keeping their *pbests* values as they are so as not to lose any useful information, if the value of σ drops under a pre-determined threshold during the search.

A different research direction is to incorporate PSO with probabilistic models that allow inter-variable dependencies. All the hybridization techniques proposed up-to-date, including the one in this chapter, use probabilistic models that assume that the problem variables are independent.

References

1. Kennedy, J., Eberhart, R.C.: Particle swarm optimization. In: Proc. of IEEE International Conference on Neural Networks, vol. 4, pp. 1942–1948 (1995)
2. Eberhart, R.C., Kennedy, J.: A new optimizer using particle swarm theory. In: Proc. of the 6th International Symposium on Micro Machine and Human Science, pp. 39–43 (1995)
3. Larranaga, P., Lozano, J.A.: Estimation of Distribution Algorithms. A New Tool for Evolutionary Computation. Kluwer, Dordrecht (2002)
4. Baluja, S.: Population-based incremental learning: A method for integrating genetic search based function optimization and competitive learning. School of Computer Science, Carnegie Mellon University, Tech. Rep. CMU-CS-94-163 (1994)
5. Iqbal, M., de Oca, M.A.M.: An estimation of distribution particle swarm optimization algorithm. In: Dorigo, M., Gambardella, L.M., Birattari, M., Martinoli, A., Poli, R., Stützle, T. (eds.) Proc. of the Fifth International Workshop on Ant Colony Optimization and Swarm Intelligence, pp. 72–83 (2006)
6. Zhou, Y., Jin, J.: Eda-pso - a new hybrid intelligent optimization algorithm. In: Proc. of the Michigan University Graduate Student Symposium (2006)
7. Eberhart, R.C., Simpson, P., Dobbins, R.: Computational Intelligence.PC Tools: Academic, ch. 6, pp. 212–226 (1996)
8. Kennedy, J., Mendes, R.: Population structure and particle swarm performance. In: Proc. of IEEE Congress on Evolutionary Computation, vol. 2, pp. 1671–1676 (2002)
9. Rudolph, S., Koppen, M.: Stochastic hill climbing with learning by vectors of normal distributions. In: First on-line Workshop on Soft Computing (WSC1), pp. 60–70 (1996)
10. Servet, I., Trave-Massuyes, L., Stern, D.: Telephone network traffic overloading diagnosis and evolutionary computation technique. In: Hao, J.-K., Lutton, E., Ronald, E., Schoenauer, M., Snyers, D. (eds.) AE 1997. LNCS, vol. 1363, pp. 137–144. Springer, Heidelberg (1998)
11. Sebag, M., Ducoulombier, A.: Extending population-based incremental learning to continuous search spaces. In: Proc. of Parallel Problem Solving from Nature, pp. 418–427 (1999)
12. Gallagher, M., Frean, M., Downs, T.: Real-valued evolutionary optimization using a flexible probability density estimator. In: Proc. of Genetic and Evolutionary Computation Conference, vol. 1, pp. 840–846 (1999)
13. Pelikan, M., Goldberg, D.E., Lobo, F.: A survey of optimization by building and using probabilistic models. Computational Optimization and Applications 21(1), 5–20 (2002)
14. Socha, K., Dorigo, M.: Ant colony optimization for continuous domains. Universitie Libre de Bruxelles, Tech. Rep. TR/IRIDIA/2005-037 (2005)
15. Socha, K., Dorigo, M.: Ant colony optimization for continuous domains. European Journal of Operationl Research 185(3), 1155–1173 (2008)

16. CEC05 benchmark functions,
    http://staffx.webstore.ntu.edu.sg/MySite/
    Public.aspx?accountname=epnsugan
17. delaOssa, L., Gamez, J., Puerta, J.: Initial approaches to the application of
    island-based parallel edas in continuous domains. Journal of Parallel and Dis-
    tributed Computing 66(8), 991–1001 (2006)
18. Bratton, D., Kennedy, J.: Defining a standard for particle swarm optimization.
    In: Proc. IEEE Swarm Intelligence Symposium, pp. 120–127 (2007)

Author Index